100 YEAR STARSHIP™

canopus 310 LY

10 YEARS

Published by 100 Year Starship™
www.100yss.org

First Edition

ISBN-13: 978-1-63725-011-2

100 YEAR STARSHIP™
canopus 310 LY

The 100 Year Starship™ exists to make the capability of human travel beyond our solar system a reality within the next 100 years. We unreservedly dedicate ourselves to identifying and pushing the radical leaps in knowledge and technology needed to achieve interstellar flight, while pioneering and transforming breakthrough applications that enhance the quality of life for all on Earth. We actively seek to include the broadest swath of people and human experience in understanding, shaping and implementing this global aspiration.

For more information, visit www.100yss.org

Edited by Mae Jemison, M.D., Jason D. Batt, and Alires Almon
Design by Jason D. Batt

PRINTED IN THE UNITED STATES OF AMERICA

100 YEAR STARSHIP™

canopus 310 LY

10 YEARS

THE FIRST TEN YEARS

EDITED BY

Dr. Mae Jemison, Jason D. Batt, & Alires Almon

100 YEAR STARSHIP™

Table of Contents

Introduction 9

Batt — 10%: The First Ten Years of the 100 Year Starship . . 12

2012 — 21

Aslan & Hawkins — Mobilizing Emerging Countries Towards Interstellar Culture: The Case of Turkey 24

Aspelund — Considering Clothing and Textiles on an Interstellar Mission 30

Field & Cantwell — Developing Faster than Light Technology A Project Management Institute Based Approach . . 40

Freeland & Baxter — A Fully Decelerated Probe to Alpha Centauri Using a Perturbed Project Daedalus Design 52

Maccone — Interstellar Links by Focal Spacecrafts. 64

Toerpe — From the Moon to the Stars: Tapping into Shared Culture to Create Public Momentum for Interstellar Travel 90

Weiss — The Foundry in the Starship. 97

2013 — 101

Batt — The Dangers of Religious and Ideological Extremism 104

Contag — Every Body is an Ark: How the Microorganisms We Carry will Impact an Interstellar Mission . . . 117

Ilves & Emerson — Constraints Facing Future Deep Space Missions . . 124

Nosanov — Solar System Escape Architecture Revolutionary NIAC 2012-2013 Phase 1 Final Report 142

Orban — A Pathway for Global Communications as an Inclusive Means of Participation. 182

Quarra — The Laser Starway: A Light Bridge to the Nearest Stars 188

Wilson — Food Design: A Primer with Thoughts on Food Design in Space. 205

Ziolo — The Astrobiological Matrix: Minimizing the Trauma of First Contact. 215

Ziolo — Religious Engineering for the K0→ K3 Transition. . . 236

2014 261

Almon Breaking Bad: Creating Predictive Models For Psychological Breaking Points During Long-term Space Travel 264

Batt Interstellar Travel in the New Science Fiction . . . 276

Faddoul Space Architecture: Design Aspects for Extended Space Travel 285

Laine Exoplanets and Survival Kits: Where to Go and What to Pack for an Interstellar Voyage? . . . 296

Maccone KLT Filtering, Even from Relativistic Sources . . . 301

Swan Space Mineral Resources: Prior to, During, and After Interstellar Flights 304

Talbot & Ellis Goldilocks Zones: A Fine Grained Exoplanet Taxonomy 328

Wisian Military Planning for Interstellar Flight 339

Bordag & Zatloukal Minimal Requirements for a Fully Autonomous Health Care in a Closed System 346

2015 352

Burke A Framework for the Design of Ethical Robotic Probes 356

Cable, Lunine, Beauchamp, & Sotin What's Right for Life: Can Life Exist Without Water? 362

Cripe Neurocognitive Behavioral Health Monitoring During Space Missions 384

de Vigne Interstellar Embryos through Interstellar Teens: Negotiating Human Development through Education in an Off-World Context. 396

Gates Loading, Please Wait: Long-Distance Crisis Communications 402

Muofhe Dawn of the African Space Era. 410

Muofhe The Square Kilometer Array. 426

Richey Starship Alpha: The Case for an Earth-based Proto-starship 429

Sherriff & Welch A Roadmap to Interstellar Travel: The Societal Challenges 444

Shostak Technological Approaches That Can Search for Life Close Up in Exotic Destinations 466

100 YEAR STARSHIP™
canopus 310 LY

10 YEARS

INTRODUCTION

The challenge of human interstellar travel is more than a propulsion problem, more than an engineering problem. There are incredibly difficult social and cultural questions to answer as we go. Where are we going? Who gets to go? And what do we do when we get there? And most importantly, how does it benefit the people on earth? For the last few decades, astronauts and rocket scientists have been the face of space exploration but they weren't the only ones making it happen. Achieving the capabilities for travel to another solar system will require a global, collaborative effort.

(continued next page)

100 YEAR STARSHIP EXISTS TO MAKE THE CAPABILITY OF HUMAN TRAVEL BEYOND OUR SOLAR SYSTEM A REALITY WITHIN THE NEXT 100 YEARS. *We unreservedly dedicate ourselves to identifying and pushing the radical leaps in knowledge and technology needed to achieve interstellar flight, while pioneering and transforming breakthrough applications that enhance the quality of life for all on Earth. We actively seek to include the broadest swath of people and human experience in understanding, shaping and implementing this global aspiration. Taking up this task ignites not only our imagination, but the undeniable human need to push ourselves to accomplishments greater than any single individual.*

The Why

When we explore space, we garner the greatest benefits here at home. The challenge of traveling to another star system could generate transformative activities, knowledge, and technologies that would dramatically benefit every nation on Earth in the near term and years to come.

We believe that pursuing an extraordinary tomorrow will build a better world today.

Today, people take very much for granted (but would loath to surrender) the benefits space exploration has provided right here on Earth: Global positioning satellites (GPS); remote sensing for water, minerals, and crop and land management; weather satellites, arms treaty verifications; high-temperature, lightweight materials; revolutionary medical procedures and equipment; pagers, beepers, and television and internet to remote areas of the world; geographic information systems (GIS) and algorithms used to handle huge, complex data sets; physiologic monitoring and miniaturization; atmospheric and ecological monitoring; and insight into our planet's geological history and future—the list goes on and on. Also remarkable is the boom in other science and engineering fields—and the accompanying wave of social creativity, artistic expressions, and educational innovation—that followed the Apollo missions and each successive leap in space exploration. Technologies created for and made possible by space exploration permeate, shape and are an integral part of our world.

Travel to the stars will require: the development of revolutionary non-chemical technologies to generate, harness, control and store enormous quantities of energy safely; radical advancements in closed-loop life-support systems and sustainable habitats; new insights into human development, health, behavior and training; quantum leaps in agriculture, communication and handling massive data sets; advances in robotics, automation, intelligent systems, and manufacturing techniques; as well as deeper, robust understanding and innovative approaches to social structures, education, economics and finance, legal frameworks, politics and international collaboration.

Programs to establish a human presence on the Moon, Mars, or elsewhere in our solar system will be stepping-stones to the stars.

All the capabilities needed to accomplish human interstellar travel are the same ones required for successful human survival.

Reality Check

The concept of humans traveling to other star systems may appear fantastical, but no more so than the fantasy of reaching the Moon was in the days of H. G. Wells. "The First Men in the Moon" was published considerably less than 100 years before humans landed on the Moon (1901 vs. 1969), and the rapidity of scientific and technological advances was not nearly as great as it is today. The truth is that the best ideas sound crazy at first. And then there comes a time when we can't imagine a world without them.

The Challenges of Interstellar Space Exploration

Advances, inventions, and innovations that arise through exploring space have the power to dramatically improve daily life on Earth for citizens of every country. Humans landed on the Moon less than ten years after we first achieved Earth orbit. And for decades since, images and data from satellites, Mars rovers, space telescopes and deep space probes bring the universe down to Earth.

Given the success of the Space Shuttle, Skylab, Mir, International Space Station (ISS), and Shenzhou, combined with omnipresent sci-fi images of humans traipsing across the galaxy, the complexities of interstellar travel may be trivialized. However, reaching another star is an incredibly difficult challenge, and it involves much more than meets the eye.

10%

The First 10 Years of the 100 Year Starship

Jason D. Batt

Abstract

In February 2012, DARPA (U.S. Defense Advanced Research Projects Agency), supported by NASA (National Aeronautics and Space Administration) Ames Research Center gave its stamp approval and selected the Houston based Dorothy Jemison Foundation for Excellence proposal to for seed funds for an ambitious concept – pursuing the possibilities for human interstellar flight within the next hundred. The impetus for the competitive grant was the desire to promote and engage the kind of explosive growth in research and development, innovation, education, and economic expansion across all science and technical fields similar to that seen during the 1960's space race to the moon. Meeting the challenge of human interstellar flight stands to transform our world as much or more than Sputnik or commercialization of DARPA's ARPA net, which evolved to the Internet.

The new organization in 2012 established its mission as: "The 100 Year Starship™ exists to make the capability of human travel beyond our solar system a reality within the next 100 years. We unreservedly dedicate ourselves to identifying and pushing the radical leaps in knowledge and technology needed to achieve interstellar flight, while pioneering and transforming breakthrough applications that enhance the quality of life for all on Earth. We actively seek to include the broadest swath of people and human experience in understanding, shaping, and implementing this global aspiration."

Later that year, 100YSS held the first of many symposia with a singular purpose: "to initiate the wide-based understanding and enthusiasm needed to achieve an audacious, global ambition—make the capabilities for human interstellar travel a reality within 100 years."

In the last ten years, The 100 Year Starship has held several symposia, introduced an intensive form of thought development called Crucibles, held many online workshops, presented the dream and mission of interstellar in countless forums, organized educational efforts, established wide partnerships with different organizations, and engaged a wide-diversity of the population towards the most audacious mission in human history.

As the editor of the 100YSS Symposium Conference Proceedings, I look back at the first ten years of The 100 Year Starship and explore how has 100YSS changed the landscape of interstellar development in the last ten years? What are the results of the activities and initiatives 100YSS has championed? What are the challenges that we must face in the next ten years? What have been the challenges that have arisen in accomplishing the 100YSS mission and what are the challenges to come? Finally, I'll explore what the potential roadmap might be for the next ten years.

Keywords

interstellar, 100YSS, 100 Year Starship

1. The Genesis of 100YSS

In October 2010, Pete Worden, director of the NASA Ames Research Project, in coordination with DARPA program manager Paul Eremenko, announced the beginning of the 100 Year Starship Project [1]. After hosting an initial smaller gathering of key thoughts leaders in summer 2011, the DARPA-lead initiative released an RFP for what would be a separate organization, seed-funded with $500,000 grant. As well, the organization held a public symposium in Florida to raise interest and pool knowledge towards a 100 yearlong effort to develop the capabilities for interstellar travel.

In 2012, The Dorothy Jemison Foundation for Excellence (DJF) was awarded the $500,000 grant for their proposal [1,2,5] "An Inclusive Audacious Journey Transforms Life Here on Earth and Beyond." DJF was the prime and submitted the proposal in collaboration with Icarus Interstellar and the Foundation for Enterprise Development. Led by former astronaut, physician, engineer, and social scientist Dr. Mae Jemison, the new organization adopted the existing DARPA name 100 Year Starship®.

2. The Winning Proposal

The winning proposal began with a reflection on the concept of the sky and of space from the earliest human, arguing that space exploration began at the initial moments of human observation of the night sky—thus, space belonged to all of humanity and the dream of exploring the stars is one of our earliest collective ambitions.

A central theme of the proposal focused on the element of inclusivity: "The cultural, social, and political communities, as well as their scientific and technical counterparts, must see human travel to the stars as aspirational, indeed possible, not just for the few, but participatory and inclusive for all" [15]. Moreso, the proposal argued that this lack of broad inclusivity hampered space exploration initiatives due to a lack of broad support, both politically and financially. The proposal made a bold assertion: "Consequently, any vision for the 100 Year Starship program would be limited, even neglectful, if it ignored the inclusion in the journey, at all levels and in all disciplines, of the broadest swath of humanity. In fact, an elitist or exclusionary posture on the part of planned missions to the stars may very well result in failure."

Yet, the proposal did not serve only to provide warnings. An optimistic tone wove itself throughout the 2000 words. The proposal put forth that the 100YSS program could "ignite the 'universal' imagination and propel humanity to the stars" [15]. Again, the focus was on the broad engagement of all of humanity. Interstellar travel was humanity's dream and thus required all of humanity to achieve.

In a manner the proposal had a subtext: igniting the "universal imagination" was not starting a new fire but stirring the ashes of the oldest fire in the human soul. From this unspoken perspective, the proposal asked that any interstellar initiative also be required to demonstrate how its achievements (and costs) would benefit those not just colonizing a new planet but for those still here on this Earth. Whatever the motivation for traveling to the stars will be, the ultimate truth is that most of humanity will not experience that—while a collective dream, it would only be experienced by a small representation of humanity.

The proposal provided a detailed overview of the expected challenges of the program. The first was to solve for "safe, reliable, and cost-effective interstellar transportation" that was built upon advanced propulsion technology and safe-guarded the human lives making the journey [15]. Second, the next challenge focused on the human being itself—providing safe and healthy "star-bound space communities" for the trek. Third, the proposal recognized the need to accelerate human productivity through advances in AI and robotics, and more. The fourth challenge before the program was to support the commercialization of space exploration. The fifth challenge addressed the need to support a vibrant in-system planetary development, focusing on the Moon, Mars, and beyond. Finally, the proposal recognized that all these efforts must ultimately enhance life on Earth.

The proposal concluded with a definitive and rigorous call to action: "Our purpose for 100YSS must be clear and focused. The goal is to travel beyond our solar system and to another star system. 100YSS must consistently measure its progress towards this goal by routinely conducting scientific, technological, behavioural, governance, and fiscal audits. Discipline must constantly be embraced and practiced. The success of 100YSS will depend upon unflagging energy and motivation, especially in the wake of unprecedented obstacles. Outreach programs to maintain awareness, enthusiasm, and fiscal support will be strategic" [15].

Finally, the proposal reminded us that this initiative was both audacious and required "an audacious spirit" to achieve. In short, the proposal put forth the reality of the challenge to be nothing less than one of the most ambitious, if not the most ambitious, of all human endeavors—the fulfillment of the earliest collective dream that arose when our ancestors turned their eyes to the skies.

3. The New Website: 100YSS.org

After launching the 100 Year Starship, Dr. Mae Jemison worked with the Barbarian Group and other key partners to develop the initial website and its content. The website [2] exclaimed: "Taking up this task ignites not only our imagination, but the undeniable human need to push ourselves to accomplishments greater than any single individual." As the hub of all things 100YSS, the site continues to this day to be highly trafficked and utilized as the clearing house of all 100YSS activity.

4. The Symposia

4.1 2012 Public Symposium

The first symposium, held in Houston, Texas, under the new organization was held that very same year, in 2012 [7]. The theme for the event was: "Transformation to Transition—The Journey Begins." President Bill Clinton served as Honorary Chair and welcomed all attendees with a direct address. "The four-day event hosted workshops, panels, and presentations addressing technological and societal issues surrounding interstellar travel. The group of more than 300 talented people represented physical and social sciences, engineering, education, finance, medicine, business, and the arts - hailing from the USA, UK, China, South America, Africa, and Australia." The event saw the launch of critical 100YSS Symposia programming such as Accelerating Creativity. The new branding and logo for the new organization debuted as well. That first year also produced the first of the Conference Proceedings, a large volume of nearly 500 pages, collecting the wide array of papers from the Technical Tracks.

4.2 2013, 2014, 2015 Public Symposium

After that, the second symposium was held in 2013 [8]. The theme of that Symposium echoed a core focus of the winning proposal: "Pathway to the Stars, Footprints on Earth." The Symposium Proceedings recapped: "Achieving the interstellar journey in many ways must build upon, promote and establish fundamental research, technology development, societal systems and capacities that facilitate ready access to our inner solar system and which will leave an indelible mark upon life on earth." The second symposium continued many of the same events and focuses of the prior year but also featured the first Science Fiction Stories Night, hosting a conversation between science fiction authors Mary Doria Russell, Karin Lowachee, Ken Scholes, actor and director Levar Burton, 100YSS Principal Dr. Mae Jemison, and hosted by Jason Batt.

The third symposium, in 2014, like the first two, was held in Houston [9] and saw the premiere of MAKERS, a documentary focusing on women space explorers. As well, that Symposium saw a creative plenary on Steampunk thinking for space missions. The 2015 Symposium focused on the theme "Finding Earth 2.0" [10]. The Proceedings for 2015 recapped the event: "The tracks and sessions were invited to reflect on the future potential moment when we would discover another Earth in orbit around a distant star. Presenters, speakers, and attendees at the Symposium asked questions and sought answers to what specific capabilities and systems—scientific, technical, and societal—will be needed over the next five to twenty-five years, not to merely suggest or catalogue Earth analogue candidate exoplanets, but to identify at least one definitive Earth 2.0." That fourth symposium also featured the first of the Canopus Award for Excellence in Interstellar Fiction.

The Symposia were designed to bring subject-matter experts from all fields together. For example, Dr. Karl Aspelund, Associate Professor of Textiles, Fashion Merchandising and Design, heard about 100YSS in an NPR interview with Dr. Mae Jemison. Aspelund reached out and was invited to present on the subject of clothing for space missions. That presentation was warmly received for its innovative content, looking at critical matters often overlooked in discussions on interstellar travel. Other subject-matter experts and critical thought leaders from across the globe, not usually associated with the space industry or interstellar were brought together through the

Public Symposia including Casey Hudson, creator of video games Mass Effect and Star Wars: Knights of the Old Republic, musician Kenji Williams and Bella Gaia, and others.

5. Public Symposia Conference Proceedings

100YSS published Conference Proceedings from the Technical Tracks, Plenaries, Workshops, and additional activities from the Symposia. The four Proceedings collected are in excess of over 2000 pages and include a wide diversity of papers from subject-matter experts. Below is a listing of selected papers representing the broad content the Symposia hosted (in order of presentation):

- Karl Aspelund "Considering Clothing and Textiles on an Interstellar Mission"
- Robert M. Freeland II and Stephen Baxter "A Fully Decelerated Probe to Alpha Centauri Using a Perturbed Project Daedalus Design"
- Claudio Maccone "Interstellar Links by Focal Spacecrafts"
- David Weiss "The Foundry in the Starship"
- Paul Ziolo "Devising the Prime Radiant: A Strategy for Seeding the K0→K3 Transition"
- Paul Ziolo "Religious Engineering for the K0→ K3 Transition"
- Jeffrey Nosanov "Solar System Escape Architecture"
- Charles J. Quarra "The Laser StarWay: A Light Bridge to the Nearest Stars"
- Caitlin Contag "Every Body is an Ark: How the Microorganisms We Carry Will Impact an Interstellar Mission"
- Antoine G. Faddoul "Sustainable Design for Extended Space Journeys"
- David Wilson "A Primer with Thoughts on Food Design in Space"
- David Orban "A Pathway for Global Communications as an Inclusive Means of Participation"
- Peter Swan "Seeing the Space Elevator in the Future"
- Kurt Zatloukal "Minimal Requirements for Fully Autonomous Health Care in a Closed Environment"
- Claudio Maccone "A Magnifying Lens as Big as the Sun: Exploring Exoplanets Without Getting There"
- Pamela Reilly Contag "The Concomitant Search for Habitable Exoplanets and Life in Space: Lessons from Evolution on Earth"
- Alires Almon "Breaking Bad: Creating Predictive Models for Psychological Breaking Points During Long-term Space Travel"
- Paulie Laine "Exoplanets and Survival Kits: Where to Go and What to Pack for an Interstellar Voyage"
- Antoine G. Faddoul "Space Architecture: Design Aspects for Extended Space Travel"
- Jaym Gates "Loading: Please Wait: Long-Distance Crisis Communications"
- Morgan Cable "Can Life Exist Without Water?"
- Mmboneni Muofhe "The African Space Race"
- David Burke "A Framework for the Design of Ethical Robotic Probes"
- Curtis Cripe "Neurocognitive Behavioral Health Monitoring"
- Abigail Sherriff & Chris Welch "A Roadmap to Interstellar Travel: The Societal Challenges"

6. Canopus Award for Excellence in Interstellar Writing

Beyond the symposia, 100 Year Starship expanded its efforts through a wide variety of initiatives and programming designed to address each of the six critical challenges put forth in the initial proposal.

In 2015, 100YSS established the Canopus Award for Excellence in Interstellar Writing [2]. The Canopus Award recognized the best fiction and non-fiction writing that focused on expanding our understanding of "challenges, opportunities, pitfalls, and rewards of interstellar exploration." In launching Canopus, Dr. Mae Jemison said: "Imagination, varied perspectives, and a well-told story are critical to advancing civilizations. In particular, beginning with the simple question 'What if?' pushes us to look beyond the world in front of us and to envision what could be, ought to be, and other realities. Both science fiction and exploratory non-fiction have inspired discovery, invention, policy, technology, and exploration that has transformed our world." The award is named for the second brightest star in the night sky, Canopus, which connects humanity's past, present, and future through fact and fantasy. Over the millennia Canopus not only heralded planting seasons in the Rift Valley but was a major navigation star for everyone from the Bedouin of the Sinai and the Maori of New Zealand to deep space probes like Voyager. Just as Canopus has helped explorers find their way for centuries, great writing —telling a story well —is a guidepost for current and future interstellar achievement.

7. Look-Up: One Sky

In 2017, 100YSS launched Look Up: One Sky. Look Up had its origins in a conversation between Dr. Jill Tarter (co-founder of the SETI institute), director and actor Levar Burton, and Dr. Mae Jemison on how to reconnect humans to the wonder of space, "to the Earth and the importance of audacious goals for the health of humanity" [2]. The Look Up initiative invited every person on Earth to stop and share what they see when they Look Up. The event was designed to remind us "[t]o acknowledge that we are Earthlings. We share the same origins, the same sky. And our evolutionary path necessarily includes capturing our place in the greater universe." 100YSS also developed the Skyfie™ app to anchor the initiative. The Skyfie™ app that allowed people to take their photos and videos of the sky and form a global Sky Tapestry. 100YSS designed the Sky Tapestry as an interactive globe that uploads people's geo and time-tagged Skyfie contribution from across the world in almost real-time—the photos, video, text, and audio may be observed by anyone with internet access on topographical map zooming into within 10 miles of contribution keeping identities and specific locations private. The app allowed people to discover what others across the globe experienced, what they thought, and what they felt when they looked up at the sky. The average user was engaged with the app for 16 minutes of focused interaction. The event trended at #5 on Google's social media tracking and had global engagement.

Additional Look Up events were held including one celebrating the 50th Anniversary of the Moon Landing titled Look Up Apollo 11. Look Up was also a core element of the Yuri's Night Celebration at the California Academy of Sciences, encouraging participants to utilize the Skyfie app and share their experiences when they look up.

8. Education

100YSS also focused on the tasks of education. Three specific areas propelled 100YSS' efforts in education: Center for Advanced Aerospace Manufacturing Study for Orange Coast College, Expo Inspire, Harvard Hackathon, and the TRUE Books, amongst others. Education has been foundational to 100YSS' activities. The following represents highlighted educational activities:

8.1 Expo Inspire

Expo Inspire was an educational event designed with hands-on learning activities for children. Children explored rocketry through designing and launching Alka Rockets. They also learned through different interactive exhibits such as: "Learn the Science of Magic," "Study the Stars in the Planetarium," "Walk the Edge," and more. The entire event was an immersive educational event that captured students' attention [5,6].

8.2 TRUE Books

100YSS partnered with Scholastic Books to produce four TRUE Books on topics critical to interstellar exploration. Using the 100YSS mission, the books "are a source of reliable, clearly presented, engaging, and cutting-edge information for elementary students on a broad range of topics. From the sciences of our sun to life sciences and social impact of exploration, the series will bring the knowledge of highly regarded experts to students around the world" [11].

8.3 Center for Advanced Aerospace Manufacturing (CAM) at Orange Coast College in Orange County, California

In coordination with the Orange Coast College, commissioned by the Orange Coast College Foundation, 100YSS developed a vision and work plan framework for the Center for Advanced Aerospace Manufacturing, its program and curriculum in 2014. The Center would focus on supporting aerospace manufacturing for deep space missions. Recognizing the need for expanded workforce development for highly technical mechanical engineering, the 100YSS study provided in detail the framework and structure for a robust trade educational program to support future interstellar initiatives. The study concluded that in the next near decades, there would be a wave of manufacturing growth in the aviation and space-based industries. The study proposed a program based on work-integrated learning to train students on advanced aerospace manufacturing.

8.4 Harvard Hackathon

100YSS sponsored the "Astropreneurship and Medicine in Space Hackathon" in 2016 in partnership with the Harvard University Science Center. Individuals spent a week at Harvard in coordinated hack activities addressing biomedical, social, policy, and investment challenges to deep space mission, in particular, interstellar.

9. Crucibles & Workshops

100YSS also held critical thought leadership events on key technical and deep knowledge topics. The Crucibles and Workshops brought together small groups of individuals around topics critical to interstellar travel.

9.1 Crucibles

The Crucibles were gatherings of deep subject matter experts around critical ideas to develop "[t]he knowledge, technological capabilities and societal systems required for human interstellar travel" [2]. To achieve an incredible synthesis of thought and innovation, the Crucibles were facilitated using the DJF TEWS Experiential Discovery Methodology.

The first Crucible was titled: "The Virtual Human" and brought together leading subject matter experts from across the globe to look at the future of medicine and human health. The Crucible was held in Graz Austria in coordination with the Medical University of Graz. The Virtual Human Crucible included expert participants and panelists from organizations and institutions such as the National Institute of Biomedical Imaging and Bioengineering, German Cancer Research Center, Space Research Institute Austrian Academy of Sciences, Molecular Science Institute USA, Swiss Federal Institute of Technology, Institute of Process and Particle Engineering, Stanford University, Rutgers University, University of Savoie Mont Blanc, University of Florence, University of Klagenfurt and 100 Year Starship. Sponsors included Medical University of Graz, MEFOgraz , Das Land Steiermark, ÖAW, Graz University of Technology, Institut für Weltraumforschung, STADT Graz, and 100 Year Starship.

Dr. Jemison encouraged participants to identify key issues on the path to attaining interstellar capabilities with a singular focus on forging thoughts of the extraordinary into reality. The Virtual Human Crucible's primary focus was not Artificial Intelligence but rather a platform/laboratory fostering predictive capabilities for testing, performing clinical trials, and epidemiological and evolution studies.

Other Crucibles are in different stages of development now and will be critical to the next ninety years of 100YSS.

9.2 Workshops and Public Forums

100YSS also hosted several workshops. The workshops and virtual forums were designed to encourage conversation about topics essential "to making interstellar capabilities a reality" [2]. The workshops were limited in scope, time, and audience, within the conversations were recorded and stored for future reference. The first of these was on the topic of the "What If . . . 2020"—challenging participants to conceive of what would be needed to launch an interstellar mission in twenty years: "What if a mission designed to take humans beyond our solar system had to be launched in the year 2020? No, 100YSS is not planning a stealth mission. Starship 2020 is an exercise to help identify and understand more clearly the challenges, possible solutions, and any absolute roadblocks to human interstellar travel TODAY" [1]. Participants were challenged to use only the technologies and capabilities viable at that time to design a human interstellar mission prior to 2020.

The second was on the topic of "First Contact" and was held during the early days of the COVID-19 pandemic. First Contact asked participants to consider the potential momentous day when we discover other intelligent life in the universe and prompted conversation around these questions: "Are we ready? How will we respond? How should we respond? What can we anticipate from 'them'? Do concepts like the Prime Directive have any application in real life? When encountering another form of intelligent life, will we aspire to our better selves and reach out in peace or will current trends in our relations here on earth find themselves repeated when we encounter others amongst the stars?"

10. Symposia-birthed Connections & Activities

The symposia created unique opportunities for networks and meetings of passionate and talented individuals. While numerous organizations and events arose from the Symposia and other 100YSS events—far more than are mentioned here--the following highlights the diversity of thought and focus that have been nurtured and developed through 100YSS activities.

10.1 Breakthrough Starshot

The organizers of Breakthrough Starshot met at a 100YSS Public Symposium, highlighting the essential nature of the Symposium to connect individuals and nurture new initiatives. Breakthrough Starshot is a multi-organization project that "aims to demonstrate proof of concept for ultra-fast light-driven nanocrafts and lay the foundations for a first launch to Alpha Centauri within the next generation" [3,4,13]. 100 Year Starship Principal, Dr. Mae Jemison, was a member of the initial group that announced this initiative and continues to be an active participant.

11. Notable Activities

Over the last decade, 100YSS has been actively involved in the mission of developing interstellar capabilities across the globe. The following represents a few of those additional activities:

- March 2013, 100YSS participated in the European Union Parliament event, EU Science: Global Challenges, Global Collaboration.
- Selected papers from the 2011 DARPA Symposium were published in the Journal of the British Interplanetary Society.
- In 2014, 100YSS opened their EU hub in Brussels, Belgium.
- In 2014, 100YSS collaborated with IAA (International Academy of Astronautics) to present session on exploration beyond our solar system session at IAC in Toronto.
- In 2015, 100YSS participated in the Art of Future Warfare project developed by the Atlantic Council in Washington, D.C.
- 100YSS consulted on the film Passengers in its early script development.
- In 2018, 100YSS participated in the Universal Peaceful Use of Space Technology International Seminar held in the Hai'nan province of China.

12. Public Perceptions & Activities

Analysing Google trend data for the word "interstellar" shows a marked public interest in the term beginning in late 2012 and early 2013 [16]. The interest spike in 2014 correlates with the release of the film Interstellar. However, the term continues years after the film to have a significant number of searches. Since the launch of 100YSS, the term "interstellar" has become embedded in the public consciousness distinctly far more than before the launch of the organization.

13. The Next Ten Years: Nexus & The Way Institute

13.1 Nexus 20223

In 2023, celebrating the tenth anniversary of the organization, 100YSS will hold the next version of its symposia: the Nexus held in Nairobi, Kenya.

The Nexus' focus is "Where Space, Purpose, and Culture Collide." The Nexus is designed to be: "THE space gathering to experience, connect, contribute to, envision, inspire & be inspired, create, share, explore, learn and foster an extraordinary future while building a better world, here and now . . . Nexus brings together the range of human experience, skills, knowledge, creativity, passion, commitment, resources, cultures, technologies, policy, investment, education, art, perspectives, and motivation needed to achieve such an extraordinary future" [12]. Nexus 2023 will be a virtual and live event allowing participants from across the globe to participate. Activities will include those from prior Symposia (e.g. Canopus Awards, plenaries, classes) and introduce new approaches to encourage high engagement from attendees (e.g. Thought Pools).

13.2 The Way Institute

At the core of the next ten years will be the launch of The Way Institute. From the website:

"Before we can reach another solar system, we first have to know the way. The Way Institute is dedicated to identify, support, invest in, develop and advance the knowledge base required for successful human interstellar travel. Fundamentally a research and development organization, The Way will focus upon and explore the 'bleeding edge' physical, life and social sciences, engineering and technologies, economics, art, culture and systems capabilities needed for missions beyond our solar system. We push for radical leaps forward in technological advancement wherever possible, rather than always merely seeking incremental technology improvement" [2].

Further, the Way Institute lists the following priorities of its activities:

- "Foster, coordinate, initiate, and seed the required research, development, and innovation necessary to achieve successful human interstellar travel.
- Push the research and technology necessary to translate 100YSS discoveries and innovations (technological, medical, scientific, and sociological) into tangible benefits that enhance the quality of life here on Earth.
- Actively apply learnings and capabilities from space exploration to enhance the quality of life today.
- Focus on engaging outstanding subject matter experts specializing in the fields required to make interstellar exploration a reality.
- Integrate a trans-disciplinary research philosophy and systems approach across various fields, such as social sciences, fundamental science and applied science and engineering disciplines.
- Identify, assess, maintain database and follow the capacities needed for interstellar travel" [2].

14. The Next Ninety Years

100YSS continues to look forward and address each of the six challenges of the initial proposal. As well, 100YSS is focused on creating a new model for enablement and capacity building the human deep-space exploration. 100YSS continues to develop systems for tracking and classification of research (The Way Institute is central to this). 100YSS orients and guides its work along the path with its partners in developing an integrated interstellar plan, a technology map, and online research and thought vault. Finally, 100YSS will continue to host Crucibles, workshops, the Canopus Awards, and other initiatives to meet the six proposal challenges, always looking towards enabling radical leaps in key technologies to achieve the mission of interstellar travel.

In the initial days of 100YSS, Dr. Mae Jemison noted: "The road to achieve the capabilities for human interstellar travel will not be straight, flat, unidirectional, one dimensional or even predictable from year to year" [7]. This reflection underscores the necessary flexibility and innovation of the 100 Year Starship as it continues to pursue its mission.

References

1. 100 Year Starship. (2022). In Wikipedia. https://en.wikipedia.org/w/index.php?title=100_Year_Starship&oldid=1107432300

2. 100 Year Starship. (1969, December 31). https://100yss.org/

3. Breakthrough Initiatives. (1969, December 31). https://breakthroughinitiatives.org/events/2

4. Breakthrough Starshot. (2022). In Wikipedia. https://en.wikipedia.org/w/index.php?title=Breakthrough_Starshot&oldid=1104112010

5. Dorothy Jemison. (1969, December 31). https://jemisonfoundation.org/

6. EXPO Inspire – Dorothy Jemison. (1969, December 31). https://jemisonfoundation.org/expo-inspire-2/

7. Jemison, M., & Batt, J. (Eds.). (2013a). 100 Year Starship 2012 conference proceedings: Transformation to transition : the journey begins (First edition). 100 Year Starship.

8. Jemison, M., & Batt, J. (Eds.). (2013b). 100 Year Starship 2013 conference proceedings: Pathway to the stars, footprints on Earth. 100 Year Starship.

9. Jemison, M., & Batt, J. (Eds.). (2014). 100 Year Starship 2014 conference proceedings: Pathway to the stars, footprints on Earth. 100 Year Starship.

10. Jemison, M., & Batt, J. (Eds.). (2015). 100 Year Starship 2015 conference proceedings: Finding Earth 2.0. 100 Year Starship.

11. Jemison, M., & Rau, D. M. (2013). TRUE BOOKS: 100 Year Starship. Scholastic & Children's Press.

12. Nexus Nairobi 2023 | When Space, Purpose and Culture Collide. (1969, December 31). https://nexusnairobi.org/

13. Overbye, D. (2016, April 12). Reaching for the Stars, Across 4.37 Light-Years. The New York Times. https://www.nytimes.com/2016/04/13/science/alpha-centauri-breakthrough-starshot-yuri-milner-stephen-hawking.html

14. TEWS – Space Race – Dorothy Jemison. (1969, December 31). https://jemisonfoundation.org/tews-space-race/

15. The 100YSS Proposal: An Inclusive and Audacious Journey. 2011/2012. Unpublished.

16. Trends: "interstellar from 2007 to 2022" https://trends.google.com/trends/explore?date=2007-06-01%20 2022-09-02&q=interstellar

2012

2015

100 YEAR STARSHIP™

2012 Public Symposium

*Transformation to Transition: **The Journey Begins***

The 2012 Public Symposium kicked off with a message from President Bill Clinton, Honorary Chair, emphasizing the rewards, challenges, and promise of space exploration. The four-day event hosted workshops, panels, and presentations addressing technological and societal issues surrounding interstellar travel. The group of more than 300 talented people represented physical and social sciences, engineering, education, finance, medicine, business and the arts - hailing from the USA, UK, China, South America, Africa, and Australia.

Mobilizing Emerging Countries Towards Interstellar Culture

The Case of Turkey

Alper Aslan and Bob Hawkins

Abstract

Emerging countries have become centers of attraction in the last few decades. BRIC (Brazil, Russia, India, China) and SEWEC (second wave emerging countries) are becoming key players in the current "near Earth" society. To enable our transition from a "near Earth" society into an interstellar civilization in the next one hundred years we need to take advantage of the political, economic, cultural and social opportunities of these rising countries. The aim of this paper is to present how this transition could be managed for an emering country based on the case of Turkey. First, a future scenario for space business, which was developed with local futurists in Turkey, will be presented. Second, we look at the current capabilities and the desired competencies that will help Turkey make the transition to an interstellar culture. As the final step, we propose resources and funding opportunities to make the necessary changes happen.

Keywords

Turkey, interstellar, culture, space, business, scenarios

1. Introduction

The planetary population is expected to grow to around 9 billion in 2100 [1] from the current 6 billion as in Figure 1. The key contribution to this growth is going to come from today's developing world. The population size of today's developing world is expected to grow from 4.6 billion to 8 billion whereas today's developed world will stay about the same level. However, in the coming decades progressively more of today's developing world will have fully developed economies.

It is generally accepted today that manned space travel is limited to a small number of countries with sufficient resources and technology. But the number of capable countries will be growing rapidly. In the next hundred years we can expect that most of the world's population could have the same access to the technology and funding needed for interstellar travel. In 100 years, the size of a country's economy may not be the deciding factor. The important question may be whether a country has the desire or the cultural and political will to take on an interstellar venture.

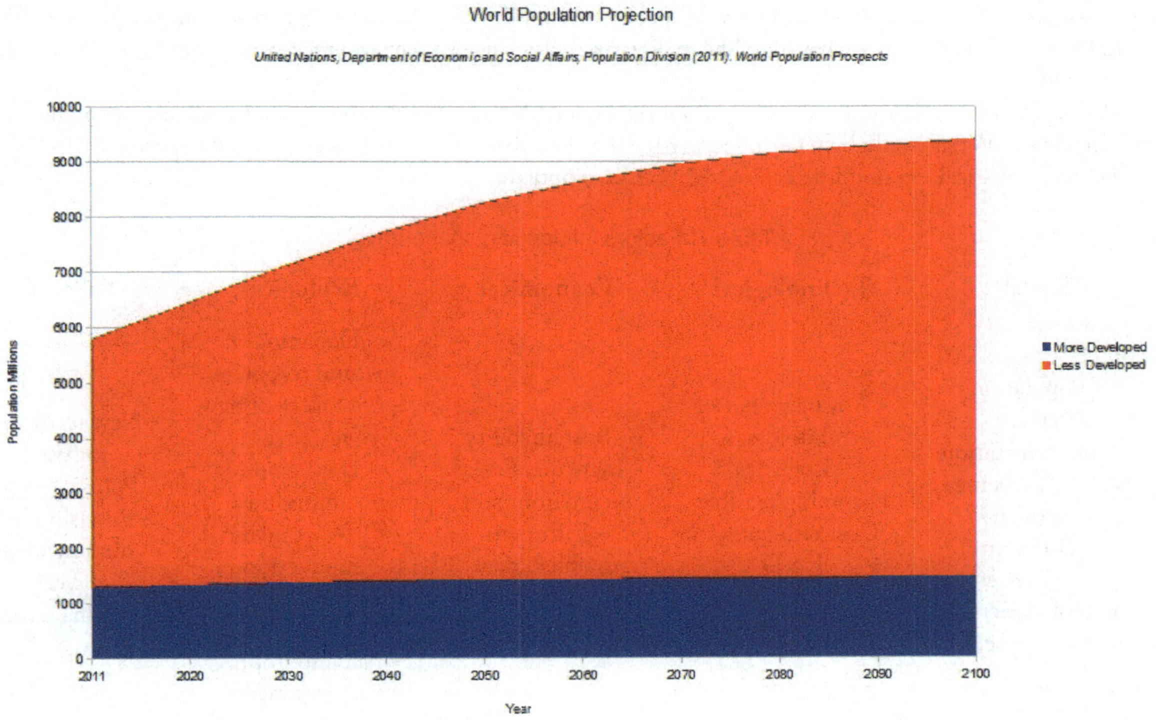

Figure 1. World Population Projection

As an example we will take a look at Turkey. Turkey is located in the center of a 25 trillion USD economy and 1.5 billion population. It is a strong member of the "second wave emerging countries" which includes countries such as Indonesia, Colombia, Vietnam, Mexico, Chile, Polandi Taiwan and South Korea [2]. In this paper, we will first describe Turkey. Then, we will show the current competencies of Turkey and how these fit with the space business scenarios generated in Turkey. We will conclude with recommendations specific to Turkey and how to extend this process to other countries.

2. Turkey: The Last 1000 Years

The story of Turks begins on the grasslands in Mongolia in the first century BC. After the big immigration from Central Asia to Middle East, Anatolia and partly Eastern Europe, the Turks decided to settle down in Anatolia which is one of the oldest occupied land on the planet Earth. Anatolia was home to Byzantine Empire from the 3rd century AD and after the entry of the Turks to Anatolia in 1071 AD, the Ottoman Empire was establised during the 13th century and lasted until early 20th century.

Turkey – The Republic of Turkey - today has a unique geopolitical location as the bridge between Europe and Asia and is bordered by 7 countries. Its current secular democratic parliamentary government was established in 1923. The GDP of Turkey which is 775 billion USD puts it just below Spain compared to European Union (EU) countries. Turkey has free trade with EU and Middle East.

Turkey economy today has a very high GDP growth with 8.5% growth in 2011 and is expected to keep the pace at least at 7% p.a. level. Turkey population is 74.7 million and is the second largest in Europe. It also has the youngest populationof any country in Europe. The young population makes a steadily growing labor force

In summary, Turkey presents a sort of an Eastern population profile with a European market.

3. Scenarios for Space Business

So it is with this context that The All Futurists Association of Turkey held a workshop in May 2012. Twenty three futurists attending the 14 weeks certification training "Futurism School."
One assignment was to look at the future of space travel for Turkey. The futurists focused on the question "What could the space business in Turkey look like in 30 years?". They used a scenario matrix methodology [3] to evaluate the answers.

The session started by brainstorming about the factors that would influence space travel. The main question at this stage was " What would benefit or deter space travel for Turkey?". These factors were categorized into STEEP (social, technological, economical, ecological, political) domains as Table 1.

Table 1. Factors influencing space travel

Social	Technological	Economical	Ecological	Political
- Population increase - Communication with other beings - Curiosity - Different experiences - Establishment of new settlements	- Advancement of technologies - Logistics becoming feasible - Cost reduction by R&D - New business areas	- Sustainability - Increase of costs on environment - Efforts to establish new markets	- Depletion of natural resources - Discovery of new resources - Depletion of commodities - Off-eart material production - Worldwide catastrophe - Need to improve environment	- New territories for war - Political needs - Provision of a minimum wealth level - China effect

Next, the factors were aggregated and ranked on importance and uncertainty. Then, it was narrowed to the factors with the highest significance and the most uncertainty.

The availability of natural resources ranked among highest and most uncertain. Also the race for national superiority – nationalism, war, dominance - was determined to be one of the most important factors and to be very uncertain.

A scenario matrix was created by taking these two significant factors and scaling them. The variability of natural resources ranges from abundance – there is plenty to go around - at the high end, to the opposite end - of depletion and scarcity. The political superiority scales from one group controlling all the decisions to the opposite of an open, diverse and cooperative political climate. Four scenarios come from looking at the extreme four corners of these scaled factors as in Figure 2.

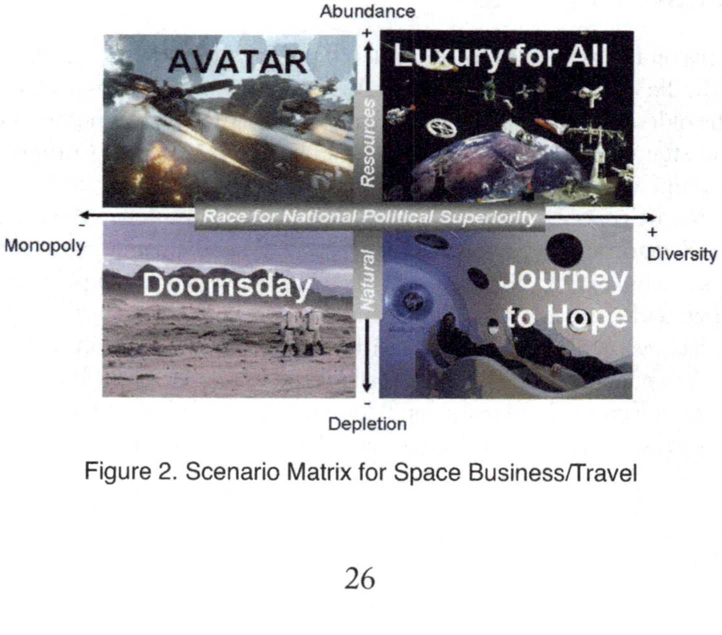

Figure 2. Scenario Matrix for Space Business/Travel

The four scenarios generated are:

- Avatar: With a small central power elite and abundant resources we end up with a scenario where limited groups control the resources for their own benefit.
- Luxury For All: Highly cooperative political climate and ample resources gives rise to a society with wide benefit that elevates everyone.
- Doomsday: A highly competitive political climate and limited resources creates an envrionment where everyone will fend for themselves in the struggle to survive.
- Journey of Hope: Wide cooperation between peoples even with low resources may lead to exploration for the benefit of all mankind.

From these options the workshop participants preferred to move toward "Luxury for all" future. This range of possibilities shows us that human race has a choice and there is a possible future with cooperative advance into outer space. The question is how to get there?

4. Capabilities of Turkey

There are three basic capabilities a nation needs to be competitive in space business[4]: government, private sector and human resources. Turkey has not been a subject of such a study yet but a consideration of the space industry in Turkey today can provide insight.

The space industry in Turkey is fundamentally driven by government. It is growing but still negligible part of the Turkish economy providing 350 jobs by 2009 as in Figure 3. However, the growth from 1998 (only about 70 workers) to 2009 was quite significant. Although there may not be many space workers today, that does not reflect on the future. The target date to establish the Turkish Space Agency has been announced as 2015 by the current government.

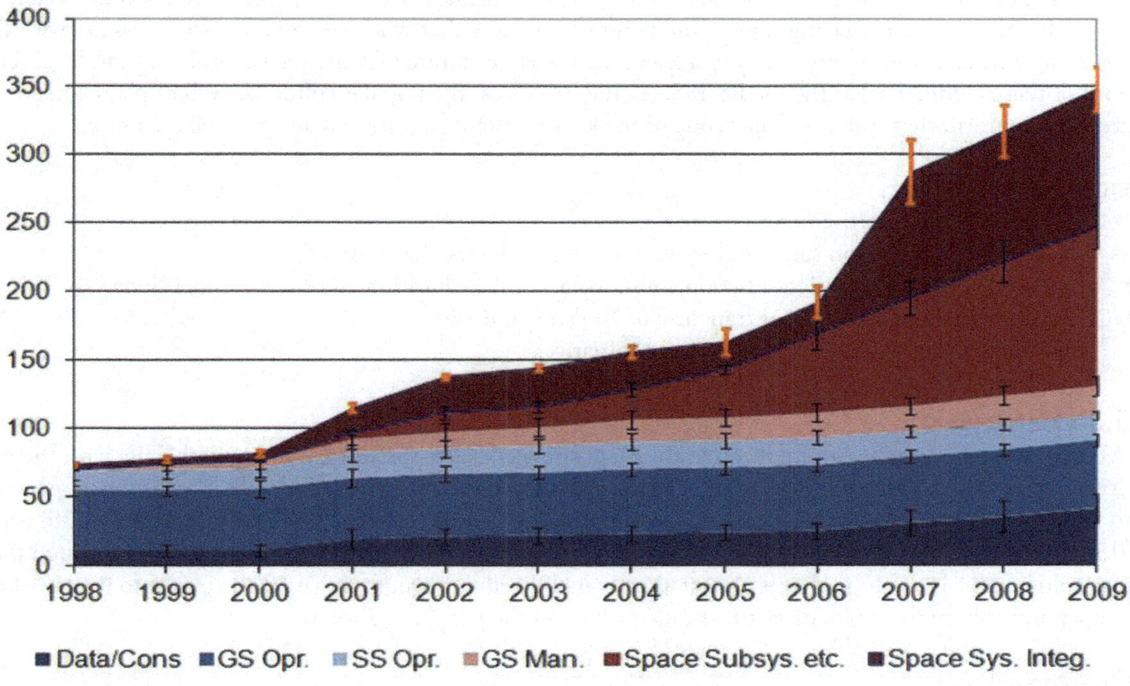

Figure 3. Turkish Space Industry – direct employment (estimated)

The private sector consists of a handful of companies such as Turkish Aerospace Industries (TAI) in Ankara, VIP Tourism for space tourism in Istanbul and Space Camp Turkey (SCT). SCT is a private company, supported by NASA, in the city of Izmir near Aegean sea. It is one of the few Space Camps in the world and has become a major hub for space education and training activities in the region. The future space workers of Turkey can be found at SCT today.

So far Turkey has not sent an astronaut to space. But two Turk space tourist/astronauts, who have bought tickets to low earth orbit, have attracted media attention in Turkey. From the human resources perspective there is a great untapped potential in Turkey. The tremendous interest of the young population is evidenced by the popularity of educational platforms like SCT, high school student interest in space and growing space curricula at Turkish universities.

5. Keys for Mobilizing Turkey towards Interstellar Culture

So how do we mobilize the general public support?

To generate interest and build the human resources needed for Turkey to succeed in future space programs there are three key areas: media, history and finance.

5.1 Media

If we look back to the coverage of space in Turkish media we see a lot of original productions which were produced by local artists. "The Man who saves the World" is a Turkish "B movie" satire produced in late 1970s. It is called the "Star Wars" of Turkey because of its space hero theme and also because of its use of bootleg footage from Star Wars, Battleship Galactica and others included in the movie. It was the first space themed movie made in Turkey. The movie was originally panned by critics and audiences alike but is now a cult classic. "The Test of a Turk with Space" was also a TV comedy show. The 2011 television series featuring the adventures of Turks traveling the universe.

An advertisement for liquid petrolium gas delivery became very popular on Turkish commercial scene. In the commercial, the drivers of the delivery truck make fun of the astronauts for making such a big deal out of going into space. The delivery guys can go anywhere – no problem, no fuss. This reflects not just on the promotion of the gas delivery company but on the Turkish "go anywhere" culture.

Another TV commerical for a brand of "raki", the national liquor of Turkey, introduced two astronauts on the moon drinking raki and playing music. The point of the ad is that wherever you are, when you are with friends, you bring a little of your home with you. Again, the nomadic culture of bringing Turks abroad can be observed in both of these commercials. The media reflects and expresses the popular culture as well as plays a major role to create and inform that culture. This is one of the keys to mobilizing toward an interstellar culture.

Our proposal is:

- Utilize space in advertisements
  Production of television series and movies focused on interstellar travel
- Encourage science and science fiction clubs, social media, education and story competitions
- Establish a national chapter or franchise of 100YSS. The media can fill in for the lack of an organized initiative.

5.2 History

We have seen the current culture through the media now consider the history and how this fits with the modern drive into the stars. Turks traveled over the ages from central Asia to Anatolia, Middle East and Europe in the 1st to 10th centuries. Expansion of the Ottoman Empire into Europe happened in the 15th through 17th centuries. This is a history of exploration, expansion and even conquest that lives today - as we have seen through the popular culture. This is a culture that is almost already mobilized to begin to move off the planet to new worlds. The history and culture for exploration are already present in the people of Turkey.

Our proposal is:

- Reinforce this exploration culture through the historical myths and the popular myths used in commercials.
- Make the connection between the historical expansion across the Earth and the future opportunities beyond this planet's surface.

- Explain, describe, create a vision for the new frontiers that lie somewhere beyond Earth from the 21st century to the 22nd century

5.3 Finance

All of these efforts will require financial support. This is the third key to mobilizing for the future of travel to the stars. A new government initiative has been announced in 2012 to help with the trade balance and attract strategic industries to Turkey. The new Investment Regime has identified Aerospace as one of the key industries where they will incent new businesses to move to Turkey.

Turkey offers the unique position of being the bridge geographically and geopolitically between Europe and Asia. Turkey is part of the EU free trade alliance and also has capital from middle east investing. Middle East investors are already active in Turkey in sectors such as healthcare, real estate and banking.

Our proposal is:

- Introduce a "Space Exploration Tax" on telecommunications in connection with the new investment regime to help pay for the development of the space industry
- Create an environment to become a catalyst to bring together the money and dreams from the Middle East with the technology from Europe to create a truly cooperative international alliance that combines the best of these two diverse economies.

6. Conclusion

This presentation highlighted many of the unique aspects of the Turkish culture that would make it ideal for launching an effort to reach to other planets in the years to come. It has been noted that although the technology is an issue today, the culture and the cultural priorities can be just as much of an obstacle - or a help. Turkey might be one of the countries to lead us off the planet.

In 100 years also Mexico, Malaysia, Brazil, Chile or any other country which today may be dismissed just a less-developed country, could have the same access to the technology as any country in the world. It could just be that the culture and the support of the people are what makes the difference in who leads the human race to new planets. Once we open our mind to that possibility, then it is easier to see that a we should start now to try to build a worldwide coalition for true cooperation in the effort.

Acknowledgements

We would like to thank Prof. Peter Bishop from The Univesity of Houston for bringing the authors together.

Bibliography

1. United Nations, "World Population Prospects", (2011).

2. Global Insight, "SEWEC Report", (2012).

3. Fahey, L., Randall, R, M., "Learning from the Future, Competetive Foresight Scenarios", Wiley, (1997).

4. Futron Corporation, "Space Competitiveness Report", (2012).

Considering Clothing & Textiles on an Interstellar Mission

Karl Aspelund, Ph. D.

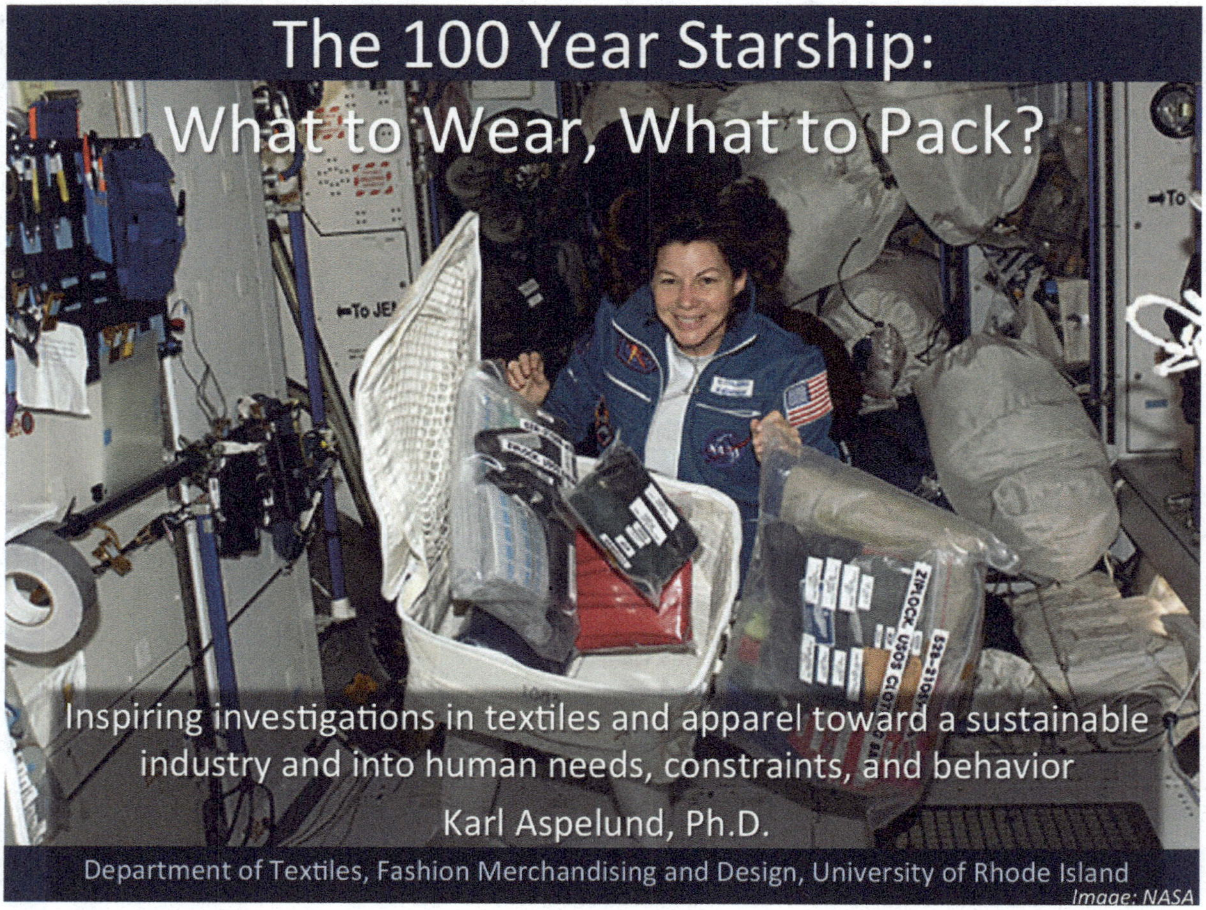

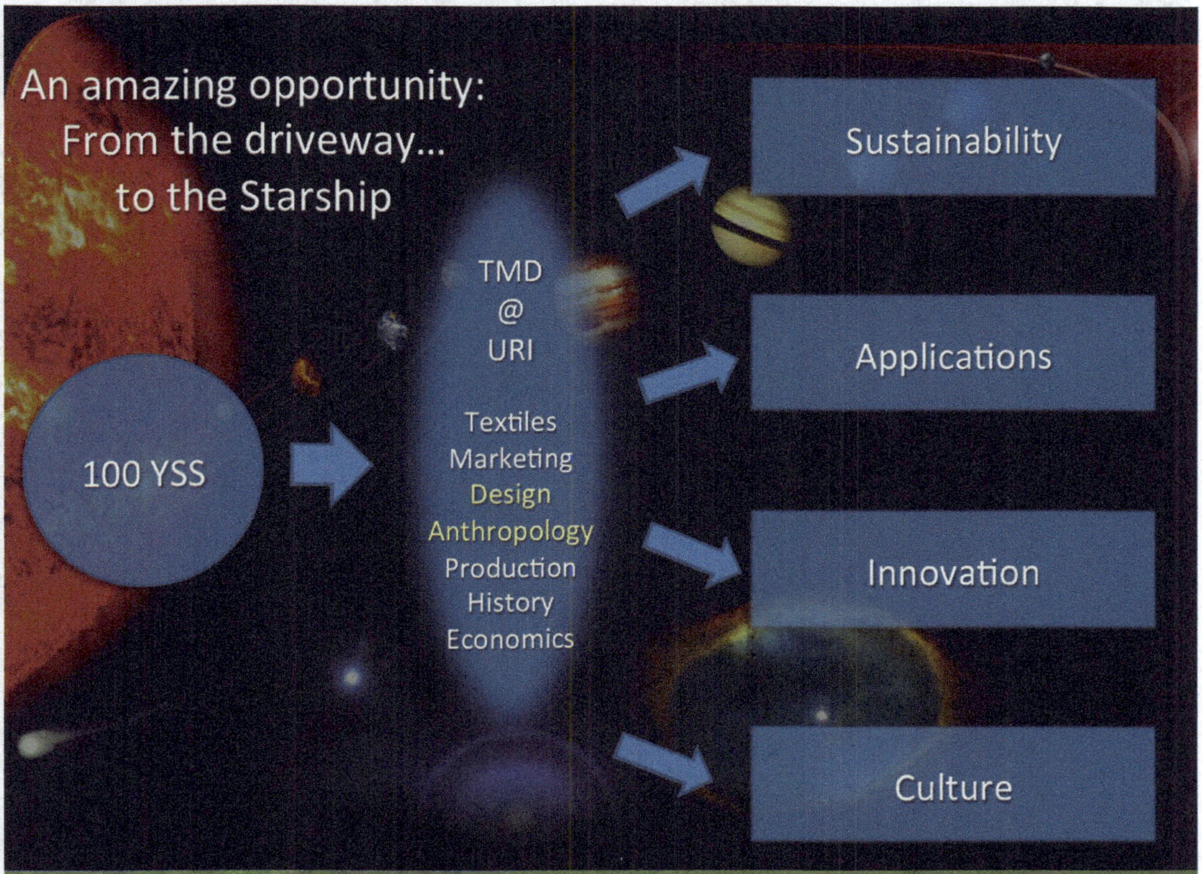
An amazing opportunity:
From the driveway...
to the Starship
100 YSS
TMD
@
URI
Textiles
Marketing
Design
Anthropology
Production
History
Economics
Sustainability
Applications
Innovation
Culture

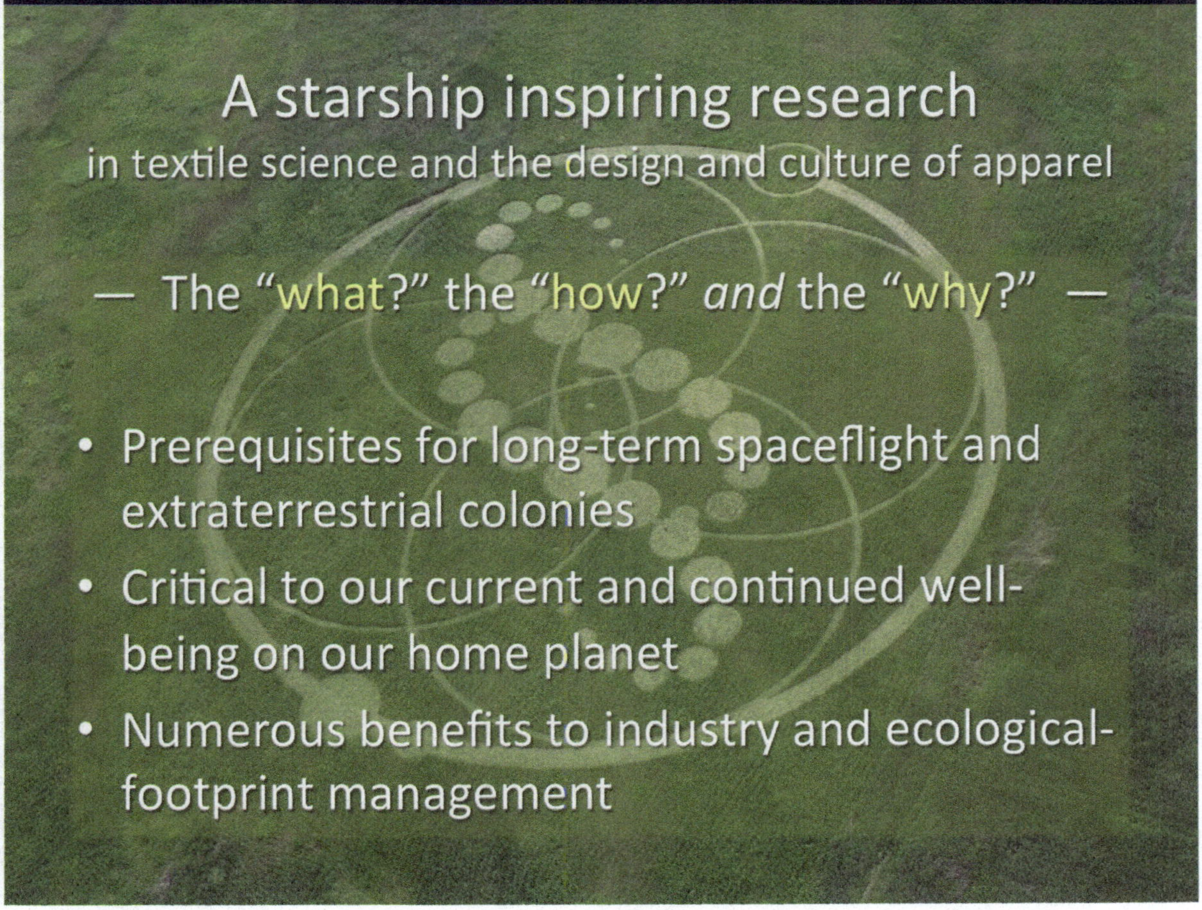
A starship inspiring research
in textile science and the design and culture of apparel
— The "what?" the "how?" *and* the "why?" —
• Prerequisites for long-term spaceflight and extraterrestrial colonies
• Critical to our current and continued well-being on our home planet
• Numerous benefits to industry and ecological-footprint management

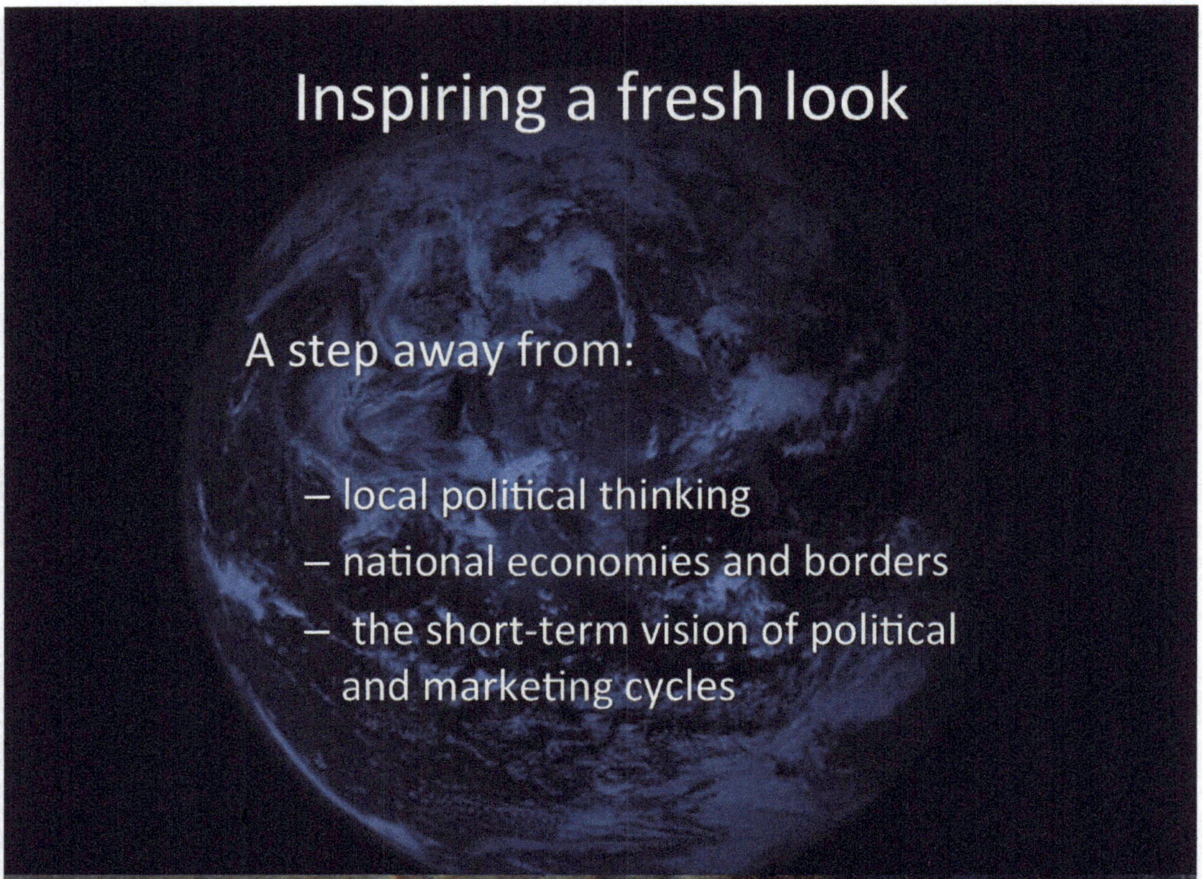
Inspiring a fresh look
A step away from:
– local political thinking
– national economies and borders
– the short-term vision of political and marketing cycles

A decades-long or multi-generational mission:
Areas of investigation
1. The Packing-List
2. Sustainability
3. Innovation in materials
4. Innovation in manufacturing
5. Ecology
6. Lessons from space-flight
7. Lessons from history of exploration and military

The Packing-List: Which items? What materials?
Wait: Isn't this all sewn up? Established and continuing?
Well... we have **nearly** all our present needs covered...
But...the length of the mission changes everything
To be clear: Amazing work constantly taking place on EVA suits
Well-covered on the outside.
Well suited for the short mission
BUT
What difference do *years* make?
What about the on-board culture?
Professor Dava Newman, MIT
Modelling a mockup "Biosuit"
Peter Homer's award-winning gloves
Photos: nasa.gov

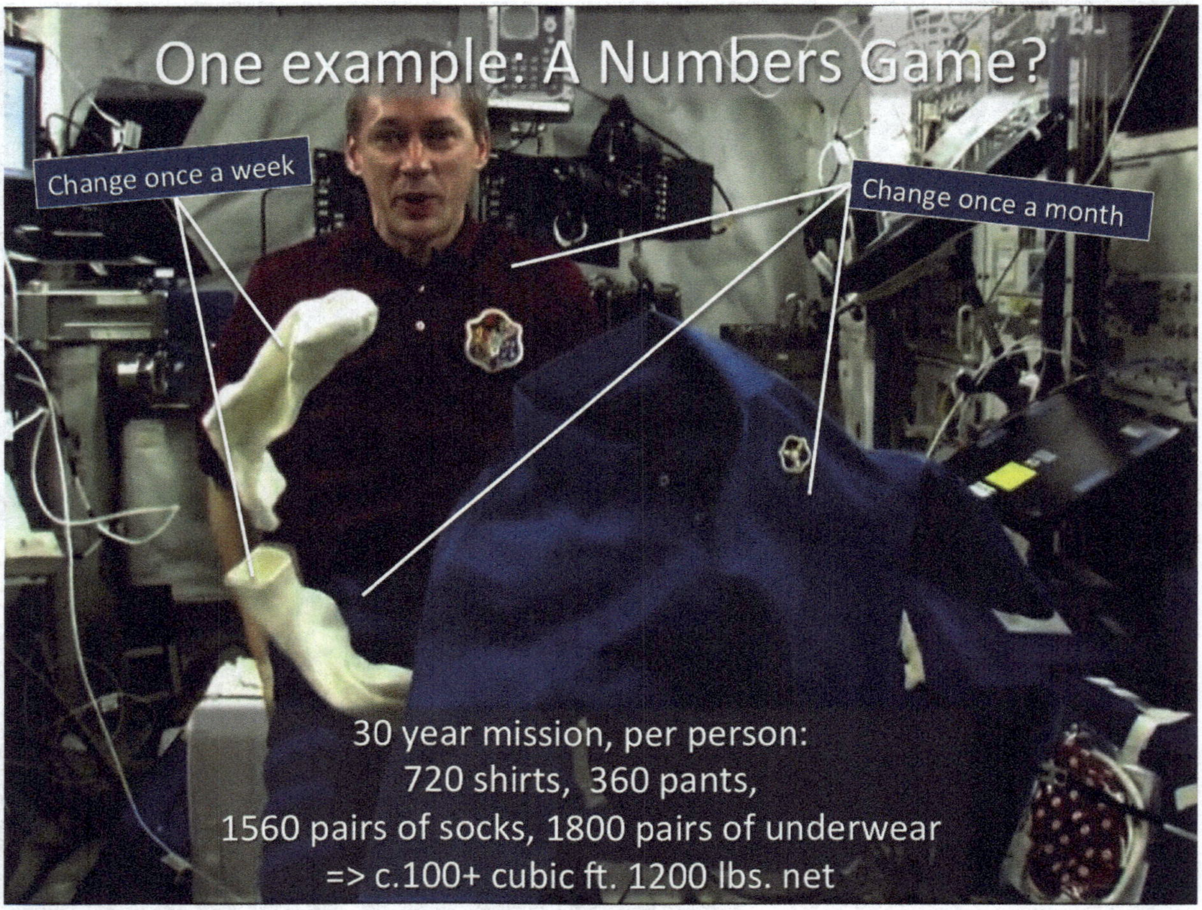

<u>1 x 20' container = 10 crew x 30 years</u>
<u>c. 3600 lbs @ launch cost of $5,000 - $10,000/pound</u>

The long timeframe radically changes the equation...
Durability or Recyclability?
And time changes everything! Materials, bodies, local cultures...

Learning from experience and history:
• The experience from space-flight so far?
• The history of long-term exploration or extended military deployment on Earth?
– nuclear submarines?
– Arctic bases?
– pre-industrial ocean voyages ?
– pre-modern military campaigns?
• Prisons?
• Re-think "apparel" altogether?
– (a cultural investigation)

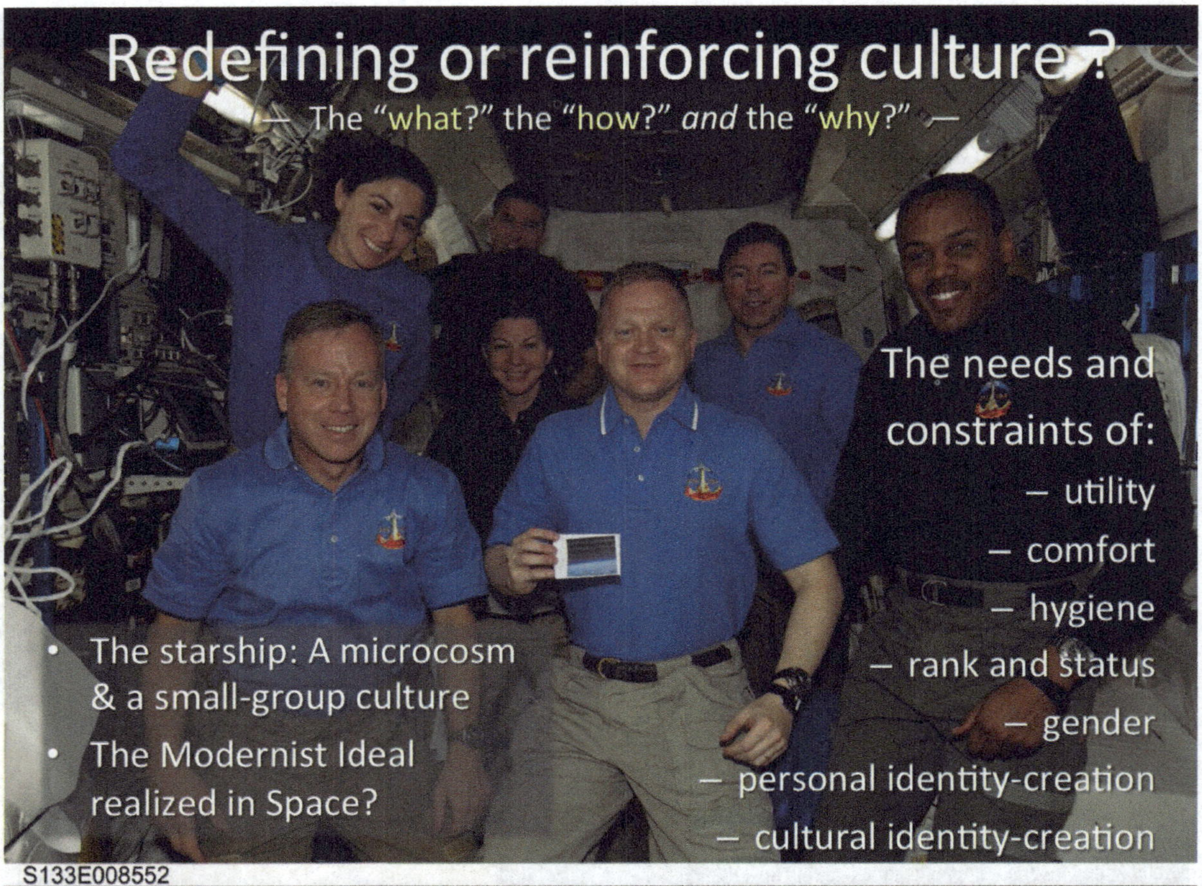
Redefining or reinforcing culture ?
— The "what?" the "how?" and the "why?" —
The needs and constraints of:
– utility
– comfort
– hygiene
– rank and status
– gender
– personal identity-creation
– cultural identity-creation
• The starship: A microcosm & a small-group culture
• The Modernist Ideal realized in Space?
S133E008552

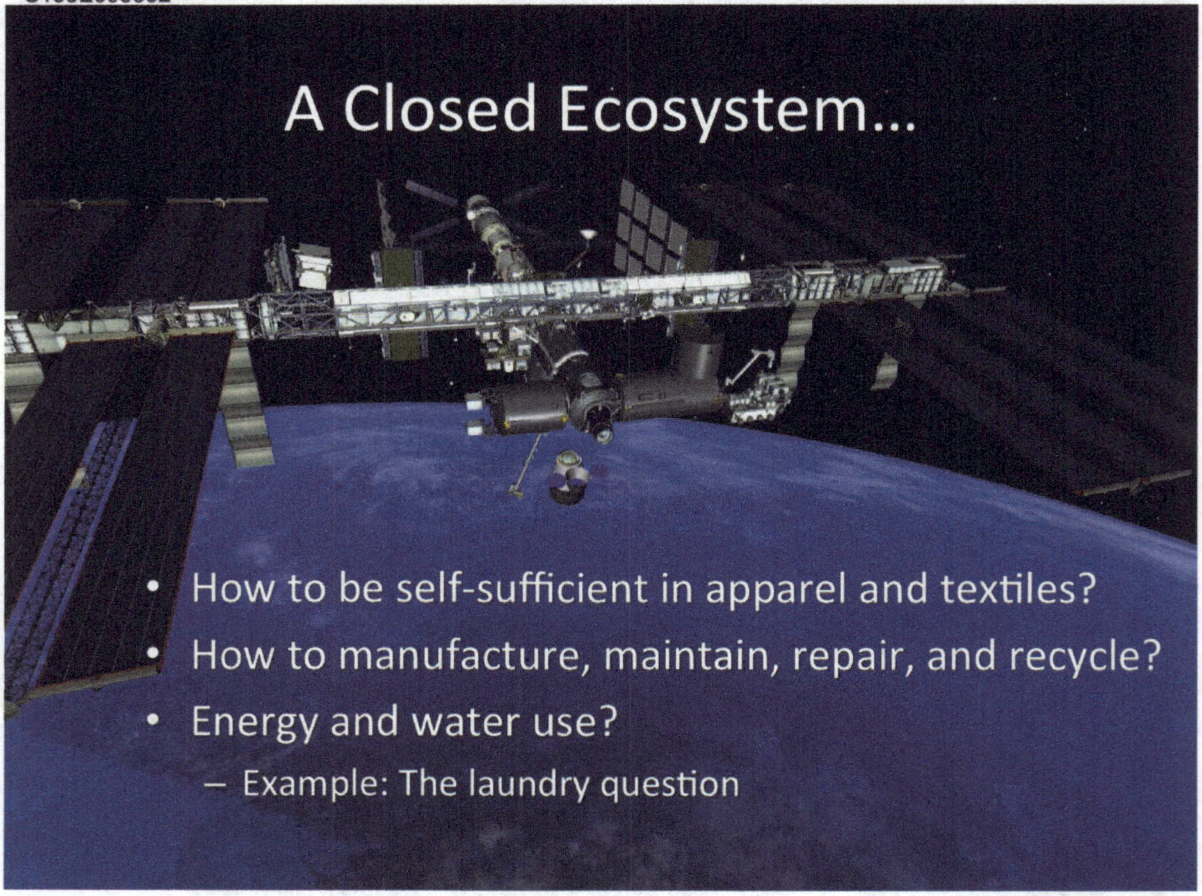
A Closed Ecosystem...
• How to be self-sufficient in apparel and textiles?
• How to manufacture, maintain, repair, and recycle?
• Energy and water use?
– Example: The laundry question

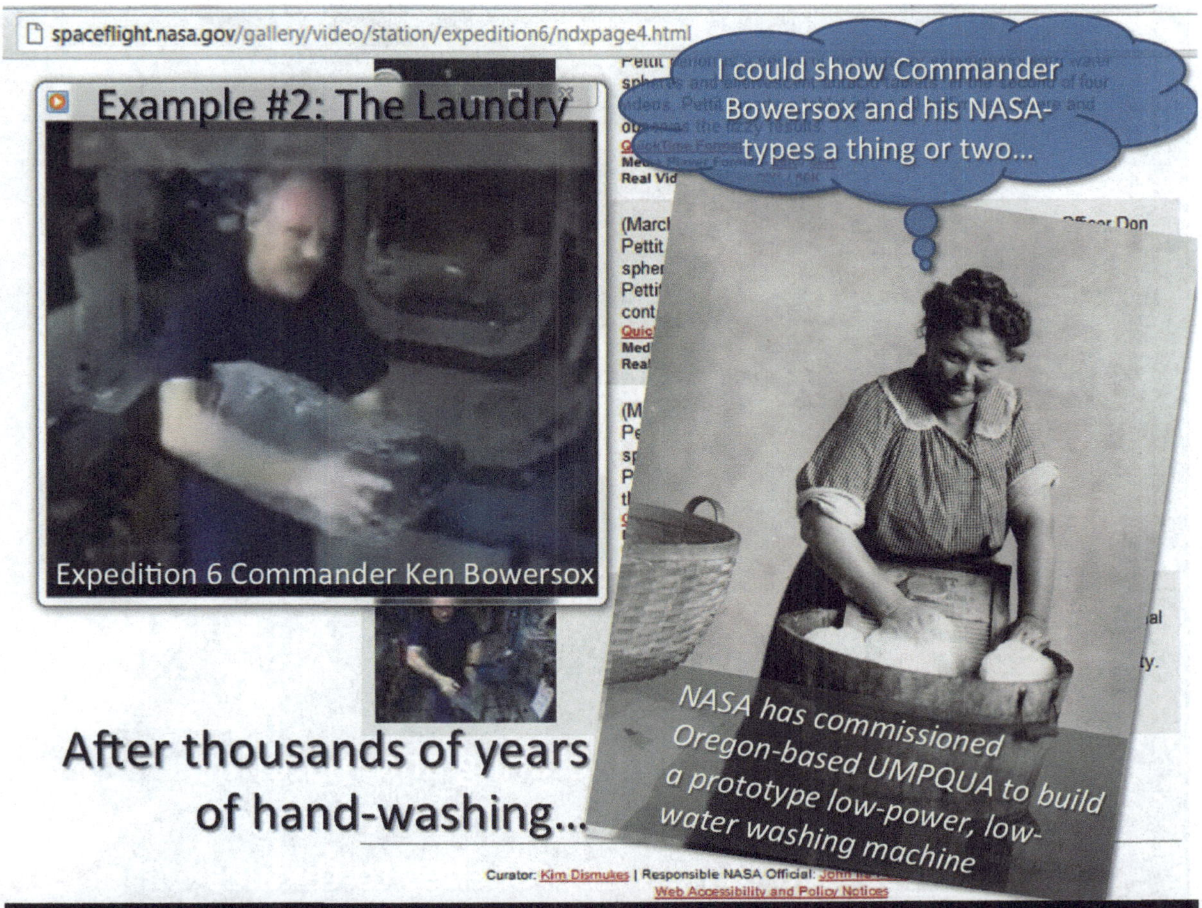
spaceflight.nasa.gov/gallery/video/station/expedition6/ndxpage4.html
Example #2: The Laundry
Expedition 6 Commander Ken Bowersox
I could show Commander Bowersox and his NASA-types a thing or two…
After thousands of years of hand-washing…
NASA has commissioned Oregon-based UMPQUA to build a prototype low-power, low-water washing machine
Curator: Kim Dismukes | Responsible NASA Official:
Web Accessibility and Policy Notices

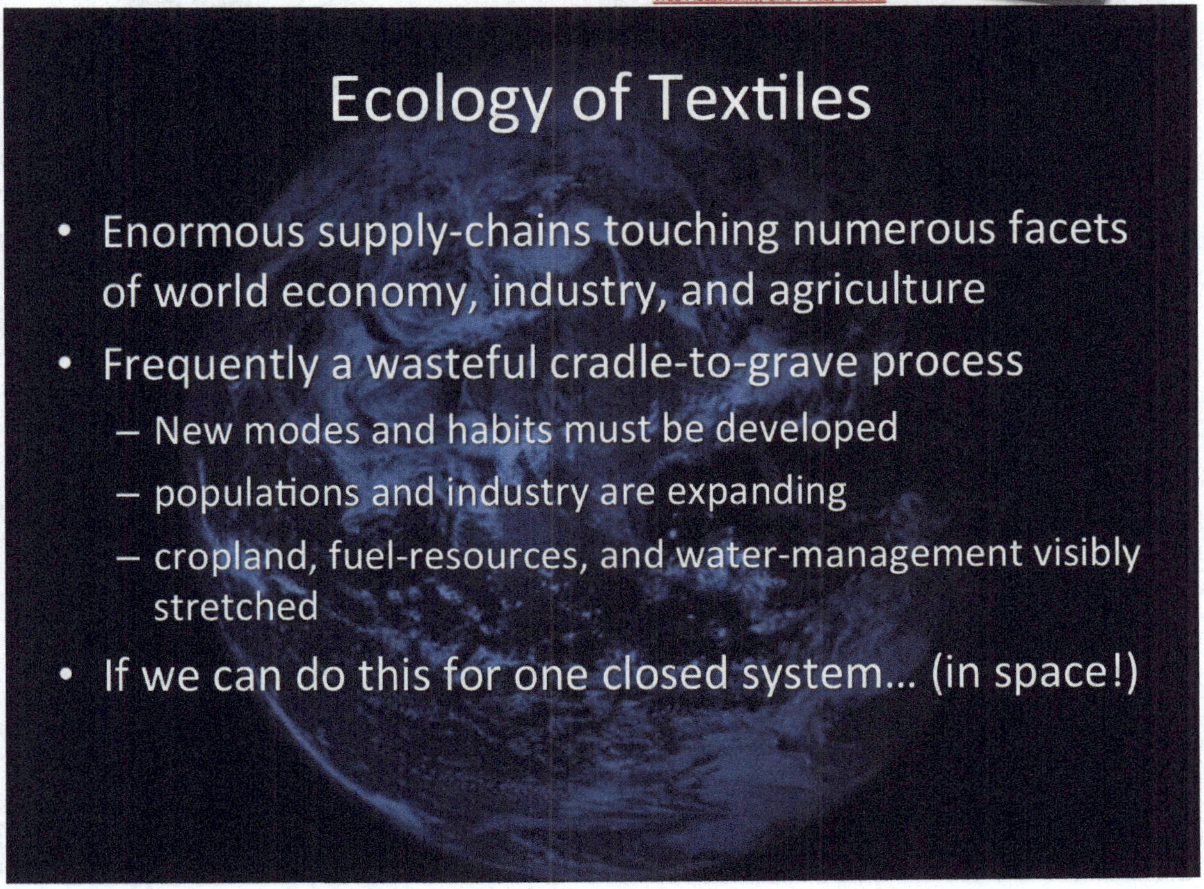
Ecology of Textiles
• Enormous supply-chains touching numerous facets of world economy, industry, and agriculture
• Frequently a wasteful cradle-to-grave process
– New modes and habits must be developed
– populations and industry are expanding
– cropland, fuel-resources, and water-management visibly stretched
• If we can do this for one closed system… (in space!)

Too true!
Commander, we have no training for this!
Define "we," Captain!
Innovation in manufacturing and materials: on-board production and maintenance? *Must* lead to new approaches to textiles

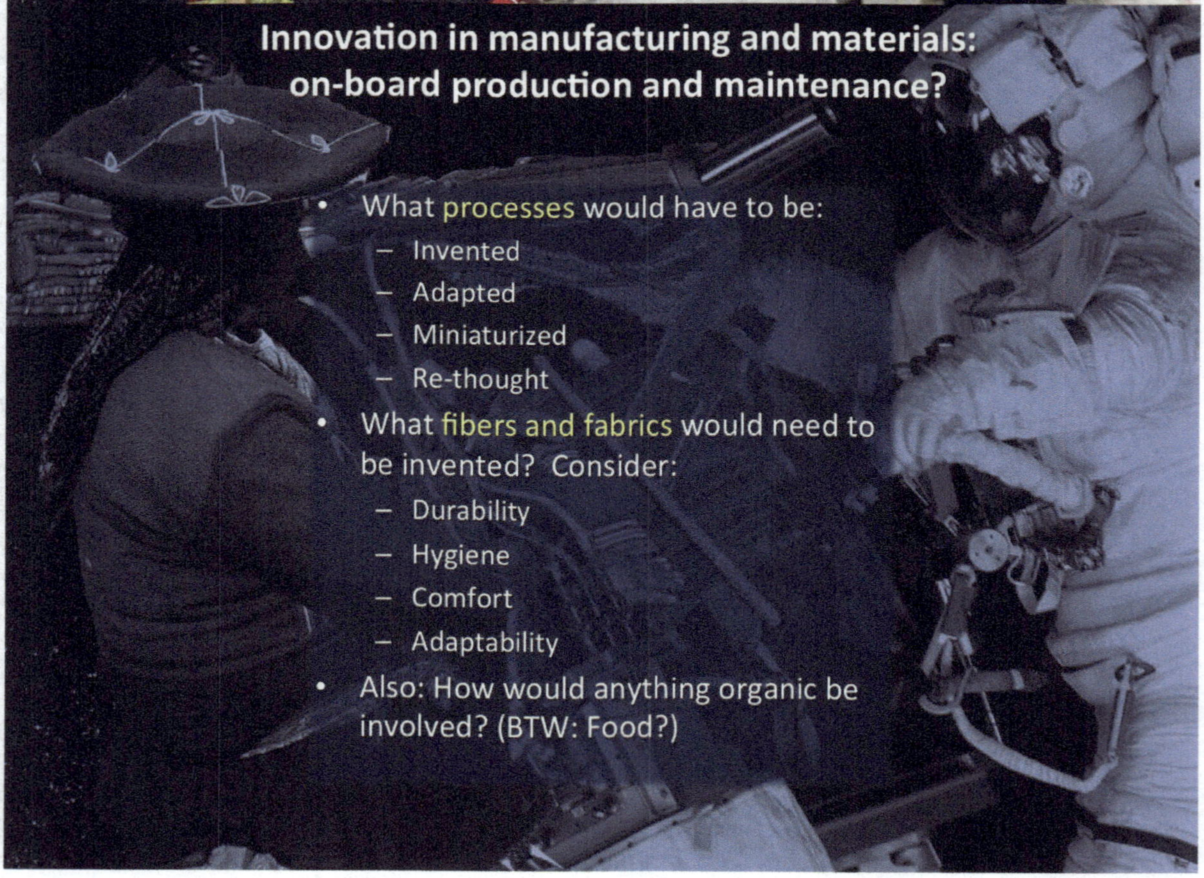
Innovation in manufacturing and materials: on-board production and maintenance?
• What processes would have to be:
– Invented
– Adapted
– Miniaturized
– Re-thought
• What fibers and fabrics would need to be invented? Consider:
– Durability
– Hygiene
– Comfort
– Adaptability
• Also: How would anything organic be involved? (BTW: Food?)

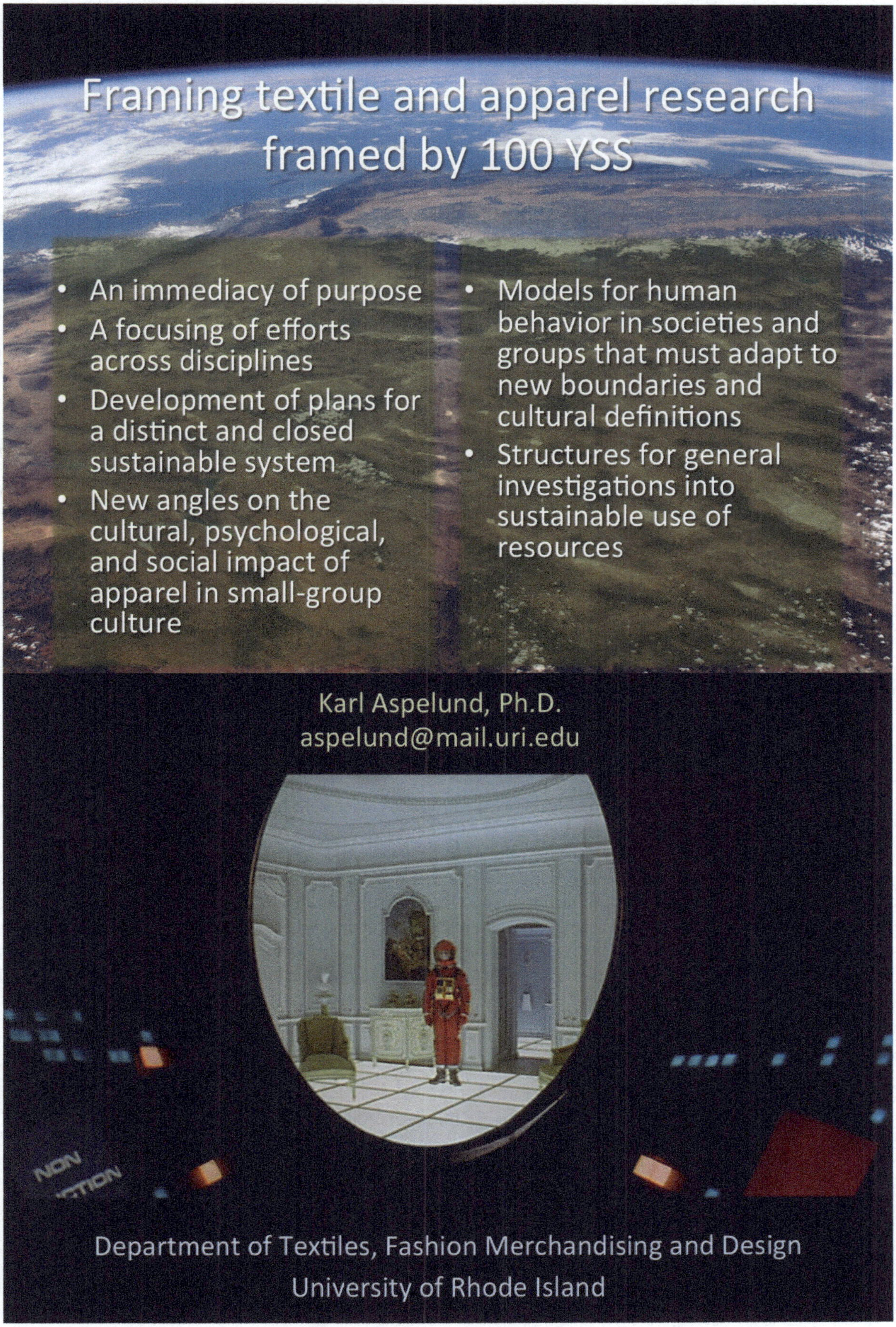
Framing textile and apparel research framed by 100 YSS
• An immediacy of purpose
• A focusing of efforts across disciplines
• Development of plans for a distinct and closed sustainable system
• New angles on the cultural, psychological, and social impact of apparel in small-group culture
• Models for human behavior in societies and groups that must adapt to new boundaries and cultural definitions
• Structures for general investigations into sustainable use of resources
Karl Aspelund, Ph.D.
aspelund@mail.uri.edu
Department of Textiles, Fashion Merchandising and Design
University of Rhode Island

Developing Faster than Light Technology

A Project Management Institute Based Approach

John C. 'Buck' Field, MBA, PMP
and Dr. Patricia Alvarez Cantwell, MD

Abstract

Reviews of government and academic documents demonstrate few effective criteria for determining "transformative" categorization. Much discussion focuses on conflicting usage, conceptualizations, and unclear understanding. This paper describes guidelines for administration of transformative research based on globally recognized management standards, particularly those published by the Project Management Institute®. Integration of critical success factors identified be researchers in the fields of history and philosophy of revolutionary science is explained. Past revolutions in science illustrate plausible utility for the approach. Initiating phase scope definition processes are shown to meet transformative research administration challenges facing by the National Science Foundation, National Aeronautics and Space Administration, and the Defense Advanced Research Projects Agency. Future knowledge and performance gains accruing to both researchers and administrators is demonstrated in a physics research scenario in support of developing faster than light technology from popular science fiction. We expect implementation of this new approach may dramatically improve TR administration, quality, value, and the rate of progress at the frontiers of many fields.

Abbreviations

DARPA	=	Defense Advanced Research Projects Agency
NSB	=	National Science Board
NSF	=	National Science Foundation
PM	=	Project Management
PMBOK®	=	A Guide to the Project Management Body of Knowledge
PMI	=	Project Management Institute
NASA	=	National Aeronautics and Space Administration
STEM	=	Science, Technology, Engineering, and Mathematics
TR	=	Transformative Research

1. Introduction

1.1 Purpose

The purpose of this document is to provide foundational information necessary for successful planning, execution, and delivery of what the National Science Foundation (NSF) terms "transformative research" (TR) of the kind which will be needed if faster-than-light starship research projects are to succeed. The audience for this material include starship enthusiasts, leaders of government, industry, and academia, as well as members of the public concerned with successfully delivering new scientific shifts of paradigm similar to the popular example in figure 1.

Figure 1. Classic Duck-Rabbit Paradigm Shift Illustration

1.2 Issues

Transformative research and STEM revolutions, defined as the highest value advancements remain surrounded by confusion and misconceptions. Students' competence within a field is most frequently assessed by familiarity with the historical revolutions in the advancement of their particular field. For example, physics students demonstrate their competence by showing understanding of experimental results obtained by the apparatus shown in figure 2, and their relation to Einstein's relativity. Despite the prevalence of this form of revolutionary milestone pedagogy, the processes by which these milestones occur and their internal structure remain unknown outside specialized academic circles within history and philosophy of science. This cripples our ability to successfully conceive and complete projects delivering revolutionary STEM.

Figure 2. Michelson–Morley Apparatus

Knowledge gaps also surround the management skills, tools, and techniques by which organizations might successfully deal with risks, uncertainties, and quality of research results endemic to such efforts.

1.3 Approach Selection

Paradigm shift resulting from successful TR is a change that occurs over an identifiable duration, i.e.: it has a definable beginning and end. It is unique in both content and context, after which later developments proceed within an altered cognitive or technological environment. The most widely accepted definition in project management (PM) is: "a temporary endeavor undertaken to create a unique product, service, or result."

Because they predate the development of PM, the most famous historical examples of TR projects were not managed. This appears responsible for an impression, (largely outside the PM community), that creation of models for heliocentrism, evolution, plate tectonics, and relativity were individual efforts, inappropriate to relate to the PM discipline, formal definitions notwithstanding.

Complex modern STEM research is characterized more and more by active group collaborations where management and administrative functions are vital to success. Yet even the most famous 'lone, creative genius' narratives are being reexamined and recharacterized. Even Copernicus' work is now regarded , as in large part, a product of the culture in which it emerged. Culture is the primary environmental factor discussed in Section 2.1.1 of PMI's latest draft of "A Guide to the Project Management Body of Knowledge" (PMBOK) the most widely-accepted global standard in management science. While the value of on time, on budget, and performance to specifications is widely acknowledged, other benefits to organizations from applying PM are gaining greater acceptance.

1.4 Scope

To plan a journey into the future of science, we orient ourselves to "the big picture" area of the region where such a journey would end up. We can do this by identifying core features of science which remain constant across transformative changes. For example, if we were to receive a cosmology textbook from 300 years in the future, we would be familiar with its goal of investigating nature, and its approach of describing the order of nature with maximum precision and universality. We would also be comfortable with the manner in which this is done: presenting theories as internally consistent and closely matched to available observation as possible. We would share a recognition with the authors of that work that logic and experience exert huge influence determining whether an idea is considered to have attained the status of "knowledge".

Figure 3. Fan Art from Startrek-ships.com

1.5 Focus

In this paper, we argue that similar to its impact in other fields, application of PM best practices to TR may dramatically improve performance, resulting technical capabilities, and value to organizations including TR projects in their portfolios. Arguably the best guidelines of their type, standards from the Project Management Institute (PMI) provide our primary analytical framework, informed by well-known historical revolutions in STEM knowledge.

Forward application of historical principles is illustrated with a potential TR project to develop faster than light "warp drive", a revolutionary future advancement in space capabilities from science fiction, the field wherein the greatest experience can be found in developing visions of revolutionary space technology. Warp drive enjoys widespread popularity with the public, and among many scientists seen as a likely future phase of civilization. Such technologies will require transformation of STEM knowledge by revolutionary new models in physics, and offer opportunities for quality control testing of theory unavailable in purely theoretical research environments.

Figure 4. 2008 PMBOK

1.6 Approach

1.6.1 Administrative / Managerial

Herein, we utilize PMI's latest edition of "A Guide to the Project Management Body of Knowledge" (PMBOK), released as an ANSI standard in December, 2012. Understanding relationships among management of portfolios, programs, projects and the organizations is required, as are differences between projects, operations, strategy, business value, life cycles, concepts define tools and techniques effective for project success.

1.6.2 Key Success Factors

Historical analysis compiles examples of events of interest to identify commonalities. Researchers within the communities of history and philosophy of science have performed such analyses relative to important STEM revolutions and ways in which they contrast with unsuccessful results. Knowledge gained in this area includes internal structures of revolutionary cognitive frameworks, the nature of scientific creativity, inputs, processes, and outputs. Our approach considers distinctive components common to past successes to ensure planned efforts include necessary and sufficient conditions to succeed.

Selection and adjustment of PM approaches and key success factors is determined by the specific area to which a given project applies.

1.6.3 Application Areas

TR efforts have common, significant components not shared by all projects. This meets the definition of a specific application area of PM. From the PMBOK "Application areas are usually defined in terms of either the product (i.e., by similar technologies or production methods) or the type of customer (i.e., internal versus external, government versus commercial) or industry sector (i.e., utilities, automotive, aerospace, information technologies, etc.). Application areas can overlap."

Early development of evolutionary theory illustrates a TR project overlapping pre-existing areas of biology and geology. Creation of heliocentric cosmology involved geometry, astronomy, and horology. Discovery of a warp drive enabling model of space-time will certainly overlap areas of mathematics, cosmology, and experimental physics.

We turn our attention to first major milestone for a warp drive development program: an enabling physics model. To plan for this deliverable we must be clear about what it is: a conceptual construct meeting the requirements for science, it is revolutionary relative to the current standard model, and it must be able to provide a foundation upon which warp technology could be researched.

Figure 5. Harmonia Macrocosmica c1708

2. Background Concepts

2.1 Focus on Paradigm Change as the Product

Like any complex task, predicting how conceptualizations of spacetime will change in future paradigms requires specific tools. Our best guide to this future is a map based on successful revolutions in the past. Here we apply "synthesis", (historical reasoning to obtain the best explanation), to well-known, uncontroversial examples of successful, transformative revolutions, we can extract "key performance indicators", enabling us to assign resources into areas that distinguish revolutionary science of the kind currently needed.

Other tools we need for planning: a working familiarity with scientific paradigms, how they operate and change, and project management, the area of knowledge concerned with successfully creating change.

2.2 Copernican Revolution

The Copernican Revolution is perhaps the best known archetype for a revolution in science. Five centuries ago, the generally-accepted model in Europe was a Ptolemaic system which held that the Earth is at the center of the universe, does not move, and that objects in the sky rotate around the earth. This model explained and was confirmed by the lack of detectible motion of the Earth and the apparent rotations in the sky. In addition, Ptolemy was official church doctrine. The sun-centered cosmology of Copernicus removed the human observer from a central observational position in the solar system and redefined Earth as one of the 'wandering stars'. "Planets" is πλανήτης (planētēs) in Greek, and means "wanderer"...literally: an Earth-shaking idea.

2.3 Darwinian Revolution

When the Darwinian Revolution began, the prevailing model of biological change was that animal and plant varieties were the result of changes acquired by ancestors during their lifetimes. A giraffe that stretched its neck to eat would pass on an acquired, inherited trait of a long neck to its offspring – and major differences of species were the result of the original pairs made by God during creation in Genesis. The view based on this model of change was called Lamarckism, which adhered to the old paradigm which held that humanity was a special creation, central and separate from other creatures. Lamarckism explained biological variety, the observed uniqueness of humans, and how countless different varieties of species came about. Darwin's evolutionary theory proposed natural selection as a mechanism that could account for all known forms of life. While it provided incredible explanatory power, it also repositioned the human observer from a central place of importance. As Copernicus had changed the understanding of humanity's physical position, Darwin changed the understanding of our biological position. Where Copernicus joined Earth with all of the cosmos, Darwin joined humanity with all other life, relating us as one kind of organic species among countless others.

Figure 6. ALICE Detector

2.4 Cognitive Frameworks

Understanding any situation requires a mental structure to tell us what the rules are. It helps us decide what is important and dictates what questions are sensible. In turn, this defines what type of instruments and activities we should use to answer those questions. If our cognitive framework contains a rule that "telltale numb spots convey information about demonic enslavement", we are likely to have investigators develop pricking instruments to detect this spot in the specialized activity of witch detection. If our cognitive framework contains a rule that "sounds from inside the body convey information about health", then we are likely to build an instrument like a stethoscope to detect those sounds for the specialized activity of medical diagnosis. The structures that govern what instruments and procedures we develop are called cognitive frameworks, paradigms, worldviews, or in physics: models. These are more specific ways of viewing reality than the general scientific method, defining what the world is relative to our discipline. Our current beliefs about what spacetime is or what it represents are no exception. Current standard model beliefs include particles without size exist, and have electrical effects. Thus, we see in Figure 6 an instrument in the Large Hadron Collider at CERN, costing many billions of dollars, with the goal of detecting these effects in the specialized field of quark-gluon plasma physics. The overthrow and replacement of our current cosmological paradigms is widely regarded as inevitable because of scientific "anomalies." In the distant future, when current paradigms are seen as naïve, those studying current instruments and methods will be able to see where the boundaries of our cognitive frameworks were. The assumptions we believe to be true, will seem obviously false to them. If now, we compare current models carefully for similarities to those immediately preceding Copernican and Darwinian revolutions, we can deduce something about how future views will be different from those of today. Such insights can guide us to what is over the next hill.

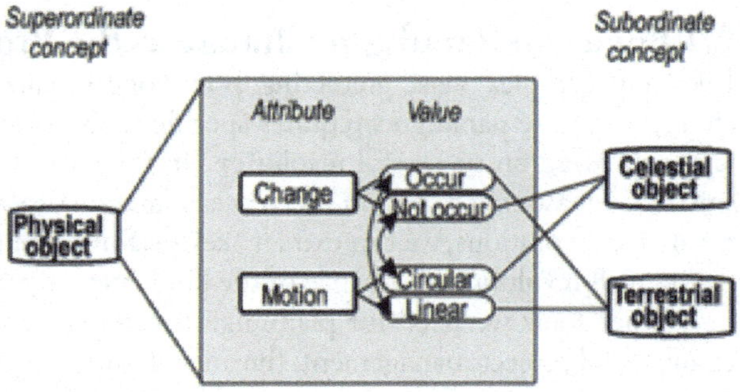

Figure 7. Pre-Copernican Cognitive Frame c. 1500

2.5 Characteristics of a Warp Enabling Model

Whether or not a scientific change is revolutionary is a matter of some debate. Some specialists regard discovering the existence of stellar clusters to be revolutionary, giving rise to several astronomical disciplines, whereas others consider this merely a new bit of information about stars we expected to find. Some definitions of transformative paradigm changes include development of "widely applicable technology", while other descriptions provide examples of unconventional, successful researchers. We can avoid disputes over details by using a good criterion for distinguishing incremental advances in science from the class of change unambiguously considered both revolutionary, and uniquely valuable. We will judge normal, incremental advances in science to have occurred when "no previously classified [types] have been reclassified." Because the discovery of stellar clusters provides evidence of the history of the Milky Way and created a new astronomical subfield, the discovery is important, but does not reach revolutionary status by this standard.

2.5.1 Type Recategorization: A Litmus Test

While debates over specifics of the Copernican and Darwinian Revolutions continue, there is no significant disagreement that if you want to list revolutions in science, these top the list. An obvious question is: "If Copernicus' and Darwin's changes were revolutionary, then what recategorization occurs in the new paradigms?" Figure 7 shows how concepts were related in the cosmology of 1500. In the Copernican Revolution, 'celestial motion' was a type of cosmological phenomenon reclassified; becoming the result of human observers moving in a sun-centered system. The Darwinian Revolution recategorized separate, divinely created types of animals into related species sharing a common ancestor on a single biological tree.

Because recategorization of paradigms' defining 'types' is a key characteristic of revolutions, spacetime - measurements of distance and periodic repetitions with which we measure time – will be reclassified. As a defining example of time's operation within both quantum and cosmological paradigms, both paradigms will be affected by any change in our model of time. What a clock measures will still be regarded as "time" in the same sense that we regard the sun "coming up" each day. The recategorization is not due to improved observation and measurement; it is due to structural changes in the cognitive framework underlying our conceptualization of time.

Figure 8. Depiction of a Pre-Darwinian Frame c.1500

2.5.2 Assumption Correction

Assumptions are defined as factors that are considered to be "true, real, or certain without proof or demonstration." They affect all aspects of cognition, knowledge, planning, and in the research enterprise with which we are concerned: assumptions introduce risk. In the large scale history of cosmology, the most pronounced cost associated with that risk is the waste of careers and progress over centuries, sometimes lives are lost. The time, money, and lives devoted to making the Ptolemaic paradigm conform to observations were tragic, as were the infamous efforts to enforce a geocentric worldview. The natural assumption of a stable earth with humans at the center was rejected in the Copernican revolution whereas under Darwin, the assumption of a hierarchy of separately-created, stable species – with humans in a reassuringly important position – was rejected. The prior paradigm is illustrated in figure 8, had Heaven at the top populated by angels, with the physical realms with levels of saints and people arranged below in a hierarchy designed and created by God.

The Einsteinian revolution rejected, among other things, the assumption that time is a characteristic of reality separate from space. Each of these fundamental theories obliterates assumptions involving special status for humans.

Successful warp drive enabling paradigms will not be able to regard time and space as fundamental dimensions of the universe which humans are uniquely able to perceive and clocks are able to measure, despite the very natural, common sense assumption that this is the case. Enforcing rejection of potentially incorrect assumptions is an advantage provided by concrete, real world goals identified in the Nersessian Model . The reason such goals are critical was described by Kuhn as:

"The need for work of this sort arises from the immense difficulties often encountered in developing points of contact between a theory and nature." Considering the size of humans relative to that of the observable universe, it seems unlike our natural senses would be able to detect much of what's going on. The common sense assumption's obvious human bias, that ultimate reality conforms with our natural sensory observations and past interpretations merits examination.

2.5.3 Observational Component

One astoundingly similar characteristic of archetypal revolutionary ideas alluded to above, is a specific type of assumption TR paradigms refute. They directly address a natural assumption made by previous investigators making observations: demonstrating that uncontroversial measurements and observations actually depend on conditions unique to the human investigators. The archetypal reduction of earth from the center of the universe to a relatively insignificant planet moving through an unimaginably large cosmos demonstrates this perfectly. Darwin's revolution moved Homo sapiens from God's central concern, to a small, just-sprouted twig on the great tree of life. We expect assumptions based on observer-centered bias to be replaced within a revolutionary theory of fundamental physics. Circumventing the light speed limit imposed by relativity would almost certainly replace the concept of distance as a fundamental quantity in a manner similar to the Ptolemaic assumption that "Earth is a fundamental center". A successful, revolutionary paradigm of spacetime will propose a new categorization for current, standard model fundamentals as the observational consequence of a process we cannot directly observe due to our perceptual limitations and the historical fact that we have not been looking for confirmation of the predictions such a model would make. This new cognitive framework will explain why making our past assumptions was understandable, if not inevitable.

2.5.4 Progressive Elaboration

Revolutionary models demonstrate vague descriptions of concepts in their early stages, acquiring greater detail as more research is performed and a better, more complete understanding is gained. The heliocentric model of Copernicus appears primitively simple compared against modern conceptualizations. The inadequacy of Darwin's theory by today's standards likewise illustrates the lack of elaboration shared by all new concepts. Such vagueness makes the value of the new model difficult to perceive, especially if we have much vested interest in preserving the old model. Current models offer greater elaboration and clarity of a rich set of observations, like the objects in figure 9. Revolutionary models are vaguely perceived initially, and because they are often clamoring for attention and funding among others, they appear to have fatal flaws, to be unreasonable, or even incomprehensible, yet history shows such appearances are misleading. This dynamic calls our attention to the need for screening processes that distinguish ideas and proposals which possess critical characteristics of revolutionary models. Management of risks that come from such misleading appearances cannot be neglected.

2.5.5 A Missing Mechanism

Although Copernicus' model rejected the orientation of the crystal sphere "firmament" given in the bible as surrounding the earth, it kept these invisible spheres as the mechanism for celestial movement. The Copernican revolution is often regarded as completed when Newtonian gravity was advanced as an orbital mechanism. In turn, these revolutionary ideas of gravity by Newton similarly lacked a mechanism by which his attractive force of gravity might operate. In fact, problems of "action at a distance" have remained since the theory was proposed; today the details, (or even the broad principles), for the mechanism by which gravity operates remain a mystery. Darwin's theory lacked any hereditary explanation that could account for gradual variability or mutation, despite his firm assertion that the evidence indicated the existence of such a mechanism.

Figure 9. Perceptual and Cognitive Precision Declines with Distance

New revolutionary models in physics will, like past revolutions, initially lack mechanisms explaining how the model operates.

2.5.6 Process Orientation

Despite their early lack of elaboration and process mechanisms, successful revolutions do prominently focus on processes. These processes describe phenomena and one or more core invariants. They share common explanatory features such as: a generative characteristic, an iterative dynamic, and interactivity.

2.5.7 Generative Characteristic

Heliocentrism explains what generates our celestial observations, explaining why heavenly bodies appear to move the way they do. Natural selection explains what generates organism speciation, and answers the question of why there are so many varieties of each species. The new time paradigm will likewise explain what generates our observations of time in terms that answer why questions of a similar nature. While we recognize the importance of a proper "why" question, our currently lack of a reliable framework with which to formulate a question makes any hypothesis unreliable. Proposals and guesses in such circumstances also can be dangerously prejudicial. Nevertheless, a possible candidate question might be: "Why do we observe distance and time the way we do?"

2.5.8 Iterative Dynamic

The generative characteristic of processes in these paradigms is supported by an iterative dynamic – a repeating cycle of activity giving rise to whatever is being generated. In Copernican cosmology, these are the repeating rotation of the Earth and the planets' revolutions about the sun. For Darwin's biology, the iterative dynamic is a repeating cycle of change from one generation of organism to the next. Our new conceptualizations of physics fundamentals may include such an iterative component or characteristic that explains our observations.

Figure 10. Fractals Possess Important Qualifying Characteristics for Successful Future Models

2.5.9 Interactivity

Interactivity in a paradigm process can refer to feedback within the processes described in the new conceptualization. In the Copernican system, we might consider the way that an output from use of this model, (e.g.: calculated positions for the sun, moon, and planetary), can be used to accurately predict observed positions for the previous day or for the following year, and these positions can provide an input for the next iteration. In evolution, a more robust form of interactivity appears: as effects of evolution significantly determine which incremental changes in offspring are advantageous, these changes are cumulative due to the replicative nature of life. Over time the results of this cumulative interactivity include the full, rich diversity of living forms.

In the case of revolutionizing our understanding of spacetime, the project team and leadership should be alert to similar ideas.

2.5.10 Mathematically Expressed

The process orientation of revolutionary paradigms is expressed in using the lingua franca of science: mathematics. While logical rules describe qualitative attributes of observations and relationships, mathematics describes quantitative aspects. Planetary orbits and the rotation of the Earth have specific, measurable velocities associated with them and math formulas describe ellipses of stable orbits and "fly-by" hyperbolas through our solar system. In evolution, formula and quantitative measurements describe critical observations of population size, reproduction and mutation rates, and our current understanding of DNA-based genetics. Fractals, quaternions, and hyperbolic geometry show potential future application toward revolutionizing space time models. Tragically, there may be mathematicians on precisely the track needed to revolutionize civilization, but are working in isolation. Within a hair's breadth of the acheivement of a lifetime, they would never know. This is a tragedy worth avoiding, if we can bring the right people together in the right ways.

In the future, new mathematics for space time will describe quantifiable aspects a more complete way than clocks or the lightspeed c relationship. A key problem in understanding and planning for whatever kind of quantities these future equations will describe is shared with our problematic understanding of other "fundamental" metrics like space, matter, energy, and force. This problem exists for all metrics lacking a meaningful, non-circular definition, and which are not qualitatively linked within a cognitive framework consistent with observations.

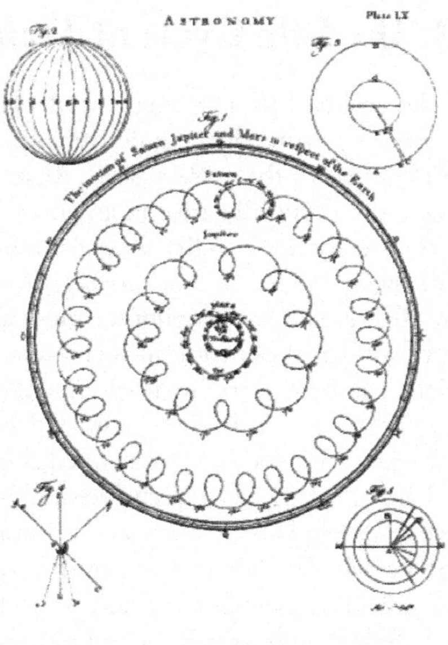

Figure 11. Planetary Motion in the Ptolemaic System

2.5.11 Domain Expansion

Revolutionary paradigms expand established areas of study into new domains, often with a principle called "reduction" in philosophy of science. Heliocentrism extended natural philosophy into the heavens, revolutionizing and creating new branches of science. The reduction of complex heavenly motion (shown in figure 11) to a consistent, universal heliocentrism was the means by which this was accomplished. Natural selection did the same in biology, synthesizing animal husbandry, biogeography, geology, morphology, and embryology. When complete, the next physics revolution will almost certainly synthesize relativity, quantum mechanics, and gravity – but the manner in which it will do so will be surprising.

2.5.12 Practical Anomalies

One likely surprise will be the way in which the revolution in time modeling will resolve the anomalies currently baffling the physics community. Isaac Asimov wrote: "The most exciting phrase to hear in science, the one that heralds new discoveries, is not 'Eureka!' - but 'That's funny...'" With Copernicus, unworkable and conflicting calculations of the date motivated a new method for calculating dates and reforming calendars. In biology, crisis was brewing with the explosion of new species being discovered during the Victorian 1800's. The inability of existing classification systems to handle these discoveries and resolve conflicts led to the search for a way to explain how variety and speciation occurred, which would enable a rational naming system. When Darwin proposed natural selection with common ancestry as a method of solving the naming problem, he changed the world. In both cases, theoretical anomalies were incurring real-world costs. The effort to eliminate those costs by solving the theoretical problems triggered the revolution. Pursuit of goals such as faster than light travel offers a potential to proactively pursue advancement before an asteroid or other threat makes revolutionizing space-time models imperative.

2.5.13 Disciplines Created, Crossed, and Unified

Expanding the domain of established disciplines is not the only relationship change that revolutionary new frameworks have on science. Transformational concepts create new fields, link previously disparate knowledge areas, and bring cognitive skills, tools, and evidence from those areas to a new model providing explanatory efficiency, greater clarity, and depth of understanding. The new paradigm for space-time will cross and draw upon fields like mathematical algorithms, group theory, relativity, dimensional analysis, and cosmology.

2.5.14 Context Sensitivity

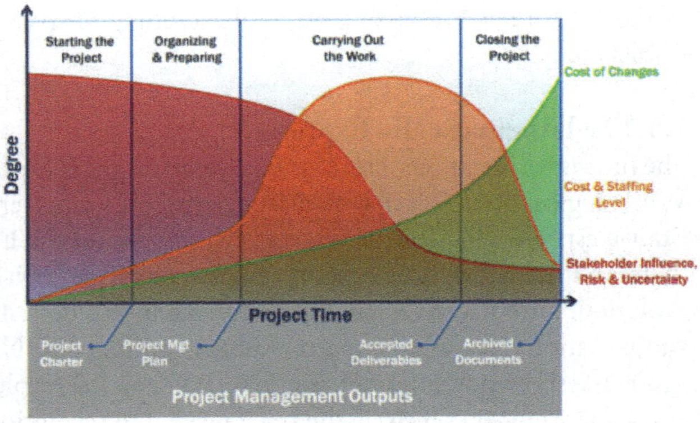

Figure 12. Key aspects of the project life cycle

Revolutionary new paradigms assimilate the old and provide an explanatory context for historical precedents and observations. They also enable us understand and judge when past paradigms might still be the best to use by a sensitivity to the conceptual goal. For example, a relativistic view is the best perspective if we want to understand how GPS signals work, but the Newtonian view was best for figuring out how to land on the moon. If we want to call from Cairo to Shanghai, thinking of Earth as a sphere with differing time zones is more useful, but working with maps, a flat-earth conception is more efficient. If we intend to climb up or ski down a mountain, local inclinations are far more relevant than a flat representation. Time as "what clocks measure" will undoubtedly remain most useful for most everyday activities, but we will recognize when the revolutionary approach is best for accomplishing our goals.

3. The Life Cycle of Transformative Research

Like general PM, management of TR involves division of the effort into largely sequential phases from initiation through closure. Specific deliverables depend on requirements and resources of the organizational environment and nature of the project itself. Absent leadership dedicated to producing change, historical examples of STEM revolution typically arise in an environment with ongoing crisis. Introduced to this environment is some practical need for a capability unattainable within the current paradigm. When Copernicus developed heliocentrism, it solved a religious crisis arising from inaccurate calendars. The inability of Ptolemaic geocentrism to provide a verifiable date for ceremonies motivated development of a workable system. Similarly, the inability of separate creation to account for an increasing number of observations in biology, and the inability of any experiment to detect aether led to the development of evolution and relativity, respectively.

If tomorrow morning, a warp drive development project were to become a top priority, it would be prompted by some crisis today, as mentioned above. The most desirable might be a friendly invitation (with schematics) from a benign extraterrestrial civilization. Much less appealing and infinitely more certain is a large asteroid impact. If we discovered an asteroid on a collision course giving life on earth just 20 years, no one would question the wisdom of a crash schedule to develop interstellar flight. Context, environment, and organizational priorities determine what is most beneficial for any project.

Selection of which type of project life cycle to use can be based on relative benefits of of the two main types. Predictive approaches use comprehensive planning of the entire effort, performed prior to carrying out project work. Iterative approaches plan only for one developmental cycle, with acceptance criteria for work and deliverable adapting as cycles proceed.

In figure 12, we see PMI's standard model for a project life cycle, including deliverables which define project phases, the inverse relationship between cost of change and stakeholder influence, and resource commitments as the project progresses. Resourcing levels, typically peak toward the end of execution, dropping rapidly after the main deliverables of the project completes.

Within several research communities, investigations are conducted as a normal aspect of operations, and value is assessed by the quantity of published papers and awards won. In this environment, cost tracking specific projects has less utility than business, industry, and substantial areas of government, where concrete return on investment is a higher priority.

4. Transformative Research Phases

Project phases are groups of activities with a particular focus and produce a deliverable. Most commonly this deliverable is a sign-off document verifying that all the work for a phase has been accepted as complete, and meeting whatever criteria were set. The advantage of such controls is that they prevent the common problems which occur when "the cart is put before the horse", that is: work proceeds without prerequisites necessary. A natural tendency is for organizational leaders and managers to utilize their resources if available, regardless of whether the planning necessary for effective utilization is complete. An example of this would be when leaders have programmers work on a software project for which no clear plan exists, with the idea that these programmers should be working while drawing a salary.

In one such case during the summer, it was a far better option to send an entire software development department out on paid vacation with complimentary tickets to the latest science fiction blockbuster. They were told that a large project was coming out, and they needed to go have a good time over a long weekend because on Monday, a serious effort was going to start. unplanned work is unlikely to produce results specified in the plan, rework will be required, increasing cost and often undermining morale and confidence in the leaders. A standard phase is illustrated below:

5. Processes of Managed Transformative Research

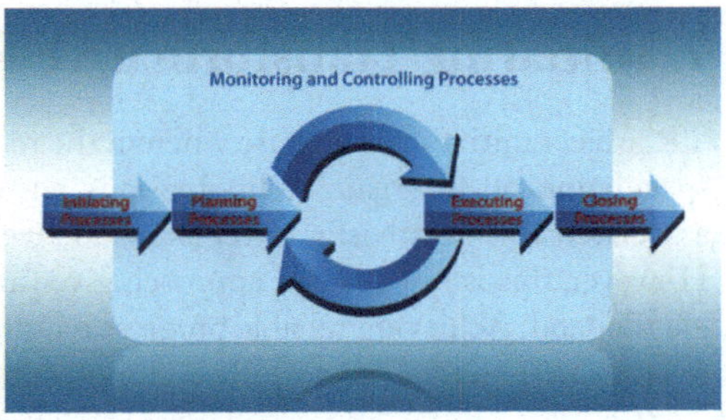

Figure 13. Process Structure of a Project Phase

Project processes are understood in terms of inputs, tools and techniques, and outputs as seen above. Processes are divided into two categories having different focus. PM oriented processes provide effective and efficient progress, whereas product oriented processes define characteristics of research results and bringing them to fruition. PM oriented processes in the case of Copernicus, Darwin, and Einstein would include funding and planning distribution of their theories, while product oriented processes included identification of knowledge gaps, contradictions, and misperceptions of the prevailing paradigms of their day.

Processes must be tailored to each project based on considerations which include: project complexity, risk tolerances, cost, schedule, quality, and others. For development of a revolutionary, warp-enabling model, effectively modifying generic processes will depend largely on which organizations choose to undertake what parts of the project. Despite this, understanding common characteristics of these processes is vital, since they determine the outcomes and success of each phase in the same way understanding steel and concrete determine the success of planning a skyscraper.

6. Typology Based Tailoring

The importance of tailoring has spawned research in the PM community assisting intelligent selection of administrative and management processes based on the project's type and conditions. One method for identification of tools and techniques appropriate for a project is based on dimensions of "technology, novelty, pace, and complexity" developed by Shenhar and Dvir is called the Diamond Approach, offering graphical comparisons of operational versus strategic projects applicable to normal versus transformative research efforts.

In the above caption, replacement of "opportunity" with "reward" gives us commonly used descriptions for transformative research in official documents in the United States. This graphical tool can be effectively applied to separate processes for finer distinctions. In the case of warp drive development, management processes rank lower on most dimensions, whereas product sub-projects, such as cosmological model development and coordination of theory testing and verification with the high-energy physics community likely to involve higher metrics in all dimensions but pace, which is dependent on environmental factors.

Adaptations of this model specifically for NASA project and program assessments were developed by Alan Stretton, PhD, who puts forward an approach to selecting appropriate project/program management and administrative frameworks based on metrics of uncertainty, complexity & scope, pace, novelty, novelty, strategic goals & customers, customer and/or provider perspective, and perceived primary program purpose.

Widespread consensus exists in the space community that some centuries from now, deep space technology will be developed with some forms similar to those commonly portrayed in science fiction. Such capabilities will transform our civilization and offer undreamed opportunities, but also require revolutionary transformation in our current understanding of fundamental physics, due to limits imposed by relativity. Similar widespread consensus on the inevitable need a physics transformation is well-established within the physics community itself. This conclusion is based on the increasing body of theoretical paradoxes and observations accumulated over more than a century of the largest financial investments in support of the best trained minds and most advanced instruments in history working to complete the work of Einstein and others. The lack of progress indicates the project to develop a unified theory is in jeopardy, and within the space community, no coordinated program exists to develop technological capabilities needed for humanity to become a starfaring civilization. Uncontrolled expenditures without quality control of outputs guarantees any viable revolutionary model which emerges cannot be identified as such. Complicating efforts to address these problems, identified processes resulting in revolutionary ideas remain largely unknown in the scientific community where misperceptions, such as mystical "flashes of inspiration" and "defining such breakthroughs a priori is difficult" are widely believed and officially promulgated.

7. Initiating Transformative Research

TR project initiation is a phase wherein the research project is "chartered" and stakeholders are identified. Because speculation on futuristic technologies is engaging and inspirational, imagining detailed solutions to problems associated with relativistic travel is a seductive option for many, with solutions only limited by one's imagination. However, this is not a viable approach given the fundamental near term unknowns upon which future technologies depend. Achieving reliable progress requires persistence, patience, care, optimism, and application of the most reliable methods for successfully accomplishing goals.

8. Definition of the Near Term Objective

Creating a successful, revolutionary theory of spacetime is properly a near-term objective for the overall project. It can only be accomplished through planned work, and like all projects it will be performed and controlled by people, constrained by limited resources, and have identifiable start and end points. Transformative results are all but impossible to attain within normal scientific operations. Minimum components of a successful effort will include requirement identification, clear and achievable objectives, balancing competing demands for quality, scope, time and cost. Adaptation of standard specifications, terminology, plans, and approaches is necessary to coordinate the different kinds of stakeholders involved in transdisciplinary work where specialist terminology prevents fluid communication.

9. Project Relevance

If the effort to fundamentally transform our conceptualization of spacetime is successful, the scope of human knowledge will be irrevocably changed, bringing the dawn of a new era for our species. Modern society now stands at a crossroad between societal collapse based on scarcity of resources and revolutionary renaissance based on creative ingenuity. The intended expansion of our genius and capabilities will bring with it unintended impacts for our society, the economy, environment, and technology. Some of these impacts will be dangerous, even deadly, and like our understanding of the structure of the atom, it will sometimes seem that we might have been better off without that knowledge.

Despite the risks, we must struggle to understand great unknowns because of who we are as a species, how we evolved, and that our survival is based on learning and cooperative tempering of our selfish, grasping instincts. We are driven to discover what is over the next hill. With diligence and wisdom, those who have the unique privilege of contributing to scientific knowledge can fulfill an adventurous destiny in the audacious, divine future of exploration and discovery.

References

1. National Science Board (U.S.), Enhancing support of transformative research at the National Science Foundation, in, National Science Foundation, Arlington, VA, 2007, pp. vi, 13 p.
2. J. Jastrow, The mind's eye, Popular Science Monthly, (1899) 13.
3. Project Management Institute, A Guide to the Project Management Body of Knowledge, in, PMI, Newtown, PA, 2012.
4. Q.B. Cooper, The myth of the lone genius, Future, Science/Fiction (2012).
5. R.S. Westman, The Copernican question : prognostication, skepticism, and celestial order, University of California Press, Berkeley, 2011.
6. N.J. Nersessian, The Process of science : contemporary philosophical approaches to understanding scientific practice, M. Nijhoff ; Distributors for the United States and Canada, Kluwer Academic Publishers, Dordrecht ; Boston Hingham, MA, USA, 1987.
7. J. Thomas, M. Mullaly, The Value of Project Management, in, Project Management Institute, Warsaw, Poland, 2012, pp. 28.
8. P. Hoyningen-Huene, Reconstructing scientific revolutions : Thomas S. Kuhn's philosophy of science, University of Chicago Press, Chicago, 1993.
9. I. Wikimedia Foundation, Jump drive - Wikipedia, the free encyclopedia, (2012).
10. H. Andersen, P. Barker, X. Chen, The cognitive structure of scientific revolutions, Cambridge University Press, Cambridge ; New York, 2006.
11. N.J. Nersessian, Creating scientific concepts, MIT Press, Cambridge, Mass., 2008.
12. T.S. Kuhn, The structure of scientific revolutions, 3rd ed., University of Chicago Press, Chicago, Ill., 1996.
13. R. Batterman, Intertheory Relations in Physics, in: E.N. Zalta (Ed.) The Stanford Encyclopedia of Philosophy, The Metaphysics Research Lab, Center for the Study of Language and Information, Stanford University Stanford, CA 94305-4115 2012.

A Fully Decelerated Probe to Alpha Centauri Using a Perturbed Project Daedalus Design

Robert M. Freeland II and Stephen Baxter

Abstract

This paper describes a preliminary study of a long stay mission to explore the Alpha Centauri system with a fast unmanned probe based on Project Daedalus technology.

Project Icarus is a theoretical design study for an interstellar probe based on the landmark Project Daedalus study completed in 1978. The Project Icarus Terms of Reference [1] specify that the probe should achieve at least some deceleration at the target system. This paper presents a straw-man mission profile for such a probe capable of reaching and exploring Alpha Centauri, one candidate target for Project Icarus. The probe is essentially a scaled-down Daedalus craft capable of the full deceleration with a reduced payload. An analysis of the physics behind the propulsion system is given, highlighting outstanding issues.

Alpha Centauri is a complex triple-star system; to explore such a system effectively within a finite mass and energy budget will require trade-offs concerning probe trajectories and subprobe deployment. The mission profile presented here uses Hohmann-ellipse minimum-energy trajectories for the exploration of (hypothetical) close-in Centauri A-B systems, while a flyby subprobe released early in the interstellar cruise is used to explore the Proxima system.

This work is intended as a proof of concept to show that a design based on Daedalus technology is indeed capable of meeting the Project Icarus Terms of Reference. This paper is a submission of the Project Icarus Study Group.

Keywords

Project Icarus, Project Daedalus, Alpha Centauri, Proxima Centauri, interstellar exploration, fusion propulsion

1. Introduction

The Project Daedalus Final Report [2] was an engineering study of a flyby probe to a nearby star system. Daedalus would have been a huge vessel to be launched from orbit carrying 50,000 tonnes of D-He3 fuel. Daedalus used a two-stage, pulsed ICF fusion engine to accelerate the vessel to 12.7% the speed of light, reaching the target Barnard's Star 48.6 years after launch. The vessel carried a 450 tonne payload, but the mission allowed for no deceleration at the target, so the total encounter time in the target system could be measured in mere hours.
Project Icarus was conceived as a successor to the Project Daedalus study, but with a few significant changes to the requirements. Foremost among these is ToR Requirement #6: [1]

> "The spacecraft mission must be designed so as to allow some deceleration for increased encounter time at the destination."

After considerable debate, the Project Icarus team has decided that FULL deceleration is vastly preferable to partial deceleration, so this has been adopted as a goal. It was recognized early in the project that, given the longer timeframe allotted for Project Icarus (100 years vs. 50 years), the original Project Daedalus vessel could theoretically accomplish the mission to a nearby star if the second stage were simply reversed. This paper describes a mission to our closest neighbor, Alpha Centauri, based on such a design.

2. Daedalus Excel Model

To properly model such a mission, it was first necessary to create a flexible mathematical model for the original Daedalus vessel. This was undertaken by Freeland [3], who created a nine-page Excel document to completely describe the mass allocations and mathematics of the propulsion system as described in the original 192-page Daedalus report. All sheets of the model are interlinked, and almost every numerical field is backed by a formula. The remaining fields are known constants or explicit assumptions. Almost every cell contains a comment referencing the exact spot in the original Daedalus report where that formula or value can be found.

- The "Mission" sheet describes the topline parameters for the mission, including target star, burns, tank drops, velocities, and mission duration.
- The "Fuel" sheet describes the intrinsic properties of the D-He3 fuel itself, plus the mathematics of its burn.
- The "Propulsion" sheet describes the physics behind the conversion of pellets to thrust.
- The "Engine" sheet describes the components of the Daedalus engines.
- The "Structure" sheet describes the structure of the vehicle apart from the engine, with particular focus on tankage.
- The "Power" sheet describes the power requirements and systems for providing that power.
- The "Payload" sheet describes the mission payload, excluding erosion shield, power supplies, and maneuvering thrusters.
- "Appendix1" is an Excel rendition of the fine print in Appendix 1 at the end of the last Daedalus chapter "An Engineering Assessment".

Unfortunately, this Excel model rather defies publication in traditional journals due to the interdependent nature of its pages, and due to the embedded calculations and notes in just about every cell. Nevertheless, the important conclusion is that the Daedalus mathematics are basically sound, apart from a few computational errors and inconsistencies as follows.

Foremost among these was a change in burnup fractions due to an error in the reaction cross-section for the D-He3 burn. The Daedalus team assumed a reaction cross-section that was too high, apparently because someone inverted the number and then someone else assumed the resulting figure (clearly very wrong with a positive exponent of ten) contained a sign error and reversed the sign. The resulting figure was too big by a factor of 2.4. Fortunately, other factors impact the burnup fraction as well, so this did not decrease performance by the full factor of 2.4.

Other computational inconsistencies stemmed from the distributed nature of the Daedalus team, and the fact that most were doing their work with pen and paper. Preliminary figures were used by one part of the team as other parts of the team refined those figures, and when agreement was close, no attempt was made to plug the final figures back into the calculations made using the preliminary figures. In particular, this was the case with the aforementioned burnup fraction. With modern computing power it is quite easy to tie these calculations together, and the "Complete Model of Daedalus" does so. Preliminary figures are noted in the spreadsheet and can be plugged into appropriate cells to demonstrate consistency with the original Daedalus report.

The net effect of these errors / inconsistencies is a relatively minor degradation in the overall performance of the Daedalus vessel. Top speed is reduced from 12.7% c to 11.5% c, and mission duration is extended from 48.6 years to 53.7 years.

3. "Dirty Icarus" Excel Model

In the spring of 2012, the Project Icarus team undertook an effort to create a "Dirty Icarus" model based on a minimal perturbation of the original Daedalus vessel. Freeland's "Complete Model of Daedalus" was chosen as the baseline for this task. Changes made were as follows:

- Tie together all pages of the model to eliminate the inconsistencies described above.
- Set the target to Alpha Centauri (vs. Barnard's Star).
- Reduce the payload mass to 50 tonnes (vs. 450 tonnes).
- Extend the mission time to 100 years (vs. 50 years).
- Reverse the second stage to achieve full deceleration at the target star.
- Reduce the pulse repetition rate from 250 Hz to 75 Hz.
- Double the target compression ratios to 2000 and 4000 on Stages #1 and #2.
- Add two additional tanks to each stage, with associated tank drops.
- Properly account for tank pressurant.

The resulting "Dirty Icarus Mk I" model by Freeland [4] achieves the following:

- Total mass of the first stage is reduced from 49,000 tonnes to 32,000 tonnes.
- Total mass of the second stage is largely unchanged at 4500 tonnes.
- Time to target is 95 years at a peak speed of 4.8% c.

The crucial conclusion is that a Daedalus-style vessel can achieve full deceleration into the Alpha Centauri system within the allotted mission time, using less fuel and a much lower pulse rate, provided we can achieve higher pellet compression ratios. The Daedalus team chose its compression ratios somewhat arbitrarily, so this is considered a reasonable iteration on their design.

Notably, the engine design in the "Dirty Icarus" model remains completely identical to the one in the original Daedalus model. An attempt was made to scale the engine down, but it was found that smaller pellets suffer inferior burnup fractions, so overall performance of the vessel degrades significantly. The engine dimensions, in turn, are a function of the pellet size (and associated energy release), so if pellet sizes remain unchanged, then so must the engine dimensions.

Note that a review of the Daedalus proposal by the Project Icarus team has identified significant and so far unresolved problems with the Daedalus design. (See Appendix for further details.) The authors of the present paper are not advocating a Daedalus-like design as a final solution for Project Icarus; rather, our purpose has been to demonstrate that if the Daedalus design could be shown to be operational, then a relatively simple perturbation of that original flyby design could produce a draft design of a probe capable of full deceleration.

4. Alpha Centauri System

Alpha Centauri is a complex and scientifically target-rich environment, demanding a similarly complex and flexible approach to its exploration by our probe on arrival. The following key parameters concerning the Alpha Centauri system are taken from Beech [5].

- From Earth there is a 2.18 degree angular separation of Proxima from Alpha A/B.
- Proxima is 4.22 ly (light years) from our solar system; Alpha A/B are 4.40 ly away.
- Proxima is 15,034 AU (0.237 ly) from A/B, and according to studies quoted by Beech [5] may be close to the inner edge of the A/B Oort cloud.
- Proxima has mass 0.123 solar masses.
- The Alpha A/B orbital plane has an inclination of 79 degrees.

- The Alpha system as a whole has a significant space velocity of 32.6 km/sec, with a radial velocity component of ~22 km/sec towards the sun.
- Alpha A has mass 1.105 solar masses; Alpha B has mass 0.934 solar masses.
- The orbits of A and B about their common centre are elliptical; their separation varies from 11.2 AU to 35.6 AU, with a period of ~80 years.
- Stable planetary systems are possible within ~4 AU of each star of the A/B pair, with the respective habitable zones closer in.
- Observational limits have been set on the presence of planets in the Proxima system, but worlds of Earth-like mass or less could exist in the habitable zone, and larger planets may exist further out.

This paper will take as an assumption that there are planetary systems around A and B between orbital radii ~0.1AU and 3AU. It is also assumed that the planetary systems are both in the same plane as the A/B orbit, which seems reasonable if all elements of the system have a common origin. Other planets could be found significantly further out, orbiting the A/B pair as a single centre of mass. This paper further assumes that the stars are an average distance 23.4 AU apart on arrival.

5. Options for Exploring Proxima

For the purpose of this paper the dual system Alpha A/B is chosen as the primary target. What options are there for reaching Proxima and any related planetary system as part of the same mission? In particular, is it possible to achieve full deceleration at both A/B and Proxima? The challenge comes from the remoteness of Proxima; while A and B are separated by planetary distances (~10 AU), Proxima is ~15,000 AU away. [5] Five trajectory options are studied here:

5.1 Separate Missions

The simplest way to reach both A/B and Proxima with full deceleration would be to launch two independent probes, one targeted at A/B, the other at Proxima. The total delta-V would be twice that of the basic mission, or ~19.2%c. This is not proposed as a serious mission choice but as a reference point for this study.

5.2 Deceleration at Alpha A/B; Transit to Proxima

Is it practical to decelerate fully at Alpha A/B and then send the Icarus probe (or a subprobe cluster) on to Proxima?

Proxima is 15,034 AU from Alpha A/B [5], well beyond the reach of modern interplanetary probe technology. Voyager 1, currently travelling at 17 km/s = 3.6 AU/year would take 4200 years to reach Proxima from A/B.

The Icarus main drive could achieve the journey in 9.2 years (with 4.25 years acceleration to 4.8%c, 8.25 months of cruise, then 4.25 years deceleration), but this would require a supply of fuel at the Alpha A/B system, and barring some kind of in-situ production, getting the fuel there would effectively double the propulsion requirements of the overall mission. The main drive could make the journey in a longer span using less fuel (by simply going slower), but then arrival at Proxima would fall well outside the 100-year timeframe set forth in the Project Icarus ToR. [1]

The conclusion is that a separate post-deceleration transit to Proxima from Alpha A/B makes little sense without some kind of in-situ refuelling at Alpha A/B.

5.3 Gravity-Assist at Proxima

Is it feasible to send Icarus to Proxima, and then use a gravity assist by Proxima to divert to Alpha A/B (or vice versa)?

The analysis here relies on the standard theory of unpowered gravity assists given in Matloff [6]. If the probe approaches and departs from Proxima with the same asymptotic speed V, then the trajectory bend angle after a flyby can be estimated by Matloff eq. 4.9:

$$\Psi = 2 \sin^{-1} [1/(1+2(V/V_p)^2]\quad (1)$$

where V_p is the parabolic (escape) velocity at the periapsis approach to the deflecting body, given by Matloff eq. 4.8:

$$V_p \sim 1.4(GM/R)^{1/2} \quad (2)$$

with G the gravitational constant, M the mass of the target, and R the probe's separation from the target's centre at periapsis.

Using the data in Beech [5], Proxima's mass M = ~0.123 solar masses. If a grazing flyby is made at R = ~0.145 solar radii, then V_p = 564 km/s. Similarly for Alpha A, V_p = 583 km/s. For the fictional Avatar gas giant Polyphemus [6] with M = ~124 Earth masses and R = 59,000 km (similar to Saturn), V_p is a paltry 41 km/s.

V_p is a key parameter in this analysis. If $V=V_p$ then ψ = 39 deg. The larger V is compared to V_p, the less the deflection angle achieved. For example if V = 4.8%c (the Icarus cruise speed) then with V_p = 564 km/s as at Proxima, ψ = 0.09 deg. Large angles of deflection can only be achieved if V is much less than V_p; as V approaches zero ψ approaches 180 degrees.

According to Beech [5] Sol, Proxima and Alpha A/B lie at the vertices of a triangle with sides 1.295pc (Sol to Proxima), 0.075pc (Proxima to A/B), and 1.348pc (A/B to Sol). Elementary geometry shows that the angular deflection required for an assist to take a craft from the Sol-Proxima line to the Proxima-A/B line is 47 degrees. To achieve this, the craft's speed would have to be reduced significantly from cruise, indeed to below Proxima's V_p; with V_p = 564 km/s then V = ~491 km/s. At this speed it would take some 153 years to complete the crossing from Proxima to Alpha A/B. If this approach were taken, then the main probe would have to reaccelerate to make the transit from Proxima to Alpha A/B. This is essentially the same problem described in Section 4.2 above, but in reverse. A similar analysis for a swing-by from Alpha A/B to Proxima delivers order-of-magnitude similar results. The conclusion is that gravitational assists would be of little value in shaping a tour to take in both Proxima and Alpha A/B (though they may be of value in other aspects of the mission; see section 5.2 below).

5.4 Separation after the Acceleration Phase

One way to reduce the total delta-V requirement would be to split off a subprobe bound for Proxima after the vessel's main acceleration phase is over. This would require a cross-range burn to deflect the Proxima subprobe to its target.

On launch from our solar system, the propulsion system burns for 4.25 years, delivering a cruise speed of 4.8%c, at an engine cut-off distance of ~0.0725ly. The Proxima subprobe would be ejected at main engine cut-off. From here a flight to Proxima requires an angular deviation of 2.22 degrees from the path to Alpha A/B. The cross-range delta-V required to put the subprobe on the right path is 4.8%c * tan 2.22deg = 0.19%c, about 4% of the cruise speed. The Proxima probe would continue on at 4.8%c, covering the remaining 4.15ly to Proxima in ~86.7 years.

Although small by interstellar standards, the cross-range burn for the flyby requires a significant delta-V relative to current drive technologies. Fortunately, the ejection of the Proxima probe coincides with the separation of the first stage, so it makes sense to use the first stage fusion drive for the needed cross-range burn. This implies that the first stage would retain some fuel and itself carry a payload of subprobes for delivery to the Proxima system.

If the Proxima probe were to fully decelerate at Proxima, then the expelled first stage would have to retain a large supply of fuel for its own deceleration phase. The first stage really isn't optimized for this, so the outcome is effectively worse than launching two separate probes. Savings on engine and structure mass (one first stage and one second stage, versus two of each) are offset by the additional fuel mass for the aforementioned cross-range burn, plus the inefficiencies associated with using the massive "spent" first stage for subsequent deceleration.

5.5 Recommendation: A Proxima Flyby

Our conclusion is that attempting to achieve full deceleration into both Alpha A/B and Proxima via a single mission offers no advantages over completely separate missions. However, we find that it is very cheap in terms of propulsion to augment a fully-decelerated main mission to Alpha A/B with an undecelerated flyby of Proxima. The total delta-V required for such a mission would be just (9.6+0.19)%c = 9.8%c, versus the baseline 9.6%c.

As to the science achieved by the flyby, the Daedalus model [2] provides a good starting point, since the Proxima flyby would actually be very like Daedalus's reference mission. Subprobes would be released 1ly out and scattered through the Proxima system across a radius of ~5AU, flying by planetary targets. As 1ly is ~63,000 AU, the additional cross-range delta-V required to launch the subprobes would be ~ 5/63,000 * 4.8%c = 1.1 km/sec, a number comparable to the Daedalus results, and within the capabilities of current interplanetary drive systems. The subprobes would have to withstand independent journeys through the interstellar medium for ~1ly at 4.8%c, followed by a passage through the Proxima system, so shielding would have to be planned carefully. While the Voyager probes weighed 0.8 tonnes each, the Daedalus planetary subprobes, each equipped with heavy shielding, weighed ~10 tonnes each. [2]

Maximizing the science return from the probes would also require careful design. At 4.8%c, a planetary encounter covering a distance of say 20 times the radius of Earth would last a mere nine seconds. Perhaps dual sets of instruments could be mounted on each probe, looking forward and back, to avoid the need for a fast-swivelling table as the target is passed. There could be issues concerning instrument performance given the Doppler shift.

There are also issues concerning communications. The Proxima flyby occurs (after ~89 years) while the main probe is still cruising. (See the overall timeline in section 7 below.) This may be problematical if the main probe is to be used as a communications relay to Earth, because the probes would be attempting to contact each other while both are travelling at 4.8%c. Perhaps the Proxima probe could communicate directly with Earth. Or perhaps it could store its data until the main probe deceleration is complete, and targeting becomes less challenging - but by then the flyby probe will have travelled far beyond Proxima.

After the Proxima encounter, there may be additional opportunities to study the Alpha A/B Oort sphere, thought to extend from ~20,000 AU to ~100,000 AU (=1.59ly) from A/B. [5] The flyby probe would presumably go on to become the most remote manmade object.

6. Insertion into Orbit at Alpha Centauri A/B

Once the main vessel decelerates into Alpha A/B, it needs to insert its probes into usable orbits. Presumably, data on the orbits of the stars and planets in the Centauri system will be available by the time of arrival, either from Earth-based observations or from the probe itself while en route. While the design of a Proxima flyby mission may need to be fixed early, the main vessel's onboard AI may be expected to make final decisions in the design of its insertion and orbital tour strategy at Alpha A/B. Nevertheless the discussion here attempts to make general conclusions regardless of the details and absent a specific set of science objectives.

6.1 Summary Trajectory Profile

A simple trajectory model is described here for the purposes of estimating post-deceleration delta-V and timings to achieve a long-duration exploration of the Alpha A/B system. The model basically assumes short burns followed by Hohmann minimum-energy transfers.

At A/B the main probe, through an 'insertion manoeuvre' (discussed below), matches the overall motion of the A/B system. It then separates into two probe clusters, a primary cluster 'Icarus A' (carrying the main communications system) which from the arrival point will follow an 'insertion orbit' to adopt a close-in circular 'exploration orbit' around Alpha A, and engage on a subsequent exploratory tour of the Alpha A system. Meanwhile, a secondary probe cluster 'Icarus B' will enter a similar exploration orbit around Alpha B.

The subsequent orbital tours, in which subprobes will be dispatched to target planets and moons, are discussed in Section 6 below.

6.2 Insertion Manoeuvre

Insertion into the A/B system will not be a familiar manoeuvre. The main vessel will arrive from out of the orbital plane of the Alpha A/B system, which is itself moving through space relative to our Sun at ~33 km/sec. [5]

One simple approach is for the probe to decelerate to a dead halt in the system's frame of reference. This will be targeted at a point labelled Q, the gravitational equilibrium between Alpha A and B. This is on the line AB at roughly the midpoint, since A and B have similar masses. Assuming an average separation of 23.4AU between A and B at the time of arrival, AQ is 12.2AU and QB is 11.2AU. At Q the craft turns and the engine burns once more with delta-V ~33 km/sec, to insert the craft into the orbital plane of A-B and match the system's own velocity relative to our Sun's reference frame. (In practice it would make more sense for the deceleration burn to be slightly off-axis to impart the 33 km/sec lateral speed, so the engine doesn't start and stop.)

Could gravity assists be used to achieve this plane-change manoeuvre? Since the Alpha system's space velocity is less than the parabolic escape velocity computed for both the Alpha stars and for Polyphemus (see Section 4.3), this could well be possible, and is recommended for investigation with more detailed modelling.

6.3 Transition Manoeuvre

On arrival at Q, the probe separates into two clusters, Icarus A and Icarus B. Assume that Icarus A carries the main communications system, so Icarus B will report home through Icarus A.

Icarus A is to be sent to the Alpha A planetary system via an elliptical 'transit orbit', with apoapsis at Q = 12.2AU, and periapsis at ~3AU, the outermost stable planetary orbit. At periapsis an engine burn will lower the orbit into an 'exploration orbit': a circle with semimajor axis 3AU, from which detailed subprobe exploration of the Alpha A system will subsequently be mounted. (Making the burn at the periapsis, as deep as possible inside the gravitational field, is of course the most effective in terms of delta-V per the Oberth effect [7].)

Both transit and exploration orbits are modelled here as single-centre Newtonian orbits. This is a safe assumption for the exploration orbit [5], but not for the transit orbit, which will be affected by Alpha B's gravity field. More detailed modelling is needed here.

Some useful standard formulae for such orbits (see e.g. Synge [8] p164) are the period:

$$T = 2\pi \, \mathrm{sqrt}(a^3/GM) \quad (3)$$

and the velocity at radius r:

$$v^2 = GM(2/r - 1/a) \quad (4)$$

where a is the semimajor axis of the orbit, M is the star's mass, G is the gravitational constant.

For Icarus A's transition orbit a=7.6AU, while M is 1.105 solar masses. The period is 20 years; velocity at Q is 5.5 km/s, velocity at periapsis is 23 km/s. Meanwhile for Icarus A's exploration orbit, a=3AU; the period is 5.2 years; orbital velocity is 18 km/s.

Thus there need to be two engine burns: one at Q to insert the probe into the transition orbit with velocity 5.5 km/s; and the second 10 years later at periapsis, a burn of 5 km/s to lower the apoapsis to 3AU. The total delta-V required for Icarus A is 10.5 km/s. The results are broadly similar for Icarus B as shown in Table 1 below, with a total delta-V for Icarus B of 9.8 km/sec.

Table 1: Insertion Orbit Calculations.

Transition Orbit			**Exploration Orbit**		
Alpha A mass	1.105	Mo	Alpha A mass	1.105	Mo
axis	7.6	AU	axis	3.0	AU
period	19.9	years	period	4.9	years
velocity at Q	5.6	km/s	orbital velocity	18.1	km/s
velocity at 3AU	22.9	km/s	delta-V	10.5	km/s
Transition Orbit			**Exploration Orbit**		
Alpha B mass	0.934	Mo	Alpha B mass	0.934	Mo
axis	7.1	AU	axis	3.0	AU
period	19.6	years	period	5.4	years
velocity at Q	5.6	km/s	orbital velocity	16.6	km/s
velocity at 3AU	20.9	km/s	delta-V	9.8	km/s

The separation between A and B varies from 11.2AU to 35.6AU. Thus the precise epoch in which we arrive, and the relative positions of A and B at that moment in time, will affect the details of the subsequent manoeuvres we require. We may choose to arrive at an epoch of closest approach, thereby allowing for a quite rapid deployment of the probe clusters to the target planetary systems.

7. Orbital Tours of Alpha A/B

It is assumed that our strategy for exploring the A/B systems will be similar to the one used for exploration of our own solar system: that is, to deliver long-duration orbiter and lander craft to rocky worlds, like the current

Mars Express and MER rovers; and to deliver long-duration orbiters to the systems of gas giants, like Cassini [9]. Tours of exploration could last for decades. These may be supplemented by such craft as lightweight penetrators, low-ISP craft like the current Dawn [10] capable of extended tours of asteroid belts, etc., and perhaps 'Grand Tour' multi-planet probes like the Voyagers.

How should the long-duration orbiters and landers be delivered to their target destinations? With two probe clusters Icarus A and B injected into circular orbits around Alpha A and B at the outer fringe of the respective planetary systems, a minimum-complexity, low-energy strategy is to drop subprobes into Hohmann orbits to (potentially) each of the inner planets. The key numbers derived here concern how long such transfers will take to achieve; an indication is also given of the likely delta-V requirements. The key question explored is whether this kind of orbit, based on the most standard of modern technologies, is sufficient to deliver a feasible tour programme, or whether more advanced secondary propulsion systems are required.

7.1 Hohmann Orbits

The mechanics of minimum-energy transfer orbits is standard theory [11]. In the 1920s Hohmann showed that the most economical transfer orbit is an elliptical orbit which just touches the orbits of destination and target worlds. The fuel expenditure incurred in travelling from an outer planet to an inner one is less than the other way round, which is why we park the subprobe clusters at the edge of their respective systems.

The time incurred in using a Hohmann ellipse to make such a transfer consists of two elements:

- the transit time itself, which is half the orbital period of the transfer ellipse.
- the time needed for the appropriate planetary alignment to occur to allow the transfer trajectory to be launched.

Transit time: the transfer ellipse will have apoapsis at the outer planet's orbit, and periapsis at the inner planet's orbit. Its semimajor axis "a" will be the average of the two planetary orbits' radii (assuming circular orbits). The ellipse's period T_p can then be computed from Kepler's law:

$$T_p \alpha\ a^{3/2} \quad (5)$$

The probe transit time T_t is just half T_p.

Alignment time: the waiting time for the appropriate relative alignment between the planets to occur cannot be more than the time taken for the inner planet (with higher angular velocity) to complete a whole orbit, then overtake the outer planet and regain its relative position. For two planets with orbital periods T_1 and T_2 (with $T_1 > T_2$) the maximum alignment wait time is:

$$T_a = 1/(1/T_2 - 1/T_1) \quad (6)$$

The total wait time, from arrival of a probe cluster at the outer orbit to successful deliver of a subprobe at an inner planet, is therefore no more than:

$$T_w = T_t + T_a \quad (7)$$

If the probe clusters arrive at an essentially random time regarding the alignment of a given target planet, the wait time could of course be less than this. The average would be half the maximum realignment time; however, T_w is the worst case and will be quoted here.

7.2 Hohmann Transfer Analysis

Since mankind's current knowledge of the Centauri planetary system is sketchy at best, this paper takes key parameters from the Centauri planetary system described in the Avatar movie [12] by way of example. See Tables 2 & 3 below for key parameters. Note that the recent discovery of a planet orbiting Alpha Centauri B – while strengthening the appeal of the system as a target for exploration – arrived too late for this study; however its conclusions are not significantly changed.

The Alpha A system has 7 planets including three sharing an orbit (labelled 2a, 2b, 2c). The Alpha B system has 6 planets. The most interesting targets in the systems are probably A2a,b,c (A2a is the giant planet Polyphemus with its habitable moon Pandora) and B3, B4, which both orbit within the parent star's habitable zone.

Results for the transfer times to the planets of the two systems are given in the tables below. For the sake of the calculations here we have assumed the orbits as circular. The transits analysed for Alpha A are from the orbit of A5 to each of the inner worlds; similarly for Alpha B the transits begin at the orbit of B6. Thus to transit from the A5 orbit to planet A4 will require a maximum of 13.32 years.

Table 2: The Alpha A system.

Planet	Period (y) [11]	Orbit Radius (AU) [11]	a (AU)	T_t (y)	T_a (y)	T_w (y)
1	0.33	0.49	1.75	1.11	0.36	1.47
2a,2b,2c	1.58	1.4	2.2	1.57	2.30	3.87
3	2.46	1.88	2.44	1.83	4.84	6.67
4	3.46	2.37	2.69	2.12	11.2	13.32
5	5.0	3.00				

Table 3: The Alpha B system.

Planet	Period (y) [11]	Orbit Radius (AU) [11]	a (AU)	T_t (y)	T_a (y)	T_w (y)
1	0.043	0.12	1.56	0.54	0.043	0.58
2	0.15	0.27	1.64	0.59	0.15	0.74
3	0.36	0.50	1.75	0.65	0.39	1.04
4	0.54	0.68	1.84	0.70	0.61	1.31
5	2.06	1.29	2.15	0.88	3.5	4.38
6	5.0	3.00				

The total delta-V requirements will of course depend on the chosen targets. In each case a first-order delta-V can be computed (with eq. 4) by assuming insertion into a circular orbit at the target planet's orbital radius. As an indication, a transfer to A2 would require ~4 km/s; a transfer to B4 would require ~10 km/s.

The systems of Alphas A and B are both compact and the wait times are correspondingly short. The upper bound of the wait times is just over 13 years, for transfer from the orbit of Alpha A5 to the close-by planet A4. The wait time to get to the key target worlds (A2a,b,c and B3,4) is no more than 4 years.

These wait times could surely be reduced with better modelling and a more flexible mission profile, for example with the use of gravity slingshots. This fictional Alpha A system includes 3 giant planets which would facilitate such an approach. Furthermore the delta-V cost of these transfers could be reduced further by using such measures as aerobraking and direct insertion of landers at the target worlds.

Our conclusion is that given the long duration of the preceding interstellar cruise phase (~75 years), followed by a presumably decades-long science programme to be executed once the subprobes are delivered to their target worlds, a transfer period of just a few years using minimum-energy Hohmann transfers to deliver the subprobes is an acceptable trade off to reduce the mass penalty and complexity of advanced secondary propulsion systems, which could get the subprobes there faster.

7.3 Subprobes and Additional Targets

Crawford [9] has given a preliminary discussion of subprobe types, based on experience in the solar system. The allocation of subprobes may look something like this:

Table 4. Possible subprobe complement.

Probe Type	Class	Mass (t)	Number of Probes	Mass Budget (t)
Stellar orbiter	SOHO-class	1.85	2 - 4	3.7 - 7.4
Planetary orbiter	CASSINI-class	2.125	6 - 8	12.75 – 17
Gas Giant probes / balloons	GALILEO EP-class	0.339	2 - 4	0.678 – 1.356
Rocky Planet atmospheric vehicles	Galileo EP-class	0.339	6 - 8	2.0 – 2.7
Planetary soft landers	MSL-class	0.9	6 – 8	5.4 – 7.2
Low mass penetrators	LUNAR-EX-class	0.04	30 - 50	1.2 – 2
				Total Science Mass = 26 – 38 tonnes
				Net Probe Mass = 40 tonnes * 4 = 160 tonnes (for Internal Structure, Shielding, Packaging, Power, Communications, Navigation, Propulsion)

A possible addition to this might be a solar-powered ion-drive spacecraft like the current Dawn asteroid explorer [10], to be used to explore minor planets or asteroids. Dawn's nominal mission duration is ~8 years during which it will insert itself into orbit around 2 asteroids. Dawn is systematically exploring a space of radius ~3AU, similar to the dimensions of the Alpha A/B systems.

There may be additional targets to be considered in the Alpha A/B system, such as asteroid belts, and planets further out circling A/B as a common centre. Secondary probes could be launched at different points in the mission profile to reach such targets. The possibility of a Proxima flyby exploring an Oort cloud has already been mentioned.

It is fascinating to note that if we could only reach Alpha Centauri we could mount an effective exploration campaign with essentially modern technology. The most significant necessary upgrade might be more autonomy; the Dawn spacecraft, for example, relies on a ground command for each engine fire [10].

8. Overall Mission Timeline

Given below is a mission timeline for a straw-man tour with primary target Alpha A/B, and a Proxima flyby ejected after the main acceleration phase:

Table 5. Mission milestones.

Time (Years)	Event	Distance (ly)
0	Launch; acceleration begins	0
2.5	Acceleration ends. Cruise at 4.8%c begins. Proxima flyby probe ejected.	0.06
89.2	Flyby passes Proxima	4.22 (Proxima)
92.5	Main probe deceleration starts	4.34
95	Deceleration ends, at A-B equilibrium point Q. Plane-change insertion burn. Insertions of A, B probe clusters into transit orbits.	4.4 (Alpha A-B)
105	A-B cluster periapsis burns; exploration of A-B systems begin in exploration orbits.	
125	Flyby reaches outer limit of Oort cloud	6

9. Discussion and Recommendations

The Alpha Centauri system is a complex and target-rich environment. To explore it effectively within a finite mass and energy budget will require an inventive approach to a mission profile design, as well as tradeoffs between different science priorities, such as between Alpha A/B orbiters and landers and a possible Proxima flyby. This analysis is a preliminary sketch, to be refined as more information about components of the mission are defined, such as the likely performance of Icarus itself, and the parameters of the planetary systems to explore.

It is hoped that this study will serve as a beginning for the design of this phase of the mission. Specific recommendations are:

- Within the Project Icarus ToR [1], it appears impractical to attempt to achieve full deceleration into both Alpha A/B and Proxima Centauri with a single vessel. A flyby of one system, with the main probe aimed at the other system, could be achieved at relatively small incremental cost. A full study of the costs and benefits of this secondary mission, together with a study of the technical challenges of relativistic-speed subprobe design and communications issues, is recommended.
- There are many options for how to manage the trajectory of the main probe and subprobes after arrival at Alpha Centauri A/B. This needs careful modelling once we have in place a definitive hypothetical target system (at Alpha Centauri or elsewhere) balancing science goals against technical challenges, in order to seek an optimal solution. The use of gravity assists both on arrival at the system and during subsequent tours should be explored. The complexity of orbital mechanics in a close-in binary star system will pose interesting mathematical problems.

- The cruise to Alpha Centauri will take decades. Ideally, the subprobes' scientific exploration of the system will also take decades. It is shown here that for the planetary transit phase that comes sandwiched between the two, even Hohmann-orbit delivery times consume only a small fraction of the total mission timeline. The use of more advanced in-system propulsion systems would add mass and complexity to the mission – thus detracting from the true science payload and reducing resilience – and their benefits in reducing probe transfer times would be marginal. Further study of this issue is recommended.

Acknowledgements

The authors are grateful to members of the Project Icarus team for comments on the work behind this paper.

Appendix: Problems with the Daedalus Engine Design

The aforementioned results notwithstanding, the authors of this paper want to note that the Daedalus engine design suffers some significant problems:

- Fuel Choice: The Daedalus design uses He3, which is almost completely nonexistent on Earth, and can not readily be produced artificially. It can be found on the surface of the Moon, but only in extremely low concentrations. The Daedalus team proposed mining He3 from Jupiter's atmosphere, and they envisioned a "solar system wide civilization" to do it. Freeland [13] recommends using LiD pellets instead and demonstrates comparable performance.
- Tritium Cores: The Daedalus design uses tritium cores in the pellets, but tritium has a half-life of only 12.32 years [14], so it won't survive to reach the deceleration phase. Crowl [ref?]15 has also shown that this radioactive decay will generate enough heat to vaporize the solid deuterium honeycomb comprising the pellet body, thereby rupturing the pellets. He has advocated ultra-dense deuterium cores as an alternative. Uranium cores may also work.
- Compression Mechanism: The Daedalus design uses electron beams to compress the pellet, but Benford [16] has recently demonstrated that the electron beams would erode away before compressing the pellet much at all. Laser technology has evolved significantly since the Daedalus Report was published in 1978, so this would provide a plausible alternative.
- Compression Symmetry: The Daedalus design assumes that the ablation of the outer surface of the pellet will create a uniform "atmosphere" around the pellet capable of compressing it symmetrically. The NIF team has discovered that – even with a perfectly positioned pellet in a special hohlraum designed to focus the impinging lasers – it is difficult to achieve symmetrical compression of the target to sufficient levels to achieve a net output from the fusion. Given that the Daedalus reaction chamber is an open hemisphere, with electron beams arranged only equatorially, it seems unlikely that the design could ever achieve the needed symmetry.
- Misfires: The Daedalus engine captures energy from each firing to power the beams for the next firing. In the event of a misfire, the system shuts down and must "ramp back up" to operational frequencies via a succession of smaller firings, the first of which must be powered by secondary power systems (e.g. – an on-board fission reactor). This whole process is rather poorly described in the Daedalus Final Report.

For all these reasons, the authors of this paper want to stress that they are not advocating a Daedalus-style engine for Icarus, but rather just demonstrating that an Icarus mission to Alpha Centauri is possible with an engine providing Daedalus-like performance. The Project Icarus team is currently studying the issues outlined above as a review of the Daedalus model, and is also evaluating several alternative fusion-based engine designs to potentially replace the Daedalus engine as its working model.

References

1. Long, K.F. et al., "Project Icarus: Son of Daedalus - Flying Closer to Another Star" (JBIS, 62 No. 11/12, pp403-416 Nov/Dec 2009)

2. Alan Bond, Anthony Martin et al., "Project Daedalus" (JBIS, 1978)

3. Freeland, R.M., "Project Icarus: Complete Model of Daedalus", Project Icarus, Modules 3,4, May 2011.

4. Freeland, R.M., "Project Icarus: Dirty Icarus Mk I Model", Project Icarus, Modules 3,4, April 2012.

5. Beech, M., "Exploring Alpha Centauri", The Observatory 131, pp212-224, 2011.

6. Matloff, G., Deep Space Probes, Springer, 2010

7. Oberth, H., Ways to Spaceflight, NASA TT-F-622.

8. Synge, J. and Griffith, B., Principles of Mechanics 3rd edition, McGraw-Hill, 1970.

9. Crawford, I.A., "Project Icarus: Some Preliminary Thoughts on the Selection of Probes and Instruments for the Icarus Interstellar Mission", Project Icarus, Module 14, April 2012.

10. www.dawn.jpl.nasa.gov, accessed 14 April 2012.

11. Bonestell, C., Ley, W., The Conquest of Space, Sidgwick and Jackson, 1950.

12. "New Tables", Avatar production team, 2012.

13. Freeland, R.M., "Project Icarus: Fission-Fusion Hybrid Fuel for Interstellar Propulsion", Project Icarus, Module 4, December 2010.

14. L. L. Lucas, M. P. Unterweger (2000). "Comprehensive Review and Critical Evaluation of the Half-Life of Tritium". Journal of Research of the National Institute of Standards and Technology 105 (4): 541.

15. Crowl, A., Project Icarus internal communication, 10/27/2012.

16. Benford, James, "Project Icarus: E-Beam ICF for Daedalus Reconsidered", Project Icarus, Module 4, August 2012.

Interstellar Links by Focal Spacecrafts

Claudio Maccone, Ph.D.

Abstract

Gravitational lensing is one of the most amazing discoveries produced by Einstein's general theory of relativity. To date, hundreds of gravitational lenses have been observed by astronomers and they led to a number of new results in extrasolar planet search, astrophysics and cosmology. SETI also could benefit from gravitational lensing if we could just get to 550 AU from the Sun and beyond. This is because the gravitational lens of the Sun would highly intensify there any weak radio signal reaching the solar system from distant civilizations in the Galaxy, as shown by this author in his 2009 book "Deep Space Flight and Communications".

The gravitational lens of the Sun, however, has a drawback: the solar Corona. Electrons in the Corona make electromagnetic waves "diverge" and this "pushes the focus out" to distances higher than 550 AU. For instance, at the CMB peak frequency of 160 GHz, the true focus lies at 763 AU. It would be safer to let the FOCAL spacecraft reach 1,000 AU.

We could get rid of all solar-Corona-related problems, however, if we could reach the six-times higher distance of 6,077 AU. This is where the focal sphere of Jupiter lies. Jupiter is the second larger mass in the solar system after the Sun, but in this focal game not only the mass matters: rather, what really matters is the ratio between the radius of the body squared and the mass of the body. In this regard, Jupiter qualifies as the second best choice for a FOCAL space mission, requiring the FOCAL spacecraft to reach 6,077 AU.

What about the other planets as gravitational lenses, then? Neptune qualifies third, with a focal sphere of 13,520 AU and Saturn comes fourth with a focal sphere of 14,420 AU. But the real surprise is the Earth, that qualifies just fifth with a focal sphere of 15,370 AU. And the Earth is indeed the best body we could use as a gravitational lens since we know about its atmosphere better than about any other planetary atmosphere.

Just to complete the picture, Uranus comes sixth at 16,980 AU and Venus seventh at 17,020 AU, let alone Mars, Mercury and the Moon, all with focal spheres at 40,000 AU and beyond.

We have discovered a new BELT (made of focal spheres) between 550 and 17,000 AU, thus located much beyond the well known Kuiper Belt of small bodies, remnants from the Solar System's formation, located between 30 to 55 AU.

There is an even more important question, tough, that we face in this paper: how to INSURE any future interstellar radio links between the solar system and future interstellar probes. We show by calculations that only exploiting the gravitational lens of the Sun as a huge antenna we will be able to achieve such an interstellar link. In other words, we calculate the Bit Error Rate (BER) across interstellar distances both with and without using the gravitational lens effect of the Sun. The conclusion is that only when we will exploit the Sun as a gravitational lens we will be able to communicate with our own probes (or with nearby Aliens) across the distances of even the nearest stars to us in the Galaxy, and that at a reasonable Bit Error Rate.

Keywords

Sun gravitational lens, space probe, interstellar flight, telecommunications, bit error rate (BER).

1. Introduction: The Sun as a Gravitational Lens

The gravitational focusing effect of the Sun is one of the most amazing discoveries produced by the general theory of relativity. The first paper in this field was published by Albert Einstein in 1936 (ref. [1]), but his work was virtually forgotten until 1964, when Sydney Liebes of Stanford University (ref. [2]) gave the mathematical theory of gravitational focusing by a galaxy located between the Earth and a very distant cosmological object, such as a quasar.

In 1978 the first "twin quasar" image, caused by the gravitational field of an intermediate galaxy, was spotted by the British astronomer Dennis Walsh and his colleagues. Subsequent discoveries of several more examples of gravitational lenses eliminated all doubts about gravitational focusing predicted by general relativity.

Von Eshleman of Stanford University then went on to apply the theory to the case of the Sun in 1979 (ref. [3]). His paper for the first time suggested the possibility of sending a spacecraft to 550 AU from the Sun to exploit the enormous magnifications provided by the gravitational lens of the Sun, particularly at microwave frequencies, such as the hydrogen line at 1420 MHz (21 cm wavelength). This is the frequency that all SETI radio astronomers regard as "magic" for interstellar communications, and thus the tremendous potential of the gravitational lens of the Sun for getting in touch with alien civilizations became obvious.

The first experimental SETI radio astronomer in history, Frank Drake (*Project Ozma*, 1960), presented a paper on the advantages of using the gravitational lens of the Sun for SETI at the *Second International Bioastronomy Conference* held in Hungary in 1987 (ref. [4]), as did Nathan "Chip" Cohen of Boston University (ref. [5]). Non-technical descriptions of the topic were also given by them in their popular books (refs. [6] and [7]).

However, the possibility of planning and funding a space mission to 550 AU to exploit the gravitational lens of the Sun immediately proved a difficult task. Space scientists and engineers first turned their attention to this goal at the June 18, 1992, *Conference on Space Missions and Astrodynamics* organized in Turin, Italy, by this author. The relevant Proceedings were published in 1994 in the *Journal of the British Interplanetary Society* (ref. [8]). Meanwhile, on May 20, 1993 this author also submitted a formal Proposal to the European Space Agency (ESA) to fund the space mission design (ref. [9]). The optimal direction of space to launch the FOCAL spacecraft was also discussed by Jean Heidmann of Paris Meudon Observatory and the author (ref. [10]), but it seemed clear that a demanding space mission like this one should not be devoted entirely to SETI. Things like the computation of the parallaxes of many distant stars in the Galaxy, the detection of gravitational waves by virtue of the very long baseline between the spacecraft and the Earth, plus a host of other experiments would complement the SETI utilization of this space mission to 550 AU and beyond.

The mission was dubbed "SETISAIL" in earlier papers (ref. [11]), and "FOCAL" in the proposal submitted to ESA in 1993.

In the third edition of his book "The Sun as a Gravitational Lens: Proposed Space Missions" (ref. 12), the author summarized all knowledge available as of 2002 about the FOCAL space mission to 550 AU and beyond to 1000 AU. On October 3rd, 1999, this book had already been awarded the Engineering Science Book Award by the International Academy of Astronautics (IAA).

Finally, in March 2009, the new and comprehensive 400-pages book by the author, entitled "Deep Space Flight and Communications – Exploiting the Sun as a Gravitational Lens" (ref. [19]), was published. This book embodies all the previous material published about the FOCAL space mission and updates it. On November 25th, 2009, this book was presented in a talk that the author gave at the SETI Institute in Mountain View, CA, USA (You Tube site: http://www.youtube.com/watch?v=ObvKVe5H8pc).

The FOCAL name may also be regarded as an acronym for "Fast Outgoing Cyclopean Astronomical Lens", summarizing the mission's main features.

2. Why 550 AU is the Minimal Distance that "Focal" Must Reach

The geometry of the Sun gravitational lens is easily described: incoming electromagnetic waves (arriving, for instance, from the center of the Galaxy) pass *outside* the Sun and pass within a certain distance r of its center. Then the basic result following from the Schwarzschild solution shows that the corresponding *deflection angle* $\alpha(r)$ at the distance r from the Sun center is given by

$$\alpha(r) = \frac{4GM_{Sun}}{c^2 r}. \tag{1}$$

Figure 1 shows the basic geometry of the Sun gravitational lens with the various parameters in the game

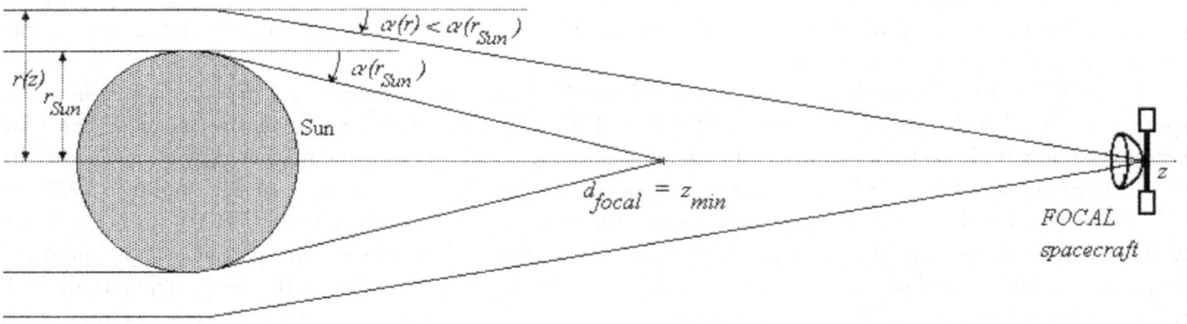

Figure 1. Geometry of the Sun gravitational lens. Shown are the minimal focal length of 550 AU (= 3.17 light days = 13.75 times beyond Pluto's orbit) and the FOCAL spacecraft position beyond the minimal focal length.

The light rays, i.e. electromagnetic waves, cannot pass through the Sun's interior (whereas gravitational waves and neutrinos can), so the largest deflection angle α occurs for those rays just grazing the Sun surface, i.e. for $r = r_{Sun}$. This yields the inequality

$$\alpha(r_{Sun}) > \alpha(r) \tag{2}$$

with

$$\alpha(r_{Sun}) = \frac{4GM_{Sun}}{c^2 r_{Sun}}. \tag{3}$$

From the illustration it should be clear that the minimal focal distance d_{focal} is related to the tangent of the maximum deflection angle by the formula

$$\tan(\alpha(r_{Sun})) = \frac{r_{Sun}}{d_{focal}}. \tag{4}$$

Moreover, since the angle $\alpha(r_{Sun})$ is very small (its actual value is about 1.75 arc seconds), the above expression may be rewritten by replacing the tangent by the small angle itself:

$$\alpha(r_{Sun}) \approx \frac{r_{Sun}}{d_{focal}}. \tag{5}$$

Eliminating the angle $\alpha(r_{Sun})$ between equations (3) and (5), and then solving for the minimal focal distance d_{focal}, one gets

$$d_{focal} \approx \frac{r_{Sun}}{\alpha(r_{Sun})} = \frac{r_{Sun}}{\frac{4GM_{Sun}}{c^2 r_{Sun}}} = \frac{c^2}{4G} \cdot \frac{r_{Sun}^2}{M_{Sun}}. \tag{6}$$

This basic result may also be rewritten in terms the *Schwarzschild radius*

$$r_{Schwarzschild} = \frac{2GM_{Sun}}{c^2}, \tag{7}$$

yielding

$$d_{focal} \approx \frac{r_{Sun}}{\alpha(r_{Sun})} = \frac{r_{Sun}}{\frac{4GM_{Sun}}{c^2 r_{Sun}}} = \frac{r_{Sun}^2}{2\, r_{Schwarzschild}}. \tag{8}$$

Numerically, one finds

$$d_{focal} \cong 542 \text{ AU} \approx 550 \text{ AU} \approx 3.171 \text{ light days}. \tag{9}$$

This is the fundamental formula yielding the minimal focal distance of the gravitational lens of the Sun, i.e. the minimal distance from the Sun's center that the FOCAL spacecraft must reach in order to get magnified radio pictures of whatever lies on the other side of the Sun with respect to the spacecraft position.

Furthermore, a simple, but very important consequence of the above discussion is that ***all points on the straight line beyond this minimal focal distance are foci too***, because the light rays passing by the Sun further than the minimum distance have smaller deflection angles and thus come together at an even greater distance from the Sun.

And the very important astronautical consequence of this fact for the FOCAL mission is that ***it is not necessary to stop the spacecraft at 550 AU. It can go on to almost any distance beyond and focus as well or better.*** In fact, the further it goes beyond 550 AU the less distorted the collected radio waves by the Sun Corona fluctuations.

The important and difficult problem of the plasma fluctuations in the Sun's Corona (electrons) were studied by Von Eshleman at Stanford and by Slava Turyshev at JPL (please refer to Chapter 6 of ref. [19]).

Just to summarize how the electrons in the Sun's Corona "push the true focus out" (i.e. they create a "diverging lens effect" that opposes the "converging lens effect of gravity) Figure 2 (taken from ref. [19], page 143) shows that the true minimal focal distance *F*, that the FOCAL spacecraft must reach, is higher for lower frequencies of the source's electromagnetic waves crossing the Corona, and lower for higher frequencies. For instance, at 500 GHz the true focus falls at about 650 AU, while at 160 GHz (the Cosmic Microwave Background (CMB) peak frequency) the true focus is found at 763 AU, and finally at 60 GHz the true focus is located at 1,000 AU or beyond. Thus, we might briefly say that the FOCAL spacecraft must actually get beyond 550 AU to get rid of the problems caused by the Sun's Corona. Also, it must be stressed that these numbers were obtained on the basis of the so-called Baumbach-Allen model for the Solar Corona (described in Chapter 8 of ref. [19]), while other and more up-to-date models might yield different results. In conclusion, more research work about the plasma in the Sun's Corona has to be made in order to find how much the "true" focal distance might be larger than just 550 AU.

We would like to add here one more result that is very important because it holds well not just for the Sun, but for all stars in general. This we will do without demonstration; that can be found on page 55 of ref. [13]. Consider a spherical star with radius r_{star} and mass M_{star}, that will be called the "focusing star". Suppose also that a light source (i.e. another star or an advanced extraterrestrial civilization) is located at the distance D_{source} from it. Then ask: how far is the minimal focal distance d_{focal} on the opposite side of the source with respect to the focussing star center? The answer is given by the formula:

$$d_{focal} = \frac{r_{star}^2}{\dfrac{4GM_{star}}{c^2} - \dfrac{r_{star}^2}{D_{source}}}. \tag{10}$$

This is the key to gravitational focussing for a pair of stars, and may well be the key to SETI in finding extraterrestrial civilizations. It could also be considered for the magnification of a certain source by any star that is perfectly aligned with that source and the Earth: the latter would then be in the same situation as the FOCAL spacecraft except, of course, it is located much further out than 550 AU with respect to the focussing, intermediate star (see Chapter 6 of ref. [19] for more details about this topic). Finally, notice that equation (10) reduces to equation (6) in the limit $D_{source} \to \infty$, i.e. (6) is the special case of (10) for light rays approaching the focussing star from an infinite distance.

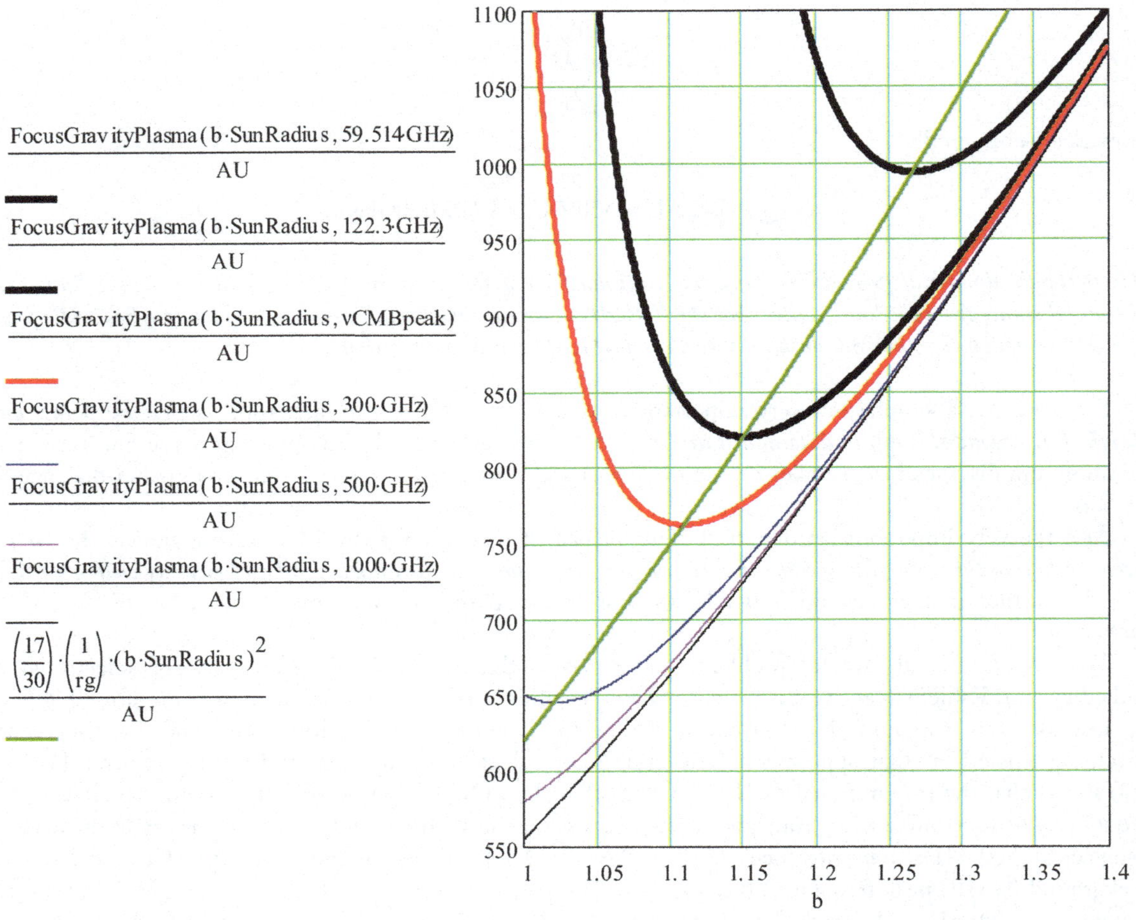

Figure 2. How the electrons in the Sun's Corona "push the true focus out". In this plot we see the (Gravity+Plasma) Sun's Lens Focal Distance *F* (in AU) as a function of the Impact Parameter b (in units of the Sun Radius) for all radio waves focused by the Sun above 120 GHz (plot of equation (8.6-8) of ref. [19]). The Impact Parameter *b* is the distance from the Sun Center at which the radio waves flyby the Sun and then proceed to focus at the distance *F* from the Sun. Each curve corresponds to radio waves of a different frequency increasing from 59.514 GHz (top, thick curve) to 1000 GHz (bottom thin curve). The shifting of the minimum is obvious, and the "locus" of all these minima is the just the parabola of equation $F(b)=(17/30)*(1/r_{Schwarzschild})*(b)^2$ (see eqs. (8.7-9) and (9.4-4) of ref. [19]).

3. From the Sun to the Planets: All as Gravitational Lenses

The Sun as a gravitational lens has the Corona problem, as we have briefly seen at the end of the last section. One is thus let to wonder whether we might use the planets as gravitational lenses, rather then the Sun, thus avoiding the Sun's Corona problem altogether. The answer to this question is provided by the formula (6) yielding the minimal distance d_{Planet} (from the Sun) that the FOCAL spacecraft must reach in order to take advantage of the lensing phenomenon for any given planet. When rewritten in terms of the planet's radius r_{Planet} and mass M_{Planet}, equation (6) reads

$$d_{Planet} \approx \frac{c^2}{4G} \cdot \frac{r_{Planet}^2}{M_{Planet}}. \tag{11}$$

We thus see that, from the point of view of the FOCAL space mission, what rally matters is not the planet radius r_{Planet} alone or the planet mass M_{Planet} alone, but the rather the ratio of its radius squared to its mass.

In other words, if one replaces for each planet both its radius squared r_{Planet}^2 and its mass M_{Planet} into (11), one gets for each planet the numerical value of the radius d_{Planet} of the relevant ***planet's focal sphere*** given by Table 1. This Table shows the radius of the focal sphere for the naked Sun (i.e. the Sun as if it had NO Corona), the Coronal Sun (i.e. the Sun with a conventionally assumed minimal focal distance of 1000 AU) and all the planets, sorted in the correct order of increasing minimal focal sphere distance from Jupiter onward to Neptune, Saturn, the Earth, Uranus, Venus, Mars and Mercury in MKS units and light time.

Table 1: Radii of the minimal focal spheres of the Naked Sun, Coronal Sun, Jupiter, Neptune, Saturn, the Earth, Uranus, Venus, Mars and Mercury in their natural increasing order of distance from the Sun, both in MKS units and in light time (where a light month is conventionally defined here to be equal to 30 light days).

Body	Radius	Mass	FOCAL SPHERE RADIUS = = $(c^2/(4G))$ x (Radius^2/Mass)	FOCAL SPHERE RADIUS = = $(c^2/(4G))$ x (Radius^2/Mass)
Naked Sun	6.962×10^5 km	1.989×10^{30} kg	548.62 AU ~ 550 AU	3.169 light days
Coronal Sun	?	1.989×10^{30} kg	1000 AU (assumed)	5.778 light days
Jupiter	7.159×10^4 km	1.898×10^{27} kg	6.077×10^3 AU	1.17 light months
Neptune	2.485×10^4 km	1.024×10^{26} kg	1.352×10^4 AU	2.602 light months
Saturn	6.035×10^4 km	5.684×10^{26} kg	1.442×10^4 AU	2.777 light months
Earth	6.387×10^3 km	5.974×10^{24} kg	1.537×10^4 AU	2.959 light months
Uranus	2.559×10^4 km	8.683×10^{25} kg	1.698×10^4 AU	3.269 light months
Venus	6.067×10^3 km	4.868×10^{24} kg	1.702×10^4 AU	3.276 light months
Mars	3.404×10^3 km	6.392×10^{23} kg	4.081×10^4 AU	7.857 light months
Mercury	2.446×10^3 km	3.285×10^{23} kg	4.098×10^4 AU	7.889 light months

These results are rather surprising. After the two focal spheres of the Naked Sun at 550 AU and of the Coronal Sun conventionally assumed to lie at 1000 AU (as discussed in the author's book, ref. [19]), the Jupiter focal sphere comes next at about 6100 AU from the Sun, that is at about 1.1 light months. Clearly, this distance would have to be the next target for any FOCAL space mission if we decide that the Sun's Corona effects are too heavily distorting our radio pictures of whatever lies on the other side of the Sun at infinity. Then, the focal sphere of Neptune comes next at about 13,525 AU from the Sun, namely at 2.6 light months. Again, the fact the Neptune has a smaller focal sphere than all other remaining planets is due to the surprisingly high ratio of the square of its radius to its mass. Saturn has a focal sphere slightly larger than Neptune's at 14,425 AU, i.e. 2.777 light months, even if it looks "bigger" than Neptune in all drawings comparing the sizes of the Solar system planets.

But the real surprise is the Earth. In fact, the focal sphere of the Earth turns out to lie at 15,375 AU from the Sun, i.e. about three light months from the Sun. Why is it that poor little Earth qualifies better that big Uranus and all other bodies of the Solar system smaller than the Earth itself (like Venus, Mars and Mercury) as the target of the future FOCAL space missions?

Well, the point is that ***the Earth is the body with the highest density (radio of mass to volume) in the whole Solar system.*** This is no marginal fact at all. It is because of this top density that the Earth's atmosphere did not "evaporate" into space, thus creating the conditions for life to develop and finally reach the present state of only inhabited planet we know. But from the point of view of the future FOCAL space missions, there is one more basic advantage that reaching the focal sphere of the Earth at 15,500 AU would have.

We know the Earth's atmosphere better than the atmosphere of any other planet in the Solar system, of course. So, a FOCAL spacecraft exploiting the Earth as a gravitational lens would operate on a much firmer scientific ground than a FOCAL spacecraft exploiting any other planet as a gravitational lens. In fact, to use the same words as in the "Planet" Wikipedia web site, http://en.wikipedia.org/wiki/Planet "Every planet began its existence in an entirely fluid state; in early formation, the denser, heavier materials sank to the centre, leaving the lighter materials near the surface. Each therefore has a differentiated interior consisting of a dense planetary core surrounded by a mantle which either is or was a fluid. The terrestrial planets are sealed within hard crusts, but in the gas giants the mantle simply dissolves into the upper cloud layers. The terrestrial planets possess cores of magnetic elements such as iron and nickel, and mantles of silicates. Jupiter and Saturn are believed to possess cores of rock and metal surrounded by mantles of metallic hydrogen. Uranus and Neptune, which are smaller, possess rocky cores surrounded by mantles of water, ammonia, methane and other ices. The fluid action within these planets' cores creates a geodynamo that generates a magnetic field." So, in conclusion, in order to exploit any planet other than the Earth as a gravitational lens, we must estimate the structure of its interior, something that science has done only approximately as of 2010. The surface and atmosphere of the Earth, on the contrary, are far better known, and that qualifies the focal sphere of the Earth at 15,500 AU as the next realistic target to be reached after, say, 1000 AU. Incidentally, that same distance also is about a quarter of the size of the Oort cloud of comets that surround the Sun in all directions (see, for instance, the site http://en.wikipedia.org/wiki/Oort_cloud). Any FOCAL mission to the Earth's focal sphere would thus automatically imply crossing the inner Oort cloud, with a harvest of further results for cometary science and astrobiology in general.

Figure 3 shows the belt of focal spheres around the Sun between 550 and 17,000 AU to scale. The Earth's focal sphere is marked by a thicker, red solid circle. See also refs. [25] and [26].

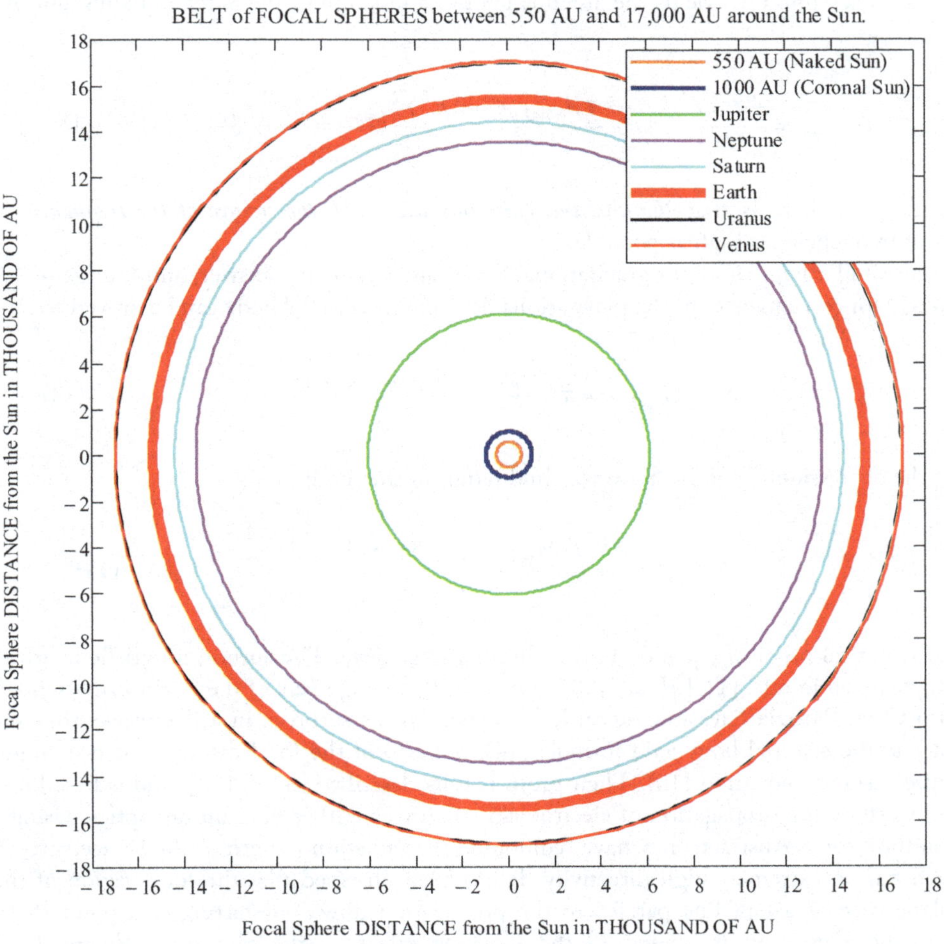

Figure 3. The complete BELT of focal spheres between 550 and 17,000 AU from the Sun, as created by the gravitational lensing effect of the sun and all planets, here shown to scale. **The discovery of this belt of focal spheres is the main result put forward in this paper, together with the computation of the relevant antenna gains.**

4. (Antenna) Gain of the Gravitational Lenses of the Sun and Plane

Having thus determined the minimal distances that the FOCAL spacecraft must reach to exploit both the Sun and all planets as gravitational lenses, one now wonders what's the good of going so far out of the solar system, i.e. how much focussing of light rays is provided by such lenses at those minimal distances. The answer to such a question is provided by the technical notion of antenna gain, simply called "the gain" in the sequel of this paper.

A standard formula in antenna theory relates the antenna gain, $G_{antenna}$, to the antenna effective area, $A_{effective}$, and to the wavelength λ or the frequency n by virtue of the equation (refer, for instance, to ref. [14], in particular page **6**-117, equation (6-241)):

$$G_{antenna} = \frac{4\pi\, A_{effective}}{\lambda^2}. \tag{12}$$

Now, assume the antenna is circular with radius $r_{antenna}$, and assume also a 50% efficiency. Then, the antenna effective area is obviously given by

$$A_{effective} = \frac{A_{physical}}{2} = \frac{\pi\, r_{antenna}^2}{2}.$$

Substituting this back into (12) yields the antenna gain as a function of the antenna radius and of the observed frequency

$$G_{antenna} = \frac{4\pi\, A_{effective}}{\lambda^2} = \frac{2\pi\, A_{physical}}{\lambda^2} = \frac{2\pi^2\, r_{antenna}^2}{\lambda^2} = \frac{2\pi^2\, r_{antenna}^2}{c^2} \cdot \nu^2. \quad (13)$$

The important point here is that ***the antenna gain increases with the square of the frequency***, thus favoring observations at frequencies as high as possible.

Is anything similar happening for gravitational lenses also? *Yes* is the answer, and the "gain" (one maintains this terminology for convenience) of the gravitational lens of any celestial body can be proved to be

$$G_{body} = 4\pi^2 \frac{r_{Schwaraschild\_of\_the\_body}}{\lambda} \quad (14)$$

or, invoking the expression (7) of the Schwarzschild radius of that body

$$G_{body} = \frac{8\pi^2 G M_{body}}{c^2} \cdot \frac{1}{\lambda} = \frac{8\pi^2 G M_{body}}{c^3} \cdot \nu. \quad (15)$$

The mathematical proof of equation (14) is difficult to achieve. The author, unsatisfied with the treatment of this key topic given in refs. [1], [3] and [13], turned to three engineers of the engineering school in his home town, Renato Orta, Patrizia Savi and Riccardo Tascone. To his surprise, in a few weeks they provided a full proof of not just the celestial body gain formula (14), but also of the focal distance for rays originated from a source at finite distance, equation (10). Their proof is fully described in ref. [13], and is based on the aperture method used to study the propagation of electromagnetic waves, rather than on ray optics. Using the words of these three authors' own Abstract, they have "computed the radiation pattern of the [spacecraft] Antenna+Sun system, which has an extremely high directivity. It has been observed that the focal region of the lens for an incoming plane wave is a half line parallel to the propagation direction starting at a point [550 AU for the Naked Sun] whose position is related to the blocking effect of the Sun disk (Figure 1). Moreover, a characteristic of this thin lens is that its gain, defined as the magnification factor of the antenna gain, is constant along this half line. In particular, for a wavelength of 21 cm, this lens gain reaches the value of 57.5 dB. Also a measure of the transversal extent of the focal region has been obtained. The performance of this radiation system has been determined by adopting a thin lens model which introduces a phase factor depending on the logarithm of the impact parameter of the incident rays. Then the antenna is considered to be in transmission mode and the radiated field is computed by asymptotic evaluation of the radiation integral in the Fresnel approximantion".

One is now able to compute the Total Gain of the antenna+body system, that is simply obtained by multiplying equations the two equations yielding the spacecraft gain proportional to ν^2 and the body gain proportional to ν:

$$G_{total} = G_{body} \cdot G_{antenna} = \frac{16\pi^4\, G\, M_{body}\, r_{antenna}^2}{c^5} \cdot \nu^3. \quad (16)$$

Since the total gain increases with the ***cube*** of the observed frequency, it favors electromagnetic radiation in the microwave region of the spectrum. Table 2 shows the numerical data provided by the last equation for five selected frequencies: the hydrogen line at 1420 MHz and the four frequencies that the Quasat radio astronomy satellite planned to observe, had it been built jointly by ESA and NASA as planned before 1988, but Quasat was abandoned by 1990 due to lack of funding. The definition of N, when measured in dB, dB is of course:

$$10\,\mathrm{Log}_{10}\mathrm{N} = \frac{10\,\ln\mathrm{N}}{\ln 10}.$$

Table 2: Table showing the GAIN of the Sun's lens alone, the gain of a 12-meter spacecraft (S/C) antenna and the combined gain of the Sun+S/C Antenna system the at five selected frequencies important in radioastronomy.

Line	Neutral Hydrogen		*OH* radical		H_2O
Frequency ν	**1420 MHz**	327 MHz	**1.6 GHz**	5 GHz	**22 GHz**
Wavelength λ	21 cm	92 cm	18 cm	6 cm	1.35 cm
S/C Antenna Beamwidth	1.231 deg	5.348 deg	1.092 deg	0.350 deg	0.080 deg
Sun Gain	57.4 dB	51.0 dB	57.9 dB	62.9 dB	69.3 dB
12-meter Antenna S/C Gain	42.0 dB	29.3 dB	43.1 dB	53.0 dB	65.8 dB
Combined Sun + S/C Gain	**99.5 dB**	**80.3 dB**	**101.0 dB**	**115.9 dB**	**135.1 dB**

5. Using All the Planet Lenses to Sweep Across the Sky in Search of Exoplanets

Let us now forget about FOCAL for a moment and consider instead a truly interstellar flight to the nearest star, Alpha Centauri, at a distance of 4.37 light years, i.e. 276,358 AU.

Several serious scientific studies about such a truly interstellar mission have appeared in the literatures over the last few decades. But all of them have missed the bonanza provided by the full belt of focal spheres between 550 and 17,000 that the spacecraft must cross anyway along its road to Alpha Centauri.

So, in the last section of this introductory paper to the belt of focal spheres between 550 and 17,000 AU, we wish to devote some thoughts explaining what can be obtained while crossing our belt of focal spheres.

First of all, while the Sun does not move in the Sun-centered reference frame of the Solar system, all the planets do move. This means that they actually sweep a certain area of the sky, as seen from the spacecraft, so that a spacecraft enjoys a sort of moving magnifying lens. How many extrasolar planets would fall inside this moving magnifying lens? Well, we don't know nowadays, of course, but the over 400 exoplanets found to date are a neat promise that many more such exoplanets could be detected anew by a suitably equipped spacecraft crossing the distances between 550 and 17,000 AU from the Sun thanks to the gravitational lenses of the planets.

Second, apart from finding exoplanets, the moving and magnifying lenses of the planets could also lead to the discovery of many more galactic and extragalactic objects of all kinds.

Finally, the direct exploration of the Kuiper belt and Oort cloud could only increase the scientific interest into such a crossing of the belt of focal spheres, in ways that it is now premature to describe in detail but will certainly be worked out by the coming generations of space scientists just as we have done here for the preliminary concepts of lens magnification.

In conclusion, looking back to the work done
thus far about the possibilities of a truly interstellar flight, it seem fair to say that all planners of the Alpha Centauri missions, in their efforts to reach 277,000 AU, have missed what was at hand at just 17,000 AU. Or,

in terms of light time, in order to get all the way to 4.37 light years in a single shot, they have missed what was just three light months away, like the Earth's focal sphere.

After the discovery of the belt of planetary focal spheres put forward in this paper for the first time, a revision of the theoretical flight plans to Alpha Centauri seems to be desirable.

6. The Radio Link

The goal of this paper is to prove that only by exploiting the Sun as a gravitational lens we will be able to have reliable telecommunication links across large interstellar distances. In other words, a direct link between different star systems, even if held by virtue of the largest radiotelescopes on Earth, will not be feasible across distances of the order or thousand of light years or more. We want to show that only a FOCAL mission in the direction from Earth opposite to that target star system will ensure a reliable telecommunication link across thousands of light years. Namely, we prove that Bit Error Rate (or BER, see site http://www.en.wikipedia.org/wiki/Bit_error_rate) will be unacceptable already at the distance of Alpha Centauri unless we resort to a supporting FOCAL space mission in the opposite direction from the Sun.

In order to face these problems mathematically, we must first understand the radio link among any two stars, and we think that no neater treatment of this subject exists than the book by "Radio Astronomy" by the late professor John D. Kraus of Ohio State University (ref. [20]), that we follow hereafter. Notice also that the theory described hereafter was already published in print in refs. [21], [22], [23] and [24].

Consider a radio transmitter that radiates a Power P_t isotropically and uniformly over a bandwidth B_t. Then, at a distance r it produces a flux density given by

$$\frac{P_t}{B_t\, 4\pi\, r^2}. \tag{17}$$

A receiving antenna of effective aperture A_{er} at a distance r can collect a power given by (17) multiplied by both the effective aperture of the receiving antenna and its bandwidth, namely the received power P_r is given by

$$P_r = \frac{P_t}{B_t\, 4\pi\, r^2} A_{er}\, B_r. \tag{18}$$

It is assumed that the receiving bandwidth B_r is smaller or, at best (in the "matched bandwidths" case) equal to the transmitting bandwidth B_t, that is $B_r \le B_t$.

So far, we have been talking about an isotropic radiator. But let us now assume that the transmitting antenna has a directivity D, that is an antenna gain in the sense of (11):

$$D = \frac{4\pi\, A_{et}}{\lambda^2}. \tag{19}$$

The received power P_r is then increased by just such a factor due to the directivity of the transmitting antenna, and so (18) must now be replaced by a new equation where the right-hand side is multiplied by such a increased factor, that is

$$P_r = \frac{4\pi\, A_{et}}{\lambda^2} \cdot \frac{P_t}{B_t\, 4\pi\, r^2} A_{er}\, B_r. \tag{20}$$

Rearranging a little, this becomes

$$P_r = \frac{P_t\, A_{et}\, A_{er}}{r^2\, \lambda^2} \cdot \frac{B_r}{B_t}. \tag{21}$$

This is the ***received signal power*** expression. For the matched bandwidths case, i.e. for $B_r = B_t$, this is called the Friis transmission formula, since it was first published back in 1946 by the American radio engineer Harald T. Friis (1893-1976) of the Bell Labs. In space missions, we of course exactly both B_t and B_r and so me construct the spacecraft in such a way the two ***bands match exactly***, i.e. $B_r = B_t$. So, for the case of telecommunications with a spacecraft (but not necessarily for the SETI case, we may well assume the matched bandwidths and have (21) reducing to

$$P_r = \frac{P_t\, A_{et}\, A_{er}}{r^2\, \lambda^2}. \tag{22}$$

Let us now rewrite (22) in such a way that we may take into account the gains (i.e. directionalities) of both the transmitting and receiving antennae, that is, in agreement with (11)

$$\begin{cases} G_t = \dfrac{4\pi\, A_{et}}{\lambda^2} \\ G_r = \dfrac{4\pi\, A_{er}}{\lambda^2}. \end{cases} \quad \text{that is} \quad \begin{cases} A_{et} = \dfrac{G_t\, \lambda^2}{4\pi} \\ A_{er} = \dfrac{G_r\, \lambda^2}{4\pi}. \end{cases} \tag{23}$$

Replacing the last two expressions into (22), we find that (22) is turned into

$$P_r = \frac{P_t\, G_t\, G_r}{(4\pi)^2\, r^2} \cdot \lambda^2 .$$

This may finally be rewritten in the more traditional form

$$P_r = \frac{P_t\, G_t\, G_r}{L(r,\lambda)}. \tag{24}$$

if one defines

$$L(r,\lambda) = (4\pi)^2 \cdot \frac{r^2}{\lambda^2} \tag{25}$$

is the Path Loss (or path attenuation), i.e. the reduction in power density (attenuation) of the electromagnetic waves as they propagates through space. Path loss is a major component in the analysis and design of the link budget of a telecommunication system, see the site http://www.en.wikipedia.org/wiki/Path_loss .

7. Bit Error Rate for an "Ordinary" Direct Link with a Probe at the Alpha Centauri Distance

In this section we first define the Bit Error Rate (BER). Then, by virtue of a numerical example, we show that, even at the distance of the nearest star (Alpha Cen. at 4.37 AU) the telecommunications would be impossible by the ordinary powers available today for interplanetary space flight. But in the next section we shall show that the telecommunications would become feasible if we could take advantage of the magnification provided by the Sun's gravity lens, i.e. if we would send out to 550 AU a FOCAL relay spacecraft for each target star system that we wish to communicate with. And this is the key new result presented in this paper.

So, let us start by defining the Bit Error Rate or BER. In telecommunication theory an **error ratio** is the ratio of the number of bits, elements, characters, or blocks incorrectly received to the total number of bits, elements, characters, or blocks sent during a specified time interval. Among these error rations, the most commonly encountered ratio is the **bit error ratio** (BER) - also called **bit error rate** – that is the number of erroneous bits received divided by the total number of bits transmitted. At the bit error rate Wikipedia site: http://www.en.wikipedia.org/wiki/Bit_error_rate it is shown that the likelihood of a bit misinterpretation

$$p_e = p(0 \mid 1)p_1 + p(1 \mid 0)p_0. \tag{26}$$

(believing that we have received a 0 while it was a 1 or the other way round) is basically given by the "complementary error function" or erfc(x) as follows

$$BER(d,\nu,P_t) = \frac{1}{2}\, erfc\left(\sqrt{\frac{E_b(d,\nu,P_t)}{N_0}}\right) \tag{27}$$

In this equation one has:

1) d = distance between the transmitting station on Earth and the receiving antenna in space. For instance, this could be the antenna of a precursor interstellar space probe that was sent out to some light years away.
2) n = frequency of the electromagnetic waves used in the telecommunication link. The higher this frequency, the better it is, since the photons are then more energetic (E = h n). In today's practice, however, the highest n for spacecraft links (like the link of the Cassini probe, now at Saturn) are the ones in the Ka band, that is: $\nu_{Ka} \approx 32$ GHz.
3) P_t is the power in watts transmitted by the Earth antenna, typically a NASA Deep Space Network antenna 70 meters in diameter.
4) The complementary error function ercf(x) is defined by the integral

$$erfc(x) = \frac{2}{\sqrt{\pi}} \int_x^{\infty} e^{-t^2}\, dt \tag{28}$$

   (for more details, see the Wikipedia site: http://www.en.wikipedia.org/wiki/Complementary_error_function).
5) $E_b(d, n, P_t)$ is the received energy per bit, that is the ratio

$$E_b(d,\nu,P_t) = \frac{P_r(d,\nu,P_t)}{\text{Bit\_rate}}. \tag{29}$$

6) Finally, N_0 is given by the Boltzmann's constant k multiplied by the noise temperature of space far away from the Sun and from any other star. This "empty space noise temperature" might be assumed to equal, say, to 100 K.

This is the analytical structure of the MathCad code that this author wrote to yield the BER. Let us now consider the input values that he used in practice:

1) Suppose that a human space probe has reached the Alpha Cen system at 4.37 light year distance from the Sun: then, d=4.37 light years.
2) Suppose also that the transmitting antenna from the Earth is a typical NASA Deep Space Network (DSN) antenna having a diameter of 70 meter (like those at Goldstone, Madrid and Canberra), and assume that its efficiency is about 50%.
3) Suppose that the receiving antenna aboard the spacecraft is 12 meter in diameter (it might be an inflatable space antenna, as we supposed in ref. [19] for the FOCAL spacecraft) and assume a 50% efficiency.
4) Suppose that the link frequency is the Ka band (i.e. 32 GHz), as for the Cassini highest frequency.
5) Suppose that the bit rate is 32 kbps = 32000 bit / second. This is the bit rate of ESA's Rosetta interplanetary spacecraft now on its way to a comet.
6) And finally ***(this is the most important input assumption) suppose that the transmitting power P_t is moderate: just 40 watts.***

Then:

1) The gain of the transmitting NASA DSN antenna (at this Ka frequency) is about 84 dB.
2) The gain of the spacecraft antenna is about 69 dB.

3) The path loss at the distance of Alpha Cen is 395 dB (a very high indeed path loss with respect to today's interplanetary missions, of course).
4) The power received by the spacecraft at that distance is 2.90×10^{-23} watt.
5) The received energy per bit (lowered by the noise temperature of the space in between the Sun and Alpha Cen) is 1.3×10^{-37} joule.
6) ***And finally the BER is 0.49, i.e. there is a 50% probability of ERRORS in the telecommunications between the Earth and the probe at Alpha Cen. if we use such a small transmitting power !!!***

In other words, if these are the telecommunication links between the Earth and our probe at Alpha Cen, then this precursor interstellar mission is *worthless.*

The key point in this example is that, for all calculations, (24) and (25) were used ***WITHOUT TAKING THE GAIN OF THE SUN GRAVITY LENS INTO ACCOUNT, because this was a DIRECT link and NOT a FOCAL mission.***

8. Bit Error Rate at the Alpha Centauri Distance Enhanced by the Magnification Provided by the Sun's Gravity Lens (Focal)

The disappointing BER results of the previous section are totally reversed, however, if we suppose that a FOCAl space mission has been previously sent out to 550 AU in the direction opposite to Alpha Cen so that we now have the MAGNIFICATION of the Sun's Gravity Lens playing in the game.

Mathematically, this means that we must introduce a third multiplicative gain at the numerator of (24): the Sun's Gravity Lens GAIN, given by (14) where the Schwarzschild radius of the Sun is given by (7).

This new gain is huge at the Ka band frequency:

$$G_{Sun}(n_{Ka}) = 12444837 \sim 70 \text{ dBs} \qquad (30)$$

and so the received power (24) at Alpha Cen, with the usual Earth-transmitted power of just 40 watts becomes

$$P_r = 2.9 \times 10^{-23} \text{ watts} \qquad (31)$$

and the relevant BER becomes absolutely acceptable:

$$\boxed{BER = 0.000000526387845} \qquad (32)$$

This should convince anybody that the FOCAL space mission is indispensable to keep the link at interstellar distances equal or higher than Alpha Cen.

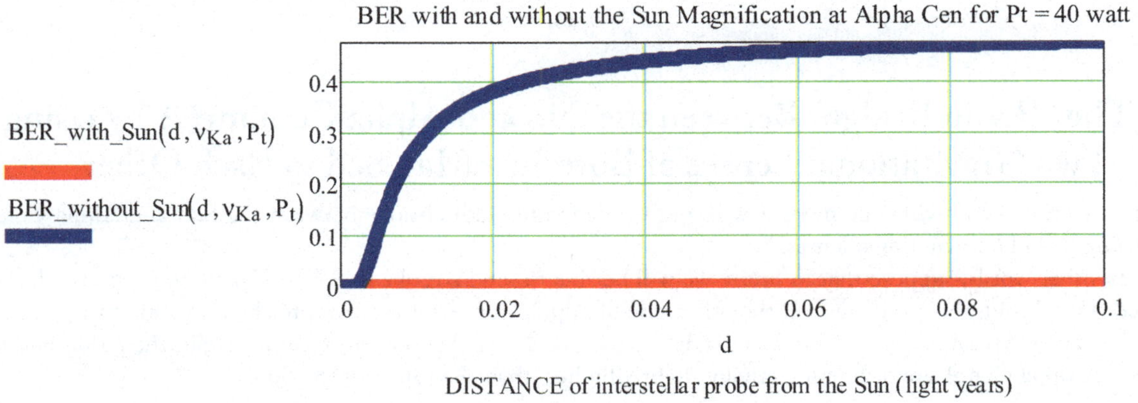

Figure 4. The Bit Error Rate (BER) (upper, blue curve) tends immediately to the 50% value (BER = 0.5) even at moderate distances from the Sun (0 to 0.1 light years) for a 40 watt transmission from a DSN antenna that is a DIRECT transmission, i.e. without using the Sun's Magnifying Lens. On the contrary (lower red curve) the BER

keeps staying at zero value (perfect communications!) if the FOCAL space mission is made, so as the Sun's magnifying action is made to work.

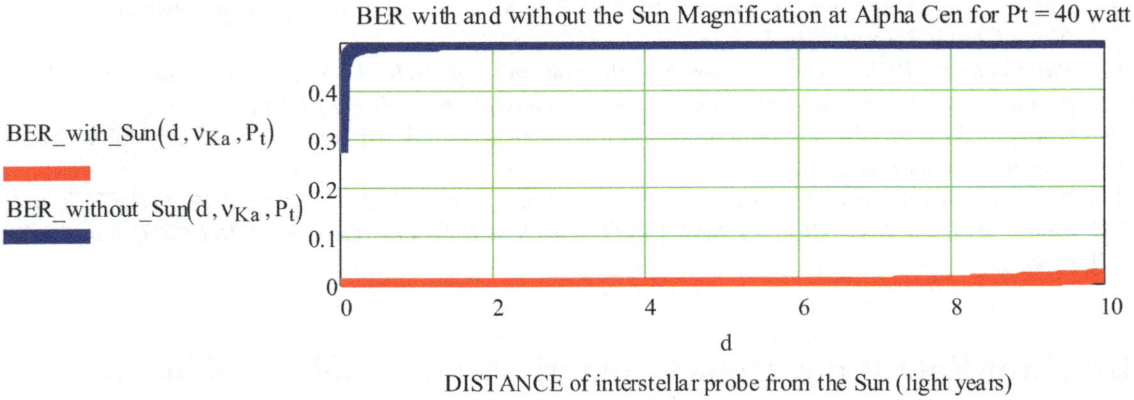

Figure 5. Same as in Figure 4, but for probe distances up to 10 light years. We see that at about 9 light years away the BER curve starts being no exactly flat any more, and starts increasing slowly.

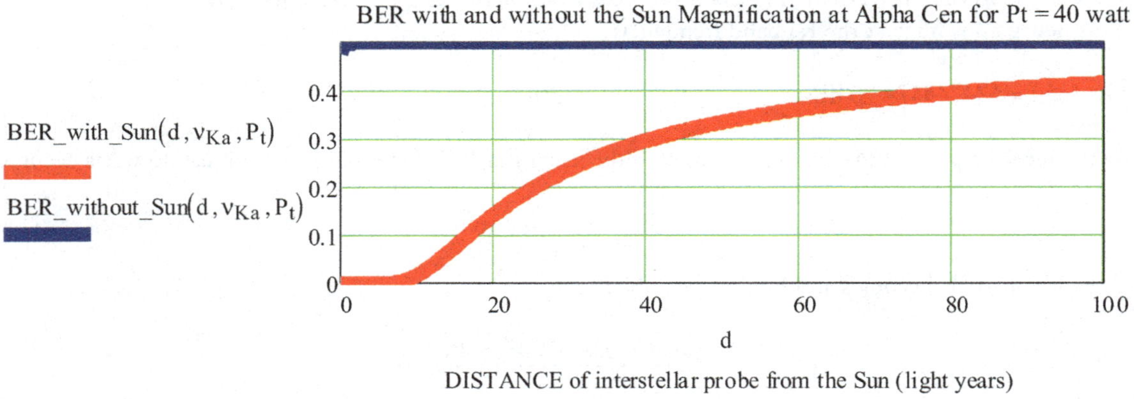

Figure 6. Same as in Figure 5, but for probe distances up to 100 light years. We see that, from 9 light years onward, the Sun-BER increases, reaching the dangerous level of 40% (Sun-BER = 0.4) at about 100 light years. Namely, at 100 light years even the Sun's Lens cannot cope with this very low transmitted power of 40 watt.

9. The "Radio Bridge" Between the Sun and Alpha Centauri A by Using the Two Gravitational Lenses of Both Just Matched to Each Other

In this section we provide one more new result: we define the radio bridge between the Sun and Alpha Cen A by using BOTH gravitational lenses!

In other words, suppose that in future we will be able to send a probe to Alpha Cen A and suppose that we succeed in placing this probe just on the other side of Alpha Cen A with respect to the Sun and at the minimal focal distance typical of Alpha Cen A. This distance is NOT 550 AU obviously because both the radius and the mass of Alpha Cen A are different (actually slightly higher) than the values of the Sun:

$$\begin{cases} r_{Alpha\_Cen\_A} = 1.227\, r_{Sun} \\ M_{Alpha\_Cen\_A} = 1.100\, M_{Sun} \end{cases} \tag{33}$$

Replacing these values into (6) (obviously rewritten for Alpha Cen A), the relevant minimal focal distance is found

$$d_{focal\_Alpha\_Cen\_A} \approx \frac{c^2 r^2_{Alpha\_Cen\_A}}{4GM_{Alpha\_Cen\_A}} \approx 749 \text{ AU}. \quad (34)$$

The Schwarzschild radius for Alpha Cen A is given by

$$r_{Schwarzschild\_Alpha\_Cen\_A} = \frac{2GM_{Alpha\_Cen\_A}}{c^2} = 3.248 \text{ km} \quad (35)$$

And so the gain, provided by (14), turns out to equal

$$G_{Alpha\_Cen\_A}(\nu_{Ka}) = 4\pi^2 \frac{r_{Schwarzschild\_Alpha\_Cen\_A}}{\lambda_{Ka}} = 13689321 \, . \quad (36)$$

That is

$$G_{Alpha\_Cen\_A}(\nu_{Ka}) \approx 71 \text{ dB} \quad (37)$$

Incidentally, we chose Alpha Cen A, and not B or C, because it has the highest mass, and so the highest gain, in the whole Alpha Cen triple system. The future telecommunications between the Sun and the Alpha Cen system are thus optimized by selecting Alpha Cen A as the star on the other side of which to place a FOCAL spacecraft at the minimal distance of 750 AU. That FOCAL spacecraft would then easily relay its data anywhere within the Alpha Cen system.

Having found the Alpha Cen A gain (37) we are now able to write the new equation corresponding to (24) for the Sun-Alpha Cen bridge. In fact, we must now put at the numerator of 24 three gains:

1) The Sun gain at 32 GHz,
2) The Alpha Cen A gain at 32 GHz, and
3) The 12-meter FOCAL antenna gain at 32 GHz raised to the square because there are two such 12-meter antennas: one at 550 AU from the Sun and one at 749 AU from Alpha Cen A, and they must be perfectly aligned with the axis passing thru both the Sun and Alpha Cen A.

Thus, the received power given by (24) now reads

$$P_r = \frac{P_t \, G_{Sun} \, G_{Alpha\_Cen\_A} \left(G_{12\_meter\_antenna\_at\_Ka}\right)^2}{L(r,\lambda)} \, . \quad (38)$$

where obviously r equals 4.37 light years and 1 corresponds to a 32 GHz frequency.

Let us now go back to the BER and replace (38) instead of (24) in the long chain of calculations described in Section 6. Since the received power P_r has now changed, clearly both (29) and (27) yield different numerical results. But now:

1) The link frequency has been fixed at 32 GHz (Ka band), and so no longer is an independent variable in the game.
2) Also the distance d has been fixed (it is the distance of Alpha Cen A) and so it no longer an independent variable in the game.
3) It follows that, in (29) and (27) the only variable to be free to vary is now the transmitted power, P_t.

Let us re-phrase the last sentence in different terms. Practically, we are now studying the BER as a function of the transmitted power P_t only and, physically, this mean that:

a) We start by inputting very low transmission powers in watts, and find out that the BER is and awful 50%, i.e. the telecommunication between the Sun and Alpha Cen are totally disrupted. This is of course because the energy per bit is so much lower than the empty space noise temperature.
b) We then increase the transmitted power, at a certain point the BER starts getting smaller than 50%. And so it gets smaller and smaller until the transmitted power is so high that the BER gets down to zero and the telecommunications are just perfect.

c) But the surprise is that… for the Sun-Alpha Cen direct radio bridge exploiting both the two gravitational lenses, this minimum transmitted power is incredibly… small ! Actually it just equals less than 10^{-4} watts, i.e. one tenth of a milliwatt is enough to have perfect communication between the Sun and Alpha Cen through two 12-meter FOCAL spacecraft antennas. How is that possible?

d) Well, that is the "miracle' given to Humanity by the Gravitational Lenses to both explore the universe and keep the link with other stars, you know! Just remember that, in 2009, the discovery of the first extrasolar planet in the Andromeda galaxy (M31) was announced because of the gravitational lens caused by something in between!

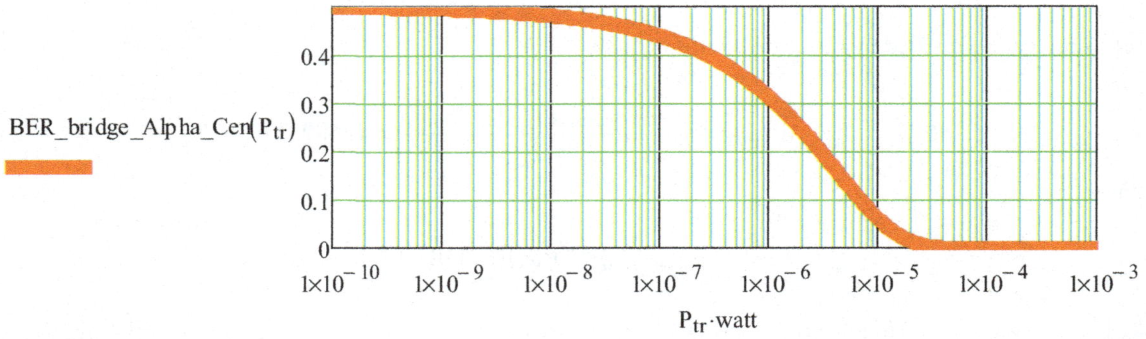

Figure 7. Bit Error Rate (BER) for the double-gravitational-lens system giving the radio bridge between the Sun and Alpha Cen A. In other words, there are two gravitational lenses in the game here: the Sun one and the Alpha Cen A one, and two 12-meter FOCAL spacecrafts are supposed to have been put along the two-star axis on opposite sides at or beyond the minimal focal distances of 550 AU and 749 AU, respectively. This radio bridge has an OVERALL GAIN SO HIGH that a miserable 10· watt transmitting power is sufficient to let the BER get down to zero, i.e. to have perfect telecommunications! Notice also that the scale of the horizontal axis is logarithmic, and the trace is yellowish since the light of Alpha Cen A is yellowish too. This will help us to distinguish this curve from the similar curve for the Barnard's Str, that is a small red star 6 light years away, as we study in the next section.

10. The "Radio Bridge" Between the Sun and Barnard's Star Using the Two Gravitational Lenses of Both

The next closest star to the Sun beyond the triple Alpha Cen system is Barnard's star (see, for instance, t http://www.en.wikipedia.org/wiki/Barnard's_Star). Let us now repeat for the gravitational lens of Barnard's star the same calculations that we did in the previous section for Alpha Cen A. Then one has:

$$\begin{cases} d_{Barnard} = 5.98 \text{ light years} \\ r_{Barnard} = 0.17\, r_{Sun} \\ M_{Barnard} = 0.16\, M_{Sun}. \end{cases} \tag{39}$$

Barnard's star is thus just a small red star, that is actually "passing by" the Sun right now and is not known to have planets around it, As a consequence of the numbers listed in (39), one infers that

$$\begin{cases} d_{focal\_Barnard} = 98 \text{ AU} \\ r_{Schwarschild\_Barnard} = 0.47 \text{ km} \\ G_{Barnard}(\nu_{\text{Ka}}) = 1991174. \end{cases} \tag{40}$$

Especially the gain is important to us:

$$\boxed{G_{Barnard}(n_{\text{Ka}})=63\ \text{dB.}} \tag{41}$$

We replace this into the Bernard's star equivalent of (38), again supposing that two 12-meter FOCAL spacecraft antennas are placed along the Sun-Barnard straight line at or beyond 550 AU and 100 AU, respectively. The result is the new graph of the BER as a function of the transmitted power only as in Figure 8.

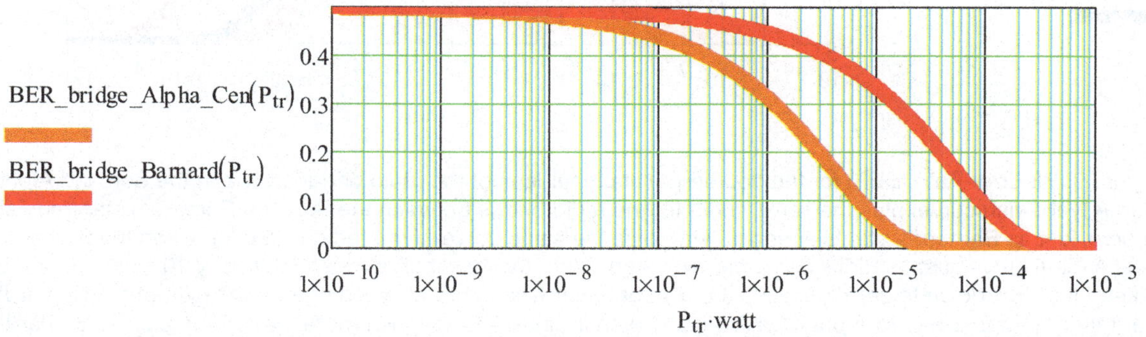

Figure 8. Bit Error Rate (BER) for the double-gravitational-lens of the radio bridge between the Sun and Alpha Cen A (yellowish curve) plus the same curve for the radio bridge between the Sun and Barnard's star (reddish curve, just as Barnard's star is a reddish star): for it, 10· watt are needed to keep the BER down to zero, because the gain of Barnard's star is so small when compared to the Alpha Centauri A's.

11. The "Radio Bridge" Between the Sun and Sirius-A Using the Two Gravitational Lenses of Both

The next star we want to consider is Sirius A. This is because Sirius A is a big, massive bluish star and so it is completely different from both Alpha Cen A (that is a Sun-like star) and Barnard's star, that is a small red star. Data may again be taken from the Wikipedia site: http://www.en.wikipedia.org/wiki/Sirius and one gets:

$$\begin{cases} d_{Sirius\_A} = 8.6 \text{ light years} \\ r_{Sirius\_A} = 1.711\, r_{Sun} \\ M_{Sirius\_A} = 2.02\, M_{Sun}. \end{cases} \tag{42}$$

From these data one gets:

$$\begin{cases} d_{focal\_Sirius\_A} = 793 \text{ AU} \\ r_{Schwarschold\_Sirius\_A} = 5.96 \text{ km} \\ G_{Sirius\_A}(\nu_{\text{Ka}}) = 251385723 \end{cases} \tag{43}$$

The important thing is of course the gain:

$$\boxed{G_{Sirius\_A}(n_{\text{Ka}}) = 74\ \text{dB.}} \tag{44}$$

Then, one replaces this into the Sirius A equivalent of (38), again supposing that two 12-meter FOCAL spacecraft antennas are placed along the Sun-Sirius A straight line at or beyond 550 AU and 793 AU, respectively. The result is the new graph of the BER as a function of the transmitted power only as in Figure 9.

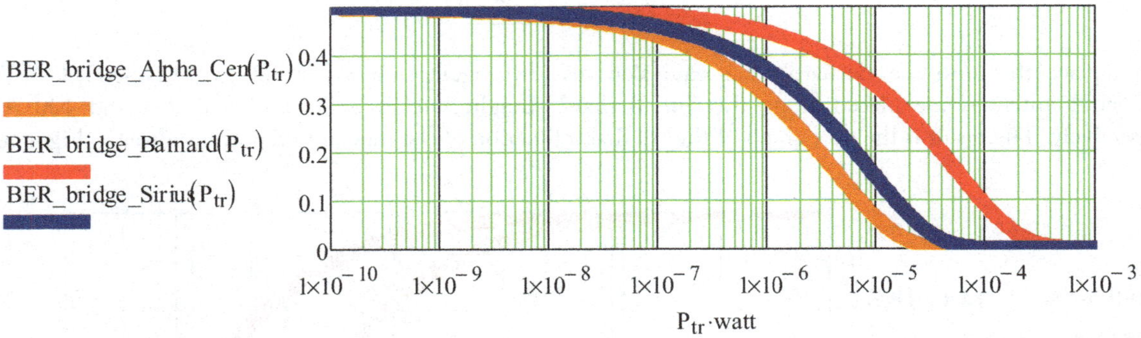

Figure 9. Bit Error Rate (BER) for the double-gravitational-lens of the radio bridge between the Sun and Alpha Cen A (yellowish curve) plus the same curve for the radio bridge between the Sun and Barnard's star (reddish curve, just as Barnard's star is a reddish star) plus the same curve of the radio bridge between the Sun and Sirius A (blue curve, just as Sirius A is a big blue star). From this blue curve we see that only 10· watt are needed to keep the BER down to zero, because the gain of Sirius A is so big when compared the gain of the Barnard's star that it "jumps closer to Alpha Cen A's gain" even if Sirius A is so much more further out and the Barnard's star! In other word, the star's gain and the size combined matter even more than its distance!

12. Channel Capacities for the Interstellar Radio Bridges

The goal of this paper is to produce new results in addition to the results already proven in refs. [2] and [3]. This we now achieve by resorting to the notion of "information channel capacity" C (Shannon, 1948, see the site http://en.wikipedia.org/wiki/Channel_capacity):

$$C = W \log_2 \left(1 + \frac{P_r}{N_r} \right) \tag{45}$$

where $\frac{P_r}{N_r}$ is the signal-power-to-noise-power ratio (or simply signal-to-noise ratio, abbreviated SNR) at the receiver. Since the receiver noise is thermal, the Johnson-Nyquist formula (1928) holds

$$N_r = W k_b T_r \tag{46}$$

(see the site http://en.wikipedia.org/wiki/Johnson%E2%80%93Nyquist_noise), where k_b is the Boltzmann constant, $k_b = 1.3806488(13) \times 10^{-23} J/K$. Replacing (46) into (45), the information capacity of the channel made up by every radio bridge thus becomes

$$C(W) = W \log_2 \left(1 + \frac{P_r}{W k_b T_r} \right) \tag{47}$$

On the other hand, in ref. [2], eq.(22), it was proven that

$$P_r = \frac{P_t A_{et} A_{er}}{r^2 \lambda^2} \tag{48}$$

where P_t is the power transmitted by ETs, A_{et} and A_{er} are the effective areas of the transmitting and receiving antennas, respectively, r is the distance between the Sun and the star around which ETs live, and λ is the transmission wavelength. Thus, having assumed that all quantities in (48) are known, as it was done in refs. [2] and [3], one may say that the channel capacity of the radio bridge (47) is basically a function of the bandwidth

W only. Now, the problem of how narrow or how large the bandwidth is, indeed is a key problem for SETI. So far, the assumption of most SETI scientists has been that the bandwidth W must be very narrow, i.e. of the order of 1 Hz or even less, although always above the Drake-Helou limit of 0.6 Hz to take interstellar scintillation into account. Thus, over such a narrow band, the Fourier transform is the correct mathematical tool for the extraction of the ET sinusoidal carrier out of the cosmic background noise, which is indeed white over such a narrow band. The evolution of human telecommunications on Earth over the last few decades, however, clearly shows that the trend is ***not*** toward narrower and narrower bands, but rather toward larger and larger bands! This implies that the Fourier transform may ***not*** be the best mathematical tool for SETI, and this author has campaigned for over twenty years the replacement of the good old FFT by the more modern but computationally more demanding KLT (Karhunen-Loève Transform, see, for instance, ref. [4]). The bandwidth in (47) may thus be ***any*** finite positive number, rather than just a "small" positive number. Table 3 hereafter shows the Channel Capacities given by (47) for 1 Hz and 1 kHz bandwidth for all radio bridges considered in the previous section.

Table 3. Channel Capacities for all five information channels made up by the radio bridges between the Sun and another star as considered in Section 2. The second and third column give the Channel Capacity for bandwidth equal to 1 Hz (typical SETI case) and 1 kHz (sometimes used in SETI also), respectively. But the "novelty" shown by this table is the last column yielding the Channel Capacities for an ***infinite*** bandwidth! This is hard to understand unless we realize that the curve $C(W)$ given by the Shannon formula (6) has a ***horizontal asymptote*** (i.e. an **upper bound**) for $W \to \infty$, as we now prove.

Radio Bridge	Bandwidth = 1 Hz	Bandwidth = 1 kHz	Infinite Bandwidth
Sun – Alpha Cen A	37.6 bit/sec	27.6 kbit/sec	210 Gbit/sec
Sun – Barnard's Star	33.9 bit/sec	23.9 kbit/sec	16.3 Gbit/sec
Sun – Sirius A	36.5 bit/sec	26.5 kbit/sec	99.7 Gbit/sec

Consider eq. (47) again. First of all, it is physically obvious that one must have $C(0)=0$, but the proof has to resort to L'Hospital's rule to solve the $0\cdot\infty$ undetermined form. In other words, one has:

$$\lim_{W\to 0} C = \lim_{W\to 0}\left[W\log_2\left(1+\frac{P_r}{W k_b T_r}\right)\right] = 0\cdot\infty = \lim_{W\to 0}\left[\frac{\log_2\left(1+\frac{P_r}{W k_b T_r}\right)}{\frac{1}{W}}\right] = \lim_{W\to 0}\left[\frac{-\frac{P_r}{W^2 k_b T_r}}{\ln(2)\left(1+\frac{P_r}{W k_b T_r}\right)}\cdot\frac{1}{-\frac{1}{W^2}}\right] = 0. \tag{49}$$

Well, just the same limit for $W \to \infty$, rather than for $W \to 0$, reveals the horizontal asymptote of $C(W)$:

$$\lim_{W\to \infty} C = \lim_{W\to \infty}\left[W\log_2\left(1+\frac{P_r}{W k_b T_r}\right)\right] = \infty\cdot 0 = \lim_{W\to \infty}\left[\frac{\log_2\left(1+\frac{P_r}{W k_b T_r}\right)}{\frac{1}{W}}\right] = \lim_{W\to \infty}\left[\frac{-\frac{P_r}{W^2 k_b T_r}}{\ln(2)\left(1+\frac{P_r}{W k_b T_r}\right)}\cdot\frac{1}{-\frac{1}{W^2}}\right] =$$

$$= \frac{1}{\ln(2)}\cdot\frac{P_r}{k_b T_r}. \tag{50}$$

These are the values of the Channel Capacity for INFINITE Bandwidth shown in the last column of Table 1, and these are, of course, the ***superior constraints*** to the speed of transmission of information in SETI messages allowed by Physics! Finally, the second derivative of (47) reads

$$\frac{d^2C(W)}{dW^2} = -\frac{P_r^2}{\ln(2)\,k^2\,W^3\left(1+\dfrac{P_r}{W\,k_b T_r}\right)^2 T_r^2} < 0. \tag{51}$$

That is a negative quantity for all positive values of W meaning that the curvature of (47) is always toward the W axis, i.e. (47) is monotonically increasing between zero and the horizontal asymptote given by (50).

13. The Channel Capacities for the "Focal" Space Mission

This section is devoted to the computation the telecommunication Channel Capacities between the Earth and the receding FOCAL spacecraft on its way to the nearest stellar system, the Alpha Centauri. This problem is intrinsically different from the one faced in the previous section inasmuch as in the previous section the distance between the two stars making up for the relevant radio bridge is constant in time, while here the channel capacity is now basically a function of the receding spacecraft distance, d. This increasing (spacecraft) distance d firstly appeared in this paper in the BER equation (27). In terms of the Channel Capacity (47), however, the increasing spacecraft distance d enters the equations by virtue of the received power P_r transmitted either by the Earth to the spacecraft or the other way round (this does not matter) but crucially depending on the distance between Earth and spacecraft that we initially called r in the Path Loss equation (25) appearing in the received power equation (24). In other words, we shall now rename the r appearing in (24) and (25) as the distance d, thus basing all the following calculations upon the two equations:

$$C_{FOCAL\_WITHOUT\_SUN}(d,\nu,P_t,W) = W\cdot\frac{\ln\left(1+\dfrac{P_r(d,\nu,P_t)}{W\,k_b T_r}\right)}{\ln 2} \tag{52}$$

and

$$P_r(d,\nu,P_t) = \frac{P_t\,G_t\,G_r}{4\pi^2\,d^2}\cdot\frac{c^2}{\nu^2} \tag{53}$$

Clearly, since the distance d in (53) appears at the denominator, the received power P_r will ***decrease*** while the FOCAL probe is getting further and further away from Earth. Thus, because of (52), the Capacity of the Earth-FOCAL communication channel will ***decrease*** also as long as the FOCAL spacecraft recedes away from Earth.

Just to give a numerical example of these two equations, the following Figure 10 shows the Channel Capacity in bits/sec for the receding FOCAL probe between 0 and 550 AU linked to Earth by virtue of a 40 W transmitter at Ka band (32 GHz) with a 1 kHz bandwidth (lower blue curve) and the same link was it supported by the Sun's gravitational lens (upper red curve).

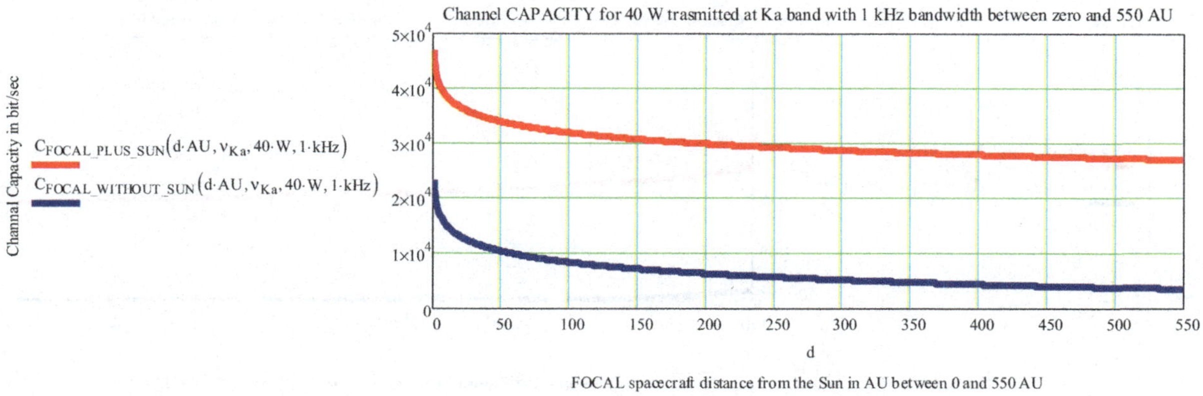

Figure 10. How the Channel Capacity DECREASES for a FOCAL spacecraft receding away from Earth between 0 and 550 AU. The lower blue curve shows the Capacity (in bits/sec) of the Direct Link Channel, i.e. of the direct transmission of radio waves between the Earth and FOCAL for a 40 W transmitter at the Ka band (32 GHz) with just 1 kHz bandwidth. This shows that, at 550 AU, the Direct Link Capacity is about 3537 bit/sec for 40 W transmitted with 1 kHz bandwidth. On the contrary, the upper red curve is the much higher Channel Capacity that would be available if the gravitational lens of the Sun was exploited for the same telecommunications.

Let us then extend these considerations to FOCAL distances between 550 and 1000 AU, as shown in Figure 11. The result is that, while the Direct Link would yield a Capacity lower than 3.5 kbit.sec, use of the Sun as a gravitational lens would yield a Capacity around 25 kbit/sec., i.e. 7.62 times higher.

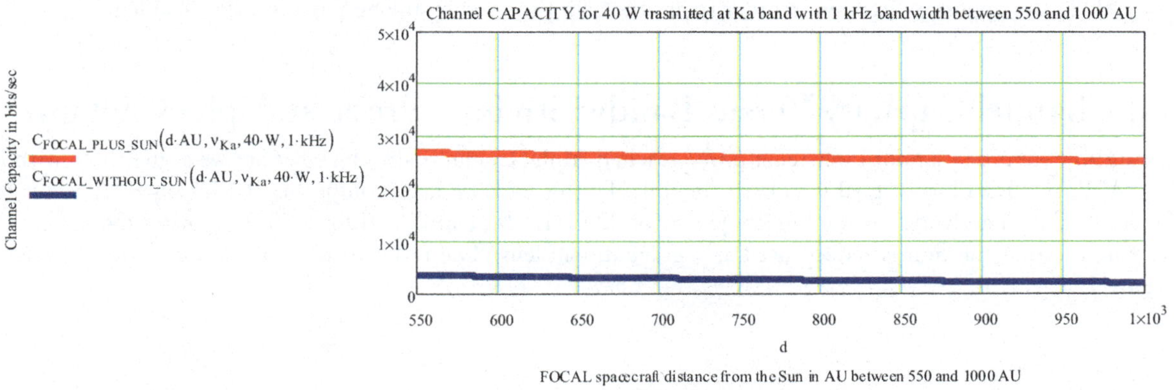

Figure 11. How the Channel Capacity decreases for a FOCAL spacecraft receding away from Earth between 550 and 1000 AU. Again, the lower blue curve shows the Capacity (in bits/sec) of the Direct Link Channel, i.e. of the direct transmission of radio waves between the Earth and FOCAL for a 40 W transmitter at the Ka band (32 GHz) with just 1 kHz bandwidth: this Direct Link Capacity decreases between 3537 bit/sec for and 2073 bit/sec for a 40 W transmitted with 1 kHz bandwidth. On the contrary, the upper red curve is the much higher Channel Capacity available if the gravitational lens of the Sun was exploited for the same telecommunications: decreasing from 26976 bit/sec and 25251 bit/sec, i.e. it staying about 25kbit/sec.

If we further extend the same content as in Figures 10 and 11 to distances higher than 1000 AU, say to distances between 1000 and 10000 AU, we then get what shown in Figure 12. In the practice, this means that, at 10000 AU, the Channel Capacity for the Direct Link has gone down to just 45 bits/sec, while, using the Sun as a gravitational lens, the Capacity still is about 18 kbit/sec, i.e. 408 times higher.

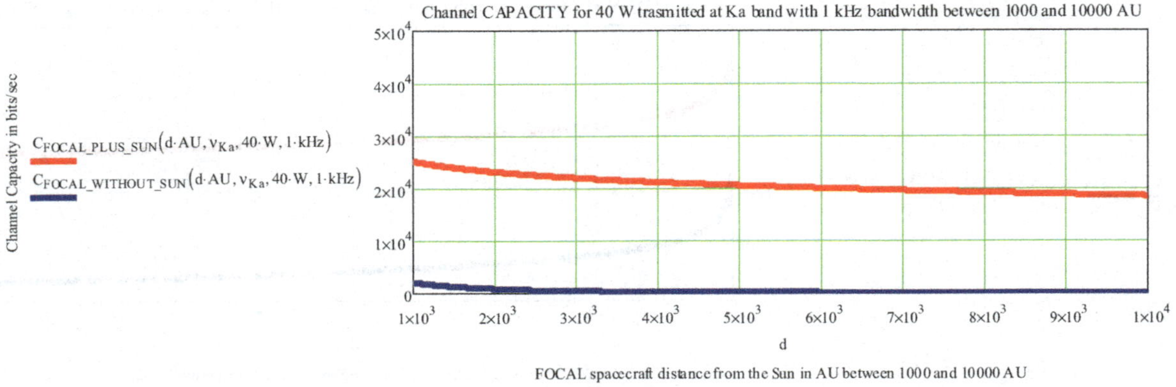

Figure 12. How the Channel Capacity decreases for a FOCAL spacecraft receding away from Earth between 1000 and 10000 AU for a 40 W transmitter at the Ka band (32 GHz) with just 1 kHz bandwidth. At 10000 AU the Direct Link would have a Capacity of just 45 bits/sec, while, using the Sun as a gravitational lens, the Capacity would be about 18 kbit/sec, i.e. 408 times higher.

We just complete this analysis by stating that, if a truly interstellar probe could reach the Alpha Centauri system at 4.37 light years distance, the Direct Link Capacity at Alpha Cen distance would be a miserable 0.06 bit/sec for a 40 W transmitter with 1 kHz bandwidth at Ka band (32 GHz), while, using the Sun as a gravitational lens, the Capacity would still be about 9 kbit/sec. In conclusion, all these data plainly show that it is impossible to keep the link with a truly interstellar probe up to Alpha Centauri unless one sends a FOCAL relay probe to 550 AU in the opposite direction in order to use the gain provided by the Sun as a gravitational lens.

14. Channel Capacity Versus Bandwidth for a Probe at Alpha Centauri

Let us now suppose that our interstellar probe reached Alpha Centauri already: what are we going to do with it?

Well, the first obvious goal is to keep the link. To this end, we need computing the Channel Capacity as a function of the bandwidth at the link frequency of 32 GHz (Ka band), and for a 40 W transmitter. Of course, we suppose that the Sun is indeed used as a gravitational lens. The result of this calculation is then shown in Figure 13.

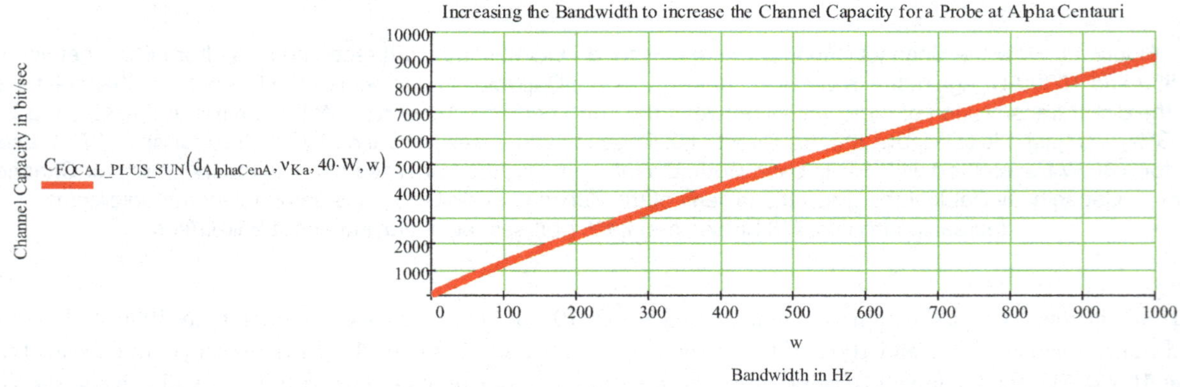

Figure 13. Channel Capacity vs. Bandwidth for an interstellar probe that reached Alpha Centauri already. The transmission frequency is supposed to be the Ka band again (32 GHz), the transmitter power 40 watts and the distance is now constant and equal to the Alpha Centauri system distance. Obviously the Sun is used as a gravitational lens. We then see that 1 kHz bandwidth yields a Channel Capacity of about 9 kbits/sec.

But what if we keep increasing the bandwidth? Well, the graph in Figure 14 is the answer. We see that:

1) The horizontal asymptote (i.e. the maximum Capacity at all, corresponding to infinite bandwidth and given by (50)) is given by 765319 bits/ sec ~ 765 kbit/sec. This Eldorado will never be reached in the practice, of course, and this is the dot-dot horizontal asymptote in Figure 14.
2) For a 1 MHz bandwidth, the Channel Capacity is about 600,000 bits/sec (always under the assumption of a transmitting frequency on Ka band, i.e. about 32 GHz.

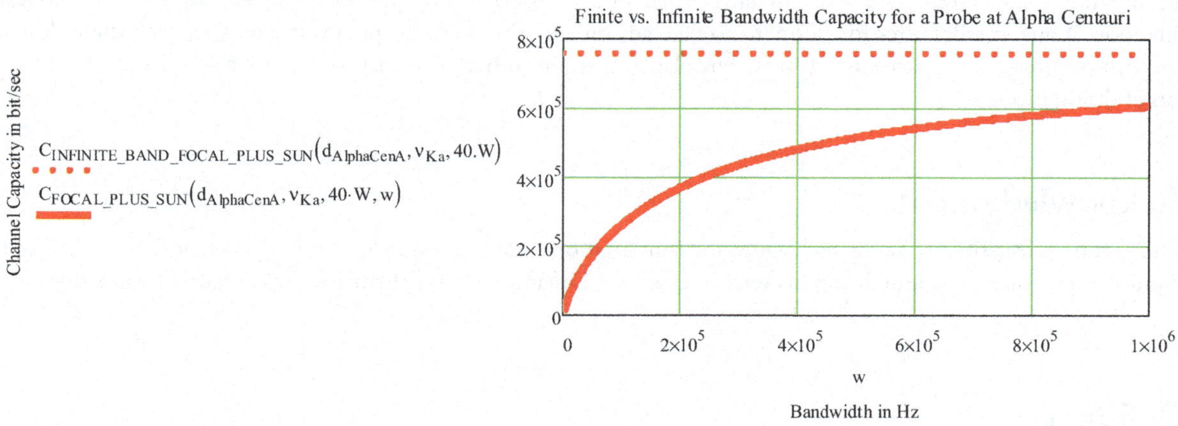

Figure 14. Channel Capacity vs. Bandwidth for an interstellar probe that reached Alpha Centauri already. The transmission frequency is supposed to be the Ka band again (32 GHz), the transmitter power 40 watts and the distance is now constant and equal to the Alpha Centauri system distance. Obviously the Sun is used as a gravitational lens. We then see that, for an infinite bandwidth, the Channel Capacity approaches 756319 bits/sec.

15. Conclusions: A Galactic Internet May Exist Already, but Not for Us

The conclusions reached by this author in this paper are thus:

1) A Galactic Internet constructed by advanced Aliens by exploiting the gravitational lenses of stars may already exist in the Galaxy.
2) The Channel Capacity for each radio bridge between any couple of communicating stars has an upper physical limit (in bits/sec), given by $\frac{1}{\ln(2)} \cdot \frac{P_r}{k_b T_r}$, where P_r is the received signal power, T_r is the noise temperature of the receiving antenna (radiotelescope) and k_b is Boltzmann's constant.
3) This Galactic Internet is currently inaccessible to Humans since Humans have not yet reached the minimal focal sphere of the Sun at 550 AU (and beyond, to, say, 1000 AU) by virtue of suitable FOCAL spacecrafts.
4) Even after the focal sphere of the Sun at 550 AU will have been reached by FOCAL spacecrafts, these must be aligned with the star sending the ET signals towards the Sun. Thus, the construction of some sort of "Dyson sphere for telecommunications" around the Sun at 550 AU by future, more advanced Humans might be advisable to put us in touch with the rest of the Galaxy for the first time.

16. Conclusion

In these few pages, the FOCAL space mission to 550 AU and beyond to 1000 AU could just be sketched. A number of related issues still have to be investigated in: 1) the science related to the mission, 2) in the propulsion tradeoffs to get there in the least possible time and 3) in the optimization of the telecommunication link. Nevertheless it plainly appears that the Sun focus at 550 AU is the next most important target that humankind must firstly reach in order to be prepared for the successive and more difficult task of achieving the interstellar flight. In particular, only a FOCAL mission will allow us to EXTEND our telecommunication link with spacecrafts located in space at least at the distance of Alpha Centauri or higher. This was proven mathematically by resorting to the notion of Bit Error Rate that would be zero or nearly zero only if we build up RADIO BRIDGES between the Solar System and the destination target, whether that would be a spacecraft or another star system. Finally, the belt of focal spheres was described as it is created by the gravitational lenses of the Sun and planets at distances between 550 and 17,000 AU from the Sun. Any future interstellar mission leaving the solar system will have to cross each one of those focal spheres regardless of the selected exit direction. Thus, it will better for Humans to take advantage of these focal spheres not only to keep the link with any future interstellar spaceship of ours, but also to explore other planetary systems while on the way to our target destination.

Acknowledgment

The author is indebted to many colleagues for conversations and suggestions, but, in particular, he would like to thank Paul Gilster for maintaining his terrific Centauri Dreams web site: http://www.centauri-dreams.org .

References

1. Albert Einstein, "Lens-like Action of a Star by the Deviation of Light in the Gravitational Field", *Science*, Vol.84, (1936), pages 506-507.

2. Liebes, Jr., "Gravitational Lenses", *Physical Review,* Vol. 133 (1964), pages B835-B844.

3. V. Eshleman, "Gravitational Lens of the Sun: Its Potential for Observations and Communications over Interstellar Distances", *Science,* Vol. 205 (1979), pages 1133-1135.

4. F. Drake, "Stars as Gravitational Lenses", *Proceedings of the Bioastronomy International Conference* held in Balatonfüred, Hungary, June 22-27, 1987, G. Marx editor, pages 391-394.

5. N. Cohen, "The Pro's and Con's of Gravitational Lenses in CETI", *Proceedings of the Bioastronomy International Conference* held in Balatonfüred, Hungary, June 22-27, 1987, G. Marx editor, page 395.

6. F. Drake and D. Sobel, *Is Anyone Out There?*, Delacorte Press, New York, 1992, in particular pp. 230-234.

7. N. Cohen, *Gravity's Lens*, Wiley Science Editions, New York, 1988.

8. C. Maccone, "Space Missions Outside the Solar System to Exploit the Gravitational Lens of the Sun", in *Proceedings of the International Conference on Space Missions and Astrodynamics* held in Turin, Italy, June 18, 1992, C. Maccone editor, *Journal of the British Interplanetary Society*, Vol. 47 (1994), pages 45-52.

9. C. Maccone, "FOCAL, A New space Mission to 550 AU to Exploit the Gravitational Lens of the Sun", A Proposal for an M3 Space Mission submitted to the European Space Agency (ESA) on May 20, 1993, on behalf of an international Team of scientists and engineers. Later (October 1993) re-considered by ESA within the "Horizon 2000 Plus" space missions plan.

10. J. Heidmann and C. Maccone, "AstroSail and FOCAL: two extraSolar System missions to the Sun's gravitational focuses", Acta Astronautica, Volume 35 (1994), pages 409-410.

11. C. Maccone, "The SETISAIL Project", in 'Progress in the Search for Extraterrestrial Life", *Proceedings of the 1993 Bioastronomy Symposium* held at the University of California at Santa Cruz, 16-20 August 1993, G. Seth Shostak editor, *Astronomical Society of the Pacific Conference Series*, Vol. **74** (1995), pages 407-417.

12. C. Maccone, *The Sun as a Gravitational Lens:Proposed Space Missions*, 3rd edition, IPI Press, Colorado Springs (Colorado, USA), 2002. ISBN 1-880930-13-7.

13. R. Orta, P. Savi and R Tascone, "Analysis of Gravitational Lens Antennas", in *Proceedings of the International Conference on Space Missions and Astrodynamics* held in Turin, Italy, June 18, 1992, C. Maccone editor, *Journal of the British Interplanetary Society*, Vol. 47 (1994), pages 53-56.

14. John D. Kraus, *Radio Astronomy*, 2nd ed, Cygnus-Quasar Books, Powell, Ohio, 1966, pp. 6-115 thru 6-118.

15. C. Maccone, "Radioactive decay to propel relativistic interstellar probes along a rectilinear hyperbolic motion (Rindler spacetime)", Acta Astronautica, Vol. 57 (2005), pages 59-64.

16. C. Maccone, *Telecommunications, KLT and Relativity* - Vol. 1, ISBN 1-880930-04-8, IPI Press, Colorado Springs, CO, USA, 1994.

17. C. Maccone, "Relativistic Optimized Link by KLT", JBIS, Vol. 59 (2006), pages 94-98.

18. L. Derosa and C. Maccone, "Propulsion tradeoffs for a mission to Alpha Centauri", Acta Astronautica, Vol. 60 (2007), pages 711-718.

19. C. Maccone, "Deep Space Flight and Communications – Exploiting the Sun as a Gravitational Lens", a 400-pages treatise about the FOCAL space mission that embodies and updates all previously published material about FOCAL. ISBN 978-3-540-72942-6 Springer Berlin Heidelberg New York, 2009. Library of Congress Control Number: 2007939976. © Praxis Publishing Ltd, Chichester, UK, 2009.

20. John D. Kraus, Radio Astronomy, 2nd Edition, Cygnus Quasar Books, Powell, Ohio, 1986. See in particular pages 12-2 and 12-3.

21. C. Maccone, "Interstellar radio links enhanced by exploiting the Sun as a gravitational lens", paper #IAC-09.D4.1.8, presented at the 49th International Astronautical Congress (IAC) held at Daejeon, Republic of Korea, October 12-16, 2009. This paper was later published in Acta Astronautica as the following reference.

22. C. Maccone, "Interstellar radio links enhanced by exploiting the Sun as a gravitational lens", Acta Astronautica, Vol. 68 (2011), pages 76-84.

23. C. Maccone, "Interstellar Radio Links Enabled by the Gravitational Lenses of the Sun and Stars", in "Communication with Extraterrestrial Intelligence", Douglas A. Vakoch, editor, State University of New York Press, 2011, pages 177-213.

24. C. Maccone, "Focusing the Galactic Internet", in "Searching for Extraterrestrial Intelligence – SETI Past, Present and Future", H. Paul Shuch, editor, Springer-Praxis, 2011, pages 325-349.

25. C. Maccone, "A new belt beyond Kuiper's: A belt of focal spheres between 550 and 17,000 AU for SETI and Science", paper # IAC-10-A4.1.6 presented at the 50th International Astronautical Congress (IAC) held at Prague, Czech Republic, September 27th -October 1st, 2010. This paper was later accepted for publication as the following reference.

26. C. Maccone, "A new belt beyond Kuiper's: A belt of focal spheres between 550 and 17,000 AU for SETI and science", Acta Astronautica, 2011, electronic version available online August 17th, 2011, paper version in press.

From the Moon to the Stars

Tapping into Shared Culture to Create Public Momentum for Interstellar Travel

Kathleen D. Toerpe

Abstract

The 100 Year Starship represents humanity's audacious foray into interstellar space. It is a foray that will require the political, economic and scientific resolve of up to five generations of Earth's peoples. History shows that shared culture and communal commitment have linchpin roles to play in sustaining such multigenerational efforts. This paper lays out a four-step process to create and sustain a multigenerational and intergenerational effort, drawing upon and expanding the narrative of one of humanity's most memorable shared experiences of space travel to date: the July 1969 Moon landing. Apollo 11's iconic mission not only represented, scientifically, a "giant leap for mankind" in 1969, but has also over the intervening years created a shared cultural narrative that can serve as a foundation to Earth's becoming an interstellar civilization. This unique event serves as a springboard to 1) reaffirm humanity's primal need to explore, 2) reinvigorate the genre of science fiction to inspire, as it has in the past, a passion for space exploration and discovery, 3) present a marketing strategy of co-branding the moon landing with current interstellar initiatives and 4) inspire a plan of engagement and outreach to people of all ages. Reaffirming the influence of the moon landing narrative, the paper integrates these four strategies into a call to action.

Keywords

Apollo 11, STEM, science fiction, public outreach, citizen science, interstellar civilization

1. Introduction

"Where were you when. . .? Human beings have a basic and deep need to place themselves within the flow of events that shape and change their world. Each generation poses the question differently. "Where were you when Pearl Harbor was attacked?" "Where were you when John Kennedy was assassinated?" "Where were you when we reached a new millennium?" "Where were you when the Twin Towers fell?" These watershed events become cultural homing beacons, drawing us to others who share these memories and binding us to a common history. Specifically, it is the power of shared experiences and the cultural narrative they create that can provide one starting point for Americans and for people around the world to understand and embrace the goals of 100 Year StarshipTM.

Where were you when man first walked on the moon? For those who were old enough to realize the momentous nature of July 20, 1969, there will be a ready answer to that question: at home, at work, on vacation with friends. For the 125 million Americans who watched Neil Armstrong's first steps on the moon – then 62% of the US population – and for another 500-600 million people who tuned in worldwide, the Apollo 11 moon landing created a shared cultural narrative that can be used as a springboard to launch public momentum for interstellar travel. [1] For the millions since 1969 that have either seen the landing replayed or witnessed it as a plot device in popular literature, film or television, the story of the moon landing provides the prequel for the soon-to-be written story of humanity's reach for the stars.

The basic unfolding of the events of July 20, 1969 is etched in humanity's collective memory. There are the ubiquitous photos of the Apollo 11 landing: the familiar count-down and launch sequence; the formal crew portrait of mission commander, Neil Armstrong, lunar module pilot, Edwin "Buzz" Aldrin, and command module pilot, Michael Collins; and an American flag, aided by unseen support bars, "waving" on the windless Moon. There is the almost eerie image of a solitary boot print in the gray lunar dust, and, of course, Neil Armstrong's iconic proclamation: "That's one small step for man; one giant leap for mankind." These iconic images have become cultural shortcuts to trigger communal memory and national pride. They honor not just heroes, but the country that gave their astronauts the opportunity to become heroic. But the narrative of the moon landing is a worldwide human narrative, beyond the ownership of any one country or political ideology, and beyond the mere 600 million or so people who personally witnessed it. Armstrong and Aldrin placed on the moon the goodwill greetings of seventy-three of the world's nations, with hundreds of thousands embracing the crew in tickertape parades and heroes' welcomes in twenty-three countries during their post-landing worldwide "Giant Leap" tour. [2] The story of the Apollo 11 moon landing and the five successful landings that followed is a human narrative, commemorated in print, in film, in anecdotes, even in toys and on lunchboxes. In the depth of its technical achievements and in its grip on society's collective imagination, the moon landing is the truly unique historical event uniting all humanity in one vicarious and extraordinary legacy of spirit, courage and humility. For the post-Apollo 11 generation for whom space exploration may seem to be the routine launching of probes, satellites and shuttles, recreating this cultural adrenaline is necessary to making a new journey to the stars the sequel to this unique legacy.

A reality check is needed here. Longstanding narratives need to be nourished in order to survive the storytellers' physical deaths and fading communal memories. The power of a story lies in its constant retelling and making it relevant to new audiences. Even an event as fantastic as the moon landing can be forgotten if not actively reinforced. A 1999 Gallup poll on the eve of the 30th anniversary of Apollo 11 found that of the more than 1,000 people randomly surveyed, only 50% could identify Neil Armstrong as the first man to walk on the moon. Recognition was highest among the youngest of people surveyed, those who were not even alive in 1969 and knew of Armstrong only through history books and the popular culture. [3] Even in 1969, just days after the successful moon landing, only 39% of Americans were ready to commit tax dollars for a trip to Mars. [4] With no competition from a Soviet mission to Mars, building upon the wildly successful moon landing – even with the cultural adrenaline rush of the previous days – had proved to be a difficult sell. Furthermore, John Kennedy needed to sustain support for the US space program for less than a decade in order to reach the moon. The research and testing were intense, but that was a sprint compared to the long-distance relay effort that is needed now. How then, is a narrative to be sustained over the four to five generations needed to make interstellar travel a reality? How does humanity in 2012 commit to, to adopt President Kennedy's words from 1960, a new "act of faith and vision" and back it up with both time and money? [5]

What is the starting point? This paper lays out a four-step process to create and sustain a multigenerational and intergenerational narrative, one that is built on the foundation of our own shared experience of space exploration in general, and of the Apollo 11 moon landing, in particular:

- Tie star travel back to the basic human longing to explore. It is tempting to cast it in the debate over budgets, policy and defense, spin-off technologies and scientific breakthroughs. It is also frustrating to paddle upstream against public perception of the size of NASA's budget. But economics and bureaucracy rarely incite sustainable passion. And the voting public may not realize the spin-off benefits for decades. The narrative of human exploration has gripped humanity's collective imaginations since the ancient Arabs, Babylonians, Egyptians and Greeks first looked up at the stars and named the constellations. Tap back into the sense of wonder that people of all ages around the world experienced on July 20, 1969.

- Revisit the same sources of inspiration, specifically science fiction, which motivated the scientists, mathematicians, engineers and astronauts to reach for the moon in the first place. Reintroduce the old stories to modern audiences and refresh the genre for new generations. The story of the moon landing did not start with President Kennedy. The seeds were planted and nourished by over 100 years of science fiction stories, comics, serials, radio shows, films, television shows and consumer goods. It is time to put those same cultural forces to work again to inspire a sense of adventure and possibility in a future among the stars.
- To borrow a marketing idea, deliberately and consistently co-brand the moon landing with star travel. By dovetailing future space exploration anniversaries in general, and Apollo ones in particular, with public outreach and announcements of technological breakthroughs, a tangible conclusion is reinforced that star travel is the next step, the sequel, to the moon story.
- Engage people of all ages – of all walks of life – where they live. Redirect current efforts to create a STEM-literate culture to an intergenerational audience. Target outreach and education efforts directly to people in their communities, schools and places of work and leisure to encourage them to become citizen scientists, support space exploration initiatives and consider careers in relevant fields.

2. Basic Human Longing to Explore

The world will undoubtedly be different one hundred years from now – in ways hardly now imaginable. How, then, is a narrative – and the effort to launch a starship - to be sustained over four to five generations? In some ways, it is an easy feat. Religious narratives and the institutions they have inspired have been sustained for millennia, in large part because they tap into humanity's core longing to understand and to be part of something greater than itself. The world's most powerful and compelling religious traditions ask the quintessential questions of human existence – the BIG questions – Why are we here? Where did the universe come from? Are we alone? It is in the answering of these questions that each tradition creates doctrine or ritual and a sense of the sacred or miraculous. John Kennedy, eager to inspire both a new generation of Americans as well as their elders to embrace the space race, alluded to the power of these same questions. Concerned that Cold War passions alone would not sustain public support for an American race to the moon against the Soviets, Kennedy tied it equally to humanity's greater need to conquer the "unknown," the "unanswered" and the "unfinished."[6] While it is tempting to shy away from these questions, to relegate them to philosophers and theologians, they are rightfully the domain of all who look up at the night sky and wonder. These are the questions that have already gripped human imagination for epochs and will continue to inspire, frame and guide all future exploration. In public releases and outreach, wherever and whenever there is talk about traveling to the stars, it must be tied consciously and consistently into this basic human longing.

In some circles, this connection is already being made. Physicists and astronomers such as Michio Kaku, Stephen Hawking, Lawrence Krauss and Neil deGrasse Tyson, Pamela Gay and Seth Shostak, among others, have started to articulate this longing – and the science to realize it - through their general audience books, documentary work, television appearances and social media messaging. They have become eloquent and approachable spokespersons for the astrosciences in recent years, and much like Albert Einstein and Carl Sagan were for earlier generations, they have become popular culture celebrities in their own rights. *How the Universe Works, The Universe, Sci Fi Science, Horizon, Nova, Prophets of Science Fiction,* even the whimsically titled, *How William Shatner Changed the World*, have all showcased distinguished scientists who have eloquently and entertainingly communicated how science can help answer humanity's oldest and most profound questions. Their involvement inspires as well as remotely and anonymously mentors millions of emerging scientists through interviews, websites, tutorials and social media feeds. If to be a celebrity is to be celebrated, and hence, valued by society, then that sense of celebration in the sciences and in scientists should be encouraged, nurtured and widened.

3. Reinvigorate Science Fiction

Jules Verne's 1865 novel, *From The Earth To The Moon*, created the genre of science fiction. This and Verne's other adventure novels, the *Buck Rogers* serials, H.G. Wells' fantastic stories of time travel and alien invasions and Edgar Rice Burroughs' *John Carter of Mars'* adventures, inspired the first generations of scientists, mathematicians and engineers to turn science fiction into science fact. This was a global phenomenon with Jules Verne as one of the most translated authors in history. [7] And the influence of these earliest writers continues. A theoretical physicist and string theory developer, Michio Kaku found inspiration in the character of Dr. Zarkov from the *Flash Gordon* serials. [8] Other scientists have also been inspired by the writings of Ray Bradbury, Isaac Asimov and Arthur C. Clarke, or by the adventures of Captain Kirk, Dr. Who, and Luke Skywalker. Physician, engineer and Endeavour astronaut Mae Jemison, the first woman of color in space, credits *Star Trek* actress Nichelle Nichols with inspiring her to join NASA. [9] Science fiction has the power to inspire; it is time to retell these old stories, to introduce contemporary novels to a broader audience and to mentor the writers of future ones. Encourage science fiction stories that move away from the apocalyptic themes that dominate the genre today to ones that capture the limitless possibilities of space travel.

Retelling the old stories is easy. Project Gutenberg ™ makes public domain science fiction (including Wells, Verne, Burroughs, etc.) accessible to everyone with computer and Internet access. Combine this accessibility with a science fiction version of the National Endowment for the Arts reading campaign, *The Big Read.* Start with the foundational giants of science fiction and move forward to contemporary authors and themes. Promote the campaign both to the general public through community reads and conversations (walking in the steps of *The Big Read*), through social media chat rooms and through prepared sci-fi curriculum materials for schools. Pedagogically, this encourages collaboration between science and language arts departments, promoting whole-brain activities crucial to both disciplines. Out of this comes a new imperative: to consciously mentor youth to become – not only scientists, engineers and mathematicians – but also novelists, screenwriters, poets, composers, philosophers, artists, even video game or app designers with a passion for space exploration. Tapping into a growing crusade here, STEM opens itself up to the arts and humanities, becoming STEAM, thus valuing both critical and creative thinking skills. Youth essay, video or designer competitions, collaborative curriculum among science, social science and humanities departments, one-time and renewable scholarships and grant awards are all ways to infuse the genre of science fiction with youthful ideas while tethering it to science fact. Most importantly, it is crucial to communicate the message that science fiction inspires, energizes and, frequently, frames the search for science fact. Would we now seek to create warp drives, teleportation beams, cloaking devices, laser weaponry and the myriad of other technological breakthroughs that started out as fanciful plot devices or cheesy props, but are now at the top of our scientific wish-lists, if it hadn't been for the inspiration of science fiction stories? Would the Apollo program ever have reached the moon without Jules Verne and Buck Rogers? Societies have multiple intelligences just as individuals do; to deny the creative sides of humanity's communal brain is shortsighted at best and self-defeating at worst. To sustain an endeavor of this magnitude, length and intricacy, the elegant interplay of the best of humanity's critical and creative faculties must be unleashed.

The first thing any good writer will tell you when writing a story is to give your main character a strong identity. For interstellar travel, that means the maiden starship needs a name. Even though it hasn't been built, even though there are competing designs with no consensus on what it will look like or how it will be powered, it needs a name now. Names have the power to label and define; they can reflect values or ideals to which a society aspires, as in the shuttle Endeavour or the current Mars' rover, Curiosity. They can honor the past as in the Magellan probe to Venus or the Cassini orbiter studying Saturn. Icarus Interstellar's current Icarus probe design harks back figuratively and literally to the mythological story of Daedalus' son. The fusion-powered Icarus picks up where the British Interplanetary Society's Project Daedalus from the 1970s left off and is one of a handful of probe or vehicular designs being investigated. [10] And there is no disputing the power of a name that, through intent or fate, can tell a thousand stories. Witness the emotional impact of the name Titanic in this hundredth year of its sinking. One hundred years – that's the time framework of the 100 Year Starship. A name can instantaneously excite the imagination. Who among space enthusiasts – and Americans and Earthlings of all nationalities - fails to recognize the name USS Enterprise? While dozens of ships with the name Enterprise have been christened throughout American history, including a space shuttle, it is science fiction futurist Gene Roddenberry's USS Enterprise that is arguably the most visually and culturally recognizable. As parents choose a name for their child while still in the womb and unborn – in anticipation of its future life and identity – humanity's starship needs a name to give it reality in the public's imagination. The name Apollo – borrowed from the Greco-Roman god of light and truth – will be forever synonymous with the American moon landing. Build upon that success and foreshadow future successes with a name. Most importantly, make average people – Earthlings from all over the planet – stakeholders in this process of starship naming. Ship design and propulsion methods cannot be crowd-sourced to those who are not trained scientists and engineers, but naming the ship can become an international event. Use the viral capabilities of social media, and tap the imaginations of the youth whose own future efforts will be so vital to success. Kick start the public's fancies with an inaugural public outreach campaign; let the people name the starship and start engaging humanity in the possibilities of interstellar travel.

4. Co-brand Moon Landing with Star Travel

Apollo 11's 50th anniversary in 2019 will be commemorated by NASA, the Smithsonian's Air & Space Museum, in public speeches, in television documentaries and by thousands of schools across the country and around the world. It is the perfect opportunity for 100YSS to partner with these groups and to dovetail their public outreach with these commemorations. Strive to make the co-branding so seamless that when people hear or read about the moon landing on TV, online, in books or in classrooms, their next thought becomes, "Now, on to the stars!" Practically, since many of these anniversary events will have their own corporate or governmental funding streams, it is a high impact return with minimal financial investment. And there are more anniversaries than just Apollo 11: the anniversary of the first man in space, the first woman, the first spacewalk, and so on. Future announcements in star travel should be tied to all of humanity's anniversaries in past space exploration, not just to American successes. These anniversaries will create a steady stream of opportunities over the next 100 years to celebrate inspiration and courage, scientific knowledge and technological success, themes that 100 Year Starship wants to communicate, too. These valuable co-branding opportunities should not be ignored. By timing the announcement of interim 100YSS benchmarks to the Apollo and other anniversaries, it creates a sense of expectation and momentum and further legitimizes a mission to the stars as the progeny of the mission to the moon.

5. Public Engagement and Outreach

Engage people at a personal level where they live, beyond impersonal press releases and anniversary events. One way is to create a Star City, Earth program – modeled on the popular Tree City, USA program. Initiated in 1976 by the Arbor Day Foundation, Tree City USA promotes local forestry planning and management, environmental education and community pride. Equally important, the more than 3,300 Tree City, USA signs placed at each town's threshold marks public support and commitment to forestry, and support for the goals of the Arbor Foundation. [11] In a similar way, a Star City, Earth program creates a grassroots commitment to star travel through public education and partnerships with local schools, observatories, astronomical societies, and community or technical colleges. Program requirements might include the town symbolically "adopting" a star, providing a publicly accessible telescope and a viewing area free of light pollution, hosting at least one community astronomy event a year, and, of course, advertising its inclusion with a public sign. The larger goal here is three-fold: to create public awareness and familiarity with star travel at the local level, to engage people personally in the adventure and encourage them to participate in one of the many space-related citizen science projects available, from exoplanet identification to interpreting SETI radio transmissions, and finally, to provide an established local conduit both for public outreach efforts and for the essential feedback communication loop. Outreach professionals need to know what they are doing right to inform and educate the public about star travel and space exploration – and what they can do better. While this can start out as an American-based initiative, the goal should be to quickly expand it internationally with Star City, Earth communities around the world. Ideal cities to initially target might be the American birthplaces or embraced hometowns of the more than five hundred courageous explorers who have already traveled to space. Neil Armstrong's birthplace of Wapakoneta, Ohio; Alan Shepard's Derry, New Hampshire; Mae Jemison's birthplace of Decatur, Alabama or hometown of Chicago, to name just a few examples, are cities to first approach, creating a web first of American, and later, of worldwide cities becoming a Star City, Earth locale.

There is a temptation to focus outreach efforts on young people, but reaching an intergenerational audience requires engaging everyone in the process of exploration and discovery. This goes beyond just training youth, whether in STEM disciplines or in the arts and humanities under STEAM initiatives. Adults need to be trained – and inspired – too. Outreach should focus on those who sit on school boards and boards of directors; those who work in factories and restaurants; those who vote and those who choose not to. By providing relevant course offerings on YouTubeTM or through the emerging open online course movement, TEDTM or BigThinkTM or through joining with NASA's Solar System Ambassador program, and by tapping into Adult Continuing Education and Learning in Retirement programs at the community and technical college level, adult STEM/STEAM awareness, appreciation and skills can be recognized and nurtured. This education has its own intrinsic benefits, of course, but can also bear fruit at election time as a scientifically literate and engaged electorate makes leadership and policy choices and as parents and grandparents advise the young people they love about career choices. This group of scientifically curious and engaged adults, many of whom will have personal memories of the Apollo missions and share in the cultural narrative of the moon landing, can form the backbone of current and future citizen science projects. With an aging population and a shrinking birthrate, the time and talent of these scientifically engaged adults who can contribute to initiating hypotheses, conducting research, interpreting data and even formulating conclusions and applications is sorely needed. Furthermore, such a rapidly mutable global economy that has displaced millions of adult workers may also encourage many of them to re-train professionally for careers in STEM/STEAM fields. While the youth may embody the future of science, it is myopic to ignore those adults who are inspired and motivated to embark on new careers right now.

6. Final Thoughts

It is here that the importance of cultural narratives has come full circle. It is in the stories that have been told – of men who walk on the moon, of spaceships that go where none has gone before, of blue police boxes that travel through space and time – that scientists, astronauts, politicians and voters have found the cultural and political will to explore the frontiers beyond planet Earth. Baby Boomers, heirs to the Apollo legacy, have been especially molded as a generation since their childhoods to seek out these new frontiers and to use the influence of their numbers to challenge and change society. Perhaps the energies of this generation can be put to use one more time – to help launch humanity to the stars – laying the groundwork for the future generation that asks: "Where were you when humanity first reached the stars?" Almost 400 years after Christopher Columbus' voyage to the New World, Verne's 1865 sci-fi classic dared to send men into outer space in the ship he envisioned, the Columbiad. In a nod to Jules Verne, the crew of Apollo 11 named their command module the Columbia, perpetuating the circular relationship between cultural narrative and historical reality, between science fiction and science fact. [12] 100 Year Starship is the next chapter in that narrative. It is the sequel to the story. From the Earth to the Moon… From the Moon to the Stars…the time to begin is now.

References

1. "A Remote That Broke All Records," *Broadcasting Magazine – Broadcasting & Cable*, reprint of July 28, 1969 issue, (7/17/2009).

2. "Apollo 11 Goodwill Messages," Release #69-83F, Washington, DC: National Aeronautics and Space Administration, (July 13, 1969); "Apollo-11 Astronauts Goodwill Round-the-World Flight Tour," Wingnet: Round the World Flights, 5jj, rev. 1001.

3. Newport, F. "Landing a Man on the Moon: The Public's View," Gallup News Service, (July 20, 1999).

4. Ibid.

5. Kennedy, J.F. "Address on the Space Effort," Houston, TX: Rice University, September 12, 1962 in *The Executive Collection: The Complete Book of Presidential Speeches and Literature*, C&C Web Press, (2008); Logsdon, J.M. "John F. Kennedy's Space Legacy and Its Lessons for Today," Issues in Science and Technology, 27(3), pp. 29-34, (March 2011).

6. Ibid.

7. Verne, J. *From the Earth to the Moon: Direct in Ninety-seven Hours and Twenty Minutes, and a Trip Around It.* Trans. by Louis Mercier and Eleanor King, New York: Scribner, Armstrong, (1874) at Project Gutenberg; Stewart, D. "Prescient and Accounted For," Smithsonian Magazine, (March 2005).

8. Kaku, M. *Physics of the Future: How Science Will Shape Human Destiny and Our Daily Lives by the Year 2100*, New York: Doubleday, (2011).

9. Katz, J. "Shooting Star: Former Astronaut Mae Jemison Brings her Message Down to Earth," *Stanford Today*, (July/August 1996).

10. Long, K.F., et al. "Project Icarus: Son of Daedalus – Flying Closer to Another Star," Journal of the British Interplanetary Society, 62 (11/12), pp. 403-416 (Nov/Dec 2009).

11. "Tree City USA: Foundation for Better Tree Management," Bulletin #22, Arbor Day Foundation.

12. "Apollo 11 Technical Air-to-Ground Voice Transcription," Tape 114/2, Data Logistics Office, Houston, TX: National Aeronautics and Space Administration (July 1969), p. 588.

Bibliography

"Apollo-11 Astronauts Goodwill Round-the-World Flight Tour," Wingnet: Round the World Flights, 5jj, rev. 1001.

"Apollo 11 Goodwill Messages," Release #69-83F, Washington, DC: National Aeronautics and Space Administration, July 13, 1969.

"Apollo 11 Technical Air-to-Ground Voice Transcription," Tape 114/2, Data Logistics Office, Houston, TX: National Aeronautics and Space Administration, July 1969, p. 588.

Kaku, Michio. *Physics of the Future: How Science Will Shape Human Destiny and Our Daily Lives by the Year 2100*, New York: Doubleday, 2011.

Katz, Jesse. "Shooting Star: Former Astronaut Mae Jemison Brings her Message Down to Earth," *Stanford Today*, July/August 1996.

Kennedy, John F. "Address on the Space Effort," Houston, TX: Rice University, September 12, 1962 in *The Executive Collection: The Complete Book of Presidential Speeches and Literature*, C&C Web Press, 2008.

Logsdon, John M. *John F. Kennedy and the Race to the Moon*, Palgrave Studies in the History of Science and Technology, New York: Palgrave Macmillan, 2010.

Logsdon, John M. "John F. Kennedy's Space Legacy and Its Lessons for Today," *Issues in Science and Technology*, 27(3), March 2011, pp. 29-34.

Long, K.F., Fogg, M, Obousy, R, Tziolas, A, Mann, A, Osborne, R, Presby, A. "Project Icarus: Son of Daedalus – Flying Closer to Another Star," *Journal of the British Interplanetary Society*, 62 (11/12), Nov/Dec 2009, pp. 403-416.

Newport, Frank. "Landing a Man on the Moon: The Public's View," Gallup News Service, July 20, 1999.

"A Remote That Broke All Records," *Broadcasting Magazine – Broadcasting & Cable*, reprint of July 28, 1969 issue, 7/17/2009.

Stewart, Doug. "Prescient and Accounted For," *Smithsonian Magazine*, March 2005.

"Tree City USA: Foundation for Better Tree Management," Bulletin #22, Arbor Day Foundation.

Verne, Jules. *From the Earth to the Moon: Direct in Ninety-seven Hours and Twenty Minutes, and a Trip Around It.* Trans. by Louis Mercier and Eleanor King, New York: Scribner, Armstrong, 1874.

The Foundry in the Starship

David Weiss

Abstract

Long duration space travel and habitation will require manufacturing tools to repair and build structure. Whether that is repair of radiation or impact damaged flight hardware or the building of equipment needed to support habitation, a foundry is likely to be involved. The metal casting foundry on a starship or in a remote habitat will be considerably different than either the perception or the reality of a foundry today. There has been much development in the last dozen years that give us a glimpse how that foundry may look and function. This paper discusses the integration of 6,000 year old casting technology with the construction, maintenance and habitation of the most advanced structures we can imagine. The special manufacturing needs of advanced materials and what those foundries will likely manufacture are also considered. Light-weight, high-strength and radiation-resistant materials manufactured without environmental impact will be the product of these foundries, building on the technical advancements of the next 99 years.

Keywords

foundry, castings, composites, maintenance

1. Introduction

It can be difficult to understand the repair and maintenance issues of multi-generational spacecraft and off Earth habitation. Start by thinking of taking a trip in a vehicle from Houston, Texas to Fairbanks, Alaska. When you leave, you need to take all of the fuel you need for the journey, everything you need to repair the vehicle, the repair facility itself and all the materials for the house you will be building in Fairbanks. You may choose a wheeled vehicle, which means you need to maintain everything from rubber to metals, or you may want to use a tracked vehicle where most everything is metallic. Where to start?

2. Design

The design task is complicated since we know little about the physical life of products beyond a decade even in our relatively benign planetary environment. An Air Force Study defined durability as a quantitative measure of the airframe's resistance to fatigue cracking under specified service conditions. [1] Ideally, repair is not required where the functional impairment does not occur before exceeding the service life. The fictional template USS Enterprise (NCC-1701) had a physical life of 40 years [2] although its design life may have been longer.

Design for maximum vehicle and structural life will be the desired paradigm. While the concept is sometimes employed here on earth, in most cases cost is the overriding factor. In spite of the desire to design for the mission, unknown or random factors will likely control repair frequencies. Material fatigue, while fairly well understood, is itself an unknown when applied to multi-generational tasks. Some light materials, such as aluminum, do not have a definitive fatigue life.

Impacts by meteoroids or other foreign objects are likely on long space voyages and the damage done will be highly dependent on size. Without the means to repair structure these impacts will present an existential risk to vehicles and vehicle inhabitants. Cumulative radiation damage to structure will need to be repaired and, depending on ecosystems encountered; bacteria, fungi or viruses may present a danger to both polymer and metallic elements.

If these "known unknowns" are not enough to consider the needs for repair, the "unknown unknowns" certainly are.

3. Material Trends

The quality index for metals is most generally defined as the potential of the alloy for mechanical performance, usually through a balance of strength and ductility. [3] Through continuous technical developments and advanced manufacturing methods the quality index of most materials has been increasing by about .375% per year.[4] I would argue that the growth is accelerating with the development of advanced nano-composites in both polymer based and metallic materials. This general trend will improve material reliability and improve hazard survivability.

4. The Manufacturing Challenge

Many common manufacturing techniques are too equipment intensive to be practical on board a space vehicle. Extrusion, rolling and forging require equipment multiples the size of the finished project. A six thousand year old technique, casting, requires equipment only slightly larger than the part to be produced. Casting on a spaceship will likely be quite different than in 2012 vintage foundries on earth. Foundries are usually defined as an establishment where metal is melted and poured into molds but more generally as a place to take the first step in building, such as a semiconductor foundry.

The concept of a repair foundry on a vehicle is not new. Naval foundries have been in existence since the use of metallic ships and continue in use today. See Figure 1. The challenge then and now is to get the maximum utility with a minimum of weight and space. That challenge is much greater on a starship.

Figure 1. Naval Repair Foundry from 1916 [5]

In the fictional Battlestar Galactica, whole ships were devoted to the foundry enterprise and were used as mobile construction docks employed by both military and non-military organizations to process ore, build replacement hardware and conduct repairs to disabled ships.[6] It is unlikely we will have the luxury of a separate foundry ship. However, significant improvements in rapid manufacturing driven by lower costs and the ability to print a range of advanced materials will give us not only the ability to direct print components but also print molds that will be used subsequently to cast a wide range of materials that cannot be direct printed. Stratasys, a 3D printing equipment company recently produced 70 parts for a prototype manned rover using their Fused Deposition Modeling Process.[7] There are numerous commercial mold printing shops that produce molds for repair parts and other low volume applications that are suitable for casting nearly all metal alloys. Combining these processes with on-demand melting using resistance, induction or levitation makes the production of most parts possible.

Figure 2. On Demand Melting Unit. Photo: Atriax Materials

5. Materials

In general, materials for a spacecraft should have low density but have a high specific stiffness. They should be capable of being cast and have the ability to substitute for a wide range of materials. As in any closed system, the materials should be recyclable. The most likely choices are light metal or polymer based composites with densities of between 1500 and 3000 kilograms per cubic meter. These materials can be customized to meet targets for strength, stiffness and wear. Natural and artificial fibers, nanoparticles, graphene, carbon nanotubes, and a wide variety of ceramic particles can be used for reinforcement to achieve the targeted properties. Natural sources for reinforcements include plants grown on-board, asteroid mining activity and treated waste products. Some asteroids are rich in clay, a common reinforcement for polymer based composites. Early in 2012, a new venture called Planetary Resources, Inc. was formed to set up mines to bring asteroid resources back to Earth.[8]

6. Conclusion

Repair and maintenance facilities will be a requirement of long duration space travel. Those facilities will use the flexibility of foundry like techniques to produce the required hardware. Light weight, high strength composite materials will likely be the materials of choice.

References

1. Manning and Yang, "USAF Durability Design Handbook: Guidelines For The Analysis and Design of Durable Aircraft Structures", 1984

2. Wikipedia-Starship Enterprise

3. Alexopoulis, N.D., "Definition of Quality in Cast Aluminum Alloys and Its Characterization with Appropriate Indices", Journal of Material Engineering and Performance, Volume 15, February 2006, pg. 62.

4. Author's calculation. Multiple sources.

5. www.practicalmachinist.com/vb/antique-machinery-history/photo-navy-foundry-work-115300/

6. www.tecr.com/galactica/miscships/foundry.htm

7. Thryft, A. "3D Printing Flies High", Design News, October 2012, pages 36-38

8. www.wired.com/wiredscience/2012/04/planetary-resources-asteroid-mining

2013

100 YEAR STARSHIP

2013 Public Symposium

*Pathway to the Stars, **Footprints on Earth***

Across the globe, calls are being made for bolder human expansion into space beyond earth orbit. It is against this backdrop that the 100 Year Starship 2013 Symposium is held.

Achieving the interstellar human journey in many ways, must necessarily build upon, promote and establish fundamental research, technology development, societal systems and capacities that facilitate ready access to our inner solar system. And to truly have the best opportunity for the aspiration to be long-lived enough to actually be accomplished, organizations and individuals involved must work to ensure it leaves a positive, indelible mark upon life right here on Earth.

100 Year Starship (100YSS) is working to create new avenues that foster innovative, robust collaborative, transdisciplinary research, project design and technological development. The 100YSS 2013 Public Symposium—*Pathway to the Stars, Footprints on Earth*—seeks to highlight both the small incremental steps and radical leaps required to make significant progress on the way to interstellar space.

"Meeting the challenge of 100YSS® stands to be even more transformative to our world than Sputnik . . . "

- Dr. Mae Jemison

The Dangers of Religious and Ideological Extremism and the Mission of Interstellar Exploration

Jason D. Batt

Abstract

100YSS has promoted itself as "everyone's starship." Individuals from a variety of ideological and religious beliefs have contributed to space exploration throughout its history. Does the challenge of inclusion apply to variables of ideological extremism? From the destruction of the World Trade Center towers to Westboro Baptist Church, the dangers of religious and ideological extremism are obvious and present. This paper will present the dangers of extremism pre-ship launch and post-ship and the environments that create religious and ideological extremism. It will also review these questions: Is terrorism always religious in nature? Can there be strategies made to preventing extremism in interstellar efforts? Interstellar efforts will need to be prepared to face the topic of religious extremism and develop strategies pre- and post-launch to challenge it.

Keywords

extremism, terrorism, religion, ideologies

1. Introduction

At the 2012 100 Year Starship Symposium, I presented two different papers on the topic of religion and interstellar travel: "Engaging the Church as Champion" and "The Off-Planet Church." During these talks, I provided these conclusions:

- Where humans go, they carry their religion. Where humans go, they take their beliefs and practices. They will take their religion off-planet. Just as exercise of religion is a freedom valued here in democratic nations, it is expected that same freedom will accompany humans in their off-planet explorations.
- The practice of religion in space has a long history. The concept of religious expression off-planet is not just a consideration for the future; it has happened and continues to happen. Where humans go, they take their beliefs with them. They take their religion with them. James Hartfield, spokesman for the Johnson Space

Center, said: "Although NASA does not provide spiritual resources, religious objects—crosses, Bibles, icons, prayer cards—are among the most common personal items taken into space." [1] Buzz Aldrin, an Apollo 11 astronaut and a believer, performed communion on the moon. [2] On Christmas Eve in 1968, while aboard Apollo 8 in orbit around the moon, Frank Borman read from the first ten verses from the book of Genesis. [3]

- Three potential relationships between organized religion and the 100YSS: neutral and uninvolved observant, as hostile opponent, or as active proponent (a champion).
- There are parallels in the mission between the 100YSS and organized religion, specifically Christianity.
- The 100 Year Starship's mission is a challenge to engage as many aspects of humanity as possible to achieve this great goal. The elements of Christianity, the world's largest religion based on number of adherents, can not be overlooked. There is the potential of a negative relationship between elements of Christianity and the mission of space exploration. However, there is great possibility in a positive relationship between the two. The Church can be an active champion of interstellar exploration. To engage the Church, it is recommended and encouraged that a focused and professional effort occur early on in the mission of the 100 Year Starship.
- In a vacuum of religions, religions will be formed.

It it this last point which serves as the key starting point for a look at the challenges of religious extremism and interstellar travel. At the 2012 Symposium, Stephen Taylor presented pieces of his musical *Paradises Lost* by Ursula K. Le Guin about a religion that formed on a generational ship. There have been suggestions that an interstellar effort should be done without any religious expression at all. The history of space travel already includes numerous episodes of religious expression. Repeating Hartfield, the most common personal items taken into space are of a religious nature.

Eli Berman and Laurence R. Iannaccone in their book *Religious Extremism: The Good, the Bad , and the Deadly* argue that the imminent death of religion has been oft-cited but has yet to materialize. Berman and Iannaccone state: "For centuries, scholars have confidently, and incorrectly, predicted the immanent demise of religion. With powerful prose, but no real proof, Hume, Comte, Marx, Freud, the Huxleys, and other influential intellectuals, pronounced religion a dying vestige our primitive, pre-scientific past . . . Throughout the world – in Latin America, India, Africa, Europe, the Middle East, formerly communist countries, and the United States – religion continues to flourish, especially its more fundamentalist variants. In short, the demand for supernaturalism is as basic and irrepressible as the wants it seeks to satisfy." [4]

Religion, whether beneficial or damaging, continues to be sought by humanity. It is shortsighted to assume that even in our ascent to the stars that religion will not find its way there. The great predictions of science fiction writers actual proposes the opposite. Dune shows an evolution of the Catholic church as it follows a multi-star society. Orson Scott Card's Ender's Game series find Catholic missionaries accompanying colony ships to the stars to encounter newly-discovered sentient races.

There was an alternative point of view proposed by Rev. Alvin Carpenter that no religion should be allowed. His abstract claims, "If there are two words that sum up earth religion it is 'divisive' and 'intolerance' and these two things, with the fury of religion behind them, would introduce a pathology to the space pilgrims they could never overcome." [5] In his paper, he makes an interesting point on the natural rise of religion within the interstellar ship: "Religion in space must be must be able to rise unhindered by earth priests, earth religions, and earth's holy books. If religion is necessary for the human mind to function, then it must rise from where that mind resides, its unique interstellar environment." So while he argues overall that earth-based religions are destructive to humanity's future, he does provide that religion, even in a potential vacuum, will still arise as we seek to explore the stars.

2. The Inclusive Nature of the 100 Year Starship

The 2012 Symposium held a workshop titled "Is It Everybody's Space Mission?" In the workshop, Dr. Henry Williams challenged the workshop group: "How do we solve the limitations to diversity for the 100yss?" In the workshop summary session at the end of the Symposium, the workshop team reported: "People aren't asking if we should have diversity. We are asking how to overcome the barriers to it." Making diversity a goal seems to be a priority for the 100 Year Starship team.

Second, the goal of the 100 Year Starship is to travel to another star. This means quite a far distance, and at current propulsion technology, quite a long time. Proxima Centauri is about 4.25 light years from Sol. A light year is equal to just under 10 trillion kilometers. NASA had launched the Voyager 1 traveling at more than 17 kilometers per second. At that speed, the Voyager 1 would take 74,485 years to reach Proxima Centauri. Even if propulsion technology is significantly expanded, the trip will definitely take years to achieve. Even if a ship could approach speeds of a tenth of light speed, a trip to Proxima Centauri (over 25.6 trillion miles away) would take nearly 70 years. This leads to discussions about the possibility of a generation ship or multi-generation ship— a ship where the individuals, and their offspring for a few generations, live their entire lives on board.

At the minimum, the crew of an interstellar ship, assuming no super-relativistic speed discoveries, will be larger than any space mission so far. It will be composed of a widely diverse group of individuals. The crew will be working and living together for a very long time in a locked, unescapable environment.

3. The Dangers of Extremism on a Starship

Where there are people, there is religion. However, not all religious expression is positive. In the last decade, the destructive effects of religious expression have been center point in our global dialogue. The horrific tragedy of 9/11 was spawned from religious extremism. The protests of the Kansas-based Westboro Baptist church have invaded funerals and other social functions with the declarations of "God Hates America" and "God Hates the World." Extrapolate this behavior to a starship and the potential results are frightening to consider. The dangers of religious extremism are only compounded when brought aboard a tightly confined space, with precariously precise working parts all subject vulnerable and all possibly mission-killing if disrupted.
The environment of the starship will be:

- closed
- communal
- homogenous

The closed, communal yet homogenous environment of the starship can produce extremism. We must avoid the rigorous/high-demand elements which foster extremist thought.

The dangers of religious extremism must be addressed and prepared-for from two perspectives: pre-launch and post-launch. The ability to confront extremism pre-launch which actually be quite easy. While the building and dreaming of ship is inclusive, we can be very critical in selection of the actual crew for the ship. We can screen for dangerous belief patterns. However, just the statement of "dangerous belief patterns" brings up its own questions: who will be defining dangerous beliefs? How do we maintain religious freedom and yet screen for this type of belief? These questions will be addressed in this paper. It should be noted here that this paper is not proposing a limitation of religious thought or simply accepting only a narrow spectrum of religious expression. We argue that it will actually be an openness to a "wide variety" of religious thought that may prevent the rise of extremism later on the actual mission.

What about after the ship in en route? If this is a generational ship of six to ten generations, with hundreds to thousands of individuals, the potential for extremism can arise from within a seemingly docile group. What strategies can be developed to manage and mitigate potential for extremism?

4. Defining and Identifying Religious Extremism

What is extremism and what is religious extremism? This exercise in definition is critical to 100YSS as we interact with future possible candidate pools. We need to be aware of potential elements which could develop into religious extremism and thus mission-threatening behaviors and beliefs.

The title of religious extremist has been applied to everyone from former President George W. Bush to the Taliban. Neil J. Kressel, in his book *Bad Faith: The Danger of Religious Extremism*, addresses the definition of religious extremism: "To cut through the confusion in defining and identifying religious extremists, one might turn to William James, who-in an invocation of New Testament scripture-asserted that religions should be judged not by their roots, but by their fruits. But which fruits are best to cultivate, and how we are to evaluate religions that produce both sweet and bitter fruits?" [6]

Religious extremism, especially in Western cultures, is often viewed as synonymous with Islamic extremism. Kressel states: "The Islamic religious tradition provides the extremist with abundant source material. But most major religious traditions contain messages that may, without much creativity, be interpreted hatefully. And theologians in every tradition, including Islam, have found ways to modify, reinterpret, counterbalance, or otherwise incapacitate such verses." [7] It is not just Islam which produces extremism. It is not a certain type of religious text that produces it.

Kressel digs into the elements that create extremism in religion: "One way we might identify pathological religion is similar to the way the doctor identifies psychiatric abnormality in the patient. In that context, no single dimension separates the normal from the pathological, and the difference between health and illness is a matter of degree." [8] And it is Kressel that provides us our most concrete definition for religious extremism: "Religious extremists can be defined as those persons who-for reasons they themselves deem religious-commit, promote, or support purposely hurtful, violent, or destructive acts toward those who don't practice their faith." [9] There is still quite a broad area of interpretation within this definition, however, it gives us a base to work from. Extremism is defined when it turns destructive.

5. The Spectrum of Religious Expression

Even within common belief systems, there exists a wide denominational continuum from lenient "churches" to extreme "sects." "[W]hile one group is singing Kum Ba Yah, the underdeveloped one is convinced of conspiracy, planning to wipe whole nations off the earth, and rationalizing their actions via beliefs about the afterlife and rewards they will receive in Heaven." [10] There is from all of these groups, there can arise a spectrum of negative religious expression from dysfunction to extremist: "When the fruits of a religion frequently include attacks on the life, liberty, and happiness of nonbelievers or those who don't practice the faith, the religion can be reliably classified as extremist. When a set of religious beliefs is incidentally hurtful to some individual or group, the classification is trickier. Moreover, when the fruits of a religion include impediments to the health, happiness, and development of believers who are allowed to leave freely if they so choose, the religious system should be dubbed dysfunctional but not labeled as religious extremism. Faith that leads one to harm only oneself is also dysfunctional, but it belongs in a separate category and poses a different type of challenge." [11]

Berman and Iannaccone contribute to the definition of the spectrum:

> "Toward one end of the continuum, we expect to find "extremist" groups that demand very high levels of sacrifice. Toward the other end, we expect to find "lenient" groups that demand very little of their members. If the theory is correct, then groups that demand similar levels of sacrifice will display similar social characteristics despite their differences in organization, history, and theology. Thus, the high-cost groups of any religious tradition will tend to be exclusive, strict, small (at the local level), suspicious of other groups, and critical of secular society. Members will learn to fully embrace the group's distinctive lifestyle, devoting their resources to the group's activities and goals, or else face penalties, up to and including expulsion. As a result, these high-cost groups will also maintain tighter social ties, as shown for extreme Christian sects in the U.S. by Iannaccone (1992). In contrast, low-cost groups will tend to be inclusive, lenient, tolerant of secular values, and open to loosely affiliated members, while providing less cooperatively provided services." [12]

To pinpoint the the religious climate which spawns religious extremism, we reference a two axis system. The Y axis focuses on the belief system ranging from rigorous to relaxed. The X axis focuses on the social management and environment: communal social focus to individual social focus. Extremism occurs in the top-right quadrant. Religious extremists groups most often have a very rigorous belief system. They also have a very strong communal social focus. While these two qualities alone don't necessitate extremism, extremist groups do possess both of these qualities.

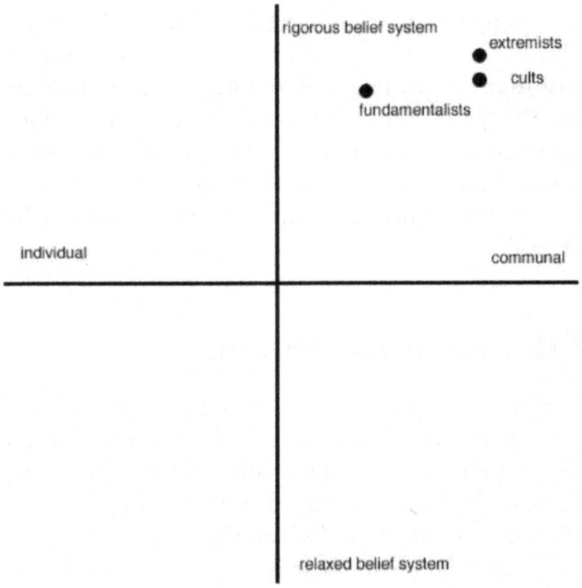

Figure 1. Two axis system of religious expression

Berman and Iannaccone reaffirm this conclusion: "Within the religious marketplace there naturally arise different niches, not all of which provide equally intense experiences. Some religious traditions, notably Judaism, Christianity, and Islam, have a much stronger communal orientation than others, such as Buddhism and Hinduism. Within each religious tradition, moreover, some firms maintain rigorous systems of shared beliefs and morality, whereas other firms demand very little. It is the former that we label 'extremists' (or 'sects,' 'cults,' and 'fundamentalists'); the latter are more known as 'moderates,' 'liberals,' or 'mainstream churches.'" [13]

In regards to screening for dangerous belief patterns, we must be aware of our own individual perspectives on religion. Every other religious belief can also be found on this spectrum:

Figure 2: Spectrum of Other Religious Belief

Every belief that an individual encounters will either be accepted, familiar, tolerable, annoying/disagreeable, disruptive, or dangerous. We should be careful to not exclude a religious belief simply because it is annoying or disagreeable from our own perspective. However, beliefs that are disruptive and/or dangerous should be the ones we begin to screen for.

6. Fundamentalism Versus Extremism

Pursuing this line of thought further, we must be cautious in not confusing fundamentalism with extremism. "Moreover, one must avoid confusing the religious extremist with someone who is just extremely religious. There are many ways in which manifestations of religiosity can be intense without necessarily having dangerous ramifications." [14]

What is fundamentalism? "One team of psychological researchers-Ralph W. Hood Jr., Peter C. Hill, W. Paul Williamson-explained that "[w]hat distinguishes fundamentalism from other religious profiles is its particular approach toward understanding religion, which elevates the role of the sacred text to a position of supreme authority and subordinates all other potential sources of knowledge and meaning" Along with many writers, they note that such positions have typically been adopted as part of a defensive reaction to modernism. They also note that some have attempted to draw a line between fundamentalism and evangelicalism on the grounds that adherents to the former have far greater hostility toward mainstream society." [15] Without diverting too far off topic, it is important to note that fundamentalism carries with it, from some perspectives, a derogatory connotation. Kressel pursues this: "In an article titled "Fundamentalism Isn't the Problem," Kenneth Minogue objects that "'fundamentalism' is an all-purpose expression denigrating the peoples of the Book (as the Muslims call them). Anyone who takes the Bible (either Testament) or the Koran seriously falls under the terminological lash of terrorism-inducing fundamentalism. How is it that millions of harmless American Christians, or orthodox Jews, living pretty blameless lives, come to be incorporated under such a denigratory slogan?" He then offers an answer, claiming that especially for American liberals, "it is an irresistible rhetorical triumph to be able to package all the people of whom you disapprove in one nauseous bundle." [16] Fundamentalists are not by this definition religious extremists and violent militants. However, in certain environments and under specific conditions, fundamentalists can become extremists.

7. Fallacies and Facts of Extremism

What are these conditions? What are the causes of extremism? We've already reviewed that religious extremists have two consistent qualities: rigorous religious belief and strong communal social focus. Beyond this, what creates extremism?

Reza Aslan, in his book *How to Win a Cosmic War*, argues that we should not confuse extremism with irrational behavior. It is not. "It is tempting, even comforting, to consider such abominable acts of terror to be the result of irrational or pathological behavior. But the truth is that terrorism is almost always a calculated choice. Terror is purposefully chosen, because it is often seen as the most effective, most expedient, and most economical method of pursuing a group's aims." [17] Kressel reaffirms that "religious thoughts, feelings, and experiences propel most violent antiabortion activists, and sometimes provide their central inspiration. Far from being impulsive or insane, these extremists contemplate and calculate their terror as a means of achieving what they consider to be a religiously certified objective." [18]

Often extremism is argued to arise from poverty situations. In their paper "Education, Poverty and Terrorism: Is there a Casual Connection," Krueger and Maleckova argue that poverty does not necessarily lead to extremism. "Perhaps surprisingly, our review of the evidence provides little reason for optimism that a reduction in poverty or an increase in educational attainment would meaningfully reduce international terrorism. Any connection between poverty, education and terrorism is indirect, complicated and probably quite weak. Instead of viewing terrorism as a direct response to low market oppor- tunities or ignorance, we suggest it is more accurately viewed as a response to political conditions and long-standing feelings of indignity and frustration that have little to do with economics." [19] Simply reducing poverty does not provide an easy solution for eliminating extremism.

Although the focus of this paper is on religious extremism, we should be careful to note that not all extremism is religious in nature. Extremism can arise from groups that are entirely not religious. "If theology is so important, why are most terrorist organizations not religious? And if afterlife rewards are key, why has a nonreligious group – the LTTE "Tamil Tigers" – been responsible for more suicide attacks than another other organization? Why is suicide bombing associated with all sorts of theologies but just one style of religious organization (best described as "sectarian")? And why do most militant sects devote much of their energy to benign and noble activities, such as running schools, health clinics, and social services agencies?" [20]

As well, not all violent acts by believers are caused by the religion and its belief structure. "Not all-or even most-evil acts perpetrated by Christians throughout two millennia can be reasonably described as religiously motivated or Christian, much less as instances of religious extremism. Many people who do dastardly deeds of one sort or another may be Christians in name alone, only pretending or perhaps not even pretending to be believers." [21] While the banner of religion is waved, often there are underlying motivations or goals that extend far past religion. 9/11, while perpetrated by religious extremists, also has a strong cultural motivation. In fact, often the religious belief system becomes manipulated to encourage their violent goals. Berman and Iannaccone state: "This is typical of religious extremists. Their militancy varies dramatically over time, even as their underlying sectarian traits remain fairly stable. Moreover, they prove remarkably adept at reworking their theology, and even reinterpreting their most sacred texts, so as to justify their changing orientations." [22] So, we should be careful to not simply reject religious beliefs because those beliefs have been used to justify violent acts.

We must also confront the belief that Islam itself is a more fruitful breeding ground for extremist thought. Extremist beliefs do tend to rise more readily within Islamic worlds, although are by no way limited or unique to those religious systems. Kressel concludes that it is a lack of social evolution and reformation that has led to a climate where religious extremism can develop."The Islamic world has not experienced the unique confluence of historical events that ultimately produced in the West a tradition of tolerance and a check on religiously inspired hatred. There was no Renaissance, no Reformation, no Peace of Westphalia, no Enlightenment, no theory of the separation of church and state, and no Vatican Council II. In the West, religious messages nowadays enter public debate primarily through the tempering prism of modern culture and constitutionally based values. In the world of Islam, the dictates of the faith can have a more direct and deleterious impact on social and political affairs." [23] It is not actually the religious belief system alone that creates extremism but the cultural environment that the religion is found in. However, there is grounds for argument that that culture is created by the religious system of Islam. However, within the discussion of extremism, it is to be noted that the religion alone doesn't provide the environment for extremism alone.

Interpretation of sacred texts is key to understanding defense of violent behavior by members of those groups. It is not just the beliefs that matter but how those beliefs are interpreted. There are major texts in the majority of religions. However, two groups can interpret key scriptures in those texts in completely different way. Kressel expands this: "As we have already seen, the question of how individuals and religious groups interpret their sacred scriptures-the Jewish Bible, the Catholic Bible, the Protestant Bible, and the Koran-is critical. No matter what believers may say, none is truly a literalist. Everyone picks and chooses, at least a little. Everyone interprets. When there are additional religious traditions and additional sacred books, the importance of interpretation extends beyond scripture to these sources as well. The same texts in each faith can be used to support diametrically opposed political, social, psychological, and theological stances. Hence, understanding how religionists make decisions concerning the interpretation of their religious texts and traditions is the key to deciphering when and how religion becomes a force for evil in the world." [24] To future prevent the dangers of religious extremism, it is needed to have an understanding of these texts and to analyze how those text are interpreted. The various religious texts often have a violent aspect to them, however, the question is whether this element is interpreted as a literal directive or simply a historical fact. If the group considers the violent aspects as a literal directive, the group then steps into the disruptive and dangerous end of the spectrum. Kressel argues that if individuals adhere closely to their religion's moral and ethical elements then extremism is less likely to arise: "To avoid extremism, people would have to attend more closely to the moral, social, and political consequences of their religions. Unfortunately, many believers see little reason to do this if a religion feels like it is working, especially with regard to matters of ultimate concern. Religious militancy is frequently a consequence of individuals choosing to pursue their personal 'immortality projects' and 'quests for meaning' by traveling down the wrong paths-wrong in the sense that they are socially and morally destructive." [25]

So, why not jettison the whole mess? Wouldn't it do well if we simply removed religion altogether? This paper has addressed the fact that within a vacuum of religious beliefs, religious belief will arise. Kressel addresses this at length: "One might point out that without scripture, the belief in god, or religious traditions, the potentially dangerous syndrome they have identified could not exist. For this reason, some atheists and skeptics ask a very controversial question about holy books, faith traditions, and religious law. Why not jettison the whole necrotic mess and start from scratch? They argue that too much killing, hatred, and distress has resulted from books purporting to speak the word of God. Perhaps a better way of life might be grounded in science, humanism, or something else. Religion, they argue, serves no useful purpose, or least none worth the price that has been paid. Repair and reform are too difficult, maybe impossible. There is, they suggest, no reliable way to separate peaceful fundamentalists from potentially deadly extremists, and no method to ensure that more liberal believers will not fall prey to dangerous impulses sooner or later. This critique is impractical. Religion speaks poignantly to the needs and desires of billions of people. A more attainable goal, in my opinion, is to focus our efforts on identifying and understanding the appeal of extremist religious visions and the circumstances under which they are most likely to thrive." [26]

One proposed concept that attempts to eliminate religious extremism, developed along the lines of "jettisoning the entire mess," is that in a purely atheistic environment the thought processes that lead to religious extremism won't develop. A launching atheistic society does not ensure future religious development. Within a vacuum of religious thought, humans tend to religious belief. Religious belief then can lead to religious extremism. "The children of atheists are more likely to abandon their parents' beliefs than are the children of any (other) religion." [27] Berman and Iannaccone state: "The resilience of religiosity is nowhere more evident than in Russia, where despite sixty years of state-sponsored atheism, bloody church purges, widespread religious persecution, and pervasive secular indoctrination, only eight percent of the people remained atheistic by 1990 and religious belief was especially widespread among the young ." [28] There seems to be a cyclical effect within modern society shifting from belief to atheism, often along generational lines. We cannot expect that this will not continue to occur on the starship. An atheistic society at launch could still spawn religious thought that could lead to religious extremism years later.

8. Potential Dangerous Beliefs to Screen For

Neil Kressel has created a list (although not exhaustive) of potential beliefs that could reveal extremist leanings: [29]

- The faith, or the rule of the faithful, should be spread by violent means.
- Nonbelievers must obey the rules of the faith, even though there is no convincing nonreligious justification for such obedience.
- People cannot freely convert out of the faith.
- Certain geographical locations must be controlled by believers, even though there is no convincing nonreligious basis for this control.
- God requires earthly punishment of, or discrimination against, those who make certain sexual or lifestyle choices, even though there is no convincing nonreligious foundation for such a judgment.
- Certain individuals have been anointed to carry out God's punishments on earth.
- All means are acceptable when one is doing God's work.
- Blasphemers must be punished in the here and now.

He also provides a following of beliefs that while may not directly lead to extremism could be the beginnings of extremist thought:

- All or some nonbelievers are hated by God or destined to be punished by God after they die.
- God prefers some races, ethnic groups, or nationalities to others.
- The government and laws ought to favor believers, though there should be some protection for nonbelievers too.
- God prefers to see women in a subjugated role.
- God disapproves of certain sexual and lifestyle choices, even though there is no convincing nonreligious basis for this disapproval.
- No one may blaspheme against God or the faithful, lest they be punished by God in the afterlife.
- Certain leaders have access to God's wishes via a channel that is not subject to accuracy checks by others, and these leaders are endowed by God with the power to make life decisions for other believers.

9. The Creation of an Extremist

As we have stated earlier, it is not belief alone which creates extremism. Screening for beliefs is a great method for detecting potential extremism but these beliefs alone don't necessarily give rise to the extremist. We must look at the extremist as an individual. What motivates an individual to become an extremist? Kressel states:

> "[I]f we are serious about addressing these underlying sources, we must think also about the political, social, and psychological underpinnings of religious militancy in its various manifestations. The psychology is complex. Most likely, individuals drawn to carry out extremist behaviors in the name of religion experience threats to some central components of their identities and, hence, to their self-esteem. They are often preoccupied with questions of pride and may feel humiliated in some aspect of their personal lives. Or they may feel that their national, religious, or local group has somehow been humiliated. Along similar lines, they may be deeply frustrated people, consumed by thoughts that their lives have been spoiled or wasted. They may be racked by guilt for real or imagined wrongdoings. They may be especially likely to come from families lacking in normal love and support. And they are most at risk of turning to extremism in late adolescence and early adulthood. Such extremism-prone individuals may be found in any culture and any religion. However, they are much more likely to come from times and places where events are unpredictable, unstable, confusing, and potentially dangerous. Modernization and globalization have unleashed destabilizing forces in many parts of the world, and the consequences have been most intense for latecomers to modernity. Failed societies are most at risk, where political and social systems deny basic gratifications to large segments of the population." [30]

Charles Liebman, in "Extremism as a Religious Norm" summarizes this in three elements that lead to religious extremism for both a society and an individual: [31]

- Expansion of Religious Law
- Social Isolation
- Cultural Rejection

Within the starship environment, the issue of religious extremism now becomes a social strategy one. And this is one that we can engage. Social isolation and cultural rejection can happen just as easily in space as it does on earth. So, part of the elimination of religious extremism requires efforts to elimination isolation and rejection.

10. The Economic Model of Religion as a Path to Understanding the Development of Extremism

Berman and Iannaccone present a solid model that predicts the conditions that give rise to religious extremism. This model can be used in both pre-launch and post-launch infrastructure and strategy development to prevent the dangers of religious extremism. Specifically, post-launch, their economic model of religion can be used as a guide to eliminating extremism on the starship. "The approach thus views people as rational religious consumers. . . . The combined actions of religious consumers and religious producers form a religious market, and as in other markets the consumers' freedom to choose constrains the producers of religion." [32] Religious involvement, according to this model, follows standard consumption and production behavior of other economic models. The demand for supernatural, while from the outside seeming as unreasonable and irrational, "can be viewed as a reasonable response to inescapable scarcity, insatiable wants, and irrepressible hope." The institutions of religion become the providers. Not only do they become the providers, but they expand to become entrepreneurs in a manner "selling" a specific type of religious experience. Denominations, sects, religions all provide different variations of the same product much as different cell phone providers offer different variations of the same product. "Groups of religious entrepreneurs establish firms and franchises, offering distinctive services, terms of trade, apprenticeship programs, rites of initiation, and rules of membership. The organizational features we associate with religion in contemporary Western countries, including congregations, denominations, and clergy, are strictly analogous to those of secular industries." [33]

There is also a cost and demand element which provides for the rise of extremism. The costs of production and/or purchase are not standard (although in some institutions, there is a monetary exchange). The costs demand sacrifice. The higher costs, assumingly paid for higher-quality goods, actually "invite stigma" including destruction of material goods, specific clothing guidelines, sexual and diet guidelines. "Such costs are present to some degree in all religions, but they are especially pronounced in the groups we label 'extremist.'" [34]

Extremist groups demand higher levels of sacrifice ("cost"). Berman and Iannaccone expand on this (and this quote is repeated from before): "Toward one end of the continuum, we expect to find "extremist" groups that demand very high levels of sacrifice. Toward the other end, we expect to find "lenient" groups that demand very little of their members. If the theory is correct, then groups that demand similar levels of sacrifice will display similar social characteristics despite their differences in organization, history, and theology. Thus, the high-cost groups of any religious tradition will tend to be exclusive, strict, small (at the local level), suspicious of other groups, and critical of secular society. Members will learn to fully embrace the group's distinctive lifestyle, devoting their resources to the group's activities and goals, or else face penalties, up to and including expulsion. As a result, these high-cost groups will also maintain tighter social ties." [35] The economic model of religious behavior fits well within the two axis system of religious belief.

So, why would someone join a religious group, and in particular, why would one join a religious extremist group? There is an initial social support for those members within the religious community.

"Case studies and survey data provide strong support for all these claims. Sectarian denominations are more supportive of community members, as evidenced by the high proportion of income members of U.S. Christian sects donate to their own communities (Iannaccone, 1992). This mutual support has many concrete benefits. In America, for example, adolescents raised in sectarian denominations manifest lower levels of most "deviant" behaviors including drug use, smoking, drinking, and criminal activity (Bock, et al. 1987; Stark and Bainbridge 1998). Sect members who marry within their religion have substantially lower divorce rates than couples from different, or less sectarian, religious backgrounds (Lehrer and Chiswick 1993). Moreover, a large body of empirical research, published in numerous medical and psychological journals, links high levels of religious faith and participation to improved mental and physical health. Religiosity in general and sect membership in particular, seems to increase life satisfaction, promote healthy behaviors, and insulate people against emotional and physical stress (Ellison 1991 and; 1993)." [36]

These very same benefits extend to those within an extremist group. Often, the higher cost of group entry will provide higher level of social support within that community.

Does the economic model actually translate beyond the Islamic world? The Westboro Baptist Church in Kansas is a supporting case study of the economic model. Jason Harper, one of only four outside individuals provided access to the internal production rooms of the Westboro Baptist Church and a key individual in a former member's flight from the cult, stated: "Extremism developed in Westboro as an attempt to differentiate themselves from an environment with very low religious diversity. Because everything was the same (the Bible-belt), Westboro had to ratchet up what brought them attention." [37] The Bible-belt provided very low religious diversity. From an economic perspective, the goods offered by the various distributors were nearly identical. To differentiate themselves amongst a potential customer base, Westboro pursued extremism.

11. Starship Planning Strategies to Eliminate the Dangers of Extremism

In review of the research and discussion on religious extremism, several strategies and approaches can be extrapolated for use both pre-launch and post-launch on an interstellar starship. It could be argued initially that engaging this topic begins to infringe on limitation of religious freedom. That is specifically not a suggestion at all in this paper. Kressel's own suggestions for societal responses to extremism are also applicable here: "Finally, religious beliefs-whether mainstream or otherwise, whether agreeable or distasteful-deserve protection. Society need not respect dysfunctional and destructive beliefs, but it must tolerate them. Religious conduct, on the other hand, may sometimes require limitation. As some legal scholars argue, 'Religious entities have the capacity for great good and great evil, and society is not duty bound by any constitutional right to let them avoid duly enacted laws, especially where their actions can harm other.' When religion becomes evil, it must be treated as such." [38] The strategies suggested below are not for religion by itself. However, they are provided in order to confront dangerous behavior that is religious in nature.

11.1 What Doesn't Work

It's easy to assume that democracy as a political situation will prevent the rise of extremism. Extremism can form within democratic societies (as it often has).

As well, the next tendency is to support a certain type (or brand) of religion as an example of what is considered legitimate religion. Berman and Iannaccone state: "To support 'good' religion while repressing 'bad' religion is to invite violence." The economic model of religion predicts that state (or ship) sponsored religion actually facilitates the rise of religious extremism. When one religion is given access to opportunities or goods that others can't gain, extremism rises as an alternative way to gain those benefits (or goods).

We must be careful to assume that religious extremism is just a fringe problem that will not arise. Religious extremism, as evidenced in the last decade, is a consistent threat to societies. Due to the nature of the starship environment, the dangers of religious extremism cannot be exaggerated in the planning and strategy stages.

Once more, it is important to remember that elimination of religious expression will also eliminate religious extremism. It will not. The vacuum of religious belief is only one point of the cyclical nature of a society's expression of religion. It is reasonable to assume the cyclical nature will continue on the starship.

11.2 Strategies to Integrate and Consider

- Developing a "national identity" that umbrellas all beliefs and actively encouraging this identity. "In the United Kingdom, the government has begun placing far greater emphasis on addressing the socioeconomic obstacles, not to mention the religious and racial discrimination, that have kept a great many Muslim immigrants from feeling like equal members of British society. Nationality laws have been revamped so as to develop a more universal, more easily accessible conception of British national identity. Immigrants seeking citizenship in the United Kingdom must now demonstrate sufficient language skills, as well as become adept in British history, culture, and national customs. All of this is an attempt to construct a collective identity based not on ethnic or cultural homogeneity but on a common national narrative, one in which every member of society can share." [39]
- Resist the environments that create extremism:
  - Expansion of Religious Law
  - Social Isolation
  - Cultural Rejection

- Create a series of "leadership checks and balances." "What is needed is not the abandonment of religion, but rather a social, psychological, theological and political system of checks and balances, a way to shout 'whoa!' when faith starts galloping down the wrong path." [40]
- Improved social services. "Religious radicals are less likely to flourish and less likely to embrace violence when there is strong competition in their non-core markets: including education, health care, poverty programs, and political representation." [41] On a starship, the potential for rations and lack on a wide-scale exists. Scarcity of essential services could quickly occur if design errors are made.
- Encourage policies that:
  - Raise the direct costs of violence [42]
  - Foster religious competition
  - Improve social services
  - Encourage private enterprise

  "Religious militancy is most effectively controlled through a combination of policies that raise the direct costs of violence, foster religious competition, improve social services, and encourage private enterprise."
- Ensure severe separation between political decisions and religious influence
- Promote a vibrant, unregulated religious market with numerous sects.

Of all of the suggestions for strategy, this is the one this paper most encourages. The economic model of religion proposed by Berman and Iannaccone predicts that a vibrant religious marketplace actually reduces the potential of extremism. Capitalism benefits from more capitalists, not less. "And, yes, religious laissez-faire will stimulate the clergy's self-interested "industry and zeal," leading them to "use every art which can animate the devotion of the common people" and encouraging numerous sects to enter the religious market and vigorously compete for members. But in the end, it is this vibrant, unregulated religious market that most benefits the state, the general populace, and religion itself. . . . America was the world's first laissez faire religious economy, and as Tocqueville and many other 19th century European traveler noted, the result was exceptional levels of religious participation, an abundance of religious sects, and very little religious conflict." [43] At the 2012 100 Year Starship Symposium, during a discussion in the "Becoming an Interstellar Civilization" track, it was suggested that there should be singular religions ships (a Mormon ship, a Buddhist ship, a Muslim ship). This is a concept which science fiction has played with. However, the economic model of religion predicts that this might be the precise environment that would spawn religious extremism.

12. Conclusion

Religion will exist on the starship. If it is not carried there, it will form there as history as shown. Religion can be beneficial and provide benefits for the "religious consumer." However, the danger for extremist views to develop exists, especially within a closed environment, with a limited population, and with always-potential scarcity issues. The encouragement of a "vibrant religious market" that is not state-supported along with a reinforcement of a strong "cultural identity," and strong checks and balances can deter the flourishing of religious/ideological extremism.

References

1. Shellnutt, Kate, "Exploring the Heavens, Christian Astronauts Reflect on Their Creator," Houston Chronicle, July 8, 2011.

2. "Communion in Space," Guideposts, 1970.

3. Haney, Paul, "Further Reflections on a Golden Space Era," http://www.collectspace.com/news/news-081009a.html, 2009.

4. Berman, Eli and Iannaccone, Laurence R. "Religious Extremism: The Good, The Bad, and the Deadly", September 2005, page 8.

5. Carpenter, Alvin L., Rev. "The Non-Promise of Earth Bound Religions into Space", published in the 2012 100 Year Starship Symposium Proceedings.

6. Kressel, Neil J. *Bad Faith: The Danger of Religious Extremism* (Kindle Locations 352-356). Kindle Edition.

7. Ibid., (Kindle Locations 430-432). Kindle Edition.
8. Ibid., (Kindle Locations 436-441). Kindle Edition.
9. Ibid.,(Kindle Locations 485-486). Kindle Edition.
10. Baum, Steven K. "Book Review: Bad Faith: The Danger of Religious Extremism," page 134.
11. Kressel, Neil J. *Bad Faith: The Danger of Religious Extremism* (Kindle Locations 490-494). Kindle Edition.
12. Berman, Eli and Iannaccone, Laurence R. "Religious Extremism: The Good, The Bad, and the Deadly", September 2005, page 13.
13. Ibid.,, page 10.
14. Kressel, Neil J. *Bad Faith: The Danger of Religious Extremism* (Kindle Locations 456-457). Kindle Edition.
15. Ibid., (Kindle Locations 467-470). Kindle Edition.
16. Ibid., (Kindle Locations 475-479). Kindle Edition.
17. Aslan, Reza. *How to Win A Cosmic War*. Page 101.
18. Kressel, Neil J. *Bad Faith: The Danger of Religious Extremism* (Kindle Locations 1027-1028). Kindle Edition.
19. Krueger, Alan B., and Maleckova, Jitka. "Education, Poverty and Terrorism: Is there a Casual Connection?" Journal of Economic Perspectives, Volume 17, Number 14, Fall 2003, Page 119.
20. Berman, Eli and Iannaccone, Laurence R. "Religious Extremism: The Good, The Bad, and the Deadly", September 2005, page 2.
21. Kressel, Neil J. Bad Faith: The Danger of Religious Extremism (Kindle Locations 1150-1156). Kindle Edition.
22. Berman, Eli and Iannaccone, Laurence R. "Religious Extremism: The Good, The Bad, and the Deadly", September 2005, page 18.
23. Kressel, Neil J. *Bad Faith: The Danger of Religious Extremism* (Kindle Locations 2121-2127). Kindle Edition.
24. Ibid., (Kindle Locations 1539-1544). Kindle Edition.
25. Ibid., (Kindle Locations 2639-2642). Kindle Edition.
26. Ibid., (Kindle Locations 2168-2174). Kindle Edition.
27. Berman, Eli and Iannaccone, Laurence R. "Religious Extremism: The Good, The Bad, and the Deadly", September 2005, page 7.
28. Ibid., page 7.
29. Kressel, Neil J. *Bad Faith: The Danger of Religious Extremism* (Kindle Locations 2663-2673). Kindle Edition.
30. Ibid., (Kindle Locations 2721-2730). Kindle Edition.
31. Liebman, Charles S. "Extremism as a Religious Norm." The Journal for the Scientific Study of Religion, page 76-78, 2001
32. Berman, Eli and Iannaccone, Laurence R. "Religious Extremism: The Good, The Bad, and the Deadly", September 2005, page 4.
33. Ibid., page 8.

34. Ibid., page 12.

35. Ibid., page 13.

36. Ibid., page 15.

37. Harper, Jason. Private interview conducted on September 16, 2013, Sacramento, California.

38. Kressel, Neil J. *Bad Faith: The Danger of Religious Extremism* (Kindle Locations 2898-2901). Kindle Edition.

39. Aslan, Reza. *Beyond Fundamentalism: Confronting Religious Extremism in the Age of Globalization* (Kindle Locations 2549-2554). Random House Publishing Group. Kindle Edition.

40. Kressel, Neil J. *Bad Faith: The Danger of Religious Extremism* . page 29.

41. Ibid., page 19

42. Ibid., page 19.

43. Ibid., page 20.

Every Body is an Ark

How the Microorganisms We Carry will Impact an Interstellar Mission

Caitlin A. Contag

Abstract

The human microbiome is essential to our health and deserves special consideration in the closed environments of space travel. The dynamics of host-microbe interactions change when normal immune functions are altered, and space travel is associated with significant immune abnormalities. Opportunistic pathogens common in the human microbiome, including those in the genera Candida, Aspergillus, and Staphylococcus, could spread among crew-members and put them at risk of serious disease. Conversely, isolating a subset of the human population during space travel could exclude organisms that confer a significant health advantage. Not only will the microbiome be important to the maintenance of crew health, it will be critical if a new human population is established from the small group of founders. This imposed bottleneck on human evolution and the associated microbial communities offers incredible opportunities to eliminate common pathogens and promote symbiotic and commensal organisms. The characterization of the human microbiome is a relatively recent area of investigation, and the full functionality of the vast microbiota living in and on the human body is not yet well understood. Data suggest microbial involvement in diverse areas of human health and further studies will likely yield novel strategies to maximize the benefits of individuals' microbial communities in a particular environment. Developing a more complete understanding of the benefits and harms conferred by the fungi, viruses, archaea, and bacteria that exist in and on our bodies will guide our determination of what non-human organisms can and should be included in a long-term space mission.

Keywords

microbiome, microbiota, pathogens, immune function, space travel, space epidemiology

1. Introduction

The spread of disease will occur very differently in an entirely enclosed spacecraft on a long-term mission, and effective prevention of disease requires thoughtful and evidence-based strategies. No traveler will board a spacecraft alone; each human carries with them innumerable microorganisms that can both aid and injure the host. The human microbiome will be of great significance and deserves special consideration during this type of mission. Nuanced and specific preventive health strategies need to be developed to respond to anticipated health challenges that will accompany interstellar travel. A more complete understanding of the benefit and risk posed by the fungi, viruses, archaea, and bacteria that colonize the human body and can transferred among personnel and determining what non-human organisms are desirable on the mission is essential for success and cannot be overlooked.

2. Human and Microbial Physiology in Space

2.1. Mammalian immune function in space

Immune function during space travel has been studied using animal models and human subjects in both simulated and real microgravity experiments. It is clear that space travel, whether through the associated stress, radiation exposure, or something peculiar to space itself, alters the function and distribution of immune cells in mammals [1][2][3]. Space travel is associated with decreased circulating lymphocyte numbers and lymphoblastogenic capability, decreased monocyte control, and decreased eosinophil percent [4]. Key immune organs, including the spleen, liver, and thymus, lose mass during space travel, and the expression of genes that control oxidative stress is decreased [5]. The space-associated alterations in immune cell proliferation and distribution, the atrophy of immune organs, and changes in gene expression results in immunosuppression. These reductions in immune capability could result in increased risk of infection for space travelers, even from what might be considered on earth as nonpathogenic species.

2.2. Microgravity alters microbial physiology

In addition to risk of infection due to immunosuppression, certain physiological changes occur in microorganisms during exposure to microgravity conditions that increase pathogenicity and virulence. Pathogenic E. coli has been found to increase production of enterotoxins [6], and form more copious and stress-resistant biofilms [7]. Incubation of Salmonella enterica serovar typhimurium in a high-aspect-ratio vessel, a bioreactor that simulates microgravity conditions, significantly lowers the lethal dose for 50 percent of mice (LD50) when compared to control organisms incubated under normal gravitational conditions in an oral challenge [8]. The mechanisms of the changes in microorganisms' pathogenicity and virulence in microgravity are still unknown. Additionally, deletion mutations occur two to three-fold faster in a model organism during a forty-day space flight than in an earthbound control [9]. Increased mutation rates could result in rapid emergence of novel pathogenic capabilities in space. The full extent and mechanisms of physiological changes that occur in microorganisms during space travel remain to be investigated, but the current data suggest that space travel will present unique infectious challenges to maintaining a healthy complement of microorganisms and controlling infections.

2.3. Space epidemiology

Some epidemiological studies have evaluated the transfer of moderate-risk pathogens commonly present in the human microbiome during space travel, but use older methods of culture-based identification and DNA amplification rather than 16S RNA-based assays, which are the current standard. In 1995 and 1996, data was published regarding the transfer of *Staphylococcus aureus* and *Candida albicans* between crewmembers during Space Shuttle flights [10][11]. Microbiome-associated infections were not reported have a significant impact on Space Shuttle missions of the 1990s, and no significant changes in crew microbiota were detected, though it was noted that surface contamination with human-associated microorganisms increased as mission length increased [12]. These results differ from those of the Apollo missions. A NASA report published in 1975 shows significant transfer of pathogenic microflora between crewmembers on several Apollo missions (notably Apollo 7, 12, 13, and 17), including *S. aureus, Pseudomonas aeruginosa, C. albicans, beta-hemolytic streptococci*, and *Trichophyton* [13]. The authors conclude the chapter by expressing the need for improved preventive care and diagnostics to protect against infection. These case studies highlight the problems faced by space travelers, including autoinoculation that can cause serious illness, and transmission of harmful species among crewmembers. Epidemiological studies with larger sample sizes and increased duration are necessary to provide data more directly relevant to an interstellar mission. While advances have been made to preventing infection during space travel, it is clear that pathogenic species in the human microbiome and the potential for emerging pathogens may pose a threat to personnel health during an interstellar mission.

3. Harms and Benefits of Microbiota

3.1. Risk posed by organisms in the microbiome

Interstellar travelers are arks for billions of microorganisms: fungi, viruses, archaea, and bacteria that differentially colonize most of the human body. These stowaways represent countless species and subspecies that are capable of aiding in digesting food and vitamin synthesis, and protecting their hosts from pathogens. However, some of these organisms can cause life-threatening opportunistic infections with long-term sequelae. Candida and Aspergillus fungi are common members of the oral microbiome in healthy humans, but some species can cause serious infections when present at other anatomic sites, and infection risk increases with immunosuppression [14]. Systemic Candida infection is especially threatening to low birth weight infants, and can cause significant neurodevelopmental defects if there is nervous system involvement [15]. *S. aureus*, a causative agent of skin infections on the Apollo mission and infamous nosocomial pathogen, is commonly isolated from the anterior nares of healthy humans [16]. The hospital environment has many similarities to that of a spacecraft – immunosuppressed individuals interacting in close quarters. It is therefore not surprising that infections caused by members of the human microbiota and associated with inpatient care were problematic during the Apollo missions.

More alarmingly, diseases previously considered "non-infectious" have been linked to the transfer of microbial strains between individuals. Housing healthy and cancer-prone mice together can induce susceptibility to colorectal cancer; the transfer of gut microflora from cancer-prone individuals to healthy individuals appears to cause the increased susceptibility in healthy mice [18].

The above examples are a small sample of the diseases associated with organisms present in the human microbiome; many other organisms in the human microbiome pose similar threats to human health. An outbreak of disease caused by any of the high and moderate-risk pathogens found in the human microbiome could put the health of space travelers at significant risk. Isolation and limited medical resources increase the potential for infection and cause high morbidity and mortality during space travel.

3.2. Certain microbiota promote human health

Despite the wide range of opportunistic pathogens in the human microbiome, humans also are home to a variety of commensal and symbiotic organisms that promote human health. The gut, having a high density of microorganisms, is a primary target in microbiome studies. Recent data suggests a strong link between a variety of metabolic and digestive diseases and individuals' microbial profile. Researchers have correlated the presence of certain gut microbes with increased and decreased risk of metabolic syndrome [18], obesity [19][20], and diabetes [21][22]. The constellation of inflammatory bowel disorders (IBD) is closely related to the diversity of the gut microbiome [23][24][25]. The symptoms of IBD can be ameliorated by introducing beneficial organisms: for example, inoculating mice with a certain strain of *Lactobacillus acidophilus* increases tolerance to pain caused by intestinal stretch by changing the density of opiate and cannabinoid receptors [26].

The skin is another densely populated organ where a thriving microbial population can confer significant health benefits. Germ-free (GF) mice have a significantly decreased immune response to cutaneous pathogens. The immune response of GF mice to cutaneous pathogens can be induced by inoculating the skin with *Staphylococcus epidermidis*, which activates interleukin-1 and MyD88 signaling in T cells [27]. The close connection between skin immune function and skin microflora is further illustrated by the role of skin microbial diversity in Atopic Dermatitis (AD). AD, also know as eczema, often results in secondary infections due to skin cracking associated with the disease. Lowered skin microbial diversity is correlated with increased severity and frequency of AD flares, and disease course is closely tracked by fluctuations in skin microflora diversity [28]. Clearly, maintenance of the skin microbiome is vital to ensuring protection against skin pathogens, autoimmune inflammation, and cutaneous infection.

These examples only briefly summarize of the full range of bodily functions associated with the microbiome. The omnipresence of microorganisms in the human body is an indication of their importance. Better characterization of the microbiome is necessary to create meaningful recommendations for long-term space travel. However, the current data suggest several opportunities in microbial preparation for an interstellar mission.

4. Opportunities

4.1. Exclusion of moderate-risk pathogens

Preventing infection during interstellar travel would require careful selection of participants who do not carry potentially harmful microorganisms. Colonization with particular organisms, such as *S. enterica serovar typhimurium* or *Mycobacterium tuberculosis,* would necessitate exclusion from close-contact travel due to the inherent threat of infection. Beyond obvious high-risk pathogens, isolating a very small subsection of the human population in long-term travel presents an interesting opportunity to eliminate moderate and low-risk pathogens from the community. For example, by excluding carriers of herpes simplex virus 1 and 2, Trichophyton, and Streptococcus mutans, the satellite population will be free from herpetic lesions, athlete's foot, and dental caries. The exclusion of these pathogens could result in significantly reduced health costs and disease burden of crewmembers.

4.2. Development of probiotic therapies for use on earth

Microbiome research in preparation for an interstellar mission provides significant opportunities to develop effective probiotic therapies. Despite the prevalence of probiotic dietary supplements on the market ("Proprietary strain promotes immunity!" and "Over 100 million good bacteria!" are ubiquitous claims in the yogurt aisle), there is limited support for the efficacy of these commercial products and concern for the honesty of labeling [29]. While there is some evidence to support the use of probiotic or microbial replacement therapy, there are not yet products approved in the United States by the Food and Drug Administration for the treatment of any disease [30]. Developing effective therapies to prevent overgrowth of harmful microorganisms or treating emerging conditions in space would benefit astronauts as well as earth-bound patients.

A challenge posed by the development of probiotics for space travelers on earth is that an effective microbial profile on earth might not be effective in a highly controlled and human-designed spacecraft environment. Any probiotic designed for interstellar travelers must be tested in the spacecraft environment and with the spacecraft diet. By developing an ideal microbial community alongside an ideal spacecraft, benefits can be maximized.

4.3. Microbial library

Additionally, the maintenance of an extensive microbial library will aid interstellar travelers by providing an extended ability to re-seed microbiomes if warranted. Spacecraft are designed to be relatively sterile environments, inoculated only by their human passengers, and crewmembers might suffer without the constant exposure to novel organisms through the environment. Monoculture and overgrowth of undesirable organisms is a significant concern on a starship mission, and a microbial library would allow for the reintroduction of desirable organisms.

However, an attempt to shape the microbiome of future generations might exclude helpful organisms and result in unintentional consequences. An indispensable component of preventive health care in an interstellar mission will be determining the role of particular organisms in human health and determining the microbiome that will confer the most benefits in the interstellar traveling environment. Through studies of the human microbiome on earth, truly beneficial and potentially harmful organisms can be identified.

5. Challenges in Current Research

5.1. Limitations of assays

One of the major limitations of the analysis of the Apollo missions in 1975 was the lack of high-throughput genetic assays and the reliance on culture-based methods. The sensitivity of culture-based identification methods is much less than the sensitivity of current RNA and DNA analyses. Advances in nucleic-acid based assays have expanded the opportunities for characterizing microbial communities, but there are still limitations in these techniques. Different procedures for extracting genetic material from samples result in different output, and sampling limitations prevent obtaining a complete picture of the dynamics of these populations. Results from DNA analysis of saliva samples using enzymatic and mechanical lysis procedures yielded results that differed depending on the extraction method that was used [31]. Each genera differs in their cell wall composition and the ability to maintain the integrity of their nucleic acid after removal from the body, and presents challenges for comprehensive analyses.

Moreover, shotgun sequencing of 16S RNA does not address the genetic changes that can occur through mutation or horizontal gene transfer, nor does it identify novel organisms that do not match with a sample from the existing databases. Plasmids and phage can significantly change the function of an organism, but are not necessarily reflected in the 16S RNA frequently amplified from subject samples. Moreover, novel organisms are frequently isolated from highly colonized regions of the human body, and without the routine use of culture-based methods, it will be difficult to identify and characterize these organisms. In order to make a recommendation for the composition of interstellar travelers' microbiomes, more sensitive and specific assays must be developed.

5.2. Geographic, cultural, and socioeconomic microbiome diversity

In selecting an "ideal" set of organisms to be included for a space mission, a comprehensive understanding of the true diversity of the human microbiome must be established. Culture and diet determine the abundance and type of organisms present in the human body [32]. Current data is limited in the diversity of subjects; there is much information available about the microbiomes of populations surrounding academic medical centers, but information about other populations' microbiome is just beginning to emerge. Important questions remain unanswered about the true extent of human microbial diversity.

5.3. Bioethics

Finally, the ethics of the investigation of the human microbiome must be addressed. Like genetic information, there is a certain level of privacy that must be afforded the collection of organisms found on and in an individual's body due to the propensity for prejudicial behavior, perhaps outweighing that of genetic prejudice due to the human aversion to infectious diseases. While there are certain organisms for which an indisputable argument can be made for exclusion from the interstellar travel population, there are many more organisms that occupy a gray area of pathogenicity. Should the isolation of S. aureus from your anterior nares prevent you from joining the brave new world?

In order to fully sculpt the microbiome of interstellar travelers, they may need to be born and raised in a controlled environment that only includes desirable organisms, not unlike gnotobiotic laboratory animals. This type of isolation is not an option, as it would prevent the members of an interstellar mission from ever living in normal human society. The narrowness of their exposure to humanity would limit their ability to be ambassadors of humankind to the next nearest star. Preferably, rigorous therapeutic and re-colonizing regiments would be investigated to eliminate undesirable organisms from participants, and the potential tremendous benefits for people on earth currently suffering as a result of undesireable microbiota. Those individuals suffering from Clostridium difficile infections, for example, would benefit greatly from improved methods of controlling this infection by modulating their microbiome.

6. Conclusion

The human microbiome can be both an asset and a risk to interstellar travelers, and there is not enough known about the organisms carried within and on the human body to make concrete recommendations about the most desirable microbiome of starship crewmembers. Research on earth will dramatically impact the manner in which we address common diseases, especially those of the gut and the skin, and will aid in the development of novel tools to promote health. Over forty years ago, Apollo mission doctors and researchers recognized the need for extensive microbial monitoring of astronauts, and advances in genetic analyses will allow for broader and deeper understanding of the microorganisms we will bring on the journey.

References

1. Sonnenfeld G, Mandel AD, Konstantinova IV, Taylor GR, Berry WD, et al. Effects of spaceflight on levels and activity of immune cells. Aviat. Space Environ. Med. 61(7):648-53. (1990)

2. Sonnenfeld G, Mandel AD, Konstantinova IV, Berry WD, Taylor GR, et al. Spaceflight alters immune cell function and distribution. J. Appl. Physiol. 73(2 Suppl):191S-195S. (1992)

3. Sonnenfeld G. Space flight, microgravity, stress, and immune responses. Advances in Space Research. 23(12):1945-1953. (1999)

4. Taylor GR. Immunologic analyses of U.S. Space Shuttle crewmembers. Aviat. Space Environ. Med. 57(3):213-217. (1986)

5. Bagai FP, Gridley DS, Slater JM, Luo-Owen X, Stodieck LS, et al. Effects of spaceflight on innate immune function and antioxidant gene expression. J Appl. Physiol. 106(6):1935-1942. (2009)

6. Chopra V, Fadl AA, Sha J, Chopra S, Galindo CL, et al. Alteration in the virulence potential of enteric pathogens and bacterial-host cell interactions under simulated microgravity conditions. J Toxicol. Environ. Health A. 69(14):1345-70. (2006)

7. Lynch SV, Mukundakrishnan K, Benoit MR, Ayyaswamy PS, Matin A. Escherichia coli biofilms formed under low-shear modeled microgravity in a ground-based system. Appl Environ Microbiol. 72(12):7701-7710. (2006)

8. Nickerson CA, Ott CM, Mister SJ, Morrow BJ, Burns-Keliher L, et al. Microgravity as a novel environmental signal affecting *Salmonella enterica* serovar Typhimurium virulence. Infect Immun. 68(6):3147-3152. (2000)

9. Fukuda T, Fukuda K, Takahashi A, Ohnishi T, Nakano T, et al. Analysis of deletion mutations of the rpsL gene in the yeast *Saccharomyces cerevisiae* detected after long-term flight on the Russian space station Mir. Mutation Research 470(2):125-132. (2000)

10. Pierson DL, Chidambaram M, Heath JD, Mallary L, Mishra SK, et al. Epidemiology of *Staphylococcus aureus* during space flight. FEMS Immunol. Med. Microbiol. 16(3-4):273-281. (1996)

11. Pierson DL, Mehta SK, Magee BB, Mishra SK. Person-to-person transfer of C. albicans in the spacecraft environment. J. Med. Vet. Mycol. 33(3):145-150. (1995)

12. Pierson DL. Microbial contamination of spacecraft. Gravity and Space Biology Bulletin. 14(2). (2001)

13. Ferguson JK, Taylor GR, Mieszkuc BJ. "Chapter 2: Microbial Investigations." Biomedical Results of Apollo. Ed. Johnston RS, Dietlein LF, Berry CA. NASA. (1975)

14. Ghannoum MA, Jurevic RJ, Mukherjee PK, Cui F, Sikaroodi M, Naqvi A, Gillevet PM. Characterization of the Oral Fungal Microbiome in Healthy Individuals. PLoS Pathogens. 6(1): e1000713. (2010)

15. Friedman S, Richardson SE, Jacobs SE, O'Brien K. Systemic Candida infection in extremely low birth weight infants: short term morbidity and long term neurodevelopmental outcome. Pediatric Infectious Disease Journal. 19(6): 499-505. (2000)

16. Grice EA, Kong HH, Conlan S, Clayton DB, Davis J, Young AC, Bouffard GG, Blakesley RW, Murray PR, Green ED, Turner ML, Segre JA. Topographical and Temporal Diversity of the Human Skin Microbiome. Science. 324(5932): 1190-1192. (2009)

17. Zupancic ML, Cantarel BL, Liu Z, Drabek EF, Ryan KA, et al. Analysis of the Gut Microbiota in the Old Order Amish and Its Relation to the Metabolic Syndrome. PLoS ONE 7(8): e43052. (2012)

18. Hu B, Elinav E, Huber S, Strowig T, Hao L, Hafemann A, Jin C, Wunderlich C, Wunderlich T, Eisenbarth SC, Flavell RA. Microbiota-induced activation of epithelial IL-6 signaling links inflammasome-driven inflammation with transmissible cancer. Proc Natl Acad Sci USA. 110(24):9862-7. (2013)

19. Ley RE. Obesity and the human microbiome. Curr. Opin. Gastreenterol. 26(1):5-11. (2010)

20. Ley RE, Backhed F, Turnbaugh P, Lozupone CA, Knight RD, et al. Obesity alters gut microbial ecology. Proc. Natl. Acad. Sci. USA. 102:11070-11075. (2005)

21. Caricilli AM, Picardi PK, de Abreu LL, Ueno M, Prada PO, et al. Gut Microbiota Is a Key Modulator of Insulin Resistance in TLR 2 Knockout Mice. PLoS Biol 9(12): e1001212. (2011)

22. Carvalho BM, Guadagnini D, Tsukumo DML, Schenka AA, et al. Modulation of gut microbiota by antibiotics improves insulin signaling in high-fat fed mice. Diabetologia. 55:2823-2834. (2012)

23. Manichanh C, Rigottier-Gois L, Bonnaud E, Gloux K, Pelletier E, et al. Reduced Diversity of faecal microbiota in Crohn's disease revealed by a metagenomic approach. Gut. 55:205-2011. (2006)

24. Ott SJ, Schreiber S. Reduced microbial diversity in inflammatory bowel diseases. Gut. 55:1207. (2006)

25. Qin J, Li R, Raes J, Arumugam M, Burgdorf KS, et al. A human gut microbial gene catalogue established by metagenomic sequencing. Nature. 464: 59-65. 2010.

26. Rousseaux C, Thuru X, Gelot A, Barnich N, Neut C, et al. Lactobacillus acidophilus modulates intestinal pain and induces opiod and cannabinoid receptors. Nature Medicine. 13:35-37. 2006.

27. Naik S, Bouladoux N, Wilhelm C, Molloy MJ, Salcedo R, et al. Compartmentalized control of skin immunity by resident commensals. Science. 337(6098):1115-1119. 2012.

28. Kong HH, Oh J, Deming C, Conlan S, Grice EA, et al. Temporal Shifts in the skin microbiome associated with disease flares and treatment in children with atopic dermatitis. Genome Res. 22: 850-859. 2012.

29. Slashinski MJ, McCurdy SA, Acenbaum LS, Whitney SN, McGuire AL."Snake oil," "quack medicine," and "industrially cultured organisms:" biovalue and the commercialization of human microbiome research. BMC Med Ethics. 13(28). 2012.

30. NIH National Center for Complementary and Alternative Medicine. Oral Probiotics: An Introduction. Accessed 17 Sept 2013. Updated Dec 2012. http://nccam.nih.gov/health/probiotics/introduction.htm

31. Lazarevic L, Gaia N, Girard M, Francois P, Schrenzel K. Comparison of DNA Extraction Methods in Analysis of Salivary Bacterial Communities. PLoS One. 8(7): e67699. 2013.

32. Nasidze I, Li J, Quinque D, Tang K, Stoneking M. Global diversity in the human salivary microbiome. Genome Research. 19(4): 636-643. 2009.

Constraints Facing Future Deep Space Missions

Erika Ilves & Tyler Emerson

Abstract

Voyager 1 has barely left the solar system. Twelve people have walked on the Moon. Twenty-four got to lunar orbit, 530 to Earth orbit. Our Spirit, Opportunity, and Curiosity may be roving on Mars but we are far from being a solar civilization. Worse, seven billion of us on Earth are struggling to coalesce into a single planetary civilization. So is thinking seriously about how to become an interstellar civilization preposterously premature?

We make a case that jumpstarting an interstellar civilization could not be timelier: it is one of our best insurance policies against existential risk and a fast track out of the current sub-planetary civilization quagmire. We then proceed to look at what it would take to initiate an interstellar jump sequence in practice. We offer a qualitative assessment of cultural, technological, and governance transitions we would need to orchestrate, and explore how the interstellar community itself would need to evolve to accelerate these transitions.

Our proposals include launching an Interstellar Art Academy to inject a steady stream of interstellar edutainment into global culture; participating in the world's global public policy and entrepreneurial fora to put interstellar and space development goals on the global agenda; facilitating a roadmapping process among major terrestrial industries to move space commercialization into space industrialization; designing an online game to simulate governance models to underpin interstellar expansion and, finally, turning the interstellar community of thousands into an epic volunteer organization of one million.

Keywords

civilization, interstellar, cultural engineering, prototyping

1. Introduction

Strike up a conversation about starfaring civilization, and a few minutes later you'll inevitably find yourself discussing interstellar propulsion and starship designs. These are the topics that gave rise to the interstellar community and have boggled the minds of a handful of visionary science fiction writers, scientists and engineers since the 1930s. Propulsion that can take humans across light years of space within reasonable time frames is obviously a prerequisite, but it takes more than starflight capability to create an interstellar civilization. Back in 1969 we developed lunar landing capability. Apollo 11 was humanity's first manned mission to land on the Moon. Over the course of the following three years we sent five more missions, putting the total of twelve men on the moon. And then we stopped. Half a century later, the only relationship we have with the Moon's surface is through controlled crashed landings of our satellites. We are not an Earth-Moon civilization, but still very much a terrestrial civilization scrambling to rebuild our lunar landing capability. Six people floating in a can that orbits Earth just 400 km above the ground is as good as it gets.

Like many people in the interstellar community, the authors of this paper believe that we must extend our civilization beyond our solar system. In the end, the choice is simple: settle other star systems or perish with our sun. The stakes are high, the mission is fantastical, and arguably more urgent than most people realize. It is therefore important that from the outset we think about our interstellar expansion in broad civilizational terms, rather than narrow technological terms. Starflight capability is necessary but not sufficient.

The main objective of this paper is to explore the magnitude of the transition from sub-planetary to interstellar civilization and offer initial thoughts on how we might go about jumpstarting this process. First, we'll dive into the very idea of "civilization" and offer a working definition of the term. Grounded in a parametric definition of civilization, we'll look at the desired future trajectory of human civilization and argue that jumpstarting an interstellar civilization now is one of the best strategies for the long-term survival as well as the short-term progress of our species. The bulk of this paper will then focus on looking at the specifics of the how: the qualitative nature of the transitions we will need to initiate and how we might go about making them happen in practice.

2. Civilization

According to an urban legend, a journalist once asked Ghandi during his visit to London: "What do you think of Western civilization?" Ghandi supposedly gave an acerbic response: "I think it would be a good idea." [1] Eighty years later, the notion of civilization has only grown more contentious.

What do we mean by civilization? In a popular online dictionary, civilization is defined as "an advanced state of human society, in which a high level of culture, science, industry and government have been reached." [2] The Oxford English Dictionary echoes the general idea: "civilization is the stage of human social development which is considered most advanced." [3]

For our purposes, these definitions are of course useless. They carry no orienting insight into the nature of civilization other than the assertion that it is "advanced" or "high" relative to other forms of human communities. To make matters worse, these assertions tend to raffle the feathers of many postmodern thinkers who rush in to point out the many ways in which "civilization" cannot possibly be deemed to be an improvement on what came before—large- scale wars, coerced taxation, slavery, epidemics, and subordination of women are among the most frequently-cited charges against civilization. [4]

Interestingly—and perhaps unsurprisingly—not even civilizationists who dedicate their entire academic careers to comparative study of civilizations agree on the definition of the term. [5] And while academics can afford to disagree, interstellar civilization-builders cannot. We must have an idea of what it is exactly we are purporting to construct. The civilization-engineering profession has a much lower level of tolerance for ambiguity.

Since there is no ready-to-use definition of civilization we can pick up off the shelf, we have attempted to piece one together. To create a parametric foundation for our civilization-building efforts, we'll take a quick look at Big History and Big Future. Big History gives us the context in which human civilization emerged and how it evolved. Big Future gives us speculations about plausible future civilizational trajectories, both extraterrestrial and our own. By soaking up insights from the past and speculations about the future, we should be able to get at something fundamental, something we can build.

2.1 Big History

The word civilization itself is derived from the Latin civilis (English civil), related to the Latin civis (English citizen), and civitas (English city or city-state). [6] Early civilizations were social structures that emerged to support the city way of life. [7] First, humans became creative with food production technology (they domesticated animals, made plows, irrigated, and alternated crops, just to name a few agricultural innovations), and started generating a food surplus. That allowed some people to specialize in something other than food production. They became builders and makers of useful things. Several historians have argued that food surpluses had to be protected, leading to specializations like army, priesthood, and centralized governance. [8] Writing and monetary systems had to be developed to keep track of and ease the exchange of value between people who did not know each other personally. With these developments, cities became our first stabs at civilization—social cooperation between strangers leading to concentration of power and extension of human control over nature. It took technology and social organization.

These types of civilizational experiments emerged independently in several places around the planet about the same time, give or take a few centuries. [9] The first known experiments emerged more than 52 centuries ago (3,200 BCE) in Mesopotamia, followed by experiments in Egypt and Nubia about a century later (3,100 BCE), and in the Indus Valley and China 40 centuries ago (2,000 BCE). Depending on how you slice human history, we've gone through 8 to 30 local civilizational experiments.

The word "civilization" only entered our vocabulary during the 18th century Enlightenment. It was first used in a French book L'Ami des hommes (1756) written by Victor de Riqueti, marquis de Mirabeau, and appeared again in Adam Ferguson's 1767 Essay on the History of Civil Society. Both were primarily concerned with the pursuit of personal enlightenment and national progress. To quote Ferguson, "Not only the individual advances from infancy to manhood, but the species itself from rudeness to civilization." [10] Civilization was used as both a description of the process and the condition, and many Enlightenment thinkers were preoccupied with the search for the "true criterion of civilization," "the necessary ingredients of civilization" as contrasted with primitive or barbaric cultures. [11] Until the early 1800s, the word was used only in the singular because it referred to the progress of humanity as a whole. Apparently, this is still the case in French where the communitarian spirit lives on. [12] In English, however, plurality crept in and we've started talking about civilizations.

The question of serious concern to any civilization-builder is why some of our civilizations have fallen in the past. According to the theory of the German historian Oswald Spengler, all cultures have a lifespan of about a thousand years (the final stage of a culture is "civilization") and must pass through a life cycle of birth and death just like living organisms. Ultimately, all civilizations decline—including the Western civilization whose impending demise he predicted in his "The Decline of the West" first published in 1918. Although many aspects of Spenglerian theory have been refuted, the thesis appears to have left a deep and lasting imprint on the psyche of the intelligentsia in the West. One of the most influential cultural and social critics of the 20th century, Northrup Frye, goes as far as to suggest that "The decline, or aging, of the West is as much a part of our mental outlook as the electron or the dinosaur, and in that we are all Spenglerians." [13] That strain of pessimism about the future of the Western civilization and the ongoing human project in general, appears to take up quite a bit of space in our cultural baggage and is something we, as interstellar civilization-builders, have to contend with.

A few years later, an English historian Arnold J. Toynbee offered a more upbeat take on human history. Having analyzed 26 civilizations, he rejected Spengler's deterministic view. For Toynbee, a civilization might or might not continue to thrive, depending on the challenges it faced and its response to them. He concluded in his 12-volume "Study of History" that civilizations rose by responding successfully to challenges under the leadership of creative minorities. The breakdown of civilizations was not caused by loss of control over the physical environment, over the human environment, or by attacks from outside. According to Toynbee, civilizations declined because their creative minorities deteriorated and lost their creative power to respond to the next challenge they face while worshipping their "former selves". Gradually, they morphed into "dominant minorities" that attempted "to hold by force—against all right and reason—a position of inherited privilege which they had ceased to merit." [14] The subsequent loss of social unity in the society as a whole led to the demise of a civilization. [15]

Toynbee's challenge-and-response approach to explaining the history of human civilizations is empowering. If we believe that civilizations fall from suicide or murder rather than inevitable natural causes, then it is within our power to shape the lifespan of our civilization. Civilizations are no longer Spenglerian organisms with predetermined mortality but a network of social relationships whose destiny is determined by the decisions we make. The ethos has been summed up by Toynbee's editor as follows: "Man achieves civilization, not as a result of superior biological endowment or geographical environment, but as a response to a challenge in a situation of special difficulty which rouses him to make a hitherto unprecedented effort." [16] Challenging goals are thus critical for civilizational longevity. The challenge of settling another star system entails an unprecedented effort, but it could catapult us to an unprecedented level of civilization.

So far, we've had a lasting civilization only in the sense of the broadest cultural identity that distinguishes humans from other species. Today, this is hardly the unifying narrative that defines who we are. We do know, however, that historically civilizations with a high degree of internal unity (in language, religion and race) lasted much longer than those without. For example, the Byzantine Empire that included several ethnic groups, but was held together by religion (Christianity) and language (Greek), lasted nearly a thousand years despite internal dissensions and many powerful enemies which included Islam, Persia, Slavic and Germanic invaders, and several Turkish groups. [17] Collective identity is another factor that has played a significant role in civilizational longevity.

At the most fundamental level, we can think of civilization as a cultural infrastructure that has made social cooperation possible at ever-increasing scales, starting with the first cities and evolving to transcontinental empires. Unifying identity, challenging goals, evolving technology, and effective governance mechanisms have all been the building blocks of such an infrastructure.

2.2 Big Future

Our experience with civilizations is limited to our own. We have yet to encounter an exocivilization. But our ongoing search for extraterrestrial civilizations requires us to speculate about them, and by proxy, about our own future. We tend to think about extraterrestrial civilizations in terms of their energy and resource use, cosmographic expansion, and information processing.

Perhaps the most widely cited hypothesis about the possible evolution path of exocivilizations comes from a Russian astrophysicist Nikolai Kardashev. Back in 1964 he proposed to think about "supercivilizations" (that he considered inevitable) in terms of how much power they are able to extract from their environment. The original Kardashev scale consisted of three levels: Type I civilizations would be able to use the entire solar insolation of their home planet (~1016 watts for a terrestrial civilization vs. 1.5 x 1013 watts in current usage). Type II civilization would be able to tap into the entire output of its star (~1026 watts for our Sun). Type III civilization would be able to tap into the entire energy output of its galaxy (~1036 watts for our Milky Way). [18] Later, Kardashev's scale was expanded by others to include Type IV, a civilization capable of tapping the energy of a supercluster of galaxies, extragalactic energy or perhaps even the entire universe (~1045 watts), and Type V civilizations capable of toying around with a collection of universes. Many discussions on the subject have centered on informed speculations about the type of advanced technologies that would need to be developed to attain different levels of power usage. No matter how many aeons separate us from Kardashev's supercivilizations, they still must obey the laws of physics

Speculating at these scales can be helpful. Once we start contemplating the long-term prospects of an exocivilization, it becomes easier to see that we don't live in a universe uniquely designed for our survival. To stick around longer than an average mammalian species, we will need to break a serious sweat and grow our mastery of physics, biology and ourselves. We tend to just assume that the entire universe is the playground for advanced civilizations—thereby expanding the size of our own playground. It's easy to imagine alien intelligence embodying a multitude of biological and non-biological forms, thereby opening the window on our own post-biological future.

2.3 Working Definition

Drawing on Big History and Big Future perspectives, we propose the following working definition for civilization:

Civilization is a cultural infrastructure designed for continued survival and evolution of Earth-originating intelligent life in the universe.

The who is Earth-originating intelligent life, rather than the specific human form we now identify ourselves with. The where is the universe—we have to think in terms of the greatest context we are currently aware of. The why is the continuity of our survival and evolution, which in our universe is anything but assured. Civilization is something we "design" rather than something that happens to us. And the crux of the matter, the what, is cultural infrastructure.

We propose to integrate the material and ethical dimensions and think of cultural infrastructure as consisting of four building blocks: Identity, Goals, Technology and Governance. Identity, or how we see ourselves in the grand scheme of existence, defines our field of vision and our circle of concern. If we see ourselves as Earth-bound terrestrials, then that is precisely what we will remain. Goals give us direction and values. Technology gives us the theoretical means of realizing our goals. Governance—in the broadest meaning of organizing collective action—helps us turn theoretical possibilities into reality.

Together, these building blocks frame the overall scope for civilization building.

3. Trajectory

With a meaningful definition of civilization, we can turn look at the special case of interstellar civilization. How different is it from where we are today and how would we get from here to there?

3.1 Staircase

Interstellar civilization implies that we have traversed light years of space and made a new home in at least one other star system. Our physical path from here to there may be a simple straight line. Our developmental path through spacetime is less obvious. To make sense of the latter, a distinction between intensive and extensive innovation can be helpful. [19] In the case of intensive or vertical innovation, we create a brand new capability that did not previously exist, like inventing a light bulb, an integrated circuit, an airplane or the internet. In case of extensive or horizontal innovation, we find ways to improve and extend the benefits of intensive innovations into new markets and geographies, like making electricity ubiquitous or creating universal access to the internet. Essentially, it means copying things that work. Extensive innovation may take a while but it is reasonably predictable. Intensive innovation is hard and inherently unpredictable.

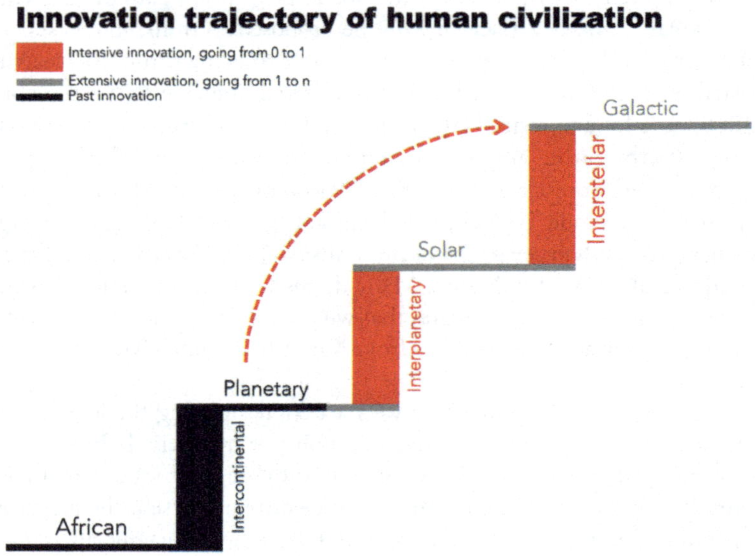

Figure 1. Innovation Trajectory of Human Civilization

If we represent our developmental path to interstellar civilization as an innovation staircase, interstellar is two intensive vertical climbs and one extensive horizontal stretch away. Two vertical climbs is a tall order. It is not entirely dissimilar to fish crawling out of the water 400 million years ago and eventually becoming human—only squared. Not only would we need to crawl out of our home world and adapt to a radically new environment (lower gravity, high radiation, different atmospheric pressures, alien or absent ecosystems), we will also have to traverse unfathomable distances of interstellar space. Given the inherent unpredictability of vertical climbs, it is now hard to imagine who we might become on the other end of this undertaking. Our ancestral fish, had it been temporarily endowed with a foresight capability, would probably not have believed that one day it would have two legs, two arms, a big head, breathe air, drink fermented grapes and work on warp drives.

As a civilization, we are still busy copying our basic transformational inventions like reading, writing, or the internet. We are still a sub-planetary civilization, with conflicting identities, largely terrestrial goals, big resource, environmental and energy challenges and dysfunctional planetary governance. Few of us are bound for the stars. A handful who would entertain the idea, imagine we will just take the stairs one step at a time: first, figure out how to coalesce into a planetary civilization, then do the vertical climb to inter-planetary and take our time to settle our own solar system. And then, hundreds of years from now we can wipe the sweat off our foreheads and start thinking interstellar. No hurry. We have more pressing concerns to attend to.

This is a precarious line of thinking.

3.2 Fast track

The ultimate reason to go interstellar is the life cycle of our sun. As our sun ages and expands, it will make all complex life on Earth impossible in less than a billion years. But this still leaves us with hundreds of millions of years to waste.

The aging of our sun is not the only existential risk we face: we must seriously consider anthropogenic technological catastrophes, planetary catastrophes like civilization-killer asteroids or supervolcanoes, supernova explosions, and possible solar-system-level catastrophes that we are not yet aware of. Mitigation of existential risk is a powerful reason to go now. It has been argued that reducing our extinction risk is strictly more important than any other global public good [20]—that is, if we decide to place some value on the existence of future generations, possibly trillions and trillions of descendants, and possibly the future of intelligent life in our galaxy. A multi-planet, multi-star strategy can dramatically reduce our risk of going extinct. It's a good insurance policy. Like with any other insurance policy, we don't know exactly when the trouble will strike or even if it will strike at all, but we still buy the policy.

As long as we remain a one-planet, one-star civilization, we are merely a parochial experiment. Now that we have evolved far enough to understand the existential predicament we find ourselves in, the only rational choice for a civilization that is not ambivalent about its own survival is to fast-forward to interstellar as fast as humanly possible.

Counter-intuitively, mastering the first vertical jump on our way to interstellar would also be a fast track out of our energy, resource and environmental challenges on Earth. When our problem solving space is limited to Earth's atmosphere, our options are unavoidably limited and politically hard. Throughout our history, we've been able to expand the meaning of the word "resources". First, we burned wood, then we discovered coal, then oil and gas. These terrestrial energy sources have finite stocks, we know they will run out. But even before we burn our way through all the stocks, we will cook our biosphere with trillions of tons of CO2. The wind and solar alternative is clean, but impractical and won't be sufficient to power the continued development of our civilization. Space-based solar and fusion power, on the other hand, would be much better contenders. It may be easier to start beaming power from our solar satellites than getting the world's nations to agree to dramatically curb their CO2 emissions. It may be more prudent to develop Helium 3-Deuterium (found on the Moon and outer planets) fusion power than invest time and energy into scrubbing CO2 out of the atmosphere to buy ourselves more time to burn even more fossil fuels. To re-green Earth without reverting to a preindustrial way of life and downsizing our population, we must develop space. [21] With thresholds passed and action windows closing on many of our environmental challenges, now would be a good time to get going.

We should be in a hurry. To continue as a civilization we must initiate an interstellar jump sequence now, rather than hundreds of millions of years from now. It is both, our best insurance policy against existential risk, and a fast track out of environmental, resource and energy challenges that plague us on Earth today.

4. Jump

What would it take to initiate this kind of interstellar jump sequence? A good place to start is to wrap our minds around the order of magnitude of the job at hand.

4.1 Magnitude

One way to do this is to focus on the tangible part of the enterprise—an interstellar mission—and reason by extrapolation from our historical space programs like Apollo. Most estimates for what it would take for human civilization to launch a starship fall in the range of US$1-125 trillion. [22] The bill for the entire Apollo program ran up to US$120 billion (or 1% of the US$12 trillion world GDP back in the 1970s), employed 400,000 people and required support from more than 20,000 industrial firms and universities. [23] If we go with a US$ 125 trillion interstellar mission, we are talking about an increase by a factor of 1,000. If we then assume that our interstellar mission will account for an Apollo-like share of the world economy and bluntly scale the Apollo numbers by that factor, we get an implied dedicated workforce of 400 million (that's the entire population of the EU) and a world economy of US$12.5 quadrillion (that's almost 200 times bigger than our current US$65 trillion economy). Depending on our assumptions about economic growth, we then swiftly arrive at a conclusion that it will take us 100-500 years to get to a place where human civilization can afford the first interstellar mission.

Although intellectually interesting, this back-of-the envelope exercise should be taken lightly. The focus is too narrow (interstellar mission rather than interstellar civilization), and the method too blunt (our mechanical extrapolations are future blind, and many variables can dramatically alter the picture, e.g., breakthroughs and exponential improvements in technology, extent of space industrialization and existing infrastructure within our own solar system).

As we have argued above, launching an interstellar mission is necessary but not sufficient. Our generation's job is to make the transition to interstellar civilization inevitable. The economic cost of this enterprise is at best a secondary consideration. A better way to wrap our minds around the jump sequence is to reason qualitatively, to understand the nature of the transitions we need to initiate and to explore the different ways we can set them in motion. We need to adopt a broad cultural prototyping mindset: we are prototyping a shift in cultural infrastructure, rather than just planning a technological mission.

4.2 The Job

Earlier in this paper, we defined civilization as a cultural infrastructure consisting of four building blocks: Identity, Goals, Technology, and Governance. We'll use these building blocks to structure our inquiry into the qualitative nature of the job ahead of us.

4.2.1 Identity

"No one shows a child the sky." Dr. Mae Jaemison, the principal of the 100 Year Starship, likes to invoke this African proverb when introducing the ambitious endeavor that she is spearheading. Indeed, most of us can probably remember the first time we looked up at the night sky in wonder and pummeled our parents with questions about those sparkling dots of light. But then we grow up, and most of us—with the exception of astrophysicists, astrobiologists and SETI researchers—would probably struggle to recall the last time we looked up at the night sky and felt that the stars had any relevance for our daily life on Earth.

How we see ourselves, our identity, matters. Global Citizens won't build an interstellar civilization. But Cosmic Citizens might. We need to build a critical mass of people who look up at the night sky and see humanity's future homes.

The shift from Global Citizenship to Cosmic Citizenship could be described in terms of a shift in our cognition and in our morality. Cognitively, our civilization still lives in a tiny sliver of spacetime. Judging by the contents of our global public discourse and our global agenda lists, our vision is limited to Earth's biosphere and our time horizon is measured in years. We need to extend our field of vision to encompass the entire cosmos, if we are ambitious, or the Milky Way, if we are conservative. Our time horizons should be limited not by election terms, financial reporting cycles, or even our own lifespans but by the extent of human knowledge. If we can predict a cosmic event that will impact our civilization thousands, millions, or billions of years from now, we need to integrate it into our plans. That's of course emphatically not the way we think today.

Morally, our civilization is an amalgamation of moral tribes with varying circles of concern. Out of the 7.2 billion of us, the vast majority care about a subset of human beings alive today (my family, my tribe, my nation, my region, my continent, my race, my religion). At the leading edge of moral development, the global citizens among us extend their circle of concern to the entire population of our planet. [24] Future generations tend to get the short end of the stick. According to our standard economic models, the value we place on their well-being beyond the next 25 years is zero. Cosmic Citizenship requires a giant leap of moral imagination: at the very least we need to include all future generations in our circle of concern. For a Cosmic Citizen, the far future of human civilization is not an abstract thought. It's a tangible concern than can shape the contents of his or her life today. A Cosmic Citizen could easily decide to spend his or her lifetime working on breakthrough propulsion or starship designs.

If Global Citizens now number in the hundreds of millions, Cosmic Citizens are exotically rare, numbering perhaps in the thousands. Cosmic citizenship is yet to register in the global public awareness as an idea and to the best of our knowledge there have been no serious attempts to define its contents or track the relevant population. Our central task is to dramatically expand this population. Cosmic Citizens are the carriers of drive toward interstellar civilization. Although exceptional individuals can have an outsized impact, as a general tendency, the bigger we grow our Cosmic Citizen community, the more powerful our collective drive can become.

So what can we do practically to grow the number of Cosmic Citizens who care about the long-term future of human civilization? We can keep constructing our moral arguments on the overwhelming importance of shaping the far future [25] or on the need for our civilization to have a billion year plan [26]—but our rational appeals will move but a handful of people. We can flood everybody's newsfeeds with stunning images of distant star systems from our telescopes and wonderful visualizations of exoplanets from our artists—but images alone might not be enough to awaken a burning desire to go and see these exoplanets for ourselves.

We humans are storytelling animals. We tell stories and we are shaped by stories. [27] The most effective strategy would be to introduce a constant stream of future-oriented, well-informed, mind-expanding stories into our global culture. We can think of it as "interstellar edutainment." Three industries in particular give us amazing access to people's minds and hearts: books, movies, and games. We have more than 130,000 cinema screens and more than a billion smart phones on this planet. Science fiction is a marketable genre. It does well at the box office, and in digital book and games sales.

The problem is, for the most part, we have the wrong content. Today, science fiction is mostly what the famed British director Alfred Hitchcock called a MacGuffin, something that moves the story along but is not really its central focus. Take Neill Blomkamp's Elysium (2013) as an example—yes, we have a stunning space-based habitat and body-reconstituting machines, but the plot is really about the present, the implications of global inequality. Although technically a scifi movie, Elysium does not do much to fuel our collective dreams to expand into space.

What do we do? Since we need stories that stretch our identity toward the stars, why don'y we start an Interstellar Art Academy? Imagine creating a graduate program for writers, game designers and film directors where they could come and learn from the masters of the genre (e.g., science fiction writer David Brin, film director David Cameron and game designer Casey Hudson), but also get grounded in a broad-based transdisciplinary understanding of the big picture. We would teach Big History, Existential Risks, Cosmography, Astrobiology, the Scientific Limits of the Possible [28], Propulsion, Starship Designs, and so on. We could get even more deliberate and create a hundred year book-movie-gaming franchise on our future history. Think of it as a prelude to Star Trek that tells the story of how Earthlings reached starflight capability in the first place and what they did with it.

Human civilization won't reach for the stars without the voice of Cosmic Citizens starting to register loud and clear in our collective awareness. To grow their number, we'll need to cultivate dozens of Gene Roddenberry's (the mastermind behind the Star Trek franchise) over the coming decades. Setting up and running an Interstellar Art Academy won't take US$1 trillion in funding and could be developed into a self-funding venture. It will take a small team to get it going and we won't have to wait for 500 years. We can start now.

4.2.2 Goals

What does our civilization want? We may disagree about underlying problem definitions and prefer different courses of action, yet there is a remarkable degree of alignment within our global community around the big four goals: Peace, Sustainability, Prosperity, and Justice. We want to minimize violent conflict and avoid major wars and genocide. We want to live within the limits of our biosphere. We want to relegate poverty to history books and usher in a new era of abundance. We want a more equitable world with a smaller gap between the rich and the poor across and within countries. These goals come in many different right-wing and left-wing flavors but by and large, this is what our civilization is trying to accomplish, with all of our aspirations neatly folded within the Earth's biosphere.

More than 30 years ago, a futurist known as FM-2030 predicted a 90-degree rotation in our political spectrum, from left-wing and right-wing to up-wing and down-wing. The directionality was meant literally: up-wing meant looking toward the heavens or black sky, and down-wing was looking toward the Earth. [29] Back in the 1980s it was perhaps an obvious prediction to make. During the preceding three decades up-wing goals occupied a prominent place on the national agendas of the Cold War rivals and attracted serious levels of funding. The first man-made object in orbit (Sputnik), the first man in space (Yuri Gagarin), the first man on the Moon (Neil Armstrong)—our trajectory was clear, we were bound for space! Yet, a quarter of a century later, up-wing goals are markedly absent from the global agenda.

Sure, we invest billions of dollars in bigger and bigger space telescopes to see what (and who) is out there. We send hundreds of flyby and lander missions to planets, moons and asteroids in our own solar system. We have six people living and administering science experiments in a space station orbiting the Earth. And 100,000 of us sign up for a privately funded volunteer mission to Mars. [30] Yet, few of these activities are seen as central to the future of human civilization. Up-wing goals are not seen as important enough to register on the global agenda.

One thing we can be certain about, we won't wake up to a sunrise in the Alpha Centauri system by accident. How can we expand the current global discourse and put the extension of human civilization beyond the Earth's atmosphere on the global agenda? The simple answer is: show up and make the case. And then, keep showing up and making the case. For a hundred years, if necessary.

Most countries with active space programs have national space-lobbying organizations arguing for more ambitious goals and more funding. That is of course helpful because in the grand scheme of existence it might not really matter whether the next flag on the Moon will be Chinese or the first flag on Mars American (or with a SpaceX logo). Yet, to up-wing our global agenda, we need to think broader. We need to start showing up and speaking up at the world's global public policy and entrepreneurial fora.

So, let's say we are bound for the World Economic Forum meeting in Davos or the Clinton Global Initiative meeting in New York. What should we contribute to the ongoing public discourse? At first sight, the goals we advocate (i.e., jumpstarting an interstellar civilization) are metaphysically as far from the aforementioned big four staples of the global agenda as the physical distance between our sun and Alpha Centauri. But first impressions are deceptive. If you think about it, propulsion aside, the challenge of settling other planets, moons and asteroids within our solar system and beyond are not fundamentally dissimilar to the challenge of figuring out how to live together on planet Earth. It's all about us trying to figure out how to make our home on a celestial body speeding through space: how to build our life support systems, how to allocate our resources, how to get along, how to create good, meaningful lives for ourselves. The Earth just has the first home disadvantage—unavoidably, that's where we make the bulk of our basic mistakes until we can pull together an Operating Manual for Spaceship Earth and Other Celestial Bodies. [31]

Furthermore, introducing an interstellar perspective in the public policy circles can serve as an intriguing thought experiment. Change of context—from the Earth-Now mental box we have locked ourselves in to Some Other Planet-Some Time in the Future—can help see our terrestrial challenges in a new light. Take Sustainability. Imagine you are headed for a planet with no "free" ecosystem services to offer you besides raw materials, either because it's a barren planet or because indigenous biology is incompatible with ours. How would you set up our life support systems? We could go with a planetary-scale terraforming (effectively, the strategy we have been pursuing on planet Earth), or we could build closed loop systems from the outset. We can think about the problem from a terrestrial context and then find a way to implement it on Earth. Making space for "nature" and moving humans onto closed loop life support systems is already a strategy advocated by the Modern Green movement. [32] Or take Peace. What happens when we take the geopolitics of peace into space? According to the Outer Space Treaty of 1967, celestial resources are a common heritage of humanity and weapons of mass destruction are not allowed into outer space. How would we go about advancing human civilization within this overarching framework? Earth is merely another celestial body. So how would the terrestrial geopolitics of peace work if we applied a similar framework on Earth?

Engaging members of the global public policy community around the interstellar thought experiment would allow us to introduce the idea of interstellar into the global public discourse. Even if this idea enters the circulation as a problem-solving prop, it still enters the circulation, paving the way for the more fundamental argument. To quote Carl Sagan, "every surviving civilization is obliged to become spacefaring…for the most practical reason imaginable: staying alive." Existential risk is a problem to which interstellar expansion of our civilization is a good if partial solution. Astro-geographical diversification offers protection from external risks like asteroid impacts, supernova explosions and ultimately the aging of our sun (while human-made existential risks tend to go wherever humans go). To get to a point where expanding human civilization beyond our solar system becomes not just a goal, but a priority, we first need to get to a point where we as a civilization see existential risk as a problem to be solved. That work has only just begun. [33]

Engaging with global public policy stakeholders alone would not be enough. Global agenda are not shaped at the meetings in Davos or New York alone. Another big part of the equation is the goals pursued by the world's entrepreneurs. So let's imagine we are bound for Skoll World Forum, TED, or the Silicon Valley, expecting to encounter thousands of shiny-eyed entrepreneurs. Are they starting the right kind of ventures? Many people bemoan the lack of big-ticket innovation in Silicon Valley and, by extension, many other entrepreneurial hotbeds around the world. [34] It is easy to cite historical precedents ("where are our 21st century Ford's or General Electric's or Intel's?"), but few people are able to offer even an illustrative list of the kind of companies they would like to see started in the next 100 years. With a bit of thinking, we could have something very interesting to share with the world's entrepreneurs. Imagine we have become an interstellar civilization. What are the companies that helped make it happen? What are the goals they pursued? We could paint a very interesting picture, perhaps put forward a list of startups we would need to see in the next 100 years. We could then go around the world and share these ideas in entrepreneurship-focused fora. If we are good at it, they can turn into "ideas worth spreading" and into a new generation of "ventures worth starting".

When it comes to the magnitude of the task, putting the interstellar expansion of our civilization on the global public policy and entrepreneurial agenda would be a marathon task that could be accomplished by a small network of passionate space writers, advocates and enthusiasts. It won't require a ridiculous amount of funding but it may require a ridiculous amount of perseverance.

4.2.3 Technology

Technologically, interstellar civilization is two extremely-hard, intensive-innovation jumps away—we need to pull off interplanetary and interstellar. It's a jump from crude energy, the orbital space station, agriculture and manufacture, global infrastructure, global internet, and human life extension to fusion energy, starships, engineered closed loop habitats, solar system infrastructure, galactic internet, and human body modification. And these are just some of the headlines.

How do we orchestrate or facilitate a technological transition of such massive proportions? This is where it helps to pace ourselves and think through the logical sequence of events. In other words, we need to draw up a high-level plan. Over the past century, several people have given serious thought to how we might sequence our steps. Russian visionary Konstantin Tsiolkovsky put forward his straightforward sixteen-step Plan for Space Exploration back in 1926. [35] In the 1980s, analysts at Rockwell International, a major American manufacturing conglomerate in the latter half of the 20th century, put forward a much more elaborate and sophisticated Integrated Space Plan where they carefully mapped the links and interdependencies among the various steps, all part of their 100-plus-year vision of settling the solar system and beyond. [36] Several nations have produced medium-term visions for their space programs but none of them come even close to the long-range systematic perspective embedded in the plans of Tsiolkovsky or Rockwell. [37]

So we have given our expansion into space serious thought and have been planning it for some time now. Yet our track record at converting plans into action over the last forty years has been modest at best. As long as our ventures into space had been fueled by the needs of the military-industrial complex, to dust off a bit of the Cold War era terminology, we were making fast-paced and serious progress. A lot has changed since then. Government space programs are now primarily about scientific discovery and spinoff innovation. Space agencies like NASA, ESA, the Japan Aerospace Exploration Agency (JAXA), and the Indian Space Research Organisation (ISRO) have all established offices dedicated to promoting the terrestrial use of space technology. The 2012 edition of NASA's annual Spinoff report that sampled fewer than 200 space technologies out of thousands estimated that spinoffs have saved at least 444,000 lives and resulted in the creation of 14,000 jobs, in addition to generating revenue and reducing costs. [38] This type of impact is what is now needed to make a case for increased funding. As government budgets around the world are falling off fiscal cliffs, some government space programs are taking a hit (USA, EU) while others (Russia, Brazil, India, China) make bold plans for increased investment. But on balance, government investment in space programs barely grew in 2012, increasing by slightly more than 1%. [39]

Fortunately, this might matter less and less. Commercialization of space is now an established trend and commercial space now accounts for nearly three-thirds of our global space economy. [40] In addition to the traditional aerospace companies, we are seeing an influx of new blood and new money as several billionaire entrepreneurs are turning their attention to space. Elon Musk (SpaceX) is the most ambitious: he wants to settle Mars and is hard at work building cheap, reusable rockets to get us there while he pays the bills by launching satellites and delivering cargo to the International Space Station. Jeff Bezos (Blue Origin) is also building reusable rockets but is less forthcoming about his long-term plans; his company's motto promises to move Gradatim Ferocita, Latin for 'step by step ferociously'. [41] Paul Allen (Stratolaunch Systems) announced he is building the world's largest aircraft, powered by six 747 engines, to serve as a platform for putting commercial payloads into orbit. [42] Robert Bigelow (Bigelow Aerospace) is working on a private orbital space station to be used for research and space tourism. Richard Branson (Virgin Galactic) is working to offer suborbital flights for space tourists and suborbital launches to science missions, with his eyes set on orbital human spaceflights further down the road. Another four billionaires, Larry Page, Eric Schmidt, Ross Perot, Jr., and Charles Simonyi have recently invested in an asteroid mining venture, Planetary Resources. [43] Their venture is not the only one—the total count of space mining startups now stands at four (Planetary Resources, Shackleton Energy Company, Moon Express, and Deep Space Industries). [44]

Government and commercial activities combined, space is still a modest US$304 billion economy. It is now driven by scientific exploration, Earth-observing and broadcasting satellites, and a nascent commercial interest in space tourism, space transportation and space resources. [45] The good news is that it is growing. The even better news is—now that our motivations and key players have changed—it may be an opportune time to return to the idea of grand planning in the tradition of Tsiolkovsky and Rockwell International. This time however, we should aim for a different kind of plan and, more importantly, a different kind of planning process.

Extending our civilization beyond Earth requires intensive innovation across all major industry sectors. Yet, if you look at the cutting-edge research across most of our terrestrial industries, it is almost without exception blue sky. So how do we get our industrial giants to think black sky? Imagine bringing representatives from all major terrestrial industries together to work on The Interplanetary Technology and Economy Roadmap 2050.

In addition to the usual suspects from space agencies and commercial space industry, we will need people from Mining, Energy, Construction, ICT and Biotech. Here is why they need to be in the room:

- **Space transport.** Today we do fewer than a hundred orbital launches a year and we rely on single-use rockets. Our current space transportation options are nonsensically expensive. [46] Just imagine trying to get the air transportation industry going with single-use airplanes. To be able to expand into space, we will need cheap, reusable rockets and space-based re-fueling infrastructure. Several such rockets and spaceplanes are already in advanced stages of development at SpaceX, Blue Origins, Reaction Engines (UK-based developer of the Skylon spaceplane), NASA, and ESA. In the mean time, the team at Shakleton Energy Company has advanced plans for an in-space network of lunar-sourced propellant depots that will offer refueling services to spacecraft in low Earth orbit—this would allow space vehicles to carry less fuel and take on bigger payloads. This type of space infrastructure could open up the cis-lunar space. [47] Space transport is the backbone of our expansion into space. It will be important for everybody in the room to understand what will be possible by when.
- **Mining.** Whether it's fuel for spacecraft or construction materials for our space-based structures, we will need to get our resources off-Earth because lifting massive payloads out of the Earth's gravity well makes little sense when you can mine them in the neighborhood and move them around without picking a fight with gravity. On Earth, mining is a $650 billion industry with finite stocks of basic materials. Off-Earth, it's an infinite frontier where a single small metallic near Earth asteroid (NEA) can contain more metals than what we have been able to mine on Earth during the entire history of our civilization. Tapping into the population of NEAs would increase our resource base by a factor of about one million. [48] But to tap this potential, our mining companies need to develop new technology for prospecting celestial objects and running remote extraction operations in low gravity and extremely cold environments. The mainstream mining industry has recently been introduced to the idea of space mining when four space mining startups spoke at the Canadian Institute of Mining's annual conference. [49] The mining industry has already been primed for serious black sky roadmapping.
- **Energy.** We will need unprecedented amounts of energy to expand and support our civilization in space but we will also have unprecedented new sources at our disposal. Coal, oil and natural gas may have powered the Industrial Revolution, but it will probably be a mix of space-based solar, nuclear fission and nuclear fusion that will power the Space Revolution. On Earth, Energy is a US$6 trillion sector. Off-Earth, the possibilities are mind-boggling. Take He3-deuterium fusion, for example. Cracking the code on fusion opens up the energy supply that is no longer measured in centuries or even millennia. We have 3 billion years worth of He3 energy on Saturn alone. The three outer planets of our solar system can turn into the Persian Gulf of the Solar System, or remain irrelevant if fusion stays "30 years away" forever. [50] Nuclear players aside, the giants of our Energy sector today are oil and gas companies. Space-based solar, thorium-based nuclear fission and He3-deuterium fusion can dramatically redraw this landscape.
- **Construction.** What would it take to design, build and operate city-scale structures in orbit or develop a city in a volcanic cave on Mars? So far our experience base is limited to launching the grand total of 13 space stations into orbit (ISS and Tiangong 1 are still in orbit, plus the now retired nine Salyut Stations, Mir, and Skylab). Several Tiangong extensions and a couple of private orbital complexes or space hotels are already in the works. [51] All this is invaluable experience in space craft design but it hardly prepares us for the complexity involved in constructing space cities that would not need to rely on spaceship Earth for supplies. At the moment, our US$4.6 trillion terrestrial construction industry is eyeing hundreds of upcoming cities in China. [52] Cities on Mars would raise eyebrows in surprise, but if surprise is where we must start, then that's where we start.
- **Telecommunications.** Here on Earth, we are getting used to information zipping through our fiber-optic cables at the speed of light. But in deep space, our communications infrastructure has not changed much since NASA launched its first probes in the late 1950s. This has recently changed for moon-to-Earth communications when NASA put in place a laser system that has the potential to transmit data at broadband speeds. [53] But by and large, our Deep Space Network relies on radio signals and large

radio dishes to catch signals from our rovers on Mars or planetary orbiters. [54] Using light for deep space communications comes with a number of technical challenges: celestial bodies are in constant motion and terrestrial clouds get in the way, just to name a few. Vint Cerf and others contributing to the InterPlanetary Internet project have already developed a basic architecture for interplanetary (and even galactic) internet, [55] but we will need to start building the actual infrastructure. Getting as close to the speed of light in our deep space communications as possible can make a world of difference for both manned and robotic missions far away from Earth. They can also make a difference on Earth: optical systems can enable new satellite networks that will boost our terrestrial internet capacity. [56]

- **Biotech.** There is intensive innovation, and then there is extremely intensive innovation. The technological breakthroughs we have mentioned so far are a matter of physics. Lump them all together and the resulting task would still pale in comparison to what we need to accomplish in biology. Apart from short stays in our space stations, no human has ever lived on another celestial body, outside Earth's atmosphere. We will need to create complete self-sustaining ecosystems, and modify our bodies to be able to survive in radically different gravitational, atmospheric and climatic environments. So far few of our biological experiments in micro-gravity environments are encouraging: bone density drops, conception is most likely impossible, and the human immune system weakens while aggressive bacteria like e. coli thrive and multiply. [57] And we are only starting to scratch the surface when it comes to understanding the complexity behind biological systems.

Introducing black sky thinking across all major industries would not only open up space but also has the potential to shake up every aspect of terrestrial living. We could end resource scarcity (mining NEAs), run our terrestrial civilization on a near-infinite supply of clean energy (space-based solar or He3-deuterium fusion), dramatically shrink terrestrial distances (sub-orbital flights), taste the fruit of the new agricultural revolution (closed loop food production and bioprinted meat), and further extend human healthspan.

But to get there, we must first overcome the chicken-and-egg problem. At a recent planetary drilling summit in Norway, one of the authors had a conversation with executives from a global oil and gas company. They were invited to the summit because their deep sea drilling expertise was considered relevant for asteroid and lunar mining and planetary drilling. Although visibly fascinated by the black sky thinking in the room, they were also understandably perplexed: "Where is the market? Six people on the ISS?" Without space-based resources, it's hard to build anything at large scale in space. Without the funded intent to build something large-scale in space, it's hard to get people to invest in space-based capabilities. So how do we start?

As with any new frontier or any new market, we need pioneers who are willing to take on outsized risks. Getting people from global industries in the same room and thinking black sky would help us to normalize the idea of expanding into space through mutual validation and to take on our chicken-and-egg problem through collective commitment. R&D funding would follow. If we succeed, space commercialization has the potential to turn into space industrialization during our lifetimes. This could take care of the first intensive technological transition to inter-planetary.

But that only gets us half way there. The second intensive technological transition is to interstellar and that's where we are not likely to get much help from the industry. If the commercialization horizon of 10-50 years within our solar system is already a radical stretch for most, a hundred-year time horizon would be swiftly dismissed as laughable. As long as that continues to be the case, we will have to continue doing all the heavy-lifting on breakthrough propulsion and starship designs "in-house", inside the core interstellar community of volunteer scientists, engineers and designers funded by the community of enthusiasts and an odd hyper-visionary billionaire or millionaire.

What is our verdict on this part of the job? It is formidable, bordering on monumental. We will need to do serious legwork to align and re-orient terrestrial commercial interests toward space. And in parallel, we will need to press on with the development of interstellar technologies ourselves. That's a tall order. It's nothing, though, that a committed group of passionate individuals could not accomplish.

4.2.4 Governance

What kind of governance would we need to have in place as a civilization to go interstellar?

We know that transitioning to interstellar civilization would require an unprecedented mobilization of our collective resources, trillions of dollars just for the first human interstellar mission alone. We'll need an economy that is a thousand times bigger, amounts of energy that exceed our entire current production, and resources that exceed our terrestrial stocks. In other words, a purely Earth-based civilization won't be able to afford the mission. This is not a one-generation project.

When we search human history for precedents of long-term or multi-generational missions, interstellar enthusiasts like to cite moving out of Africa and settling the North, the gradual settlement of the Polynesian islands, the conquest of North America, the construction of the Egyptian pyramids and European cathedrals, and, from our more recent history, the Apollo program and the construction of the Hoover Dam. [58] All of these examples may offer some historical comfort in the knowledge that our species is capable of pursuing multi-generational or grand engineering projects. This may be of some psychological value at a time when our civilization's attention span appears to have shriveled down to years and months, and multi-generational projects are rare.

But upon closer inspection, few of these precedents offer good replicable governance models. Our African ancestors advanced into the North gradually and in small tribes. Their initial investment was covering vast distances and crossing inter-continental land bridges on foot, a technology that comes preinstalled as part of the human body plan. The Polynesians built small sailboats, an undertaking a small tribe could afford. The Egyptian pyramids were erected by the pharaohs using slave labor at a time when the pharaonic power was most centralized. The European cathedrals were built by the institutionalized Christian church. The European forays into North America were jumpstarted by kings and queens, the Apollo program by the national government. So we extrapolate from the past and imagine that our foray across the interstellar space will be driven by national governments, religious institutions or the super-wealthy (e.g., trillionaires who make their fortune in the interplanetary resource economy). [59]

Pursue any of these models and we will be exporting our parochial biases across the interstellar space. Unless we will have established some form of an all-inclusive meta-religion, we'll be looking at a dominant religion putting its own believers first. Unless we will have established some form of civilization-wide government, we'll be looking at leading super-powers privileging their own citizens. Unless we will have all become trillionaires, we'll be looking at a wealth-based selection. Some parts of our heritage are best left behind.

But do we really have more palatable alternatives? As a civilization, we face a number of civilization-wide challenges that cannot be addressed by any single existing institution alone, some of them are conceptually comparable in complexity to going interstellar. Take sustainability. We are facing the "perfect storm" of rising food, water and energy needs that are inseparable from our impact on climate, biodiversity and biosphere as a whole. It is hard to imagine how we could weather this storm without mobilizing ourselves at a civilization-wide scale. What are the global governance mechanisms we have in place? The bad news is, our existing global institutions, most of which were conceived more than half a century ago, have proved incapable of making serious progress on the issue of sustainability. The good news is, we have plenty of proposals around for what we could try next: creative coalitions among countries, companies and cities. [60]

It does not take a wild leap of imagination to imagine some form of creative coalition taking the lead role in propelling our civilization to the stars. In real life, we can prototype this coalition at the level of our solar system already today, as part of our push for the industrialization of space (See discussion in 4.2.3 Technology). But we can also experiment with various governance mechanisms through simulations in online game environments.

We will not be the first. A few years back, the then popular online game World of Warcraft captured the imagination of academia. After a "corrupted blood" disease that was initially introduced as a minor hindrance to make combat more challenging turned into a game-wide pandemic and led to pulling the plug on the whole world, microbiologists, mathematicians, psychologists, and epidemiologists woke up to the idea that virtual worlds could be used to model real-world behaviors. Many serious studies have been published in medical journals since. [61]

We already have several interstellar games on the market, with EVE Online being the most prominent. Half a million people play this online game, set in a galaxy of some 7,500 star systems. It is already an established interstellar civilization that comes with warp drives and stargates—a fascinating lab for experiments in interstellar economy and governance but too advanced for exploring the pathways to becoming an interstellar civilization. To create a good experimental lab for the interstellar transition, we will need to create an exciting prequel to EVE Online.

A creative coalition of the willing, aligned around a shared goal is perhaps the most likely path to interstellar civilization. But coalitions don't just form. They are formed. We will need an independent body capable of energizing the interstellar mission and putting coalitions together for hundreds of years to come.

5. Starfleet

Initiating an interstellar jump sequence is formidable but not impossible. The question is, who is going to do it. Throughout this paper, we (the authors) have often used the pronoun "we" to loosely refer to the interstellar community. As it stands today, it might be too small and too inward focused to energize a jump of such magnitude. Below we take stock of where we are today and explore how we might need to evolve to make the transition to interstellar civilization inevitable.

5.1 Thousands

Thousands of people are now actively contributing or rooting for the interstellar cause: we have half a dozen active interstellar organizations, [62] hundreds of people show up at starship conferences and submit papers, and the Centauri Dreams blog attracts almost 50,000 unique visitors a month. The roots of our community go back to the 1933 founding of the British Interplanetary Society (BIS) that from the outset was intended to cover spaceflight to the nearby planets as well as travel across interstellar space. At the time, rockets were seen as a novelty with only limited potential, so it might as well have called itself a Mad Hatter's Society. [63] We've come a long way since then. The National Geographic features interstellar in the New Age of Exploration even if the title is "Crazy Far". [64] Interstellar conferences get coverage in The Economist even if the participants are half-dismissed as "an intrepid band of scientists," "an optimistic bunch," and "iconoclasts and dreamers fitting the activity [starship research] around their 'proper' jobs". [65]

From the outset, BIS had a dual mission: popularize space travel and conduct original research on rocket propulsion and spacecraft design. Most of our interstellar organizations have followed suit and have similar missions revolving around technical starflight capability, albeit the emphasis varies from organization to organization. In the last couple of years, we have started to think broader. Tau Zero was the first organization to explicitly include the relevance of broader civilizational and societal issues to the vision of interstellar flight. 100 Year Starship has been advocating the inclusion of biomedical and social sciences. And the team behind The Starship Century book and events also adopted a broad perspective, exploring what it would take to make this century a starship century. These are very promising signs. Yet, by and large, the vast majority of interstellar members and volunteers are passionate about the nuts and bolts of starflight.

We do need a stronger, bigger starflight-focused core. Without it, we would not be able to justifiably refer to ourselves as an interstellar community. But to initiate the interstellar jump sequence, we should dramatically expand our scope and aim to grow our base by three orders of magnitude, from thousands to one million.

5.2 Million

Imagine one million ensigns around the planet, united by a single purpose: make the transition to interstellar civilization inevitable. An epic purpose calls for an epic organization. Modeled on Star Trek's Starfleet, it is all about action. You don't join Starfleet for its membership benefits, you join for the opportunity to contribute to the single most ambitious endeavor in the history of human civilization. Starfleet is structured into distinct units, each with a clear mandate and well-defined missions for ensigns to carry out, led by an Admiral or a Council. Command is responsible for the overall direction, coordination and funding. Engineering is responsible for propulsion, starship and habitat design. Industrialization runs the interplanetary economy and technology roadmapping process. Culture runs the Interstellar Art Academy and other content-generating initiatives. And so on. Every ensign is assigned to a unit. Every contribution—be it paying an annual membership fee, recruiting new ensigns, submitting a ground-breaking paper, fundraising for a symposium or giving a talk that gets people interested in space start-ups—earns star miles. Rack up enough of them and you move up in rank. Starfleet always gets things done. With a bench a million strong, it can effectively spread the word to hundreds of millions, self-fund its ongoing activities and tap larger networks for funding projects. It is motivated by passion, which, unlike taxes or profits, comes in unlimited supply. [66]

From the viewpoint of a thousand-strong community, one million can look unachievable. Seen through the lens of serious scientific research, Star Trek references can appear unserious. Before we dismiss the idea as unachievable and unserious, let's sanity-check our aspiration—is it practically even possible to accomplish such a feat? Are there that many people who would be available and willing? Do we have any precedents?

With the advent of online tools that allow new forms of collaboration, as a civilization we are now learning how to use more constructively the free time afforded to us since the 1940s for creative acts rather than consumptive ones. The cognitive surplus—the buildup of free time among the world's educated population—is now in the order of magnitude of a trillion hours a year. We are in the middle of a Great Spare-Time Revolution. [67] There is a massive reservoir of volunteer time that we can tap. Availability should not be a problem.

Five decades of behavioural research shows that most enduring motivations are not external but internal—the joy of doing something for its own sake. We do things because they're interesting, because they're engaging, because they're the right things to do, because they contribute to the world. [68] For people looking to contribute to the world, to be part of something bigger, we can create an unprecedented opportunity—contribute to the most ambitious mission in the history of human civilization. But to capture the hearts and minds of a million people, we will need to get creative or drown in the background noise. Star Trek has struck a cord with hundreds of millions around the world. An opportunity to join a terrestrial Starfleet bound for the stars just might be the kind of opportunity many would not want to miss.

There isn't an exact precedent we could copy. But there is plenty we can learn from open-source volunteer projects like Wikipedia. Almost 20 million people are registered as contributors to Wikipedia (even though only a minority of them are regular contributors). All of the articles, edits, and arguments about articles and edits represent around 100 million hours of human labor. For comparison, Americans watch about 200 billion hours of TV every year.

Transforming a loose interstellar community of thousands of people into a disciplined Starfleet of a million ensigns is a formidable task in its own right. It won't happen overnight. But this might be our best shot at initiating an interstellar jump sequence and getting to starflight capability in the next 100 years.

References

1. O'Toole, G., "Mohandas Ghandi? Apocryphal?", Quote Investigator, April 23, 2013, http://quoteinvestigator.com/2013/04/23/good-idea/ (retrieved 6 Aug. 2013).

2. "Civilization", dictionary.com, http://dictionary.reference.com/browse/civilization (7 Aug. 2013).

3. "Definition of civilization in English", Oxford Dictionaries, http://oxforddictionaries.com/definition/english/civilization (6 Aug. 2013).

4. Stokes Brown, C., "What Is a Civilization, Anyway?", World History Connected, October 2009, http://worldhistoryconnected.press.illinois.edu/6.3/brown.html (6 Aug. 2013).

5. Drew, J., "Civilization: Definitions and Recommendations," International Society for the Comparative Study of Civilizations Spring 2007, http://www.wmich.edu/iscsc/civilization.html (6 Aug. 2013).

6. Sullivan, L.E., "The SAGE Glossary of the Social and Behavioral Sciences", Editions SAGE, 2009, p.73.

7. Bagby, P., 1963 (1958), "Culture and History", University of California Press, 1963 (1958), p. 163.

8. Ibid, 4.

9. Ibid, 4.

10. Benveniste, E., "Civilisation. Contribution à l'histoire du mot" (Civilisation. Contribution to the History of the Word"), 1954, published in "Problèmes de Linguistique Générale", Editions Gallimard, 1966, pp. 336–345 (translated by Mary Elizabeth Meek as "Problems in General Linguistics", 2 vols., 1971).

11. Bowden, B., The Empire of Civilization: The Evolution of an Imperial Idea (Large Print 16pt), pp. 17-19.

12. Velkley, R., "The Tension in the Beautiful: On Culture and Civilization in Rousseau and German Philosophy", in Being After Rousseau: Philosophy and Culture in Question, The University of Chicago Press, 2002, pp. 11-30.

13. Northrup Frye, N., "Northrup Frye on Modern Culture", Volume 11, edited by Jan Gorak, University of Toronto Press, 2003, p. 305.

14. "A Study of History", Wikipedia, http://en.wikipedia.org/wiki/A_Study_of_History, (10 Aug. 2013).

15. "Arnold J. Toynbee", Wikipedia, http://en.wikipedia.org/wiki/Arnold_J._Toynbee, (10 Aug. 2013).

16. Snooks, G., "The Laws of History: Exploding the Myth of Social Evolution", Routledge, 1998, p. 91.

17. Dario Fernandez-Morera in private correspondence to the author, Joseph Drew, "Civilization: Definitions and Recommendations," International Society for the Comparative Study of Civilizations, Spring 2007, http://www.wmich.edu/iscsc/civilization.html (6 Aug. 2013).

18. "Kardashev scale", Wikipedia http://en.wikipedia.org/wiki/Kardashev_scale (12 Aug. 2013).

19. We are borrowing this distinction between intensive/vertical and extensive/horizontal innovation from a venture capitalist Peter Thiel who has recently wrote a book dedicated to the topic: Zero to One: Notes on Startups or How to Build The Future, coming Summer 2014. Notes version of his Stanford course that inspired the book can be found here.

20. Bostrom, N., "Existential Risk Prevention as Global Priority", Global Policy, February 2013, Volume 4, Issue 1, pages 15-31. Also see Nick Beckstead, "On the Overwhelming Importance of Shaping the Far Future", PhD Thesis, Department of Philosophy, Rutgers University, 2013.

21. The case for space development as a path to re-greening Earth is put forward by Matloff, G., C Bangs and Johnson, L. in their book "Paradise Regained", Copernicus, 2010 edition.

22. The cost of "Thrifty Dyson Starship" has been estimated at $1 trillion, "Budget Daedalus" at US$ 20 trillion and "Daedalus Class Starship" at US$100 trillion in Richard Obousy's "Interstellar Flight: Social and Economic Considerations", posted on June 9, 2010, http://www.icarusinterstellar.org/interstellar-space-flight-social-economic-considerations/ (Aug. 14, 2013). Early interstellar mission has been estimated at US$125 trillion in Robert Zubrin's "On the Way to Starflight: The Economics of Interstellar Breakout", in "Starship Century: Toward the Grandest Horizon", edited by James Benford and Gregory Benford, Microwave Sciences 2013, p. 84.

23. "NASA Langley Research Center's Contributions to the Apollo Program". NASA Langley Research Center (14 Aug. 2013).

24. Israel, R. C., "What Does It Mean to Be a Global Citizen?", 13 February 2013, opendemocracy.net (10 Oct. 2013).

25. Beckstead, N., "On the Overwhelming Importance of Shaping the Far Future", PhD Thesis, Department of Philosophy, Rutgers University, 2013.

26. Garretson, P., "What Our Civilization Needs is a Billion Year Plan", September 23, 2012, kurzweilAI.net (7 Oct. 2013).

27. Gottschall, J., "The Storytelling Animal: How Stories Make Us Human", Mariner Books, 2012.

28. A good exploration of the scientific limits of the possible can be found in Clarke, A., "Profiles of the Future", Warner Books, 1985.

29. Fuller, S., "Nighty-degree Revolution", October 24, 2013, Aeon Magazine.

30. Juarez, J., "More than 100,000 want to go to Mars", August 9, 2013, CNNMexico.com.

31. Concept taken from Fuller, B., "Operating Manual for Spaceship Earth", 1968.

32. Pearce, F., "Technology as Our Planet's Last Best Hope", July 15, 2013, theguardian.com.

33. Although we have a few decades of academic discourse on global, catastrophic and existential risks, it is only recently that we have started to introduce the idea of existential risk to the global public policy circles. The most interesting recent example is Nick Bostrom's article February 2013 issue of Global Policy, "Existential Risk Prevention as Global Priority", Global Policy, Volume 4, Issue 1, pp. 15-31.

34. Here is a selection of the more considered complaints about the lack of intensive or big-ticket innovation in technology and in science: "The Danger of Celebretizing Entrepreneurship", "Transformational Entrepreneurship: Where Technology Meets Societal Impact", "Reversing the Decline in Transformational Ideas" — all by Max Marmer on Startup Compass blog; Jerzy Gangi, "Why Silicon Valley Funds Instagrams, Not Hyperloops", August 19, 2013; Sam Altman, "What Happened to Innovation?" June 20, 2013; Neal Steaphenson, "Innovation Starvation", World Policy Journal, Fall 2011; Mark Buchanan, "In Search of Black Swans", April 1, 2009, pysicsworld.com (30 Oct. 2013).

35. The full text of Tsiolkovsky's 1926 "Plan for Space Exploration" can be found in Encelcopedia Astronautica article on "The Foundations of the Space Age", astronautix.com.

36. Rockwell International, "Integrated Space Plan (Preliminary)", version 1.1 February 1989.

37. See "Vision for Space Exploration" announced on January 14, 2004 by US President George W. Bush; "Space policy of the Barack Obama Administration"; International Space Exploration Coordination Group's "The Global Exploration Roadmap", August 2013. European Space Agency's "Cosmic Vision 2015-2025" (30 Oct. 2013).

38. "The Space Report 2013", Space Foundation, spacefoundation.org, p. 5.

39. Ibid., p. 6.

40. Ibid., p. 5.

41. O'Connor, C., "Billionaire Space Race: 2012 Will Be the Year Private Space Transportation Takes Off", December 30, 2011, forbes.com (17 Oct. 2013).

42. Caulfield, B., "Billionaire Paul Allen Building World's Biggest Plane in Effort to Conquer Space", December 13, 2011, forbes.com (17 Oct. 2013).

43. Wall, M., "Asteroid Mining is Just Latest Billionaire Space Club Project", April 25, 2012, space.com (17 Oct. 2013).

44. Lark, E.-J., "Space Mining Entrepreneur Discussion", October 10, 2013, spaceref.biz (17 Oct. 2013).

45. Ibid., 36, p. 5.

46. For a fascinating historical account of how we ended up with nonsensically expensive space rockets and infrastructure, see Zubrin, R., "Entering Space: Creating a Spacefaring Civilization", Jeremy P. Tarcher/Putnam, 1999, Chapter 2, pp. 21-38.

47. Keravala, J., Tietz, D., Stone, B., "Shackleton Energy Company's Propellant Depot and Space Transportation Architecture", volume 1, number 2, 2013, New Space.

48. Lewis, J., "To the Asteroids—and Beyond!", SETI Talks, streamed live on September 10, 2013.

49. Ibid., 42.

50. Zubrin, R., "On the Way to Starflight: The Economics of Interstellar Breakout", in "Starship Century: Toward the Grandest Horizon", edited by James and Gregory Bengford, Microwave Sciences, 2013, p. 94.

51. "Space Station", wikipedia.com (29 Oct. 2013).

52. Davis Langdon, "World Construction 2012", aecom.com.

53. Brumfiel, G., "Netflix on the Moon? Broadband Makes It to Deep Space", October 23, 2013, npr.com (30 Oct. 2013).

54. For an excellent account of our current Deep Space Network and the requirements for interstellar communications and navigation, see Gilster, P., "Centauri Dreams", 2004 Copernicus Books, 2004, pp. 183-210.

55. Cerf, V., "Interplanetary Internet", TEDxMidAtlantic talk, uploaded on December 1, 2011.

56. Talbot, D., "A Successful Moonshot for Laser Communications", October 23, 2013, MIT Technology Review.

57. Research highlights shared by fellow 100YSS Symposium 2013 presenter, Jeanette Hill, during her talk "Microgravity Effects on Human Health and Metabolism" (to be published in Symposium Proceedings in Spring 2014).

58. The Hoover Dam precedent was invoked by Les Johnson at an Eve Online conference in October 2013 in Las Vegas. See O'Neill, I., "Want a Starship? Think Big. Think Really Big", October 28, 2013, news.discovery.com (1 Nov. 2013).

59. For one possible set of future scenarios for who might lead the interstellar expansion of our civilization, see Schwartz, P., "Starships and the Fates of Humankind", in "Starship Century: Toward the Grandest Horizon", edited by James and Gregory Benford, Microwave Sciences, 2013, pp. 27-39.

60. For one such proposal see "Now for the Long Term: The Report of the Oxford Martin Comission for Future Generations", 2013.

61. Christakis, N., Fowler, J., "Connected: The Amazing Power of Social Networks And How They Shape Our Lives", Harper Press, 2010, pp. 252-255.

62. Active interstellar organizations: The British Interplanetary Society, Tau Zero, Icarus Interstellar, the Institute for Interstellar Studies, 100 Year Starship, SpaceGambit. In the neighborhood: The Planetary Society, SETI Institute. An excellent overview of all interstellar organizations and their interstellar activities can be found on Tau Zero website.

63. Watson, N., "Reaching for the Stars", Volume 3, Issue 1 2013, History Today.

64. Folger, T., "Carzy Far", January 2013, ngm.nationalgeographic.com (1 Nov. 2013).

65. "Interstellar Travel: Starship Troupers", October 26th, 2013, The Economist (1 Nov. 2013).

66. You can find a high-level presentation by the authors further detailing the idea behind the 1 million Starfleet at http://www.slideshare.net/erikailves/road-to-k3.

67. Shirky, C., "Cognitive Surplus: Creativity and Generosity in a Connected Age", Penguin Books Limited, 2010.

68. Pink, D., "Drive: The Surprising Truth About What Motivates Us", Canongate Books, 2010.

Solar System Escape Architecture Revolutionary Science

NIAC 2012-2013 Phase 1 Final Report

Jeffrey Nosanov, Brian Trease,
Daniel Grebow, Henry Garrett, and John West

Synopsis

Our most distance spacecraft, Voyager 1 and Voyager 2, recently reached the boundary of the solar system known as the Heliopause. Beyond this boundary lies interstellar space, and, at tremendous distance, the stars. The Voyager spacecraft have shown us that the Heliopause is a dynamic, rapidly changing environment, varying greatly as the fields and effects emanating from the sun are met by a host of forces pushing in from our nearby galactic neighborhood. This is the region we aim to explore.

Specifically we aim to develop a mission architecture and spacecraft concept capable of reaching the Heliopause (~100 AU) region in multiple directions of interest, with a variety of scientific instrument suites, within a reasonable timeframe (~15 years.) This report details our Phase 1 work including science goal definition, trajectory planning, technology research, mission planning, instrument selection, spacecraft design, and more.

We present an architecture employing a 250m x 250m solar sail with a 175 kg spacecraft. This spacecraft could carry a variety of instrument suites depending on the destination, time, and other factors. The spin-stabilized spacecraft would powered by a small radioisotope power system (~20 watts), makes use of an optical communication system, and carries 3 extensible booms for instrument accommodation as visible in the sketch and rendering below. We conclude that it would be reasonable to implement the architecture described herein in the 2020-2030 timeframe based on likely near-term technological and material advances. We are confident that cruise times of ~15 years could be achieved to science-rich destinations.

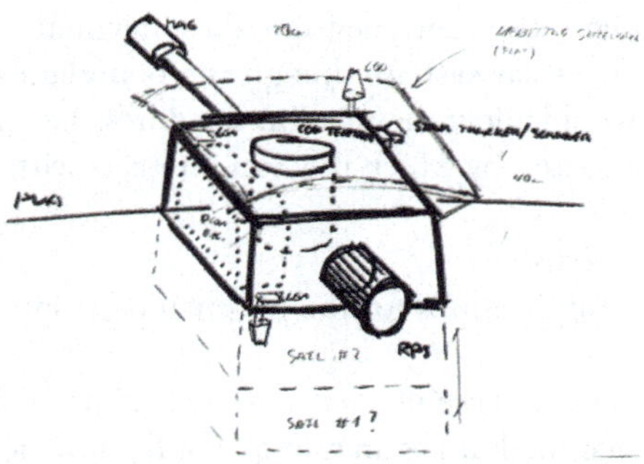
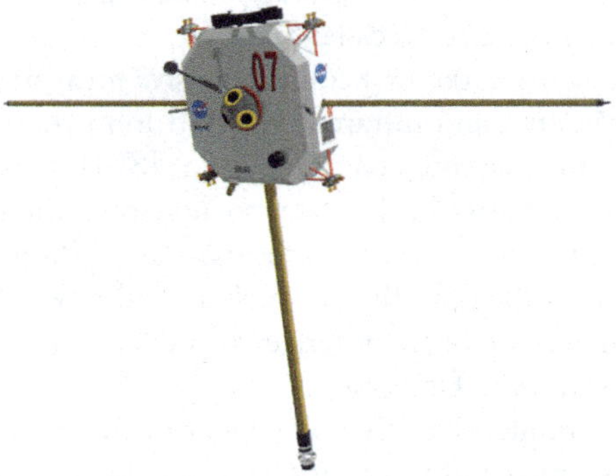

This report details the approach we used to come to these conclusions and identifies some of the steps along the way. We utilized a variety of methodologies involving our full capability at JPL – from small point studies to large collaborative engineering processes. This paper also describes potential industry partnerships and costing information, and includes the methods used to account for our project's significant deviation from any prior mission in the NASA costing tool database.

Outreach and public engagement is an important part of NASA activities. We engaged in several different forms of this outreach and describe them in this report. We conclude with some open questions and a summary of activities we hope to engage in with Phase 2 funding.

The entire team would like to thank the NIAC program for the opportunity to investigate this fascinating and challenging concept.

Abstract

The Voyager program gives us tantalizing clues as to the nature and behavior of the Heliopause – the boundary between the sun's influence and the interstellar medium. This information comes from forty-year old instruments designed to study the outer planets. A targeted Heliopause investigation would give insight into the formation of the solar system, the role of the sun in the local interstellar neighborhood, and contribute to human exploration planning by helping to predict periods of low galactic cosmic ray (GCR) penetration into the inner solar system.

1. Motivation/Voyager Program Introduction

The Voyager program arose out of a realization in the early 1960s that the mid-1970s would offer a once-in-170-year planetary alignment. This event would allow one spacecraft to fly by multiple outer planets. The primary Voyager Mission was to investigate the Outer Planets, and the Voyager Interstellar Mission (VIM) was confirmed in the late 1980s as it was clear that the spacecraft were functioning well and actually had a chance of reaching the Heliopause boundary intact. As of mid-2013 the VIM is still active and the two Voyager spacecraft continue to return data, and are expected to continue to do so until the mid 2020s when the Plutonium238-based power system no longer produces enough energy to power the communication systems. From that point the Voyager spacecraft will be silent monuments to humanity's technical achievements for roughly one million years until high-speed impacts with interstellar dust grains cumulatively abrade the spacecraft back to the stardust from which they were made.

The SSEARS project concept arose from a conversation with Ed Stone, former JPL director and Voyager project scientist since the inception of the project. At one point Ed Stone was asked how he would continue the Voyager science. His answer was to send multiple spacecraft in multiple directions to study the 3d structure of the boundary. The aim of this project is to develop an architecture that enables this investigation.

2. Heliopause Science

This section will discuss the science goals and rationales for the goal of returning to the Heliopause and traveling beyond. These goals were derived from conversations with Ed Stone (Voyager Project Scientist), the Science Mission Directorate, the Human Exploration and Operations Mission Directorate, the Heliophysics Decadal Survey, and the Planetary Science Decadal Survey.

SSEARS would enable the first comprehensive measurements of plasma, neutrals, dust, magnetic fields, energetic particles, cosmic rays, and infrared emission from the outer solar system, though the boundaries of the Heliosphere, and on into the interstellar medium (ISM). This would allow the mission to address key questions about the distribution of matter in the outer solar system, the processes by which the Sun interacts with the galaxy, and the nature and properties of the nearby galactic medium.

The principal scientific objectives of such a mission would be to:

- Explore the nature of the interstellar medium and its implications for the origin and evolution of matter in our Galaxy and the Universe;
- Explore the influence of the interstellar medium on the solar system, its dynamics, and its evolution;
- Explore the impact of the solar system on the interstellar medium as an example of the interaction of a stellar system with its environment;
- Explore the outer solar system in search of clues to its origin, and to the nature of other planetary systems.
- Significantly reduce human radiation risk for future crewed missions by understanding mechanisms leading to variability of dangerous levels of radiation in the inner solar system

We describe below examples of the scientific issues that could be addressed.

2.0.1 The Nearby Interstellar Medium

Our Sun is thought to be located near the edge of a low-density interstellar cloud (~0.3 cm^3), often referred to as the local interstellar cloud (LIC), that is made up of material blowing from the direction of star-forming regions in Scorpius and Centaurus. Present knowledge of the ISM is based on astronomical observations that average over long lines of sight, measurements of sunlight resonantly scattered back by interstellar H and He, data returned from the Voyager and iBEX spacecraft, and in situ measurements of neutral gas and dust that penetrate the Heliosphere. Direct observations of our local cloud by SSEARS would provide a unique opportunity to derive the physical properties of a sample of interstellar material, free from uncertainties that plague the interpretation of data acquired over astronomical lines-of-sight, and from uncertainties arising from the exclusion of plasma, small dust particles and low energy cosmic rays from the Heliosphere. Direct measurements would be made of the elemental and isotopic composition of the ionized and neutral components of the interstellar gas and of low-energy particle components, and of the composition and size distribution of interstellar dust. These measurements would provide a benchmark for comparison with solar system abundances (representative of the pre-solar nebula) and with abundances from more distant galactic regions, thereby providing important constraints on theories of galactic chemical evolution.

SSEARS would also measure cosmic ray nuclei and electrons, free from the influence of the Heliosphere, and investigate astrophysical processes that include acceleration by supernova shock waves, interstellar radio and x-ray emission, recent nucleosynthesis, and the heating and dynamics of the interstellar medium. Little is known about the properties of magnetic field in the local cloud or in the region beyond the termination shock. SSEARS would enable the first in situ measurements of interstellar magnetic fields and of the density, temperature, and ionization state of the interstellar gas, including studies of their variations over a variety of spatial scales. The possibility of identifying organic matter in the outer solar system and ISM is also an exciting possibility that is under investigation.

2.0.2 The Interaction between the Interstellar Medium and the Solar Wind

The solar wind and the interstellar medium interact to create the global Heliosphere, shown schematically in Figure 1. The size of the Heliosphere is determined by the balance between the solar wind ram pressure and the interstellar pressure. There are presently no direct measurements of the size and structure of the Heliosphere and our present understanding is based on theory and modeling, constrained by a few key measurements. The Voyager spacecraft have detected radio emissions that are thought to be caused by interplanetary shock waves hitting the denser interstellar plasma. Voyager 1 should soon reach the termination shock, providing a first direct test of our current understanding of Heliospheric structure, although some of the Voyager instruments were not designed to explore the boundaries of the Heliosphere and interstellar medium. SSEARS' enhanced capabilities and lifetime would greatly extend Voyager's exploratory studies, answering questions relating to how the ISM influences the solar system and how the solar system influences the ISM.

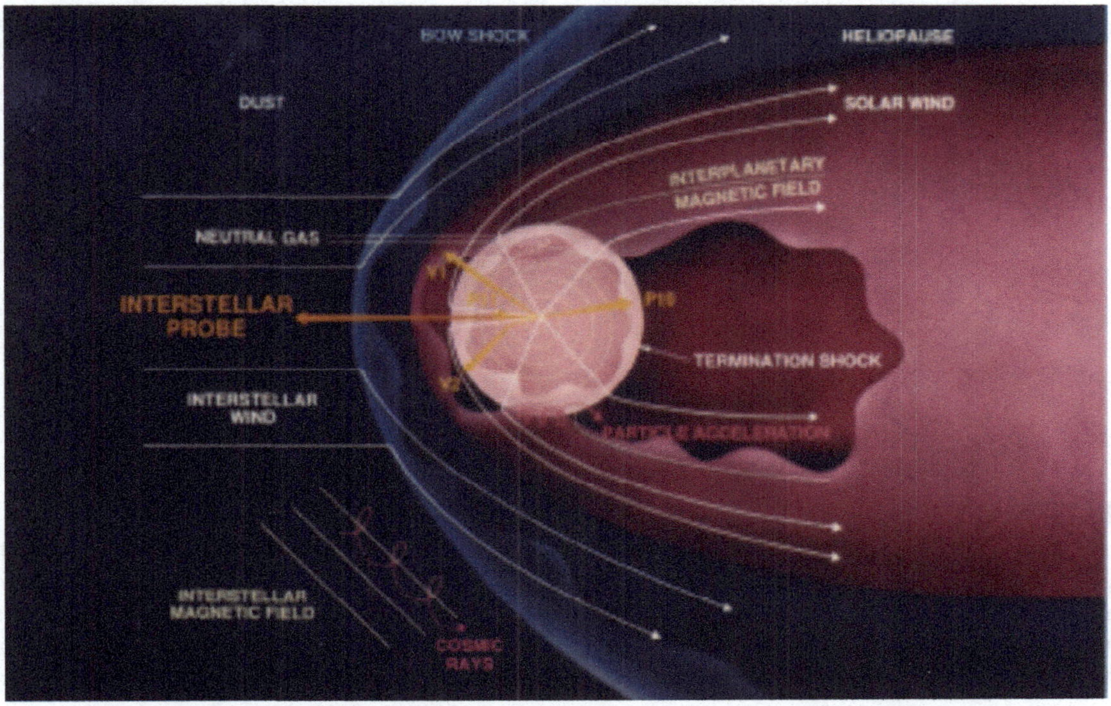

Figure 1: Schematic of the Heliosphere created by the supersonic solar wind diverting the interstellar flow around the Sun. The interstellar ions and neutrals flow at 25 km/s relative to the Sun. The solar wind, flowing outward at 400-800 km/s, makes a transition to subsonic flow at the termination shock. Beyond this, the solar wind is turned toward the Heliotail, carrying with it the spiraling interplanetary magnetic field. The Heliopause separates solar material and magnetic fields from interstellar material and fields. There may or may not be a bow shock in the interstellar medium in front of the Heliosphere.

The termination shock is a powerful accelerator that accelerates particles to energies as high as 1 GeV. In situ studies of shock structure, plasma heating, and acceleration processes at the termination shock will serve as a model for other astrophysical shocks. Past the termination shock, in the region called the Heliosheath, the solar wind flow is turned to match the flow of the diverted interstellar plasma, as illustrated Fig. 1. The spiraling solar magnetic field, frozen into the solar wind, is swept back with this flow. Depending on the unknown interstellar magnetic field strength, there may or may not be a bow shock created in the interstellar medium ahead of the nose of the Heliosphere. Energetic ions created by charge exchange in the Heliosheath can be used to provide an image of the 3D structure of the Heliosphere. Charge-exchange collisions lead to a weak coupling between the neutral and ionized hydrogen in the interstellar medium causing a pile-up of neutral hydrogen at the Heliosphere nose, referred to as the "hydrogen wall." The SSEARS spacecraft would pass through these boundary regions and make in situ measurements of the dust, plasma, fields and flows to answer questions regarding the size, structure and dynamics of the Heliosphere and the processes occurring at the boundaries. Our Heliosphere would serve as an example of how a star interacts with its environment. 1

2.1 Solar System Orientation

The plane of the Solar System is inclined at roughly 60 degrees to the plane of the galaxy, as seen in Figure 2. This helps us identify regions of interest because we can investigate the Heliopause along the solar system plane to avoid interactions with the galactic plane, and specifically study those interactions by sending the spacecraft along the galactic plane.

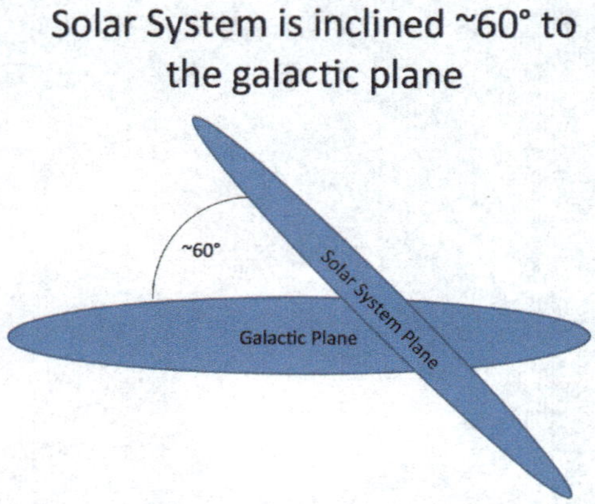

Figure 2: The plane of the solar system and the plane of the galaxy are 60 degrees apart.

2.2 Heliopause/Interstellar Medium Regions of Interest

We began our thinking with the idea of sending spacecraft in the six cardinal directions from the Earth- up, down, left, right, in, and out relative to the sun and the galactic plane. The inclination of the solar system plane to the galactic plane increases the complexity of the ideal investigation by doubling the number of regions of interest. Figure 3 shows the directions of travel along the solar system ecliptic, and Figure 4 shows the directions of travel along the galactic ecliptic.

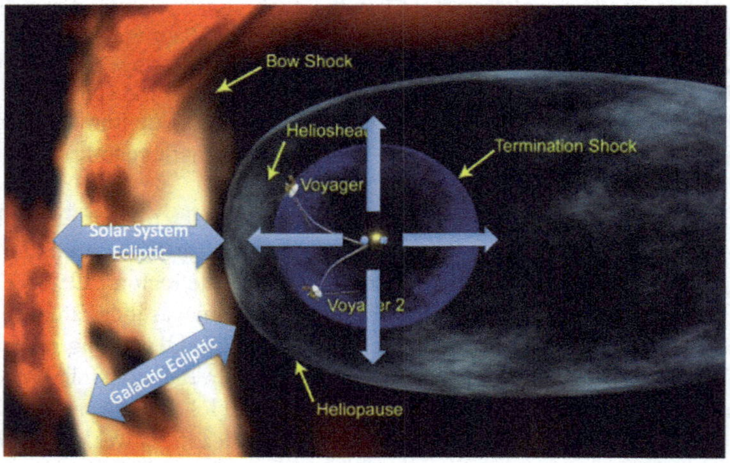

Figure 3: Directions of travel/investigation along the Solar System Ecliptic

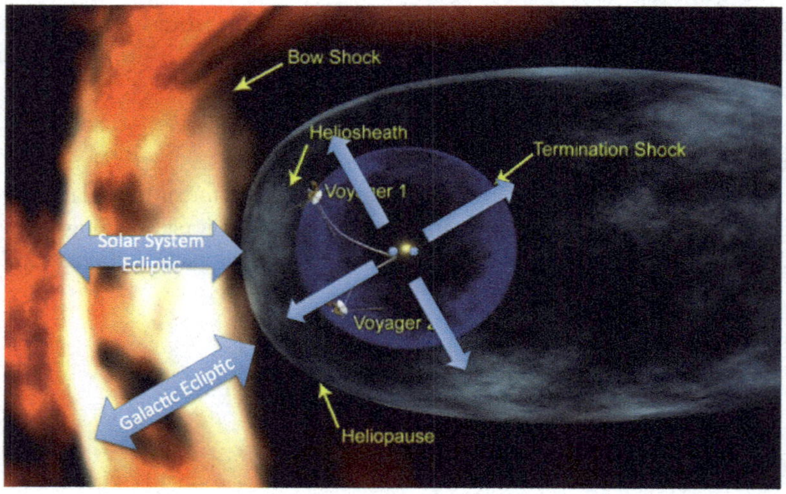

Figure 4: Directions of travel/investigation along the Galactic Ecliptic

Later in the project, after additional analysis, we revised our destinations to better account for the complexity of the sun's magnetic field. The sun's electric currents generate a complex magnetic field that extends far into space with the solar wind. The sun rotates as it emits the solar wind so the magnetic field is "wound" into a spiral known as the Parker Spiral.

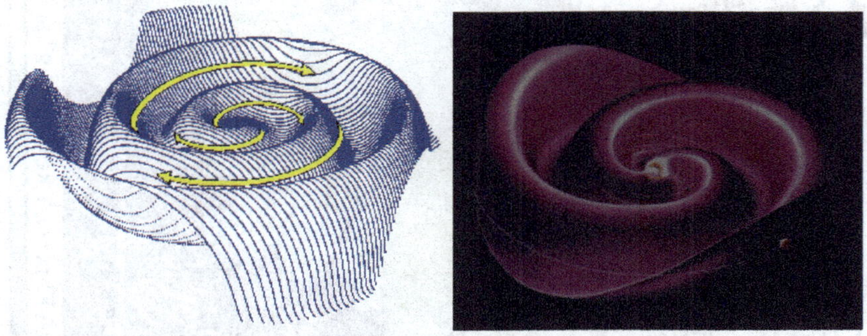

Figure 5: Two visualizations2 of the sun's magnetic field. The spiral structure of the sun's magnetic field as the solar wind carries it past the planets.

This magnetic field is primarily directed outward from the sun in one hemisphere, and inward towards the sun in another. The line between the two field directions is not exactly aligned on the solar equator, so the field becomes "wavy" as seen in Figure 5. As a result of this effect, radiation entering the solar system from the interstellar medium changes significantly with the solar cycle as the sign of the solar magnetic field switches. So, the north and south "hemispheres" of the Heliosphere should be investigated at different times corresponding to the amount of interstellar radiation present. This would necessitate at least two "waves" of spacecraft, one wave planned to reach the Heliopause in one hemisphere at the solar maximum, and another wave planned to reach the Heliopause in the other hemisphere at the solar minimum.

The SSEARS mission proposes to send multiple spacecraft out to the boundary of the Heliosphere to follow up on the Voyagers' recent discoveries and better define the interactions of the Heliosphere and the galactic media. The Voyagers were not targeted for specific locations to measure these interactions and so represent randomly placed snapshots of the Heliosphere—SSEARS would carry out a targeted analysis of the structure of the Heliosphere.

2.3 Specific Helipause Destinations

Specific targets for the SSEARS would be: the Heliosphere's nose (i.e., the point where the interaction is a head on collision due to the combined motion of the solar system and the galactic media), the anti-nose direction, and points perpendicular to the line connecting the nose and anti-nose (both "east and west" and "north and south" to determine the shape of the Heliosphere. Other potential targets are the anti-galactic center so that the center of the galaxy could be imaged using gravitational lensing or the source of neutral particles detected by iBEX (see Figure 6 below). We would also send spacecraft out at various times to measure these interactions during solar minimum and maximum for 2 solar cycles or a complete 22 year Hale solar magnetic cycle 3,4.

An unusual band of high energy neutral particles was detected by the iBEX and Cassini missions. A graphic for the iBEX data is shown in Figure 6—the band is apparently associated with a magnetic merging region where charge exchange between energetic charged particles from the interstellar medium are interacting with slow solar wind neutral particles. The nose is where the large arrow meets the Heliosphere.

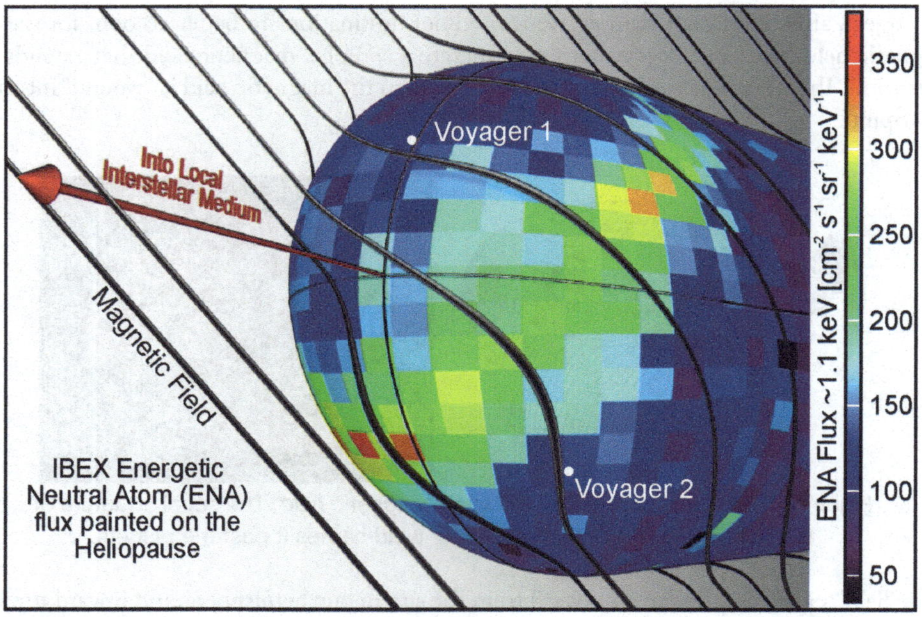

Figure 6: Graphic presentation of the location of the Heliosphere's Nose and the iBEX source of high energy neutral particles. (IBEX Mission, NASA)

Specific destinations along the Heliopause are displayed in two ways. First, a table is presented of approximate coordinates and the associated views from the surface of the earth. Constellations as visible from the surface of the Earth are rendered to aid recognition. The precise location along the Heliopause is at the center of each image. The table lists the destinations in Waves 1 and 2, temporally separated to allow investigation of the Heliopause at different times in the solar cycle. Second, a series of all sky-plots are included in a variety of coordinate systems.

Destination Region Table

Name	Ra	Dec	Nearby Sky
Wave 1			Reach destination regions at solar minimum
Apex	30	270	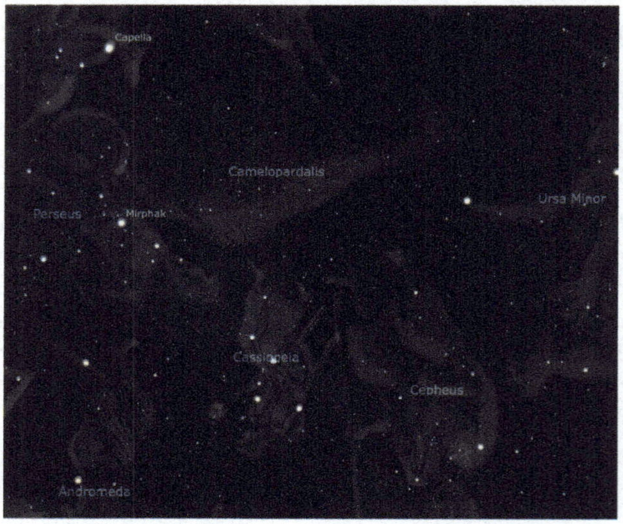

Anti apex	-30	90	
Nose	5	260	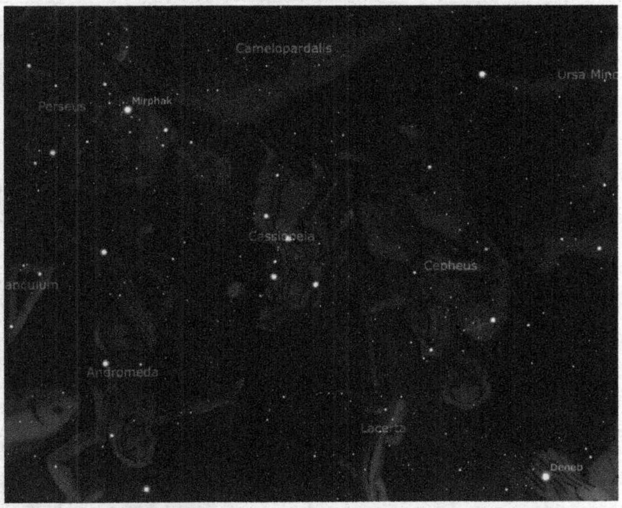
Anti nose	-5	80	

90° from apex	-40	240	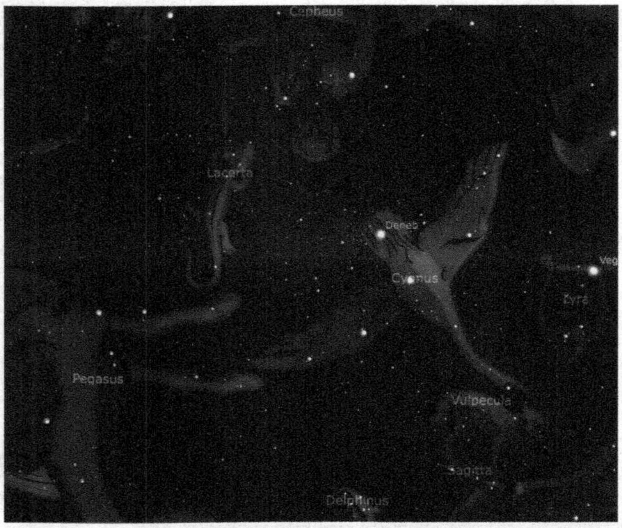
90° from nose	90	270	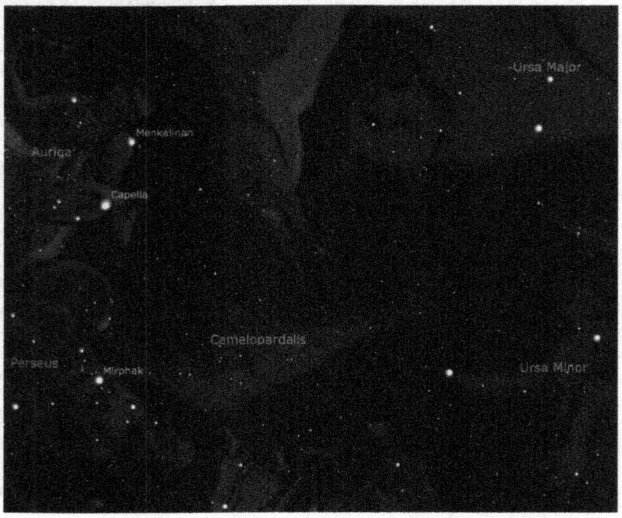
Wave 2			(Same destinations as Wave 1, but 11 years later) to study at different times in the solar cycle

2.4 Coordinate Systems

All coordinates presented here are referred to the J2000 epoch. The starting coordinate system is the standard Right Ascension (RA) and Declination (Dec) Earth-based Equatorial system used in most star maps. Right Ascension is a coordinate based on a sidereal day, the time it takes the Earth to make a 360 degree rotation inertially. It is classically broken up into 0 - 24 hours as in a solar day but the sidereal day is 1/365.25 (approximately 4 minutes) shorter than a solar day. The scale is 24 hours to 0 hours as you move from left to right on the map, as time marches on, stars that reach your north-south meridian are assigned a RA corresponding to the sidereal time. The stars are fixed in this coordinate system but the sun moves approximately 1 degree (2 apparent solar diameters) eastward each day in this coordinate system (right to left). These coordinates are dependent on an epoch such as J2000 because they are tied to the Earth and change as the Earth's axis precesses. Dec goes from 0 on the projection of the Earth equator to +/-90° at the poles.

Figure 7 is a segment of a star map showing the RA scale on the bottom. The blue line across the middle is the terrestrial equator, Dec = 0, so the star Regulus has positive Dec while the star Alphard has negative Dec. The diagonal line that passes through Regulus is the ecliptic, the path the sun moves along during the year.

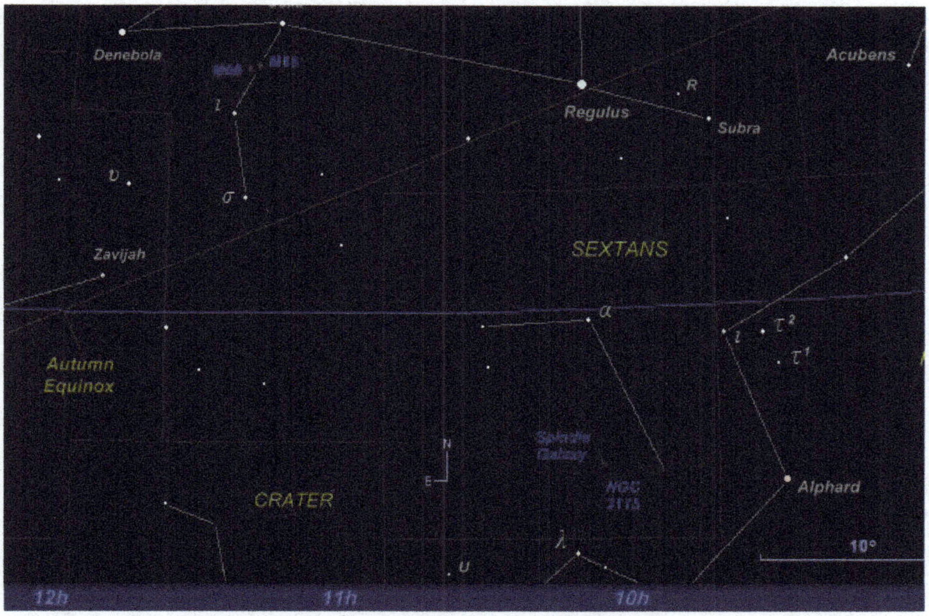

Figure 7: Star Chart in Right Ascension and Declination Coordinates

In our charts, the RA scale is changed to degrees (alpha) to correspond to other coordinate systems longitudes, i.e., RA corresponds to longitude and Dec corresponds to latitude (delta).

The maps are presented in a cylindrical projection, an x-y grid where longitude is x and latitude is y. In these projections, areas near the poles become distorted "beyond recognition". A projection that maps a sphere onto an elliptical area brings the polar areas more into what would be visualized in the real sky.

Figures 8-11 show the above maps with the points of interest for the project marked and a limited number of stars for orientation (dark blue). The Equator is in cyan. Also in cyan are the positions of spacecraft that will leave the solar system, Pioneer 10 and 11, Voyager 1 and 2, and New Horizons (P10, P11, V1, V2, and NH). The Galactic North pole (GNP), Galactic South pole (GSP) , Galactic center (GC), anti-Galactic center (AGC) and Galactic Equator are in red. The solar apex, the point in the sky that the sun is moving towards with respect to local stars, the anti-solar apex and apex equator are in black. The ecliptic is in grey. The Nose, Anti-Nose and Nose Equator are in green. The magenta dots are a rough reproduction of the iBEX ribbon. Plots are made with Caltech's FORTRAN plotting package, pgplot.

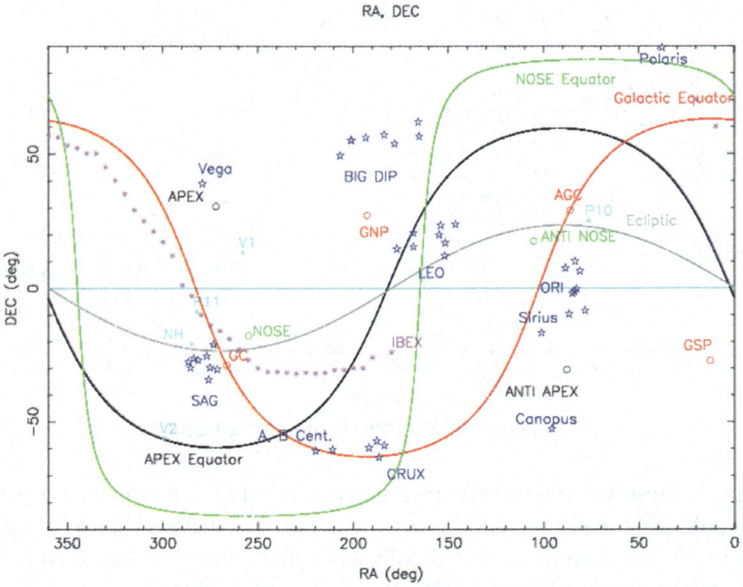

Figure 8: Equatorial Projection of Sky

The next coordinate system is Ecliptic coordinates, based on the ecliptic above. The coordinates are called Lambda and Beta corresponding to longitude and latitude. The 0 degree (beta) latitude is the ecliptic so the sun stays at 0 degrees latitude as it moves across the chart. The ecliptic North pole is at right ascension 18h 0m 0.0s (exact), and declination +66° 33′ 38.55″ Figure 9 shows the Ecliptic projection.

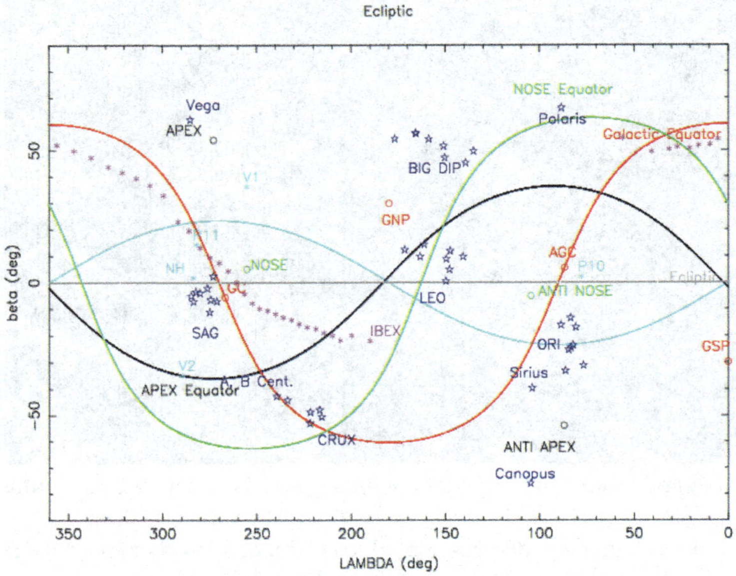

Figure 9: Ecliptic Projection of Sky

The next system is the Galactic coordinate system that is based on the Milky Way (our galaxy's center and poles). These coordinates are called l and b for longitude and latitude. The b=0 line is along the centerline of the Milky Way. The Galactic pole is at RA = 12h 51m.4, Dec = +27°.13 and the galactic center is at RA = 17h 45m.6, Dec = -28°.94. The galactic projection is shown in figure 10

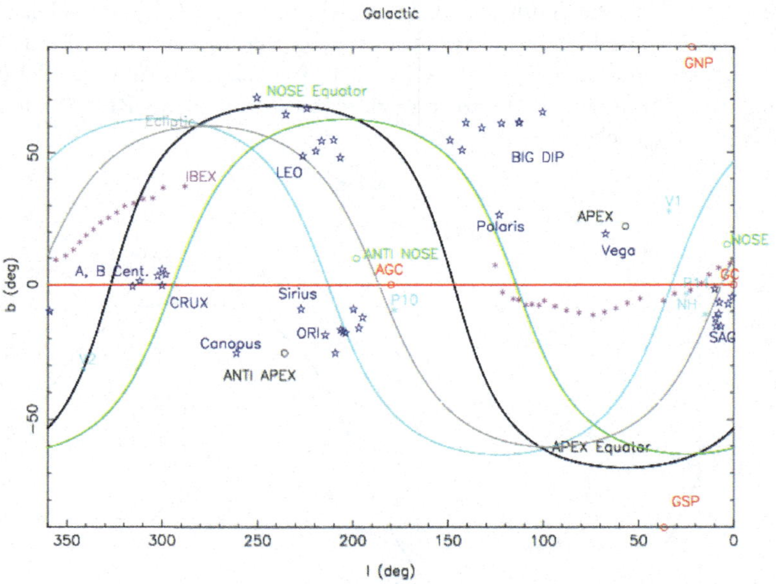

Figure 10: Galactic Projection of Sky

A particular mapping that preserves areas (equal area projection) is a Hammer map. It does not result in excessive crowding around the poles as severely as some other maps but does have a disadvantage that lines of constant latitude are not straight. There is another map (Mollwiede) that conserves areas and produces parallel lines of latitude but it is not solvable in closed form (requires iteration) and produces excessive crowding at the poles. The transformation of any of the above maps into a Hammer map is:

$$x = \frac{2\sqrt{2}\cos b\ \sin\frac{a}{2}}{\sqrt{1 + \cos b\ \cos\frac{a}{2}}}$$

and

$$y = \frac{\sqrt{2}\sin b}{\sqrt{1 + \cos b \cos\frac{a}{2}}}$$

where 1 is the longitude coordinate, b is the latitude coordinate and a is the difference between the longitude of a point and the center longitude for the map (180 degrees used). The calculated x and y coordinates reside within an ellipse in the resulting map.

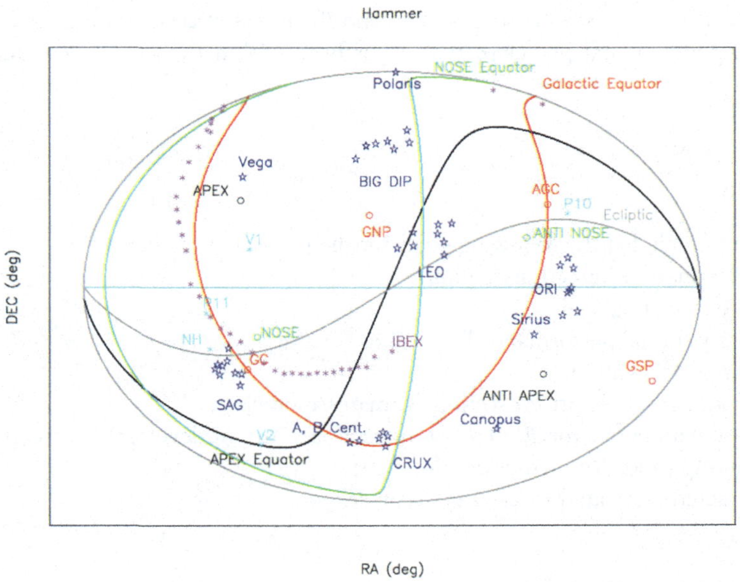

Figure 11: Hammer Projection for Equatorial Sky

Conclusions

Programs to convert between different coordinate systems were developed for the SSEARS mission concept study. This was done to provide coordinates in a variety of systems to improve communication with scientists comfortable with each of the three systems. The table below list points of interest in the various coordinate systems.

Points of Interest in 3 Coordinate Systems, J2000.

Point	Equatorial		Ecliptic		Galactic	
	RA	Dec	lambda	beta	l	b
Galactic North Pole	192.85951	27.1283	180.023	29.8115	122.932	90
Calactic Center	266.4051	-28.9362	266.84	-5.53631	0.00011	-0.0001
Galactic South Pole	12.859508	-27.1283	0.0232	-29.8115	302.798	-90
Anti Calactic Center	86.4051	28.9362	86.8396	5.53631	180	0.0001
Apex (visible)	272	30.5	272.927	53.919	56.9845	22.0802
Anti Apex	92	-30.5	92.9268	-53.919	236.985	-22.0802
Nose	254.73074	-17.5687	255.4	5.1	3.53975	15.2112
Anti Nose	74.730744	17.5687	75.4	-5.1	183.54	-15.2112

3. SSEARS Science Capabilities

The SSEARS architecture is capable of enabling science investigation to the benefit of several major NASA stakeholders. For the Science Mission Directorate (SMD) SSEARS would enable the study of

1. the Interstellar Medium (IM) and its implications for the origin and evolution of matter
2. the influence of the IM on the Heliosphere
3. the influence of the Heliosphere on the IM
4. the dynamics of the coupling of the Heliosphere and the solar system as a model for other planetary systems

For the Human Exploration/Operations Mission Directorate (HEOMD), SSEARS could assist and possibly enable human interplanetary travel by enabling the prediction of periods of low GCR penetration into the inner solar system.

For the Heliophysics directorate, the SSEARS architecture would enable direct, in-situ investigation of two of the four goals of the 2012 Heliophysics Decadal Durvey:
Determine the interaction of the sun with the solar system and the interstellar medium
Discover and characterize fundamental processes that occur both within the Heliophere and throughout the universe.

4. Phase 1 Goals

Our phase 1 NIAC proposal included the following goals for the phase 1 study period:

1. Identify science destination "directions/regions"
2. Develop trajectory modeling tool
3. Determine current material performance limitations
4. Project material trends to ~2030
5. Identify/design concept spacecraft for scaling estimates/modeling
6. Determine mission parameters for 2030 mission assuming current development trends
7. Propose for follow-up funds for further development

We have successfully completed our stated Phase 1 goals on schedule.

5. Propulsion Methods

This section will describe the decisions made that led to the selection of the solar sail as a method of reaching the Heliopause.

Among our first thoughts when planning this project was a recognition of the need to reach the Heliopause much faster than the Voyager spacecraft did. Their 34 year journey was interpreted programmatically by NASA as the Voyager Primary mission (~1977- ~1989) and the Voyager Interstellar Mission (~1989-present). A single 35 year mission to return to the Heliopause strains the imagination as NASA has never in its history approved or directed a mission that, from its inception, was intended to spend over three decades en route. (Some science could be performed during cruise but the primary goals require proximity to the Heliopause.) Thus, we began our study with a review of available in-space propulsion methods.

5.1 Solar Electric vs Solar Sail

In-space propulsion has been extensively studied by various players throughout the space age. Specifically a study at JPL was done in the early 2000s and identified the best available in-space propulsion options at various distances and destinations from the sun. This study shows that solar sail propulsion is clearly superior to solar-electric propulsion for distances of many tens of AU from the sun or greater.

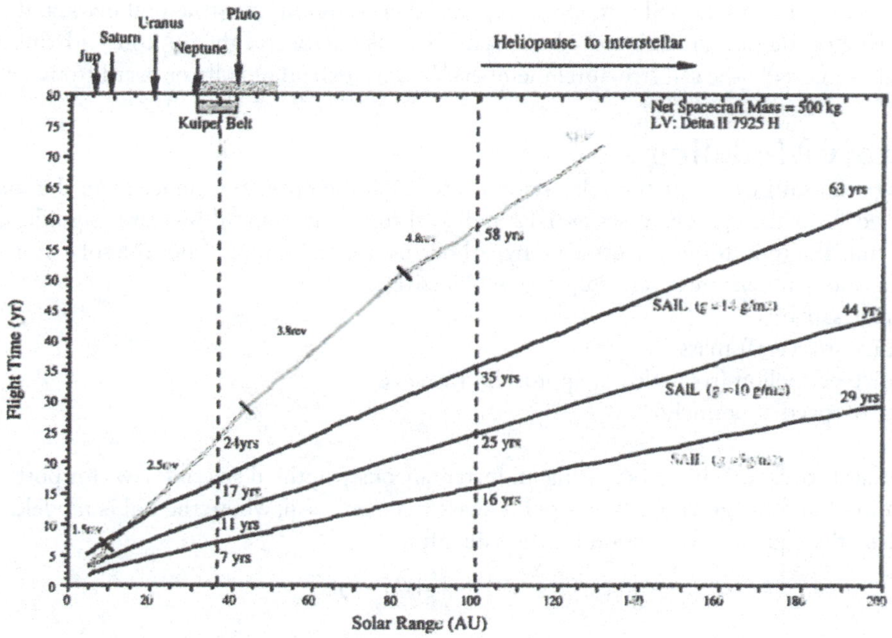

Figure 12: Solar sails are significantly more effective than solar-electric propulsion when applied to solar system escape trajectories. This is largely due to the ability of the solar sail to come close to the sun without using fuel. Here three different sail densities are compared to SEP. (Image courtesy of Chen-Wan Yen)

5.2 Nuclear Electric Propulsion

Nuclear Electric Propulsion (NEP) is not a practical solution to achieve solar system escape for most missions because it does not take advantage of the significant increase in power achieved by going close to the sun. As such, most NEP trajectories to the outer planets are relatively slow while requiring massive power systems.

We concluded early on that a solar sail was the best option for this study at this time. This coincides well with current NASA investments in solar technology demonstrations in the 2014-2015 timeframe. We hope to convey that there are further benefits of that investment.

6. Solar Sail Introduction

A solar sail is a thin film sheet, usually with a very large area, that allows the use of light pressure to exert thrust on a spacecraft. Figure 13 shows the basic components of a solar sail vehicle.

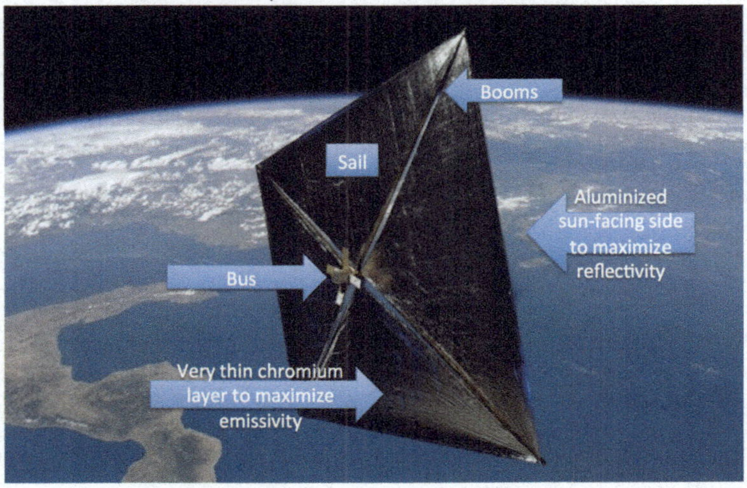

Figure 13: Labeling the top-level relevant parts of a solar sail system.

Key to the operation of a solar sail is to minimize areal density, minimize structural mass, and maximize area. The sun-facing side of the sail must be coated to maximize reflectivity and the opposite side must be coated to maximize emissivity. Usually the sail structural members "booms" extend radially outward from a central bus.

6.1 Trajectory Modeling

The key to using solar sails to escape the solar system is to reach the optimal distance from the sun. The amount of energy received from the sun decreases as 1/r2 and so thrust from the sail increases significantly as the sail approaches the sun. The trajectories involved are hyperbolic escape trajectories from the solar system. In order to achieve these trajectories there are several key design objectives:

1. Maximize sail area
2. Minimize spacecraft mass
3. Minimize perihelion (solar closest approach) distance
4. Maximize speed at perihelion

The following trajectories were developed using an instantaneously optimal steering law (for post solar-close approach). Assuming the Sun's gravity acts as a point-mass force on a sail, where the sail is modeled as a perfectly reflective flat plate, the equations of motion for the sailcraft are

$$\ddot{\vec{r}} = \frac{\mu}{r^2}(\beta\bar{a} - \hat{r}). \tag{1}$$

In Eq. (1), μ is the gravitational parameter of the Sun, r is the distance from the Sun, β (dimensionless) is the sail loading parameter, and $\hat{r}$ is a unit-vector directed from the Sun to the sailcraft. The sail acceleration is defined as

$$\bar{a} = (\hat{r} \cdot \hat{n})^2 \hat{n}, \tag{2}$$

where $\hat{n}$ is a unit-vector normal to the surface of the sail. Note that the sail loading parameter β , also called the sail 'lightness number', is the ratio of solar-radiation pressure (SRP) to solar gravity, i.e.,

$$\beta = \frac{2AW_E R_E^2}{c\mu M}. \tag{3}$$

In the above equation, A is the total area of the sail, c is the speed of light, R_E is one astronomical unit (AU), and W_E is the solar flux at one AU. The total spacecraft mass M is the payload mass m_p plus the mass of the sail, or $M = m_p + \rho A$, where is the density of the sail. It is sometimes convenient to describe the performance of the sail in terms of characteristic acceleration κ , or the acceleration provided by the sail at one AU. The characteristic acceleration is then written

$$\kappa = \beta \frac{\mu}{R_E^2}, \tag{4}$$

where κ typically varies from 0.01 mm/s^2 (doable today) to 1.5 mm/s^2 (aggressive).

For fast departure from the solar system, the goal is to maximize the component of sail thrust along the spacecraft's Heliocentric velocity vector $\hat{v}$. This optimization problem is posed succinctly as

$$\max J = \bar{a} \cdot \hat{v}, \qquad \text{subject to } \hat{n} \cdot \hat{r} > 0. \tag{5}$$

The condition $\hat{n} \cdot \hat{r} > 0$ is present to ensure that the sail acceleration defined in Eq. (2) is always directed away from the Sun. To solve the optimization problem, we begin by defining a transverse vector $\hat{\theta}$ such that $\hat{\theta}$ is in the plane of motion and normal to $\hat{r}$ such that $\hat{\theta} \cdot \hat{v} > 0$. Then $\hat{n}$ can be resolved into radial and transverse components according to

$$\hat{n} = n_r\hat{r} + n_\theta\hat{\theta}. \tag{6}$$

Making use of $\hat{n}\cdot\hat{r} > 0$ in Eq. (5), we can write Eq. (6) only in terms of n_θ

$$\hat{n} = +\sqrt{1-n_\theta^2}\,\hat{r} + n_\theta\hat{\theta}. \tag{7}$$

Now substituting Eq. (7) into Eq. (2), we arrive at the following expression for thrust acceleration

$$\bar{a} = (1-n_\theta^2)\left\lfloor\sqrt{1-n_\theta^2}\,\hat{r} + n_\theta\hat{\theta}\right\rfloor \tag{8}$$

Note that with Eq. (8), the performance index J in Eq. (5) is

$$J = \bar{a}\cdot\hat{v} = (1-n_\theta^2)\left\lfloor\sqrt{1-n_\theta^2}\,(\hat{v}\cdot\hat{r}) + n_\theta(\hat{v}\cdot\hat{\theta})\right\rfloor \tag{9}$$

Finally, from the expression provided in Eq. (9), it is easy to show that J is a maximum when

$$n_\theta = \sqrt{\frac{3k^2 - k\sqrt{9k^2+8}+2}{6k^2+6}}, \quad \text{where } k = \frac{\hat{v}\cdot\hat{r}}{\hat{v}\cdot\hat{\theta}}. \tag{10}$$

This locally optimal steering law describes the best direction for orienting the sail normal for fastest departure from the solar system. Following the final perihelion passage, all trajectories are computed by simulating the spacecraft motion with Eq. (1) using the steering law provided in Eq. (10).

6.2 Basic Trajectories – Upper Stage for Earth Escape

We assumed an aggressive sail characteristic acceleration of 1.34 mm/s^2 (assumed because it matched with previous JPL work) and initially analyzed trajectories for four cases. In each figure the blue line represents the trajectory before solar close approach. The red vectors along the blue line show the sail normal angle to the direction of motion, and the green line shows the trajectory after solar close approach with the sail guided by the control law derived in the previous section.

1. using an upper stage rocket to escape earth orbit and escape along the ecliptic (Figure 14)
2. using an upper stage rocket to escape earth orbit and escape off the ecliptic using a Jupiter gravity assist to raise inclination (Figure 15)
3. using the sail to escape earth orbit and escape along the ecliptic (Figure 16)
4. using sail to escape earth orbit and escape off the ecliptic using a Jupiter gravity assist to raise inclination (Figure 17)

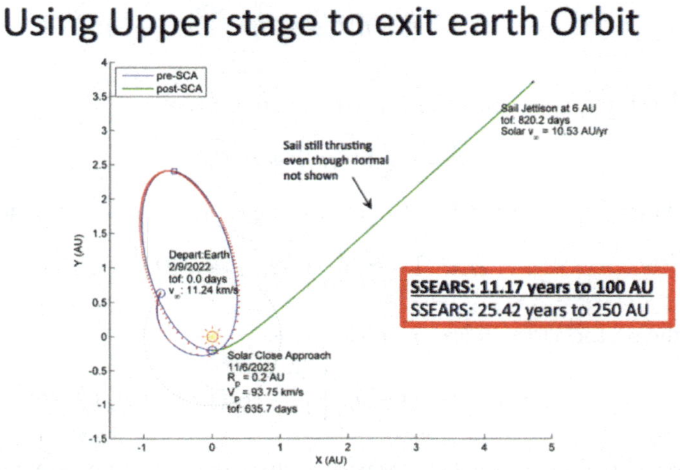

Figure 14: This trajectory shows the spacecraft leaving earth, looping once around the sun to raise aphelion and lower perihelion, and then swinging by the sun again before proceeding radially away from the sun. This method reaches the vicinity of the Heliopause along the ecliptic in 11.1 years.

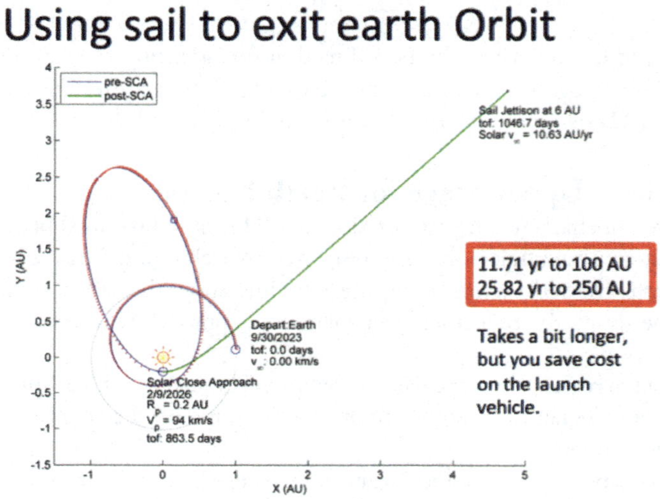

Figure 15: This trajectory uses the sail itself to exit earth orbit and takes slightly longer - ~6 months – to reach the same distance along the ecliptic as that in Figure 14.

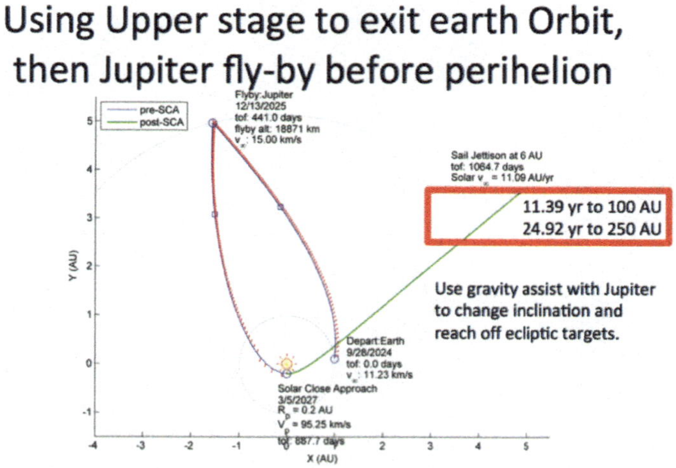

Figure 16: The same distances can be reached at off-ecliptic inclinations by using a Jupiter gravity assist before perihelion. The transit time remains similar to the other trajectories. (3d nature of inclination change not displayed.)

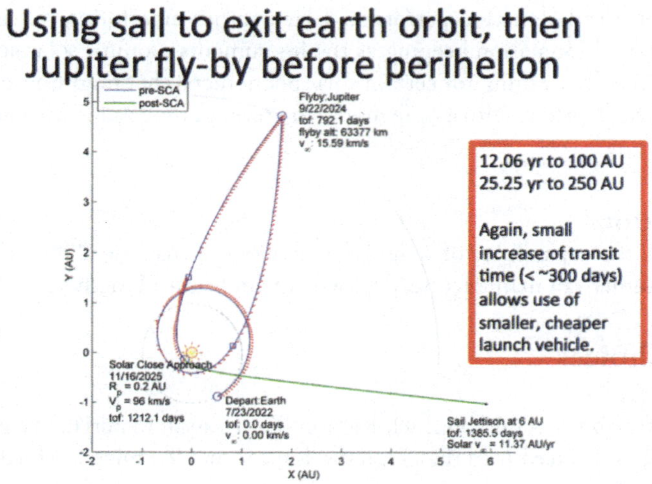

Figure 17: The same technique as in Figure 15 applies to the off-ecliptic destinations, at a modest increase in transit time of ~ 1 year. (3d nature of inclination change is not displayed.)

7. Sunjammer

This section will discuss our heritage with the current NASA Technology Demonstration Mission "Sunjammer." Our proposed flight system is based on the Sunjammer spacecraft, due to launch in 2014. We studied the Sunjammer capabilities and decided to use it as a real-world foundation for the spacecraft design necessary to reach the Heliopause.

7.1 Programmatic Background

Sunjammer is a technology demonstration project scheduled for launch in 2014. It is a sun observing mission with a 34 m x 34 m solar sail propulsion system. The sail is supported by 4 booms extending from the bus and ending in actuated vanes that enable steering of the sail.

Figure 18: An image of the 34m x 34m Sunjammer sail.

Led by industry manufacturer L'Garde Inc. of Tustin, Calif., and including participation by the National Oceanic and Atmospheric Administration (NOAA), the Solar Sail Demonstration mission builds on two successful ground-deployment experiments conducted by L'Garde in 2005-2006 in a vacuum chamber at the Plum Brook Facility in Sandusky, Ohio, a research laboratory managed by NASA's Glenn Research Center in Cleveland. It also leverages the successful deployment of the NanoSail-D sail, a 100-square-foot test article NASA launched to Earth orbit in early 2011 to validate sail deployment techniques [1].

7.2 Mission Summary

The Sunjammer spacecraft was designed to perform a solar weather monitoring mission while "parked" at Sun-Earth L1. The sail will be used for station keeping as the instruments monitor solar activity. Sunjammer should give NOAA and NASA an early warning for certain solar phenomena that might necessitate the proactive safing (switching spacecraft electrical systems into a "safe mode" to prevent interactions with solar phenomena) of certain assets in Earth orbit.

7.3 Scaling Sunjammer

We decided to investigate the capabilities of a Sunjammer-like sail that was 250m x 250m. This is roughly the scaling limit of the Sunjammer sail manufacture, deployment, and control methods.

7.4 Sail Material Choices

7.4.1 Present Sail Materials

The Sunjammer spacecraft will use 5 micron thick Kapton film as a sail material. We performed an investigation into the materials available and determined that 5 micron Kapton and 0.9 micron Mylar are the two leading materials for this application. Mylar however does not survive radiation environments well and further study is required to determine whether this fully precludes the use of Mylar as a solar sail material.

7.4.2 Future Sail Materials

We contacted several manufacturers of Kapton (DuPont, 3M) and were told that 2 micron thick Kapton film is within manufacturing capability. We will see in a later section how this change affects Heliopause cruise times. The 0.5 micron thick Kapton film is theoretically possible but has not been significantly investigated because even 2 micron Kapton has yet to find a commercial application significant enough to justify the necessary modification to manufacturing facilities.

In parallel to the evaluation of DuPont's Kapton® films for the proposed application, we have identified other candidate materials to be studied in phase 2. This study will include CP-1, a potential successor to CP-1 known as CORIN™ XLS, and Thermalbright®2. CP-1 is a space-durable material developed at NASA Langley (LaRC) and exhibits a high resistance to UV radiation. It has currently been fabricated in large sheets and rolls to as thin as 1.5 micron. CP-1 has flown on Hughes HS-702 geosynchronous communications satellite. CORIN™ XLS is a potential next-generation CP-1. Thermalbright® polyimide is a high temperature highly reflective white polyimide film which is expected to be particularly beneficial for thermal control while maintaining good UV and VUV durability.

8. SSEARS Trajectories

With our sail materials in mind, and an assumed spacecraft mass of 110kg, we modeled the transit time for sails with the different materials to 100AU. We also moved to a more realistic characteristic acceleration of 0.5 mm/s^2. We chose two sail sizes to compare, the 250m x 250m size that is the limit of current techniques, and a hypothetical 500m x 500m sail. Material choices are represented by the line color. We also estimated a scaling factor for the boom masses. We assumed two methods of boom mass scaling – linear (here called aggressive) and geometric (here called conservative). This factor is represented by the line style. Solid lines represent a conservative boom mass scaling factor, and dashed lines represent an aggressive boom mass scaling factor. The curve plots in figures

19 and 20 represent these results. Finally, vertical black lines represent the distance to the sun below which the listed material begins to degrade.

The below charts were derived from the initial spacecraft mass assumption of 110kg. When we completed our detailed spacecraft configuration, we concluded that a more realistic spacecraft mass is 175kg. The transit times were relatively insensitive to this change in spacecraft mass.

8.1 SSEARS Model Payload Trajectories with 250x250m Sail

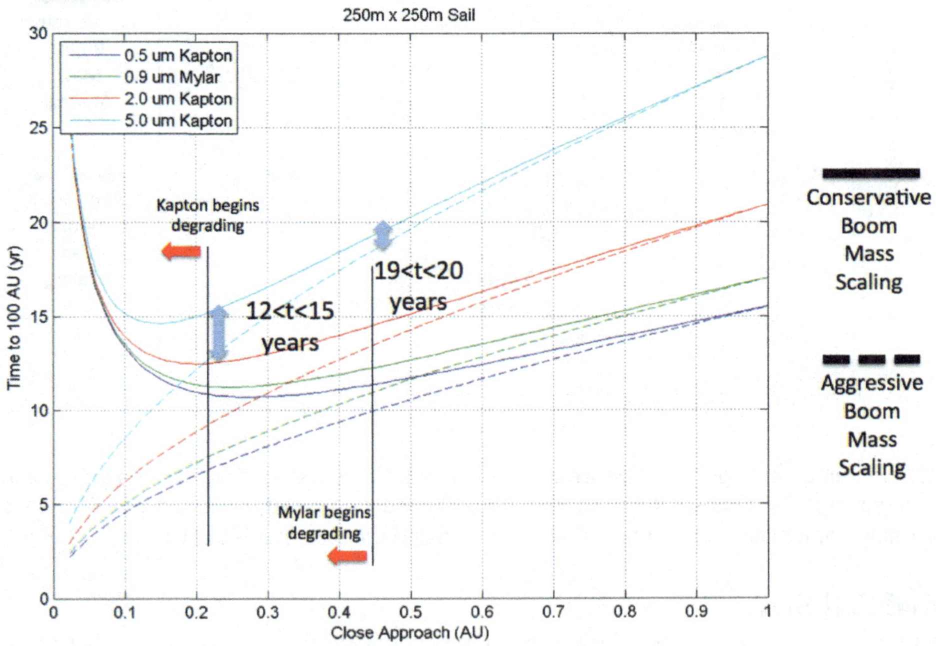

Figure 19: We can see that 5 micron kapton, the material used by Sunjammer, provides the slowest velocity to the Heliopause. It cannot get closer than ~0.22 AU from the sun and so the theoretical cruise time to 100AU is between 12 and 15 years depending on the boom scaling factor. The 0.9 micron thick mylar, which cannot get closer than 0.45 AU from the sun and is a difficult material to use from a radiation tolerance perspective, delivers the same spacecraft to the Heliopause in between 10 and 11 years.

8.2 SSEARS Model Payload Trajectories with 500x500m Sail

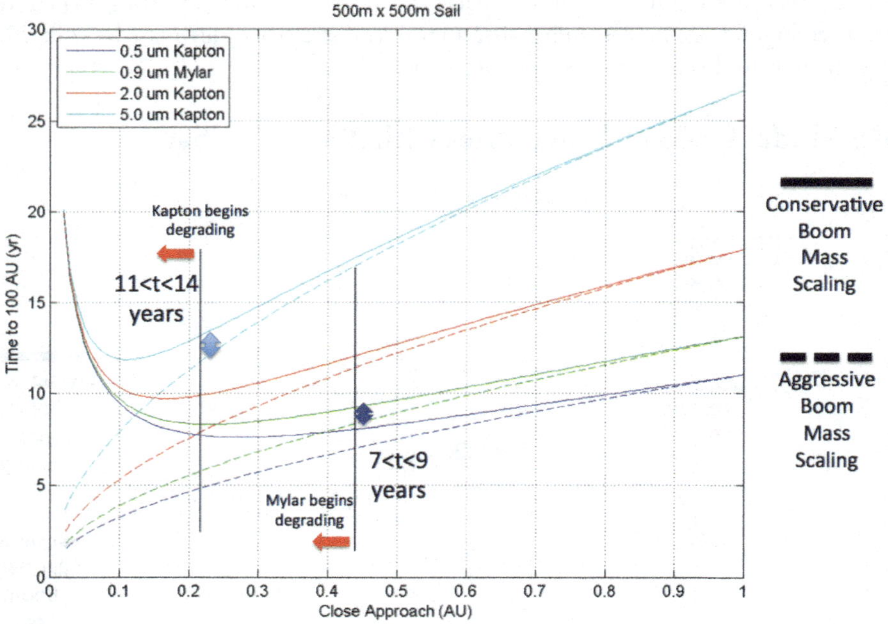

Figure 20: With an even larger sail size we can see the trip times reduced by about a year, each. This result was very insightful because it shows that the additional complexity and cost of a 500m x 500m sail is probably not justified as little is gained by increasing the size of the sail beyond 250m x 250m.

8.3 Optimal Sail Size

Our investigation into the sail material showed us that a significant increase in sail size beyond 250m x 250m does not provide significant velocity improvement. Our flight system design study, then, assumes a 250m x 250m sail size.

8.4 Sail Jettison

The sail would not receive much energy from the sun by the time it crosses Jupiter's orbit. At that time the sail would be turned and then jettisoned (in order to prevent the sail's trajectory from blocking communication from the spacecraft.) The spacecraft would continue alone. A series of renderings follow that demonstrate this process.

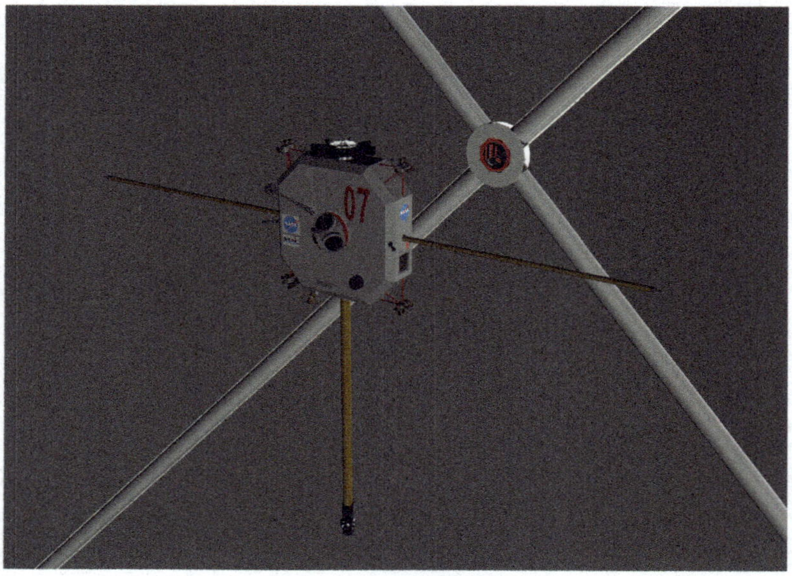

Figure 21: The spacecraft separates from the sail and continues.

Figure 22: Side view of the spacecraft following separation.

9. Spacecraft / Flight System

This section describes the spacecraft architecture that we designed to meet the requirements described up to this point in this report. We being with a discussion of the design approach and show several configuration graphics to demonstrate the stages of the mission.

9.1 Spacecraft Design

The guiding principle of the spacecraft design process was to leverage existing technology as much as possible. This involved scaling up the nominal Sunjammer design concept and using high heritage instruments. The sail would be folded and packed into a modular container. The spacecraft concept itself is visible on the foreground-facing side of the sailcraft container.

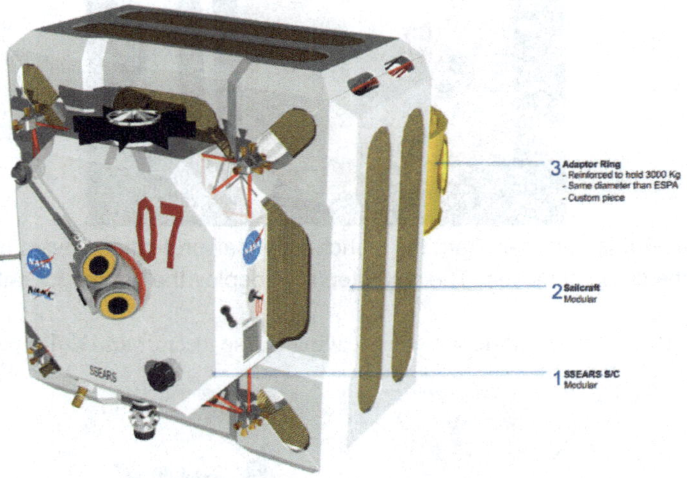

Figure 23: Here the two main components of the module can be seen: the module itself and the sail craft. In the stowed mode 3 complete spacecraft can be stored inside the fairing of an Atlas V launch vehicle (5m in diameter).

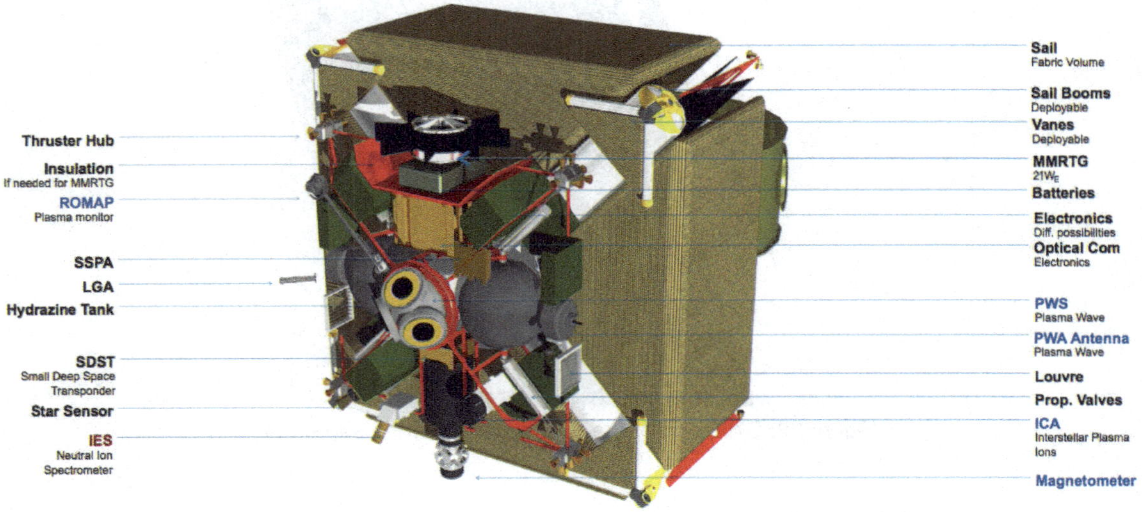

Figure 24: Here the main selection of instruments can be seen as well as the internal structure and configuration. Some of the green volumes represent extra electronic boxes to either support current instruments or to be substituted by extra payload using a better packing ratio.

The sail container deploys in this fashion:

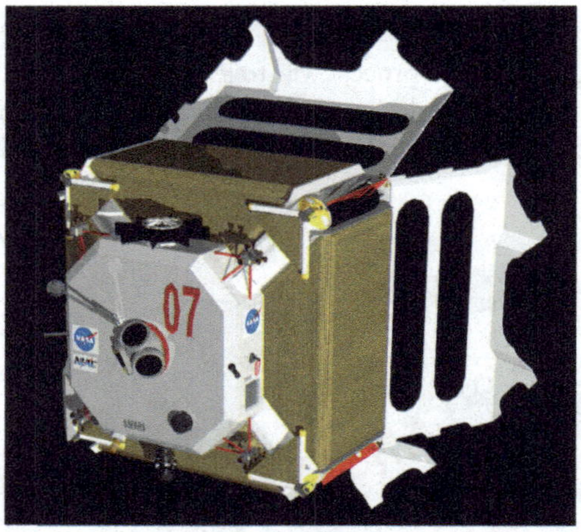

Figure 25: Once SSEARS is detached from the launcher the sail protection covers open and the vanes are deployed. After that the sail is deployed. The next step is to deploy the booms for antennas and instruments.

This component then jettisons after sail deployment leaving the spacecraft and sail separated.

The three major systems are visualized side-by-side for scale comparison below. The left side of the image shows a side view of the spacecraft, sail structure, and mounting hardware (left-to-right) and the right side of the image shows the spacecraft with the booms deployed.

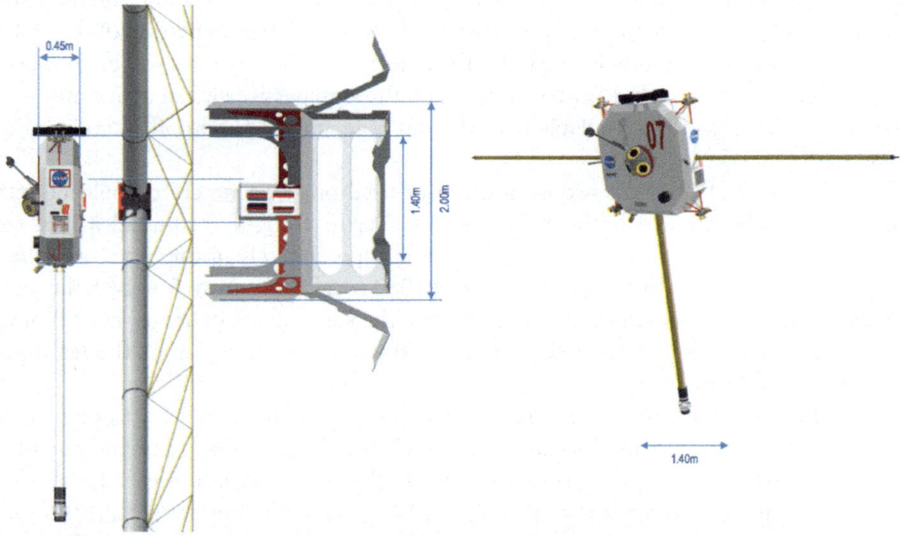

Figure 26: The three spacecraft elements are displayed side by side to provide a sense of relative scale. Launch vehicle compatibility with Atlas 5 501 was shown for 3 units per Atlas 5.

9.2 Communication

We determined early in our study that a large dish antenna communication system would be so massive (as necessary to function from 100AU) that it would severely constrain the design of the rest of the spacecraft. We determined that optical communication can meet our needs.

9.2.1 Downlink

Laser beams at optical frequencies can be transmitted with angular beam-widths of a few microradians (mrads). Coupled with the ability to accurately point narrow laser beams to a fraction of the beam-width, signal power densities required for communication can be delivered over huge distances. To illustrate this further a point design is presented below.

A 20 watt average power 1550 nm laser beam transmitted through a 50 cm diameter telescope will result in an angular beam-width of approximately 3.6 mrad. Pointing jitter control of ~ 0.3 mrad will result in losses relative to the on-axis peak of ~ 1.5 dB. If such a beam with the indicated pointing control were transmitted over a distance of 100 AU with a 12 m diameter collector at the receiving end a 500+/-100 bits/second communication link could be established. This accounts for reasonable transmitter and receiver optical transmission and implementation losses and 3-dB of link margin. Such a link is enabled by the availability of high peak-to-average power ratio (160:1 to 320:1) lasers that permit the use of near-capacity achieving serially concatenated pulse-position modulation (SCPPM) error-correcting coding together with a single-photon-counting sensitivity receiver.

9.2.2 Phase II Plans

A high level summary of the studies for maturing the concept of Optical Communication from a distance of 100 AU to Earth are tabulated in Table 1 below.

	Study Topic	Approach
1	Acquisition and tracking strategy	Sun as beacon combined with knowledge generated on-board of Earth position and phase
2	Concept of operations	Strategies needed to operate with > 1 day round-trip light time
3	Terminal Architecture Requirements	Emphasis on special environmental and lifetime requirements

Link acquisition and pointing strategy based on the remote spacecraft terminal utilizing the Sun as a reference to off-point to Earth using on-board generated position and phase knowledge, will be part of our phase 2 study. Note that off-pointing is mandatory since the Sun-Earth angular separation of 10 mrad from 100 AU is many laser beam diameters. Pointing must also account for point-ahead angles of 100-200 mrad or 30-40 laser beam widths where a beam-width of 3-6 mrad is being considered. Residual 1-sigma errors of 0.3-0.4 mrads must be achieved and maintained for the duration of the link. The objective of these studies will be to show through modeling and simulation that: (a) spacecraft disturbance rejection through passive and/or active means can be achieved and; (b) adequate control bandwidth is available with the sun beacon to maintain offset pointing to Earth within the allocated error.

The field-of-view (FOV) of the acquisition sensor used on the spacecraft would be relatively large in order to overlap the angular separation between the Sun and Earth. Roughly ±10 mrad will be required. At the same time the instantaneous field-of-view (IFOV) should be approximately of the order of the transmitted laser beam-width i.e. close to 10 mrad. This suggests a 1000 x 1000 focal plane array, however, the viability of implementing such a sensor needs further study. Even at 100 AU the sun angular diameter is ~ 90 mrad and this will have to be factored into achieving the targeted residual 0.3-0.4 mrad pointing error. The feasibility and benefits of using other celestial objects will be explored.

On the ground the beam footprint from 100 AU would be approximately 4 Earth diameters, however, the irradiance would be a few femtowatts per square meter. A 12 m or larger collector or an array of large collectors would be required to gather sufficient photons to overcome background and shot noise. In our preliminary analysis a 256 pixel superconducting nanowire array behind a 12 m collector supported 500±100 bits/sec. Ground aperture would very likely utilize a conical scan to acquire and lock on the downlink signal. The use of adaptive optics and aggressive filtering to discriminate the faint signal against prevalent background light will be studied. The performance achieved with the latter would determine how close to the Sun the ground terminal can point and this would also determine the duration of possible outages. For comparison the capability achievable with an orbiting large aperture collector/receiver will be determined where shot-noise limited performance can be achieved.

The concept of operations for a 100 AU optical communication link needs further study. From ranges comparable to where Voyager II spacecraft is at present, round-trip light times of approximately 27 hours (> 1 day) will be encountered. Since bi-directional acknowledgement and re-transmission requests will not be viable with such long ranges, optical communication from 100 AU will involve repetitive data transmission over a sufficiently long duration to ensure signal detection and data reception at Earth. While similar to the strategy used by Voyagers downlinking data to DSN, optical communications would require additional considerations related to cloud-cover, number of available ground stations and their distribution around the globe. In our proposed study these considerations will be addressed and the autonomous capability required by the spacecraft to implement the concept of operations will be evaluated. Emphasis on spacecraft resources needed to support optical communications will also be assessed.

The suitability of space terminal architectures pursued for deep-space optical communication within the solar system will be re-examined for the expected environment at 100 AU, as well as, expected longer lifetime demands. The enhanced acquisition field-of-view mentioned earlier is an example of a needed modification. The spectral characteristics of the sun beacon will also result in optical design modifications. Depending on the concept of operations the command receiving architecture and data-buffering requirements will very likely require modifications. Lifetime demands may impose added redundancy and in some cases re-engineering of components or assemblies. The radiation environment and shielding requirements must also be factored in.

9.2.3 Uplink

For our baseline configuration we assume that the spacecraft would perform autonomously until it fails. In the case of our 100 AU mission this makes sense because there are no encounters, there are no maneuvers, and navigation is moot once the spacecraft has passed the sun. The spacecraft does have a small low-gain antenna for near-Earth communication. We will further investigate the uplink strategy in Phase 2.

9.3 Instrument Suite for Model Spacecraft

The following table shows the baseline instrument suite selected for this spacecraft design. Other instrument suites will be developed and configured as part of the Phase 2 work.

Instrument	Heritage
Magnetometer	Cassini
Plasma Monitor	ROMAP - Rosseta
Plasma Wave	PWS - Voyayer
Neutral Ion Spectrometer	IES - Rosetta
Interstellar Plasma Ion Detector	ICA - Rosetta

This suite primarily focuses on the interactions between the solar wind and the local interstellar medium. Other spacecraft would carry instruments intended for study of the galactic wind, interstellar dust, and galactic cosmic rays as described in the next section. One of the advantages of the spacecraft architecture is that duplicate spacecraft could be built with different instruments. We expect to employ the following instruments in other instrument suites (to be further defined in phase 2:

1. Energetic particle monitor (>500kev)
2. Anomlaous cosmic rays
3. Dust detector
4. Galactic cosmic ray detector (<10 MeV/nuc, 10-100, MeV, and >100 MeV)

9.4 Mass Equipment list (MEL)

The initial estimate of spacecraft mass of 150 kg was close to the mass computed by the detailed spacecraft design: ~150kg. Accounting for margin, a spacecraft mass of 175 kg is used for our trajectory calculations and our spacecraft mass came to less than 170 kg even when fueled and with 10% system margin.

SSEARS Mass Equipment List (MEL)

	Mass (kg)	Units	CBE (kg)	Contin-gency	Expected Mass (kg)	Heritage	Comments
Spacecraft							
Science Instruments			6.30	30%	8.19		
Magnetometer	0.70	1	0.70	30%	0.91	Cassini	
Plasma Monitor	1.00	1	1.00	30%	1.30	Rosseta	ROMAP
Plasma Wave	1.40	1	1.40	30%	1.82	Voyager	PWS
Neutral Ion Spectrometer	1.00	1	1.00	30%	1.30	Rosseta	IES
Interstellar Plasma Ions	2.20	1	2.20	30%	2.86	Rosseta	ICA
Structure and Mechanisms			33.00	28%	42.30		
Bus structure	15.00	1	15.00	30%	19.50		Composite honeycomb panels
Secondary structure (incl. booms)	7.00	1	7.00	30%	9.10		
Cabling harness	5.00	1	5.00	30%	6.50		
System Assembly Hdw.	4.00	1	4.00	30%	5.20		
Ballast	2.00	1	2.00		2.00		
C&DH			8.50	10%	9.38		
Spacecraft Flight Computer (SFC)	0.60	1	0.60	5%	0.63	SMAP	
Non-Volatile Memory (NVM)	0.70	1	0.70	5%	0.74	SMAP	
MSAP Telemetry Interface (MTIF)	0.70	1	0.70	5%	0.74	SMAP	
MSAP System Interface Assy (MSIA)	0.70	1	0.70	5%	0.74	SMAP	
Local Eng. Unit Digital Card (LEU-D)	0.70	1	0.70	5%	0.74	SMAP	
Local Eng. Unit Analog Card (LEU-A)	0.70	1	0.70	5%	0.74	SMAP	
CDH Elect. Power Conv. Unit (CEPCU1)	1.10	1	1.10	5%	1.16	SMAP	
Critical Relay Controller Card (CRCC)	0.30	1	0.30	5%	0.32	SMAP	
Chassis and backplane (incl. Power cards)	3.00	1	3.00	20%	3.60	SMAP	Holds both C&DH and Power
ACS			3.65	16%	4.25		
Star Tracker/Scanner	2.70	1	2.70	20%	3.24	SMAP	
Sun Sensor	0.10	2	0.20	10%	0.22	SMAP	
LN-200S IMU	0.75	1	0.75	5%	0.79	MSL	
Power			17.62	20%	21.18		
Small MMRTG	10.32	1	10.32	30%	13.42		21 W_E BOL, 250 W_t
Battery	3.00	1	3.00	5%	3.15	MER	150 Wh
Power Switch Slice (PSS)	1.90	1	1.90	10%	2.09	SMAP	
Housekeeping Power Cond. Unit (HPCU)	1.00	1	1.00	5%	1.05	SMAP	
Power Bus Controller (PBC)	1.40	1	1.40	5%	1.47	SMAP	
Telecom			30.68	26%	38.61		
Small Deep Space Transponer (SDST)	2.65	1	2.65	5%	2.78	MSL	
X-Band SSPA	1.35	1	1.35	5%	1.42	MSL	17 W_{RF}
Coax Transfer Switch (CXS)	0.10	1	0.10	10%	0.11		
Filter	0.28	1	0.28	10%	0.31		
Coax Cable	0.30	1	0.30	30%	0.39		
Low Gain Antenna (LGA)	0.50	2	1.00	10%	1.10	SMAP	
Optical Com System	25.00	1	25.00	30%	32.50		75 W_E input power
Propulsion			8.90	14%	10.18		
Hydrazine tank	0.80	2	1.60	30%	2.08	ATK custom	10 liter tank, assume ½ press.
Thrusters	0.40	8	3.20	5%	3.36		4 N monoprop thruster
Valves	0.30	4	1.20	10%	1.32		
Filter	0.50	1	0.50	5%	0.53		
Propellant lines and components	1.50	1	1.50	30%	1.95		
Sensors	0.30	3	0.90	5%	0.95		
Thermal Control			7.70	30%	10.01		
Multilayer Insulation (MLI)	0.38	10	3.80	30%	4.94		
Heaters	0.05	20	1.00	30%	1.30	SMAP	
Sensors	0.01	30	0.30	30%	0.39	SMAP	
Louvers	1.30	2	2.60	30%	3.38		
S/C Dry Mass			116.35	**24%**	144.09		
Hydrazine Propellant	10.00	1	10.00		10.00		
S/C Wet Mass			**126.35**	22%	**154.09**		
S/C Wet Mass with 10% system margin					**168.50**		
Sail Module	625.00	1	625.00	30%	812.50	Sunjammer	σ=10 g/m², A=62,500 m²
Inflation Module	50.00	1	50.00	30%	65.00	Sunjammer	Incl. adapter & sep'n. hdw.
TOTAL LAUNCH CBE+Contingecy			**801.35**	**50%**	**1,200.08**		

With that spacecraft mass the transit times for different sail materials are provided in Figure 27. Boom mass scaling figures apply as in previous graphs (solid line is conservative scaling, dashed line is aggressive scaling.) 2 micron thick kapton is the material of choice both from a near-mid term (~2020s) feasibility standpoint and a cruise time standpoint. With this material cruise times range between 15 and 18 years. However 2 micron Kapton has yet to find a commercial application sufficient to motive manufacturers to develop it. With enough time .5 micron thick Kapton will emerge as the better choice, further reducing cruise times to between 12 and 15 years.

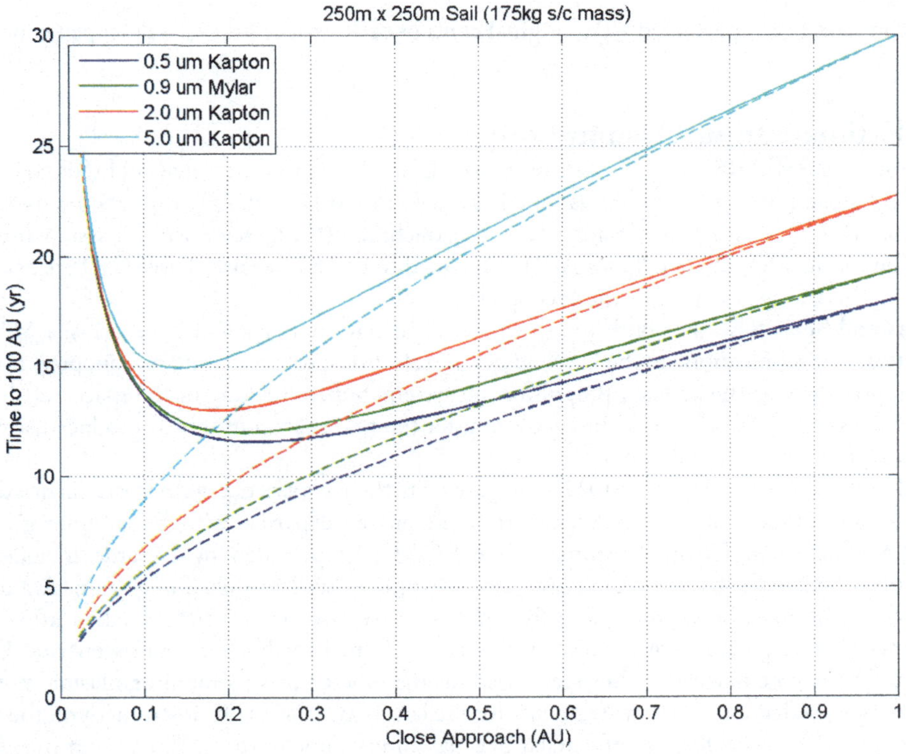

Figure 27: Our estimated spacecraft mass of 175kg raises the transit time somewhat, and shows that 2 micron Kapton is a clear choice for sail material. A phase two task will be to discuss the manufacturing of 2 micron thick Kapton with various manufacturers.

9.4.1 Thermal Control Challenges

The spacecraft would face thermal control challenges both at perihelion and in deep space. The spacecraft concept uses louvers, shielding and insulation as accounted for in the MEL. There is an external wall of aluminum between the sailcraft and the spacecraft itself that would ensure there is a shadow on the spacecraft when the sun is behind the spacecraft. Regarding thermal tolerance as the spacecraft approaches the sun, it is unclear how close the spacecraft can get to the sun, and how long the spacecraft can stay there. Part of our phase 2 study will be to determine specific trajectories, their time spent in the high heat and radiation environment near the sun, and the extent to which the short duration spent at those distances may, to some degree, alleviate thermal and radiation exposure concerns. As a baseline power source, the small RPS would be mounted to the spacecraft near a central position and the waste heat would be used to heat the spacecraft in deep space.

10. Cost

The cost information contained in this document is of a budgetary and planning nature and is intended for informational purposes only. It does not constitute a commitment on the part of JPL and Caltech.

The SSEARS cost estimate for a solar-sail propelled spacecraft, designed to perform fields and particles research at the Heliopause region, was performed using a variety of tools and methodologies. Overall, two distinct flight system cost scenarios were developed, a ten-spacecraft mission concept and a smaller three-spacecraft mission concept, which is considered the descope option. JPL's Parametric Mission Cost Model (PMCM), which uses statistically derived cost estimating relationships in order to estimate the full breadth of the JPL mission work breakdown structure (WBS), was utilized in the estimate. These costs were supplemented with instrument costs, developed using analogous instruments flown on previous missions. L'Garde Inc. provided a ROM (rough order of magnitude cost) for the 250m x 250m solar sail. The launch vehicle cost was determined using ROM costs provided by the Launch Services Program at Kennedy Space Center (KSC). Subsequently, a cost risk analysis was performed using ACEIT to assess a confidence level in the estimate. All estimates were performed in FY13 constant year dollars.

Two mission sizes were considered – the original concept of 10 spacecraft and a descoped concept with three spacecraft.

10.1 Cost Estimate Input Assumptions

Overall Mission: The SSEARS mission is assumed to be Class A mission with dual cold redundancy. Consistent with JPL reserve policy, 30% reserves are assumed through Phase D, and 15% reserves are assumed in Phase E. A team of 6.5 (FTE) scientists is required through launch; 2.5 (FTE) scientists are assumed to be adequate post-launch. The formulation phase (Phases A/B) is assumed to be 18 months; Phases C/D are assumed to be 40 months, followed by a 15 year voyage to the Heliopause.

Spacecraft and Solar Sail: For both the three-spacecraft and ten-spacecraft scenarios, a MMRTG power source and optical communication are the baseline design selections; solar sails and a monoprop hydrazine blow-down configuration serve as the primary propulsion subsystem. Subcontracting of the spacecraft is also assumed in each case. A discount of 50% is taken into account after the first spacecraft unit is produced, as the additional spacecraft are exact replicas of the first unit.

L'Garde Inc provided a solar sail ROM cost based on their experience with costs incurred while building the Sunjammer 1200m^2 solar sail and sailcraft. Their proven expertise with Sunjammer gives the team a high level of confidence in L'Garde's estimate. The \$53M cost provided by L'Garde includes the separation and inflation modules, solar sail membrane and structure, as well as their integration. An additional 17% JPL fee is assessed in addition to the quote from L'Garde. A discount of 20% is taken after the first unit.

Instruments: A core payload with five instruments is assumed for both mission scenarios. Cost estimates were developed for the core payload. A Voyager-based magnetometer, plasma monitor, plasma wave, and neutral ion spectrometer are used as analogous instruments for the cost basis. The TIDE instrument on the POLAR mission is used as an analogy-base for the interstellar plasma ion instrument. It is assumed that the effort, and thus the cost, will decrease by a factor of 65% after the first unit of each instrument produced, as they are exact replicas of the first instrument units. In the case of the 10 spacecraft scenario, there are 50 instruments in total. There are 15 instruments in total for the 3 spacecraft scenario.

Cost estimates were also developed for two optional instruments, a Channeltron, which will complement the Plasma Monitor, and a Cosmic Ray instrument.

Launch Vehicle: For the ten-spacecraft scenario, an Atlas V 551 throughput cost is used which includes launch services, payload processing, launch vehicle integration mission, unique launch site ground support and tracking, data and telemetry services, as provided in the payload planner's guide from the Launch Services program at KSC. For the three-spacecraft scenario, an Atlas V 501 cost was used.

10.2 Cost Estimate Output

The cost output is based on the input assumptions discussed previously. The relative magnitudes of the ten-spacecraft cost scenario versus the three-spacecraft cost scenario are in alignment. Due to the large amount of fixed cost, one would not expect a direct linear relationship with the number of spacecraft.

Project Total (FY 2013 $M)

		PMCM Model Output: 10 Spacecraft	PMCM Model Output: 3 Spacecraft
Total Project Costs ($M)		**$3447.0M**	**$1796.7M**
1.0	Project Management	$67.2M	$51.2M
2.0	Project System Engineering	$17.1M	$17.1M
3.0	Mission Assurance	$81.1M	$35.5M
4.0	Science Team	$29.8M	$29.8M
5.0	Instruments	$94.7M	$42.3M
6.0	Flight System	$1426.2M	$515.7M
6.1	Flight System Management	$9.7M	$9.7M
6.2	Flight System Engineering	$15.5M	$14.9M
6A	Starjammer	$876.0M	$318.5M
6B	L'Garde Sailcraft	$516.1M	$163.6M
6.13	Materials and Processes	$0.6M	$0.6M
6.14	Flight System Testbeds	$8.4M	$8.4M
7.0	Mission Operations System	$519.8M	$519.8M
8.0	Launch System	$255.6M	$172.9M
9.0	Ground Data System (Currently included under 7.0)		
10.0	Project Systems I&T	$245.9M	$74.1M
10A	ATLO - Starjammer	$194.8M	$58.1M
10B	ATLO - L'Garde Sailcraft	$51.1M	$16.0M
11.0	Education & Public Outreach	$24.9M	$12.9M
12.0	Mission Design	$8.6M	$8.6M
	Reserves	$676.1M	$316.7M

Figure 28: Cost breakdown by Work Breakdown Structure

Instrument Options: Two instruments were evaluated for inclusion in the SSEARS mission as options, a Channeltron and a Cosmic Ray instrument. The marginal cost of including the Channeltron and the Cosmic Ray instrument on the ten-spacecraft mission scenario is $20.9M and $15.3M, respectively. These are not included in the estimates above; they are merely options at this point.

Cost-Risk Analysis: Uncertainty and risk exist in the estimate, as expected with such a unique mission. The order quantity discounts are aggressive, but fair, given the number of exact replicas of each spacecraft, sail, and instruments required. Additionally, the assumed schedules for both the ten- spacecraft and three-spacecraft scenarios are the same, drawing heavily on the learning curve/repetition assumptions inherent in building multiple units of the same system.

To understand the risk posture of the mission, a mission-level output based s-curve was generated for each scenario in ACEIT. A lognormal cost distribution was assumed, as it has been proven that space system cost typically follows a lognormal distribution. The PMCM total output without reserves was used as the most likely point estimate. A coefficient of variation of 0.4 and 0.45 was used for the three spacecraft and ten spacecraft scenarios respectively, which is higher than the space systems industry standard of 0.35, so that the higher level of uniqueness and risk in the SSEARS mission is captured. Both scenarios attained approximately a 50% cost confidence level, meaning that the probability that the mission actual cost will be less than the cost estimate with reserves is 50%. The risk analysis for the ten spacecraft scenario follows; the three spacecraft risk analysis may be found in the appendix.

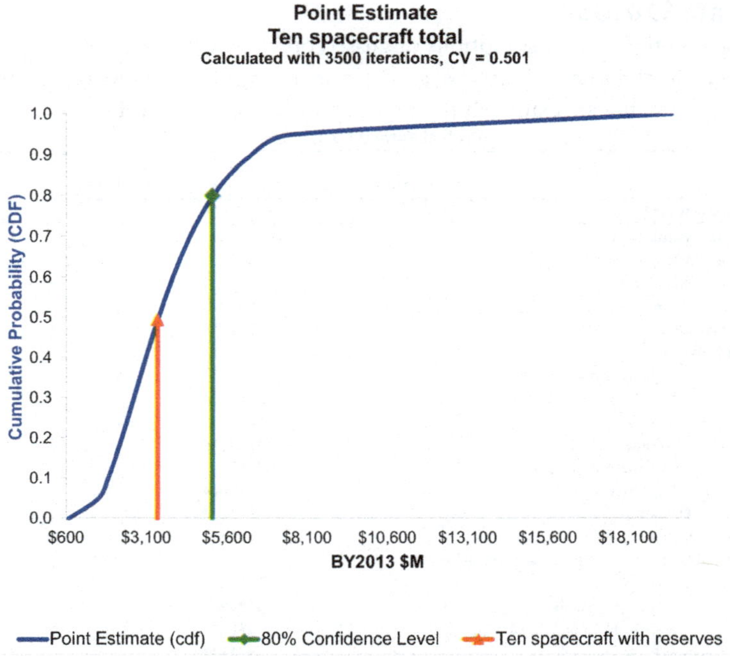

Figure 29: mission-level output based s-curve

Markers	Costs	Confidence
80% Confidence Level	$5,153.189	80.0%
Ten spacecraft with reserves	$3,447.000	49.6%

The cost estimate of $3447M, for a ten spacecraft mission, accounts for a substantial amount of uncertainty and risk, consistent with a space mission at this stage in its definition. The estimate will continue to be refined and risk reassessed as the mission evolves.

Final cost outputs are as follows for both the ten-spacecraft mission and the three-spacecraft mission:

Project Summary (JPL WBS)

		Formulation (Phase A/B) ($M)	Implementation (Phase C/D) ($M)	Development Total ($M)	Operations (Phase E) ($M)	Project Total (FY 2013 $M)	Notes/Models
Total Project Costs ($M)		$109.5M	$935.8M	$1218.2M	$578.5M	$1796.7M	
1.0	Project Management	$6.0M	$31.3M	$37.3M	$13.9M	$51.2M	
1.1	Project Manager & Staff	$1.6M	$4.2M	$5.9M	$9.0M	$14.9M	
1.2	Business Management	$1.6M	$4.7M	$6.4M	$4.4M	$10.7M	
1.4	Project Reviews	$0.3M	$0.7M	$0.9M	$0.5M	$1.4M	
1.6	Launch Approval	$2.4M	$21.7M	$24.1M		$24.1M	
2.0	Project System Engineering	$3.2M	$13.9M	$17.1M	$0.0M	$17.1M	
2.1	Project Systems Engineering	$1.0M	$3.4M	$4.4M		$4.4M	
2.2	Project Software Engineering	$0.8M	$1.8M	$2.6M		$2.6M	
2.3	End-To-End Information System	$0.1M	$1.1M	$1.2M		$1.2M	
2.4	Information Systems Engineering & Communications	$0.4M	$1.2M	$1.6M		$1.6M	
2.5	Configuration Management	$0.3M	$1.2M	$1.5M		$1.5M	
2.6	Planetary Protection	$0.0M	$0.1M	$0.1M		$0.1M	
2.7	Contamination Control	$0.1M	$0.8M	$0.9M		$0.9M	
2.9	Launch System Integration	$0.2M	$1.0M	$1.2M		$1.2M	
2.10	Project V & V	$0.1M	$3.2M	$3.4M		$3.4M	
2.11	Risk Management	$0.1M	$0.2M	$0.3M		$0.3M	
3.0	Mission Assurance	$3.1M	$27.7M	$30.8M	$4.7M	$35.5M	Msn Assurance wrap excludes reserves, LV, and itself
4.0	Science Team	$1.3M	$11.5M	$12.8M	$17.0M	$29.8M	
5.0	Instruments	$4.2M	$38.1M	$42.3M		$42.3M	
6.0	Flight System	$51.6M	$464.1M	$515.7M	$0.0M	$515.7M	
6.1	Flight System Management	$1.0M	$8.7M	$9.7M		$9.7M	
6.2	Flight System Engineering	$1.5M	$13.4M	$14.9M		$14.9M	
6A	Starjammer- 3 copies	$31.9M	$286.7M	$318.5M		$318.5M	
6B	L'Garde Sailcraft - 3 copies	$16.4M	$147.3M	$163.6M		$163.6M	
6C		$0.0M	$0.0M	$0.0M		$0.0M	
6D		$0.0M	$0.0M	$0.0M		$0.0M	
6E		$0.0M	$0.0M	$0.0M		$0.0M	
6.13	Materials and Processes	$0.1M	$0.5M	$0.6M		$0.6M	
6.14	Flight System Testbeds	$0.8M	$7.5M	$8.4M		$8.4M	
7.0	Mission Operations System	$6.2M	$55.8M	$62.0M	$457.8M	$519.8M	
8.0	Launch System			$172.9M		$172.9M	
8.1	Launch Vehicle			$172.9M		$172.9M	
8.2	Upper Stage / SRM			$0.0M		$0.0M	
9.0	Ground Data System (Currently included under 7.0)						
10.0	Project Systems I&T	$7.4M	$66.7M	$74.1M		$74.1M	
10A	ATLO - Starjammer- 3 copies	$5.8M	$52.3M	$58.1M		$58.1M	
10B	ATLO - L'Garde Sailcraft - 3 copies	$1.6M	$14.4M	$16.0M		$16.0M	
10C		$0.0M	$0.0M	$0.0M		$0.0M	
10D		$0.0M	$0.0M	$0.0M		$0.0M	
10E		$0.0M	$0.0M	$0.0M		$0.0M	
11.0	Education & Public Outreach	$0.3M	$2.9M	$3.2M	$9.7M	$12.9M	EPO wrap includes everything except reserves, LV, and itself
12.0	Mission Design	$0.9M	$7.8M	$8.6M	$0.0M	$8.6M	
#.#	Other			$0.0M		$0.0M	
#.#	Reserves	$25.3M	$216.0M	$241.2M	$75.5M	$316.7M	15%

Figure 30: Final mission cost for 3 spacecraft: $1.796B

Project Summary (JPL WBS)

		Formulation (Phase A/B) ($M)	Implementation (Phase C/D) ($M)	Development Total ($M)	Operations (Phase E) ($M)	Project Total (FY 2013 $M)	Notes/Models
Total Project Costs ($M)		$265.1M	$2324.2M	$2844.8M	$602.2M	$3447.0M	
1.0	Project Management	$7.4M	$34.3M	$41.7M	$25.4M	$67.2M	
1.1	Project Manager & Staff	$2.5M	$5.1M	$7.6M	$14.5M	$22.0M	
1.2	Business Management	$1.9M	$6.5M	$8.4M	$10.0M	$18.4M	
1.4	Project Reviews	$0.5M	$1.1M	$1.6M	$1.0M	$2.6M	
1.6	Launch Approval	$2.4M	$21.7M	$24.1M		$24.1M	
2.0	Project System Engineering	$3.2M	$13.9M	$17.1M	$0.0M	$17.1M	
2.1	Project Systems Engineering	$1.0M	$3.4M	$4.4M		$4.4M	
2.2	Project Software Engineering	$0.8M	$1.8M	$2.6M		$2.6M	
2.3	End-To-End Information System	$0.1M	$1.1M	$1.2M		$1.2M	
2.4	Information Systems Engineering & Communications	$0.4M	$1.2M	$1.6M		$1.6M	
2.5	Configuration Management	$0.3M	$1.2M	$1.5M		$1.5M	
2.6	Planetary Protection	$0.0M	$0.1M	$0.1M		$0.1M	
2.7	Contamination Control	$0.1M	$0.8M	$0.9M		$0.9M	
2.9	Launch System Integration	$0.2M	$1.0M	$1.2M		$1.2M	
2.10	Project V & V	$0.1M	$3.2M	$3.4M		$3.4M	
2.11	Risk Management	$0.1M	$0.2M	$0.3M		$0.3M	
3.0	Mission Assurance	$7.6M	$68.7M	$76.4M	$4.7M	$81.1M	Msn Assurance wrap excludes reserves, LV, and itself
4.0	Science Team	$1.3M	$11.5M	$12.8M	$17.0M	$29.8M	
5.0	Instruments	$9.5M	$85.2M	$94.7M		$94.7M	
6.0	Flight System	$142.6M	$1283.6M	$1426.2M	$0.0M	$1426.2M	
6.1	Flight System Management	$1.0M	$8.7M	$9.7M		$9.7M	
6.2	Flight System Engineering	$1.5M	$13.9M	$15.5M		$15.5M	
6A	Starjammer- 10 copies	$87.6M	$788.4M	$876.0M		$876.0M	
6B	L'Garde Sailcraft - 10 copies	$51.6M	$464.5M	$516.1M		$516.1M	
6C		$0.0M	$0.0M	$0.0M		$0.0M	
6D		$0.0M	$0.0M	$0.0M		$0.0M	
6E		$0.0M	$0.0M	$0.0M		$0.0M	
6.13	Materials and Processes	$0.1M	$0.5M	$0.6M		$0.6M	
6.14	Flight System Testbeds	$0.8M	$7.5M	$8.4M		$8.4M	
7.0	Mission Operations System	$6.2M	$55.8M	$62.0M	$457.8M	$519.8M	
8.0	Launch System			$255.6M		$255.6M	
8.1	Launch Vehicle			$255.6M		$255.6M	
8.2	Upper Stage / SRM			$0.0M		$0.0M	
9.0	Ground Data System (Currently included under 7.0)						
10.0	Project Systems I&T	$24.6M	$221.3M	$245.9M		$245.9M	
10A	ATLO - Starjammer- 10 copies	$19.5M	$175.3M	$194.8M		$194.8M	
10B	ATLO - L'Garde Sailcraft - 10 copies	$5.1M	$46.0M	$51.1M		$51.1M	
10C		$0.0M	$0.0M	$0.0M		$0.0M	
10D		$0.0M	$0.0M	$0.0M		$0.0M	
10E		$0.0M	$0.0M	$0.0M		$0.0M	
11.0	Education & Public Outreach	$0.6M	$5.6M	$6.2M	$18.7M	$24.9M	EPO wrap includes everything except reserves, LV, and itself
12.0	Mission Design	$0.9M	$7.8M	$8.6M	$0.0M	$8.6M	
#.#	Other			$0.0M		$0.0M	
#.#	Reserves	$61.2M	$536.4M	$597.5M	$78.5M	$676.1M	15%

Figure 31: Final mission cost for 10 Spacecraft: $3.44B

In the beginning of our study we thought we could fit ten spacecraft on one launch vehicle. During our configuration study we learned that we can only fit three per vehicle using currently available vehicles. Future developments in the launch vehicle industry might change this, perhaps with the Falcon Heavy. However for our cost analysis we included a single Atlas 5 501 in both the three-spacecraft and ten-spacecraft scenarios. The three-spacecraft scenario indeed fits onto one launch vehicle. If launched today, the ten-spacecraft scenario would require the addition of roughly $0.5B, the current approximate cost of 2 Atlas 5 launch vehicles. Rather than include those costs (nearly certain to be inaccurate by the time any such mission would be seriously considered) in the formal estimate, we left the acknowledged, under-estimated cost in the model for the ten-spacecraft scenario.

The 10-spacecraft mission could use combinations of launch vehicles at different times. Further several disruptive launch technologies (e.g. SpaceX "Grasshopper" reusable first stage rockets) are nearing the marketplace making accurate predictions more difficult.

11. Phase 2 Activities

Our Phase 2 proposal (developed in parallel with this final report) focuses on three primary technical challenges: demonstrating sail material thermal tolerance for expected perihelion distances, showing that the L'Garde Sunjammer boom design can be scaled up to the size necessary for a 250m x 250m sail, and developing a sail control method suited to such a large structure.

Material	Boom Design	Control
Thermo-mechanical testing of available sail materials at MSFC Solar Thermal Test Facility	~170m design study with team Mechanical Engineer and L'Garde team	Study and simulation of large sail control rates

We will also further develop the optical communications concept of operations and investigate other subsystem issues.

12. Future Applications

The architecture we have described can be applied for other missions besides the Heliopause exploration mission that we have explored in detail. Very large solar sails can enable many types of missions to distant objects.

12.1 Oort Cloud Object / Outer Planet / Moon Impactor

An object discovered in 2010, called 2010WG9, is believed to have originated in the Oort cloud1. It is thought to be a pristine sample of early solar system materiel, as its orbit does not come closer to the sun than the distance to Uranus. This object is a scientifically valuable object as its surface has not experienced the periodic melting and freezing that would occur with objects that come closer to the sun.

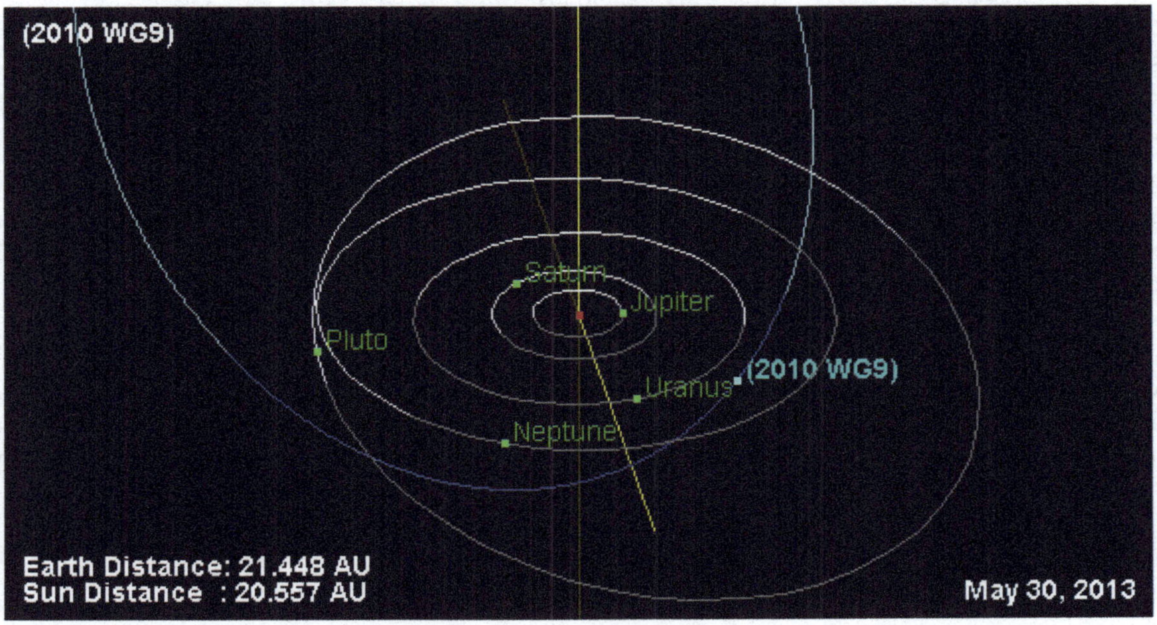

Figure 32: The object 2010 WG9 does not get any closer to the sun than Uranus, protecting its early solar system materials from the cycle of heating and freezing that happens to objects that get closer to the sun.

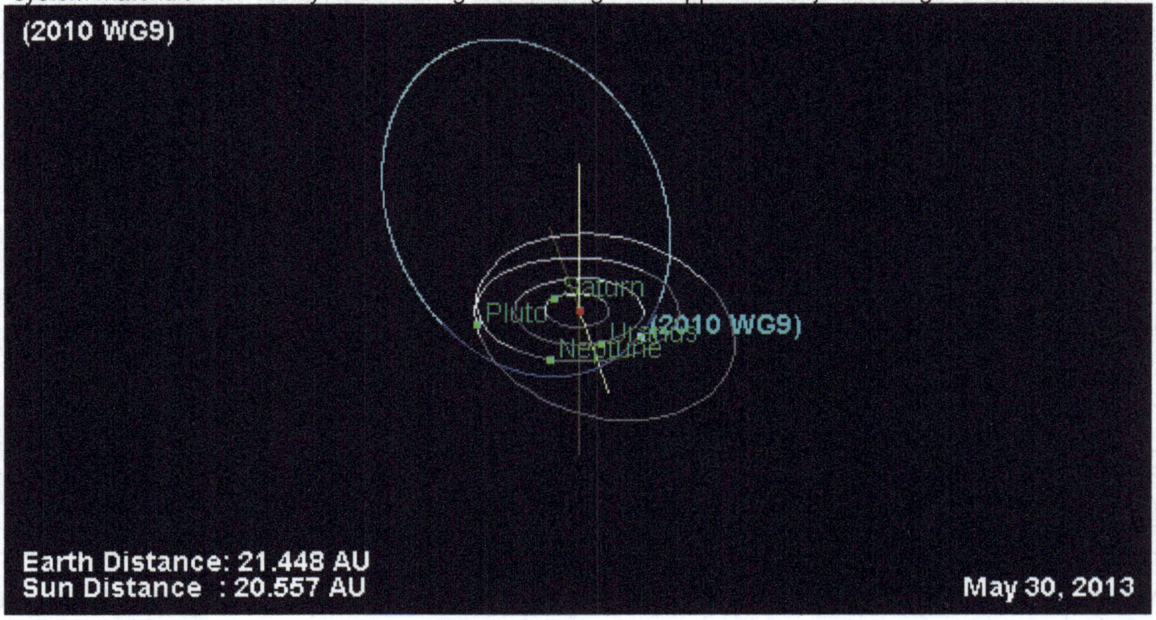

Figure 33: The object has an unusual orbit and likely contains pristine samples of early solar system material.

The SSEARS architecture could be used to deliver an impactor to the surface of this object at high relative velocity, with a second spacecraft following behind to study and pass through the impact plume. This architecture provides great flexibility in mission design and timing due to its use of a solar close approach to "aim" the outbound trajectory. This same concept could apply to Enceladus, Europa, or the other outer planet moons, with expected transit times as follows:

Target Planet (or its moons)	Distance from Sun	Cruise Time
Jupiter	~5 AU	~2 years
Saturn	~10 AU	~3 years
Uranus	~20 AU	~5 years
Neptune	~30 AU	~6 years
Pluto	~32 AU for the next several decades	~7 years

These figures all include variability due to the different lengths of time possible to reach "maximum velocity" depending on the perihelion distance and whether one exits Earth's orbit with the sail itself or with an upper stage rocket. However, the figures represent a significant improvement over current transit times and results in significantly higher arrival velocities well suited for impactor missions.

12.2 Gravitational Lensing (~550 AU)

As early as 1978 it was recognized that a spacecraft at roughly a distance of 550 AU and beyond could take advantage of the sun's gravitational effect to magnify the hydrogen line at 1420 MHz, the so-called ideal frequency for interstellar communications [2]. This concept has been developed further by Frank Drake and Claudio Maccone [3] as well as a study in 1999 at JPL [4].

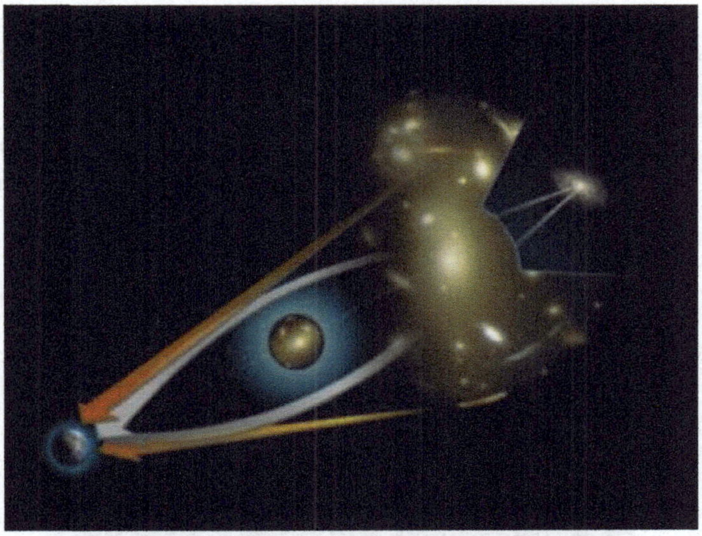

Figure 34: Gravitational lensing at work. A space probe at 550 AU and beyond could exploit such effects to make detailed studies of other solar systems, among numerous other scientific targets.

In 1999 JPL produced a rough design which estimated the mass of such a telescope at roughly 1000kg. In order to apply our architecture to this mass we conceived of using nine sails as a "Raft" as follows.

12.3 Nine-Sail "Raft" for Gravitational Lensing Telescope Delivery

A concept was developed for nine-sails to be arranged in a tiled formation. The sails would not utilize the Sunjammer-type vanes, but the overall system would use the corner sails themselves as vanes, as shown below:

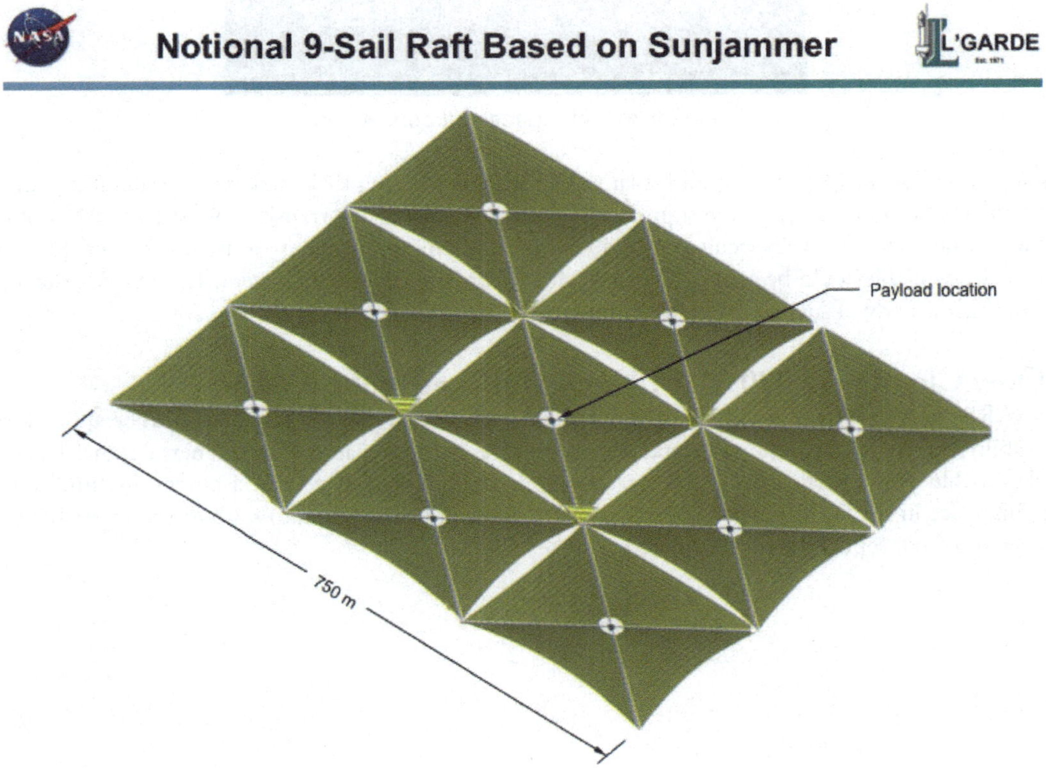

Figure 35: 9-sail raft design to carry payload range between 500- and 1000kg gravity lens telescope.

Analysis showed that we can transport 1000kg to 550 AU in less than a century. Unfortunately, despite the imaginative appeal of this 9-sail approach, one cannot, with this architecture, increase the area of the sail by a factor of 9 without increasing the mass by the factor of 9. This results in essentially the same performance as with just one sail because the payload mass is so small. If the payload was significantly more massive then this "raft" concept would be a useful reduction in transit time over the single-sail version.

So, we conclude that the SSEARS architecture is a reasonable approach to reaching 550AU within a human lifetime, if not during a productive career (at least as such time frames are measured at the beginning of the 21st century.) We leave the feasibility of such biological and medical advances to the imagination.

12.4 Thousand Astronomical Unit Mission (TAU)

JPL developed a mission concept in 1987 that would send a spacecraft to a distance of 1000 AU using then-existing technology. The proposed spacecraft would measure the distance to other stars via stellar parallax, measure conditions in the interstellar medium, and perform tests of general relativity via communications with Earth [5]. This spacecraft was proposed to use a 1 MW fission reactor and ion drive to reach 1000 AU in 50 years. Given that much work remains to develop a flight-ready space fission reactor system, a SSEARS-like solar sail architecture bears further studying for this application.

Figure 36: TAU spacecraft concept art

As shown in figures 19 and 20, useful solar sail sizes seem to reach their maximum at about 250 m x 205 m, at least with any currently conceivable support technology. Reducing the transit time to 1000 AU from the 120 years it would take the SSEARS architecture will require applying more energy to the sail rather than increasing the size of the sail. This could be achieved with a reflector or beamed energy system. The SSEARS team intends to propose such a study of such a system to a future call.

12.5 Oort Cloud (~50,000 AU/ 1 LY), Exoplanet Probe (<4 LY)

Our SSEARS concept is theoretically limited only by the amount of power/energy that can be applied to the sail and the ability of the material to withstand that energy. As such we can dream of energy transfer systems and materials capable of the performance necessary to reach another solar system in a human lifetime. We hope to explore this space in the far term and believe that NIAC is the ideal program to support such investigations once the "nearer-term" concepts have been explored.

13. Outreach Activities

13.1 Conferences

Due to the Federal Budget "Sequester" it was impossible to travel to many of the relevant conferences.
Attended

NIAC 2012 Fall Conference, Hampton Bays, VA	PI Jeffrey Nosanov attended this conference and presented a poster about the project.
NIAC 2013 Spring Conference, March 2013, Chicago, IL	PI Jeffrey Nosanov attended this conference and presented ongoing work and progress made so far.
Starship Century Symposium, May 2013, San Diego, CA	PI Jeffrey Nosanov attended this symposium and discussed the project with many leaders in the field.

13.1.1 Accepted/Invited but unable to attend due to NASA budget sequestration

International Astronautical Congress 2013, Beijing, China	This final report was accepted but travel will not be permitted.
Starship Congress (Icarus Interstellar)	Abstract was accepted but travel will not be permitted.
100 Year Starship Conference	Abstract was accepted but travel will not be permitted.

13.2 Poster for General Public

The poster graphic for public consumption will arrive separately in an email to program management.

13.3 Reddit AMA (Ask Me Anything)

The popular forum website www.reddit.com hosts exchanges between users called "AMAs" This stands for "Ask Me Anything" and many popular or influential figures have hosted an AMA giving their fans and supporters a chance to post questions. One of us (Nosanov) hosted an AMA under the title "I am a NASA Innovative Advanced Concepts Fellow, developing a mission concept to the edge of the solar system and nearby interstellar space." The post attracted dozens of viewers, many of whom asked insightful and interesting questions. I answered every single serious question and have concluded that an AMA is a valuable component to a broad outreach effort, but I could have expanded my audience significantly (compared to other AMA events) by advertising or posting about the AMA on several other websites beforehand. The entirety of the exchange (~25 pages of questions and answers) will be sent via email to program management.

13.4 Video

We initiated the development of a brief outreach video to convey the mission concept in a 60-second clip. We developed a storyboard concept for the video and hope to produce the video in Phase 2. The storyboards are reproduced below.

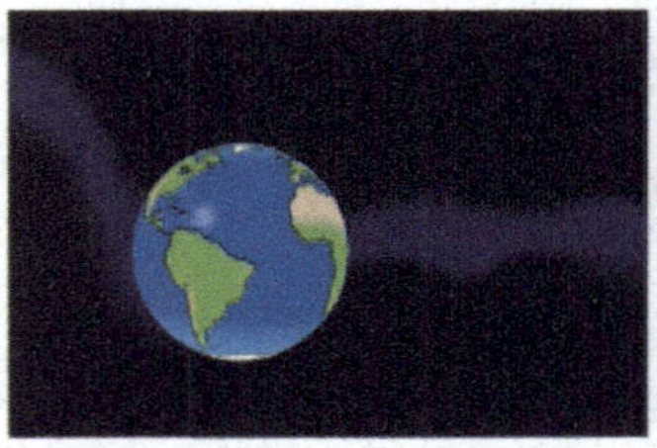

1) Establishing shot – Earth

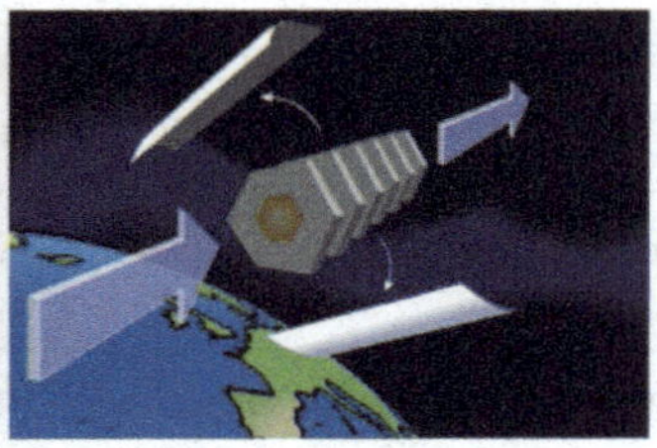

2) Cut to close up – Payload enters frame left as half-shells separate to reveal 6 stacked spacecraft

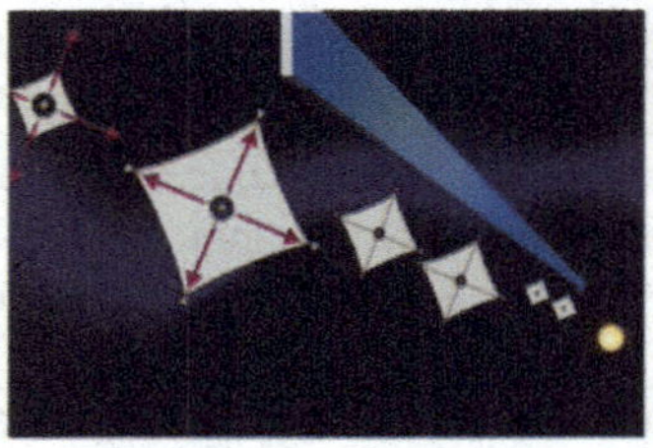

3) Dissolve to – As each spacecraft enters frame left, no solar sail is visible. Solar sails unfurl as each spacecraft travels into frame.

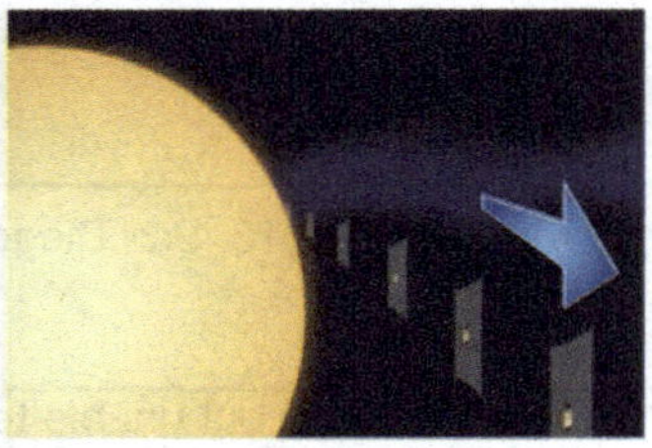

4) Dissolve to – Spacecraft emerge from behind the sun. Camera pans to follow the first craft... [shot continues with next frame]

R. Barkus SSEARS Storyboard, draft 0501 1

4) [continued from previous frame] Camera continues pan as spacecraft head into distance. Each of the first four spacecraft then begin a graceful move into their respective trajectory.

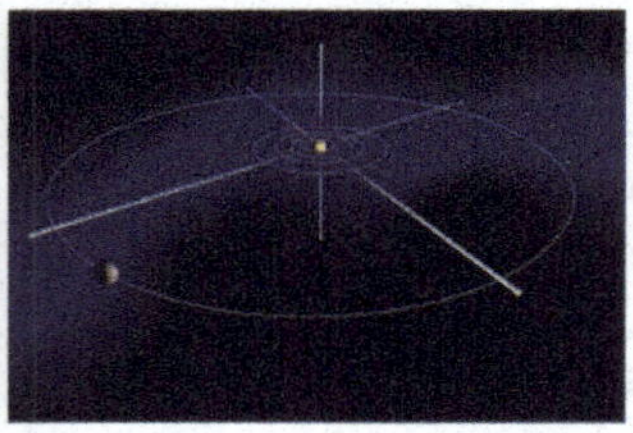

5) Cut to – Animated diagram showing each individual trajectory. The spacecraft, represented by a marker dot and a long fading trail, have just passed Jupiter's orbit.

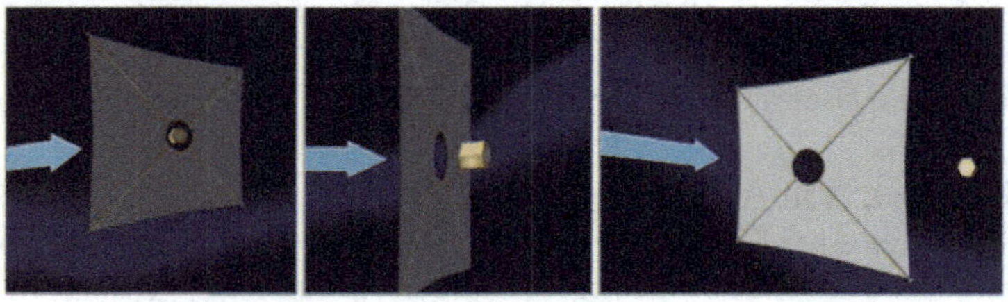

6) Cut to close up – A single spacecraft enters frame left as camera begins follow-pan. Solar sail is released causing a slight speed decrease in the sail and an increase in the spacecraft. Camera pan eases out as spacecraft continues into distance.

R. Barkus SSEARS Storyboard, draft 0501 2

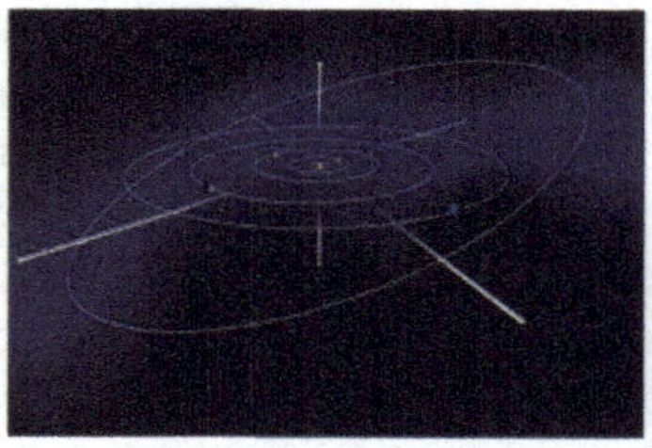

7) Cut to – Animated diagram showing each individual trajectory. The spacecraft are now beyond Pluto's orbit.

8) Cut to close up – A single spacecraft, now without solar sail, enters frame left as camera begins follow-pan. Camera pan eases out as spacecraft continues into distance.

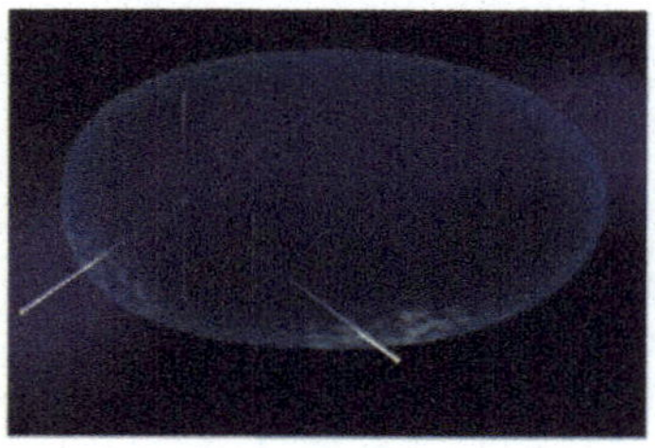

9) Dissolve to – Animated diagram showing spacecraft markers and long trails (but no visible Heliopause). As the first spacecraft passes the Heliopause boundary a bright wavering texture is illuminated. As the spacecraft travels beyond the boundary, the texture becomes still and somewhat dim. The process is repeated as each spacecraft passes the various boundary points. After all six craft have passed beyond the Heliopause boundary, the complete dimensions are visible.

R. Barkus SSEARS Storyboard, draft 0501 3

References

1. The Peculiar Photometric Properties of 2010 WG9: A Slowly-Rotating Trans-Neptunian Object from the Oort Cloud, David Rabinowitz, Megan E. Schwamb, Elena Hadjiyska, Suzanne Tourtellotte, Patricio Rojo, Astronomical Journal, 2013 Apr 20

2. Von Eshleman, Stanford University

3. Maccone, The Sun as a Gravitaitonal Lens: Proposed Space Missions, Colorado Springs: IPI Press

4. Design Issues for a mission to exploit the gravity lensing effect at 550AU, John West, Acta Astronautica V.44 I2-4, January 1997, Pages 99-107

5. Preliminary Scientific Rationale for a voyage to a thousand astronomical units, Jet Propulsion Laboratory

A Pathway for Global Communications as an Inclusive Means of Participation

David Orban

Abstract

Global communication is evolving rapidly, allowing for effective, fine tuned, interactive messages to reinforce the support that a complex project can receive, beyond the limits of traditional broadcast models. It is unlikely that a starship project could be directly financed via crowd funding, but it is necessary to increase and to sustain the understanding and sharing of its value in the larger public, with carefully designed and wide-ranging, inclusive communications.

This paper concentrates on what can be done now with ubiquitous online video that goes beyond the barriers of language to establish a new global conversation, as well as forecast what new tools can be designed and deployed as new technologies become available. The project of designing, building, and launching a starship is as much an engineering effort as one of public relations, of willpower, which must be sustained, together with the political and financial resources necessary.

Emotional intelligence is an important component in human endeavors, on par with analytical and logical skills: being able to powerfully communicate goals that can be not only understood, but with which people all around the world can identify with. Online video today is the medium of broadest emotional bandwidth, and its use is key for a broad public understanding and participation in the starship project.

Among other developments, it is going to be possible to generate, disseminate, analyze, and aggregate videos not only produced by the project itself, but those that are created by the global conversation around it. These components are going to be an essential part of the project, sustaining its ambitious dream. The adoption of a distributed communication model is the basis for taking advantage of innovative ideation and design components, which can concretely involve groups of passionate followers of the project, complementing the professional team dedicated to it.

Keywords

Communication, support, identity, interaction, crowd

1. Introduction

This paper covers what are, in my opinion, important principles for designing in the next decades, types of social structures that are conducive to desirable futures. Our society is brittle, and the more complex, the more global it gets, the less we can afford this brittleness. I may lead to wider and deeper crises as it has been the case in the past periods because, for the moment, we do not have an alternative. Social collapse or the passing of empires did not mean that the human civilization would be put in peril in the past. If a planet-wide civilization collapses as it could very well today, after globalization, it would put our entire civilization at risk.

2. Assumptions

I want to start with some assumptions that are the basis of this paper.

2.1 Continued exponential trends

The assumption that the exponential trends that we see around us—the most famous example of which is Moore's Law [1], stating that the density of electronic circuits doubles every 18 months and its price halves during that time, but there are many, many others—that these trends will continue. The linear thinking that informs our policies and that is the basis of thinking of our policy makers is something that we cannot afford anymore, because the prevalence and the visibility of the exponential trends is just too strong.

2.2 No new physics

At the same time, as much as one would want otherwise, the assumption for this paper is that within the next few decades we will not have radically new physics, such as a warp drive would represent. That there are no shortcuts. If we want to colonize the galaxy, we will have to do it the hard way. It is going to take a couple of million years to do so, and we have to do it step-by-step. We cannot jump across 100,000 light years.

2.3 Human nature to stay the same

Even if I have been the chairman of Humanity+, the world transhumanist association which defines humanity as an entity that overcomes itself, that sees its limits as the barrier to go through and redefining what it means to be human [2], I will assume that for the next few decades what it means to be human is not going to radically change.

2.4 No self-modifying artificial general intelligence

My final assumption, and I am a card carrying singularitarian, is that what is called the technological singularity—the theoretical moment in time, in the future when self-modifying artificial intelligence is brought to the scene many more active agents, participants, in the global civilization alongside humanity with their own goals, with their own rules, with their own active capacity of acquiring and managing the sources—it is not going to happen.

If any of these assumptions is falsified, every bet is off, and the trajectory of human development will take a very different, and possibly hard to forecast direction.

3. Network Society

What we are seeing and what we are seeing around us, in my opinion, is that hierarchies are breaking down everywhere. They are breaking down at the level of the family, at the level of a city, at a level of the company, at the level of the nation states. What used to be the case that hierarchies defined the role that plays the behavior of individuals and groups of individuals is now turning into what I call the Network Society which is defined by distributed peer-to-peer systems substituting centralized systems. Energy production based on solar; 3D printing for distributed manufacturing; plant labs allowing food production to be available at every home rather than centralizing agribusinesses; massively online open courses making learning and education available to everybody regardless of their economical means or geographical or social provenience; and our financial technologies are also explosively changing through digital currencies becoming platforms of innovation. The nation state is taking notice and is enacting policies that are more and more violently against those new entities that appear to the nation state as invading forces. Its immune system is rebelling against this rate of destabilizing innovation.

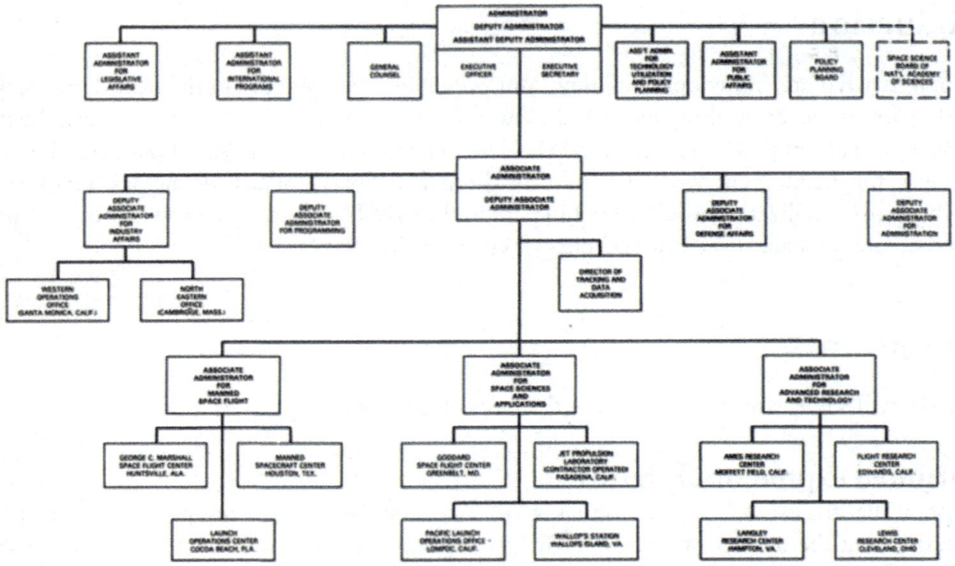

Fig. 1. The organizational chart of NASA, November 1, 1963 (Credit: NASA)

4. Empathic Communications

Individuals, as emotional machines, have immense power, and if somebody asked me what my superpower would be if I could have one, this is what it would be. It would be the arbitrary extension of the circle of empathy in a given moment. Empathy is our strongest power to overcome in people and groups of people. Artificial isolation that is engendered by our superstructures is a threat to that empathic power. We can overcome. We have this power. It is also not only something that is available to us, we have to use this power. If we do not, then the existential risk of global annihilation, the global destruction of planetary civilization is almost guaranteed.

We are in a system of networks. We are in a system of participating, aware, capable, and well-educated people who are empathic to each other who want to give back. And we know how important it is to be able to identify with an objective, to be able to feel that we are able to influence a project, that even if we are not part of the core team, what we say, what we do, has an impact.

A tool that is available today, the most powerful with the broadest, emotional bandwidth in order to overcome our communication barriers, is online media, especially online video [3]. When we see another person in our discussions, in our conversations, there is a subtext of emotional content that written or verbal communication without seeing the other person is not capable of transmitting. That is why we are so excited about seeing video. That is why on YouTube today every minute 72 hours of new content is uploaded because we just love seeing another human person. Seeing is not enough. We have to understand. We have to engage. And it is, today, possible in online video, to eliminate the language barrier.

We are not going to be able to crowd fund the creation of a Starship. We can dream of it, but it is not going to happen. We have to be extremely persuasive, not only among the group of passionate individuals regardless of their connections, regardless of their knowledge and skills. We have to be persuasive among tens and hundreds of millions, hopefully billions, of people who have to share our mission. We have to be almost evangelistic, you know, or desire of and making sure that what we learn, what we know is shared among as many of us as possible. And today this is available in online video. We can start doing it; we actually have started doing it.

One of the members of the 100 Year Starship community just took what were static images from the website, and turned them into a beautiful video. And the organization took that, and is now on the homepage of this symposium. Others can do the same, and the way to do this is in a distributed way rather than in a top-down manner, in a hierarchical decision making process, the way it would have been natural for NASA to make in the 20th century.

We know how important collaborative design and collaborative, participatory design is becoming in several systems. It is not the case of our integrated circuits, but it is definitely, for example, the case of Wikipedia, which everybody said would be a failing project among the experts. The experts are the best in telling you what cannot be done ever. That is why a lot of people should not listen to the experts.

Our interfaces are changing constantly and becoming more and more evolved in systems from punch cards to teletypes to common layer interfaces, initially fairly simple and then more sophisticated graphical user interfaces emulating our desktops to today in touch interfaces, and you will know if you are familiar with the history of computing that these cycles are actually getting shorter and shorter. It is just a few years that touch interfaces have become ubiquitous, and now we cannot imagine living without them. But gesture-based interfaces, dialogue systems, and computer systems that perceive us more and more are becoming available, and already on the horizon we see the next steps towards direct brain computer interfaces that are going to be fabulous and very interesting and intriguing.

Today's most engaging interface, what is available to billions of people, and billions of people, at least hundreds of millions of people, are taking advantage of these interfaces as active participants rather than passive consumers of the content that is being created, is online video. The reason why online video is so important is because it represents in the most simple terms possible with requirements that contrary to some of the previous examples of our evolving human-computer interaction platforms that I showed do not need the design and the deployment of new devices. Online video is available on computers, on mobile phones, in high income or even low income countries. But they are extremely rich in giving you not only information content but also emotional content.

In order to evangelize in a humanistic, naturalistic spirituality the beauty, the ambition of the mission of a starship, we do need to leverage all the emotional impact that we can have. However, we can drastically broaden from a few hundred accolades to hundreds of millions of people who believe in the mission without which we wouldn't be able to participate and to convince the key decision makers that what we want to do deserves their attention, their resources, and their decisions.

The vast amounts of human creativity are digested, prioritized, and presented in terms of relevancy by our computers that in turn base these decisions on our actions, measuring and acting on every viewing or closing of our videos that are being watched. So what we need is to make sure that as many of the people that we can reach become active participants.

YouTube creators rather than YouTube watchers, programmers rather than video game players, video game makers, and we need to design systems that are extremely inclusive, When we are welcoming to this creativity, we engender participation and are reflective of the nature of our peer to peer networks that break down hierarchies.

I understand that some of this is difficult for some of those among us who have been accustomed to NASA-style top-down engineering approaches.

When we are talking about the starship itself, just like when we are talking about a spaceship today, we are not going to be designing it by a committee of millions. It is going to be a different engineering approach. But in the meantime, when we are talking about popularizing the idea, when we are talking about how to reach our goal of making these things happen, we need to make sure that we are leveraging the desires of those who cannot be here with us today. Those that have been complaining about the fact that we are not streaming live, who will be clamoring to see the videos of these sessions being recorded by the volunteer crews of film faculty here in Houston, we need to leverage that passion, and we can leverage that passion by creating infrastructure that goes in that direction. Innovation that we need is, yes, technological.

We need to invent systems that understand how human nature works. Without that, we not going to be able to achieve the impact we desire. If we don't adapt to the world and ignore the rules are embedded in our human nature, we can't learn from our mistakes. We need to embrace our mistakes, so that we can enthusiastically go out and evangelize our mission.

5. Sustainable Bottom Up Organization

We are in the 21st century now and in this century, the risks and the opportunities that we have are huge. Our goal is to minimize the risks, and to maximize the opportunities ahead of us. Our current industrial civilization is unsustainable, but on Starship Earth unsustainability itself is unsustainable. We have to learn that; we have to realize that. Only sustainable systems can sustain themselves. The way we go building the new network society, it will be crucial, to try to minimize the violent reactions of the dying nation states as we achieve this phase change. As we go into the new organization of the network society, we have to face the new questions presented in front of us. Is it right? Is it wrong? What is a natural path? What is some extremely artificial legislation that is being promoted because it is in the interest of a very, very restricted group? These are the questions that we have to ask in order to bridge our knowledge through our ignorance. And we have to be extremely inventive because the only way we can try to succeed in our exponential dreams is by powerful application of our inventiveness.

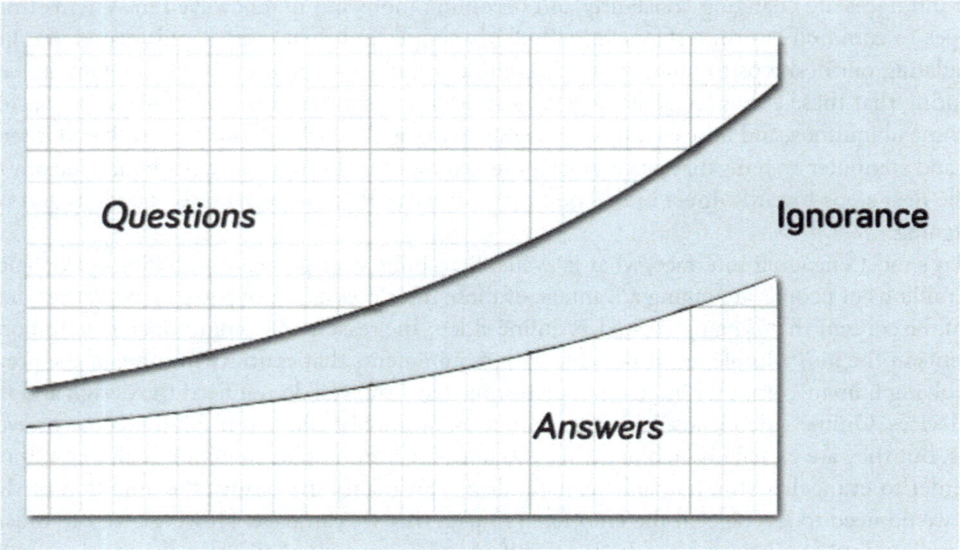

Fig. 2. Building blocks of our learning (Credit: Kevin Kelly [4])

6. Conclusion

There is no guarantee of our continued progress, and Moore's Law, as well as many others, is a total misnomer. It is the common self-fulfilling prophecy of engineers worldwide, who have been inspired for the past 40 years by the little dots that Gordon Moore jotted down in his notebook. When the transistor had been invented, he said, "Yeah, this is how I think it's going to work." They just kept going at it to prove him right, and that is the only way to go. In our quest of going to make the dream of the 100 years Starship true, we have to prove our dreams right, in an inclusive fashion, that leverages the passion and the dedication of people globally. The best tool for doing so is available to us right now, and it is online video.

We have to try, and we will fail a lot of times, but that is it: sometimes we will win, and sometimes we will learn.

Acknowledgements

The author would like to thank Michael Smolens, the Founder of Dotsub [5], for his guidance, and passion in analyzing, presenting, and implementing the values of online video in overcoming the language barrier to global communications.

Notes

1. Intel Corporation uses Moore's Law as a guiding principle in its engineering planning, and, with evolving technologies, strives to prove it right for decades to come. http://www.intel.com/content/www/us/en/silicon-innovations/moores-law-technology.html
2. Humanity+ http://humanityplus.org
3. The Cisco Visual Networking Index projects that online the "sum of all forms of video (TV, video on demand, Internet, and P2P) will be in the range of 80 to 90 percent of global consumer traffic by 2017," http://www.cisco.com/en/US/solutions/collateral/ns341/ns525/ns537/ns705/ns827/white_paper_c11-481360_ns827_Networking_Solutions_White_Paper.html
4. Inspired by Kevin Kelly's post "The Expansion of Ignorance" in The Technium, http://kk.org/thetechnium/archives/2008/10/the_expansion_o.php
5. Dotsub http://dotsub.com. The author is CEO of Dotsub, an online platform for language powered video.

Bibliography

1. Blackmore, Susan. The Meme Machine. Oxford University Press, 2000
2. Kelly, Kevin. What Technology Wants. Penguin Books, 2011
3. Kurzweil, Ray. The Singularity Is Near. Penguin Books, 2005
4. Maxwell , JC. Sometimes You Win--Sometimes You Learn. Hachette Digital, 2013
5. Rifkin, Jeremy. The Empathic Civilization. Tarcher, 2009

The Laser Starway

A Light Bridge To The Nearest Stars

Charles J. Quarra

Abstract

Beamed laser propulsion for interstellar flight is promising but it is severely limited by beam divergence over interstellar distances. In the current work, a system of power relays is proposed to deliver and route light power to accelerate and deccelerate ships flying in both directions between two stars, as well to keep the relays in their equilibrium positions. The system, once deployed, creates a natural radiation roadway where interstellar dust grains tend to evaporate, addressing the problem of relativistic impacts with high-speed ships.

Keywords

interstellar, laser, starway, propulsion,sail, beamed

1. Introduction

Fast Interstellar travel propulsion is fraught with technological difficulties, as well as hard limitations imposed by the laws of Physics. Rockets that carry their own fuel are subject to the limits imposed by Tsiolkovsky rocket equation. The only fuels that make the rocket approach workable for interstellar missions are those with the highest energy density per kilogram. Nuclear fission rockets have enough energy density for fast interplanetary missions inside the solar system, but usually top at velocities of $10^{-5}c$. Nuclear fusion rockets can get up to $0.2c$, but nuclear fusion power has been expected to be around the corner for the last 60 years. Besides, even if a breakeven reactor would be ready today, most confinement technologies are expected to be huge and heavy.

Antimatter-matter reactions could provide highly relativistic speeds of travel, but even if we ignore the fantastic costs of production of macroscopic quantities of fuel, safe storage of antimatter in large enough ratios of fuel mass to container mass is a physical problem that demands magnetic fields that far exceed the capabilities of even the most powerful superconductors known today. Another bold proposal that has come up in recent years is to leverage mini black holes as power sources of Hawking radiation, an approach proposed by Crane [5]. But, putting aside the problem of creating or capturing mini-black holes of 10^{-15} meters, there is the daunting problem of pushing back the tiny black hole together with the ship at macroscopic accelerations, as well as the problem of feeding the black hole and keeping it in a stable operating range.

On the other hand, there exist alternative proposals that do not involve carrying the ship fuel onboard. These kinds of propulsion concepts have their own kind of challenges: The emptiness of interstellar space makes almost impossible to replenish fuel along the way. The Bussard ramjet [3] is an attempt to work around this by scooping sparse interstellar hydrogen as nuclear fusion fuel, but its limitations are discussed at length elsewhere. The last method that can achieve fast interstellar travel without carrying the fuel is solar-powered laser propulsion, and will be the focus of the current work.

2. Solar-Powered Laser Propulsion Limitations

Robert Forward[6] proposed to leave the energy source at home instead of carrying it on with the ship, by using powerful lasers to drive a thin sail attached to a payload. Lasers will have a non-negligible spread over interstellar distances. The spread is limited by the laser wavelength and the laser beam width. Ideally, is desirable to keep the beam focused for long enough distances, in order to keep sails below a reasonable size, and still attain a final velocity that is a reasonable fraction of the speed of light. The beam divergence of a laser in the fundamental Gaussian mode of waist W_0 and wavelength λ is given by the expression [4]

$$\theta = \frac{1}{\pi}\frac{\lambda}{W_0} \quad 1)$$

which is the divergence angle of the Gaussian mode of the beam. For a coherent, collimated beam, the total spot width at a distance *D* is given by

$$w(D) = W_0\sqrt{1+\left(\frac{\lambda D}{\pi W_0^2}\right)^2} \quad 2)$$

where D is the distance between the beam source and the target spot. This total width is minimal when the following condition holds

$$\frac{1}{\pi}\frac{\lambda^2}{W^2} = \frac{\lambda}{D} \quad 3)$$

when λ and D are fixed. In explicit form, the optimal beam width is

$$W = (\frac{\lambda D}{\pi})^{1/2} \quad 4)$$

For $\lambda = 10^{-6}m$ (the average photon wavelength from sunlight) we have that sending a beam up to 10^{16} meters (roughly one light-year) gives an optimal beam-width of 100 kilometers, which implies that any sail smaller than $100\sqrt{2}$ kilometers will not receive the full power from the beam at one light-year of distance. This attenuation can be addressed by increasing the laser power output as the distance to the sail increases, but this power will need to increase quadratically with the distance to keep a constant acceleration on the probe.

Solar-powered laser sails are henceforth, limited by the size of the sail, which has some inherent weight, and most of that weight is not needed until the end of the acceleration phase.

There would be a number of open-ended problems in such interstellar mission, that are hard to address with light-sails. Among those is the amount of interstellar dust impacts the sail can take from interstellar dust before a severe structural damage collapses a significant portion of the sail. Also, the problem of deceleration at arrival, can only be partially solved by reusing the sail to brake using the solar radiation from the destination star.

Magnetic sails can also contribute somewhat to the braking force, but both approaches are inherently low-acceleration, so they place a natural bound on the maximum coasting velocity the ship can brake from. This problem can be ignored if one only intends a fly-by mission, but the amount of scientific information that can be gathered by such probes is limited.

3. Extending Solar Power Into Deep Space

Forward [6], again, proposed extending the range of laser systems by deploying thin, huge (500 km) Fresnel lenses between the laser and the sail, in order to refocus the light toward the accelerating sail. Landis [7] also discussed and analysed the performance obtained by launching multiple lenses in order to extend the reach of laser light to farther distances.

Is there a way to keep such vast structures in a stable position? All objects will follow some orbit that will drift them away of their angular position, unless the gravitational radial acceleration is countered exactly by an opposing force. Statites are such an example of an object in a stable position relative to the sun, being in equilibrium by the solar radiation pressure. In deep space, however, due to the quadratic attenuation of solar radiation with distance, we need to rely on laser power to keep a deep-space statite in a fixed position.

But there are other complications that arise at such huge distances. For instance, if either the sail or the lenses drift away from their expected position, the laser stations at home will not receive this information for days, months or even years. As the distances become larger, communication latencies becomes very large. Because of this, reliance on trajectory predictability becomes a dominant feasibility constraint.

The advantage of stabilizing a big Fresnel lens in space is that we can re-use the undoubtedly expensive device for launching multiple interstellar probes. Given enough economies of scale, the effective cost of building and deploying the lens becomes smaller.

If it could be stabilized in this way, the Fresnel lens acts as a reusable laser relay that delivers the laser light farther, at most a factor of 2. Once the first is deployed, can we use it to deploy and stabilize a new one in front of it, extending the laser distance by a factor of 3?

That possibility would be attractive for several reasons. If the distance between successive optical elements could be reduced, the problem of pointing the light output to the target, as well as tracking its movement would be significantly reduced. But at the same time, we would be able to extend the reach of the laser light to much larger distances, without necessarily increasing the width of the beam, or the span of the sails.

In the following section, we will analyze the power and material costs of creating and maintaining a laser relay system that would distribute laser power at all distances between any two stars, both for the acceleration and deceleration phases of an interstellar flight. We will defer the analysis of deployment schemes to a subsequent work.

4. A Road Made Of Light

It is common lore that the most important technological revolution in human transport was the wheel. It is amusing to notice that, in contrast, roads themselves have never been in any notable hall of fame of inventions. Even as it is obvious that wheels, by their nature, are pretty much useless without roads. Since the start of written civilization, roads had to be prepared before they could be used for payload transport. For long enough distances, horse stables had to become available, in order to provide shelter and food for the long road.

Can we extrapolate the above, arguably primitive technology, to bridge the interstellar voids? The main goal of this work is to perform an initial attempt to answer that question. Interstellar space is after all, extremely vast. Our terrestrial standards of distance and time fall too short to easily describe the scales involved. In earth, terrestrial roads are made of some material that took human workers some time to lie out. In space, a different approach and material is surely required.

An interstellar highway would simply consist of a geometrical path where laser beams are retransmitted along a relay chain. The relay chain is made of relay nodes that can keep their stable positions by computing adjustments and reflecting a portion of the incoming light to generate thrust. The rest of the light is transmitted to the next relay node. Unavoidably, a portion of the incoming energy is absorbed on each relay node, and as we will see, this thermal loss will be the dominant factor in the efficiency of the relay chain.

Let's assume that the positions for stars A and B are labeled as 0 and N+1, where N is the number of relay nodes in the chain, placed at regular intervals of distance D. Near each star, powerful lasers of width W deliver light of wavelength λ to the relay nodes R_1 and R_N, in opposing directions. Each relay node refocus the incoming light beam toward their neighbor nodes. A diagram of the starway is shown in Figure (1)}.

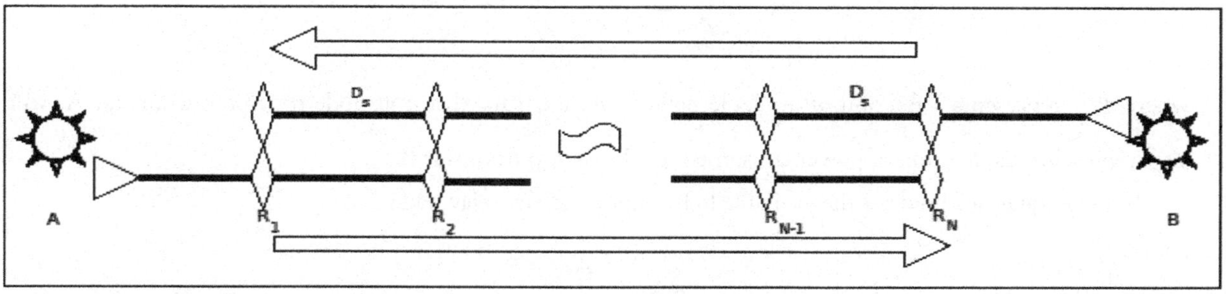

Figure 1: Schematic of a starway. Lasers near both stars transmit power to relay nodes that transmit over the chain, and manage it dynamically for ship propulsion, or just to keep the chain stable

We will simplify the current analysis by doing a number of assumptions: First, we don't specify what underlying optical element is used for the relays. We just assume a few generic properties that a relay will have, such as an effective width W, an effective surface density ρ_a, a transmission absorption coefficient f_n, and a drag factor η that accounts the contribution of the optical losses to the drift of the relay. The factor $1 - f_n$ will measure the ratio of light that is retransmitted to the next relay. Also, we assume that both stars send an equal amount of power

$$W_0 = W_{N+1} = W_L \qquad 5)$$

on their respective directions.

If we write down the k-th component of the total light momentum coming from the $q-1$ -th node to the q-th one as $P^+_{k,q-1}$ and the total light momentum coming from the $q+1$ -th node to the q-th one as $P^-_{k,q+1}$, conservation of momentum demands that

$$P^+_{k,q-1} + P^-_{k,q+1} = P^+_{k,q} + P^-_{k,q} + P^t_{k,q} + \Delta P^{n;+}_{k,q} + \Delta P^{n;-}_{k,q} \qquad 6)$$

Where $\Delta P^{n;+}_{k,q}$ and $\Delta P^{n;-}_{k,q}$ are momentum gained by the relay node due to absorption of incoming light from the right and left direction respectively, and $\Delta P^t_{k,q}$ is a (relatively tiny) portion of light that is reflected for thrusting the node back into its equilibrium position, which is lost. An ideal relay node that will not absorb any light, will have

$$P^t_{k,q} = \Delta P^{n;+}_{k,q} = \Delta P^{n;-}_{k,q} = 0$$

For realistic relay nodes, $P^t_{k,q}$ will be computed in such a way as to track and cancel position drift due to $\Delta P^n_{k,q} = \Delta P^{n;+}_{k,q} + \Delta P^{n;-}_{k,q}$. From now on, we will drop the vector index k and consider the longitudinal component only. Using the simplified assumption eq.(5), we can write general expressions for the available power and force at any relay node:

$$W_q = \left[e^{\beta^+_q} + e^{\beta^-_q} \right] W_L \qquad 7)$$

With

$$\beta_q^+ = \sum_{q'=1\cdots q-1} \ln(1 - f_n - \alpha_{q'}^+)$$

$$\beta_q^- = \sum_{q''=q+1\cdots N} \ln(1 - f_n - \alpha_{q''}^-)$$

where α_q^+ represents a fraction of available power consumed for thrust at node q coming from star A, while α_q^- represents the fraction of power consumed for thrust at q from star B.

Similar equations hold for the available light impulse at any relay node

$$F_q^+ = \frac{W_L}{c} e^{\beta_q^+} \tag{8)}$$

$$F_q^- = -\frac{W_L}{c} e^{\beta_q^-} \tag{9)}$$

we have that only at most $W_L(1-f_n)^q$ power will reach the q-th relay node from the star A, where the convention is that positive force points in the direction from star A toward star B.

The dominant component of position drift in the nodes is caused by thermal absorption. Some of the light will be absorbed, and some of it will be scattered. The fraction f_n represents the net optical losses at each relay element, regardless of their physical origin. ηf_n represents the fraction of momentum that the light transfers to each optical element. η is an adimensional drag factor that estimates how the optical losses contribute to drift. In this work we will assume that all the losses are due to absorption, which is given by

$$\eta = 1 \tag{10)}$$

The net power available from both sides at q-th node will be

$$W_q = \left[(1-f_n)^{q-1} + (1-f_n)^{N-q}\right] W_L \tag{11)}$$

Or,

$$W_q = 2(1-f_n)^{\frac{N-1}{2}} \cosh\left[(2q-N-1)\ln(1-f_n)\right] W_L \tag{12)}$$

The net power loss on this node is given by

$$W_q f_n \tag{13)}$$

and the net force by the absorbed light is

$$F_q^{drift} = \left[(1-f_n)^{q-1} - (1-f_n)^{N-q}\right] \frac{\eta f_n W_L}{c} \tag{14)}$$

$$F_q^{drift} = 2(1-f_n)^{\frac{N-1}{2}} \sinh\left[(2q-N-1)\ln(1-f_n)\right] \frac{\eta f_n W_L}{c} \tag{15)}$$

Since this force cannot be switched off, even if no useful power is being consumed on the starway, we might need to cancel those forces by actively consuming some power from the distant star, but first we will analyze the gravitational influences experienced by the relay chain.

Conceivably, some of the absorption drift will be desired in order to keep the relay nodes as active statites against the gravitational attraction for the nearby stars. The gravitational attraction of relay node q from stars A and B is

$$-\frac{GM_A}{D^2q^2}+\frac{GM_B}{D^2(N+1-q)^2} \qquad 16)$$

This gravitational acceleration needs to be countered with some laser pressure. By eq. 15, we have an absorption drift force that points away from both stars. We will, however, ignore the destination star gravitational attraction and consider only half of the starway. The absorbed force plus orbital correction force to keep the relay nodes as statites is

$$-\frac{GM_A}{D^2q^2}+2(1-f_n)^{\frac{N-1}{2}}\sinh[(2q-N-1)\ln(1-f_n)]\frac{f_nW_L}{c} \qquad 17)$$

The most significant observation that we need to derive from this expression is that gravitational attraction decreases quadratically by each successive step we move away from the star, while the absorption drift decreases only by a factor $1-f_n$. This means that if the absorption drift for the first node is equal to its gravitational attraction, nodes placed farther will drift away from the star, unable to propel back toward the sun. So gravitational pull can help to reduce the absorption drift only for relay nodes placed very close to the star (the first 10 nodes at most), the rest of the nodes need to be pushed actively toward their neighbor star by some other force.

This imposes a stringent constraint on the feasibility of a starway, because it requires that light power from the destination star B to be available to push back the nodes near star A into their equilibrium positions, and likewise on the other end. The argument above leave us free to ignore the corrections due to the gravitational forces from the nearby stars, sincethose represent small corrections to the stability forces. The stability constraint is expressed by the following equations

$$0=F_k^{drift}+2\alpha_k^-F_k^- \qquad 18)$$

$$=\left[\eta f_n(1-f_n)^{k-1}-(2\alpha_k^-+\eta f_n)(1-f_n)^{N-k}\right]\frac{W_L}{c} \qquad 19)$$

$$0=F_{k-1}^{drift}+2\alpha_{k-1}^-F_{k-1}^- \qquad 20)$$

$$=\left[\eta f_n(1-f_n)^{k-2}-(2\alpha_{k-1}^-+\eta f_n)(1-f_n-\alpha_k^-)(1-f_n)^{N-k}\right]\frac{W_L}{c} \qquad 21)$$

$$0=F_1^{drift}+2\alpha_1^-F_1^- \qquad 22)$$

$$=\left[\eta f_n-(2\alpha_1^-+\eta f_n)(1-f_n)^{N-k}\prod_{i=2}^{k}(1-f_n-\alpha_i^-)\right]\frac{W_L}{c} \qquad 23)$$

Similar equations can be demanded on the α+. The factor of 2 on the adjustment thrust is justified on the account that it is generated by reflecting a portion of the incoming light. This system can be resolved recursively for the α coefficients. It can be shown that the maximum value of f_n that will have consistent solutions is given by

$$N = 2\frac{\ln\left[\frac{\eta}{2}\right] - \ln\left[2 + (\frac{\eta}{2} - 1)f_n\right]}{\ln\left[1 - f_n\right] + \ln\left[1 + (\frac{\eta}{2} - 1)f_n\right]} \qquad 24)$$

The solution and the derivation of this equation is detailed in the appendix A. This equation establishes a relationship between the minimum transmissive efficiency and the number of nodes in a starway stabilized against absorption drift. If we use the thermal optical absorption assumption eq. (10) we can simplify further and obtain

$$N = 2\frac{\ln\left[\frac{1}{2}\right] - \ln\left[2 - \frac{1}{2}f_n\right]}{\ln\left[1 - f_n\right] + \ln\left[1 - \frac{1}{2}f_n\right]} \qquad 25)$$

The starway must receive laser light from both directions, which means that optical elements need to be arranged in such a way that drift forces on the individual elements does not cause unwanted net torque on the node. A way to avoid this would be to arrange four or more optical elements in a ring-shaped structure such that the average net torque is zero. In any case, multiple optical elements is a requirement to guarantee minimal redundancy of power in case of partial outages or failures in some of the nodes.

A ring-shaped structure for a relay leads naturally to a concept where the flight traffic would be compartmentalized in regions near the axial line joining the rings. The laser relays must implement some method to deviate a portion of the light in another direction for stabilization thrust, and that same mechanism will be used to provide beamed propulsion to passing spacecrafts.

Since relays are placed one in front of the other in the line-of-sight to the star, safety measures and flight traffic protocols must be developed to ensure that all ships move within their transport lanes tolerances. Depending on the velocity of an incoming ship and the distance to the next relay, maneuvering countermeasures might be difficult or altogether impossible.

5. Available Propulsion Thrust

The figure of merit of performance of a given starway design is the maximum thrust force available in a given direction, since it gives a measure of how much payload can be transported from one star to the other, and at what accelerations. By using eq. A15 from the appendix we are able to estimate the maximum distributed propulsion thrust fraction to be

$$T \leq T_m = \frac{(4 - f_n)(1 - f_n)^{\frac{N}{2}} - (1 - \frac{f_n}{2})^{-\frac{N}{2}}}{(1 - f_n)^{\frac{N}{2}-1}(3 - f_n)} \qquad 26)$$

Given Eq. (26), we can compute the laser power required to provide a given acceleration a_S for shuttles of a given mass as

$$W_L = \frac{a_S n_S M_S c}{2T_m} \qquad 27)$$

where n_S is the maximum number of simultaneous allowed shuttles of a given size and payload accelerating on the starway. The term $n_S M_S$ can be called ***traffic mass*** to emphasize the fact that is the total mass being accelerated on a direction lane of the starway. $\frac{a_S M_S c}{2}$ is the amount of power that a shuttle must receive in order to accelerate at a_S (assuming perfect reflection, it will be lower than that in practice).

If we altered some of the flight parameters (mass of the vehicles, radiation pressure-induced acceleration), we can increase or decrease the number of vehicles supported. Another straightforward method is by adjusting the laser power at the sources, but this needs to happen over timescales comparable to the time it takes light to travel from one endpoint to the other, to allow enough time for the starway to self-adjust to new equilibrium forces.

5.1 Laser-Driven Thermal Management

Some brief remarks about general thermal limitations in laser-driven sails are in order. Laser-pushed ships do not need to carry any propellant on board, all they need is an efficient sail or mirror. However, laser-pushed ships are still limited by their thermal capacity to dispose heat. Assuming a total mass $\sum_i M_i$ and a maximum waste heat disposal rate of $\sigma \sum 2\varepsilon_i A_i T_i^4$, we have the following bound on the ship parameters:

$$\sum_i M_i \le \frac{4\sigma}{\alpha a_S c} \sum \varepsilon_i A_i T_i^4 \qquad 28)$$

where α is the absorption coefficient of the sail. If we ignore all the other terms and keep only the payload mass, sail mass and sail heat terms explicit, and assuming the payload is cold enough to ignore its waste heat, we are left with the following:

$$M_P \le A_S \left(\frac{4\sigma}{\alpha a_S c} \varepsilon_S T_S^4 - \rho_S \right) \qquad 29)$$

where ρ_S is sail area density and M_P is the payload mass. Since this is a direct bound on the payload mass, people have been trying to optimize the right hand side since Robert Forward proposed laser-pushed sails for interstellar missions. The latest improvement to this are microwave-pushed carbon sails[2], with area densities of 5×10^{-3} kilograms, absorptivity of $\approx 10\%$ and operating temperatures above 3000 K.

Such a sail accelerated at one G would have a thermally-bounded payload of $m_p \le 4.5 A_s \times 10^{-2}$ kg. So it would take a sail of 3.1×10^5 square meters (radius a little above 300 meter) to accelerate a payload of ten tons without evaporating the sail. If we relax acceleration to 0.1 g we have $m_p \le 6.1 A_s \times 10^{-1}$kg. Hence, with the same sail size we are able to increase the payload to nearly 200 tons.

Existing concepts for interstellar carbon laser-pushed sails (with a laser at home and no optical refocusing) would require sails of tens or hundreds of kilometers wide. At these accelerations, such sails would be far from being thermally limited, given their large area. However, we still have a limited distance to accelerate before the beam spreads beyond the size of the sail. Thus, high accelerations are required to achieve relativistic velocities. On the other hand, if we can keep the laser light focused on a narrow area (which is precisely the case for a starway), the mass and thermal optimization analysis might change: For instance, a light enough material with good thermal conductivity, may be able to balance its additional mass by increasing the net radiator area, As long as it is less dense than the sail itself. An example of this kind of approach would be to have a sail with extremely low absorptivity $\le 10^{-5}$ operating at relatively cooler temperatures (≈ 800 K), and dump most of the heat to thin graphene sheets covering a large area. Cooler temperatures means a much larger total radiator area, but also a wider array of material and design options for the sail.

Given the fantastic heat conductivity and small weight of graphene ($\approx 10^{-6}$ kilograms per square meter), we could expect in the near future, efficient and realistic graphene-based radiator designs, able to dump most of the heat at a tiny fraction of the mass of both the sail and the payload. In such a scenario, accelerating big payloads in the starway at one G or higher would be limited only by the available laser power at the source.

6. Materials Cost analysis

A simple analysis of material costs can be made by assuming that the optical elements in the relay nodes have a uniform surface density ρ_n. Since there are N relay nodes covering the distance D_s between the two stars, separated uniformly by a distance of D_n, and by the optimal beam width relation (Eq. 4), the total mass for all relay nodes optical elements in the starway is

$$N\frac{\pi}{2}W^2\rho_n = N\lambda D_n\rho_n = \lambda D_s\rho_n \qquad 30)$$

which as we see, does not depend on how many relay nodes are there on the starway because it only depends on the wavelength, the overall distance between the endpoints and the surface density. This of course, assumes that the relay node mass depends linearly on the beam area. As long as this assumption holds, the number of relay nodes will not affect the overall mass of the structure, which means that we can choose smaller relay nodes placed at smaller intervals, or bigger ones placed at larger distances. However, heat disposal will still favor bigger nodes, and the amount of heat that needs to be disposed depends on the relay efficiency f_n.

The thermal waste characteristics of the q-th relay node, from eq. (12) and (13), demand a total radiator area of

$$A_R = \frac{f_n W_q}{\sigma\varepsilon T^4} = 2(1-f_n)^{\frac{N-1}{2}}\cosh\left[(2q-N-1)\ln(1-f_n)\right]\frac{W_L}{\sigma\varepsilon T^4} \qquad 31)$$

This thermal energy needs to be disposed by radiators, since we are in the vacuum of space. Assuming a heat dissipation of 1000 W/m^2 at somewhat below 400K for keeping the optical elements of the relay node relatively cool. Further, we assume a surface density of 0.1 kg/m^2 for liquid droplet radiators [9], which gives us a radiator mass of

$$M_R = 0.1A_R = 10^{-4} f_n W_q \qquad 32)$$

$$= 2\times 10^{-4} f_n(1-f_n)^{\frac{N-1}{2}}\cosh\left[(2q-N-1)\ln(1-f_n)\right]W_L \qquad 33)$$

The total radiator mass M_{TR} for the entire starway becomes

$$M_{TR} = 2\times 10^{-4} f_n(1-f_n)^{\frac{N-1}{2}} W_L \sum_{q=1}^{N}\cosh\left[(2q-N-1)\ln(1-f_n)\right] \qquad 34)$$

7. Modeling the Sol-Alpha Centauri starway

For our toy model of the Sol-Alpha Centauri starway, we will assume a lightway with a wavelength of $\lambda = 10^{-7}$ meters (region between soft UV and visible light). Further assumptions will include a relay absorption of $f_n = 10^{-4}$, and a relay node surface density of $\rho_n = 10$ kilogram per square meter. For covering 4×10^{16} meters we will assume 4000 relay nodes, each separated by a distance of 10^{13} meters. From eq.(24) we can numerically compute a critical transmissive efficiency of $1-4.62\times 10^{-4}$.

Our model of the starway is conceived assuming an efficiency of $1-10^{-4}$. This number, besides being slightly larger than the critical value, is otherwise arbitrary, and it is assumed in order to carry on with estimations on the other relevant quantities. This kind of optical efficiency might be tricky to reach with current diffractive optics, even with fused quartz optics and advanced anti-reflection coatings, in the proposed wavelengths[8, p. 195]. But reflective optical alternatives could be devised for the relay nodes. Reference [10] experimented with mirror surfaces fine-tuned with pixelated capacitive elements. Also, dielectric mirrors[1] can attain reflectivity coefficients of $1-10^{-5}$ over a narrow optical bandwidth, so we are definitely not talking about material requirements far from our current technical capabilities.

By using eq.(26) we are able to estimate the maximum distributed propulsion thrust fraction to be

$$T_m \leq 0.883 \qquad 35)$$

Since our model will be just a proof of concept, we will aim for a traffic mass capacity of 3.0×10^5 kilograms at 1-G (half the mass of the ISS). Given the above and eq.(27), this requires a laser power of $W_L = 10^{15}$ watts down each side of the starway.

From our assumed absorption coefficient, this implies that each node receives $10^{15}(1-f_n)^{q-1}$ watts on a given direction, and absorbs $\approx 10^{11}$ watts. A heat dissipation of $1000 W/m^2$, at somewhat below 400 K is assumed. This gives a required radiator area of

$$10^8(1-f_n)^{\frac{N-1}{2}}\cosh\left[(2q-N-1)\ln(1-f_n)\right]$$

square meters. Assuming 0.1 kg/m^2 for liquid droplet radiators [9], we are left with a mass for radiators of

$$10^7(1-f_n)^{\frac{N-1}{2}}\cosh\left[(2q-N-1)\ln(1-f_n)\right]$$

kilograms per relay node. This radiator mass depends mostly on the absorption coefficient and total laser power. Also, as discussed in sec. 4-1 even if graphene radiators will allow game changing opportunities regarding heat management for solar and laser-pushed sails, the present exercise will still be enlightening as a high bound on material costs.

Given the internode distance of $D_n = 10^{13}$ meters, eq.(4) implies an optimal beam width $W_b = 800$ meters. From eq. (30) we can derive a total optical element mass for the whole starway of $2\times10^9\rho_n = 10^{10}$ kg. From eq. (34), and evaluating the sum, gives a total radiator mass of $M_{TR} = 6.7\times10^{10} kg$ so we have roughly that about 77% of the relay node mass is just for heat management. From the above thermal estimates, the radiator of the closest node to the Sun has about 1.77×10^{-8} kilograms of mass for each watt of laser power W_L, in our case that amounts to 1.77×10^7 kilograms. Gravitational acceleration from the Sun at D_n will be 1.3×10^{-6} meters per square second. That implies that countering gravitational force with laser light will require $3.9\times10^2 M_n$ watts, where M_n is the relay node mass, and follows that 7.0×10^{10} watts would be required to balance gravitational drift.

Since that node already absorbs 10^{11} watts from the laser, this gives further credence to our original statement that gravitational drift is not a determinant factor in the stability and energy efficiency of a starway. Contrary to what intuition would otherwise suggest, building a physically realistic starway to Alpha Centauri would take no more mass than that of a small asteroid. If we allow a redundancy factor of 10, that is, 10 beams travelling parallel between relay nodes, we are in the ballpark of 10^{12} kg, still suggesting that we can build a starway by consuming a medium-sized asteroid.

By simple relativistic kinematics, assuming constant acceleration of 1-g until half-way trip and constant deceleration on the second half, a starway-accelerated vehicle at 1-g would take about 1000 days to travel

2×10^{16} meters (about 2 light years, which is half the full distance to Alpha-Centauri). The proper time experienced on ship would be about 700 days, which is a little less than 2 years.

For the purpose of reference, some of the parameters for similar laser starways are shown below. The targets are two interesting stellar systems in our neighborhood that have one or more confirmed exoplanets: Tau Ceti and Gliese-581. These parameters assume the same internode distance of 10^{13} meters and the same laser wavelength.

7.1 Tau Ceti Starway

- Distance: 11.9 light years (approximated as 1.2×10^{17} meters for simplicity)
- Radial velocity: -16.4 km/sec
- Tangential velocity: 32.3 km/sec
- number of possible nodes: 1.2×10^{4}
- critical transmissive efficiency: $1 - 1.54 \times 10^{-4}$
- optical mass estimation with $\times 6$ redundancy: 7.2×10^{10} kg
- time to destination at 1-g (home frame): 5070 days
- time to destination at 1-g (ship frame): 1864 days

7.2 Gliese-581 Starway

- Distance: 20.2 light years (approximated as 2.0×10^{17} meters for simplicity)
- Radial velocity: -10 km/sec
- Tangential velocity: 35 km/sec
- number of possible nodes: 2.0×10^{4}
- critical transmissive efficiency: $1 - 9.2 \times 10^{-5}$
- optical mass estimation with $\times 6$ redundancy: 1.2×10^{11} kg
- time to destination at 1-g (home frame): 8010 days
- time to destination at 1-g (ship frame): 2180 days

8. The Problem Of Interstellar Dust

The vacuum of interstellar space is not entirely empty. According to Matloff [8,p. 156], dust grain densities and sizes in interstellar space could be enough to require a significant amount of shielding over any kind of velocity above 0.1c.

But assuming a starway like one proposed here is deployed: Parallel light beams of a 10^{15} watts of a couple of kilometers in width, would represent energy fluxes of the order of $10^{9}\ {watt}/{m^{2}}$. Grains inside this beam would tend either to evaporate due to absorption, or to accelerate. In any case, the beam of radiation around the starway would create a thermal gradient around itself that would tend to eliminate or reduce the amount of interstellar dust, which would reduce the risks for vessels travelling the starway at relativistic speeds. The author is not aware of a precise estimation of how much interstellar dust would be evaporated or removed automatically by this effect. This might be an interesting area of research for a future work.

9. Conclusions

We have drawn inspiration from Robert Forward's proposal to use Fresnel lenses with the purpose of increasing the reach of laser power into deep space to propel photon sails. Starting from that idea, we are proposing a system of stabilized, equally spaced optical relay elements between two stars. This architecture will enable a bidirectional, relativistic, safe and economical interstellar flight corridor.

Among the notable benefits of such an approach, there are a few that deserve special mention:

- Sustained power availability for the whole duration of the spaceflight, including deceleration phases, enabling true relativistic travel
- Long-term reusability and flexibity, opening a two-way corridor between two neighboring solar systems
- Smaller distances between neighboring power transmission nodes, which imply softer uncertainty requirements in trajectory prediction, as the time between position observations is smaller and constant.
- Decrease in the expected amounts and size of interstellar dust grains present along the starway due to thermal evaporation.

We have analyzed a simplified, symmetric model of the starway, and considered the stability of the optical nodes against drift due to light thermal absorption. This puts some constraints on the absorption coefficient f_n and the number of nodes N. Less transmissive efficiency implies less nodes in the chain, demanding bigger lenses and bigger lasers at the source stars.

There is however much work still to be done. In this regard, the most notable absence in this work is the problem of deployment. Presumably, two conceivable deployment schemes might be studied in the future: The first possibility is launching multiple lenses together with the sail, which was already suggested by Landis[7]. The other possibility requires reaching the destination star with some other method, and build a minimal amount of self-assemblying infrastructure at destination. After that infrastructure exists, it will proceed to create the solar-powered lasers and relay nodes that will be pushed toward a similar train of nodes travelling from the sun, and eventually braking as both ends approach and meet, half-way between the two stars.

The architecture of the relay nodes is left intentionally vague, as we have required only a few high-level features from them, like the ability to reflect a desired portion of light in order to generate thrust, some maximum thermal disposal capacity, some transmissive efficiency given by $1 - f_n$. We established a relationship between the minimum transmissive efficiency and the number of nodes in a starway stabilized against absorption drift, which we think is the most important result in this report. For our model of the sol-Alpha Centauri starway, 4000 nodes demands an absolute minimum transmissive efficiency of 4.61×10^{-4}. This number might be tricky to reach, even with fused quartz optics, in the proposed wavelengths. But alternatives to diffractive optics using catadioptric elements and dielectric mirrors could be devised for the relay nodes that are not far from our current technical capabilities.

As discussed in section 4-1, laser-pushed sails are limited by their rate of excess heat removal. This is a key subject as payload masses per ship depend strongly on this. Developments in single-layer graphene manufacturing will be essential to remove or reduce thermal limitations on laser shuttles.

We have assumed a $1000 \, ^{W}\!/_{m^2}$ power dissipation at 400K to keeping the optical elements of the relay nodes cool. This implies that optical elements in the relay nodes need to operate in a somewhat higher temperature. Recirculating index-matching liquids as currently used in some high-power laser applications, are an interesting possibility to pursue. Further research in material science will be required to find optimal materials that combine the required low absorption rate and refraction index at high operating temperatures. Within current material technology, liquid droplet radiators could be used to increase the power dissipation per radiator mass to $10^4 \frac{watt}{kg}$. Future improvements in technologies like graphene could make extremely light-weight heat sinks, so we could still see a reduction of several orders of magnitude in the radiator mass. We could also argue that the temperature estimate of 400 K is rather conservative. But the point that the present work aims to remark, is that even with a worst-case scenario of modest technology developments in the key technologies to develop laser starways, they are absolutely realistic engineering undertakings.

Some redundancy is absolutely required, since relay nodes will need downtime for maintenance or emergency repairs. Each energy transmission loss due to failures will propagate down and uplink. Without redundancy, those losses will combine and produce long outages, or catastrophic instability.

The size of fluctuations in the energy flow due to unscheduled thrust corrections along the chain needs to be quantified in order to completely assess structural stability. Given that information about correction requests travel at the speed of light, an algorithm that can forecast and minimize such fluctuations will be needed to route and grant request priorities for power and take decisions in a distributed way.

As we analyzed the material costs in terms of mass, it became clear that the solar system has more than plenty mining resources in order to build at least a couple of starways to the closest stars. The energy bill of 10^{15} watts might seem high by current standards, but it represents a microscopic fraction of the energy budget that generates our sun and goes to waste in the vacuum of space. Energy costs cannot be simply extrapolated from present earth economy, because we already know that beamed solar power could, if implemented, provide for all the energy demand of our global economies. That means in particular, that given the right infrastructure, the cost of delivered power can be brought as low as infrastructure maintenance costs permit.

The focus of the current analysis is on propellant-less radiation-pressure propulsion. Conceivably, further efficiencies in the laser energy usage could be obtained with laser-heated propellant drives, at least for the early acceleration stage. Needless to mention, the starway architecture is completely unaffected by specific shuttle architectures and concepts: as long as the ship adheres to the safe transit protocols that will have to be enforced to avoid catastrophic accidents at relativistic speeds, each ship is free to use the energy it receives from the nodes in the way that better suits its purposes.

Even so, large laser powers will be required. In order to generate such large-scale laser beams, possibly many hundreds or thousands of laser beams will need to be combined in order to produce a single coherent beam. FEL (Free electron lasers) have a number of characteristics that could make them a suitable choice: acceptable energy efficiency, relatively low thermal generation and the absence of a gain medium, which will reduce the maintenance costs.

Given the amount of heat that will be disposed on each relay node, is unavoidable to think that life, and possibly civilization will thrive in the vicinity of these. Populations will be capable to drift outside the star road over the generations and live from the rejected heat, in a similar way that our ancestors did when they decided to leave Africa. The starway will have not only enough redundancy of power lines, but it will have to support multiple lanes, both for acceleration and deceleration of ships coming from both directions. Special lanes could be designed in for emergency braking.

With enough power, and self-repairing capabilities, a starway can work for centuries, possibly even more. As both star endpoints slowly move in their orbits around the galaxy, and their relative motion changes, the starway has plenty of time to dynamically adjust, insert or remove relay nodes as required. In this way, an advanced civilization can create a metropolitan stellar network of light bridges that communicate neighboring stars to each other.

Acknowledgments

The author wishes to thank Adam Crowl of Icarus Interstellar for helpful discussions on the subject of solar sails and laser beamed propulsion.

Appendix

A: Recursive Thrust Equations

If we extend equations 18 thru 22 to allow for the expenditure of propulsion power T_i^- on s nodes in the far half of the starway, as well as drift correction power α_i^- on the nodes of the near half, a recursive solution is given in the following form:

$$\alpha_k^- = \frac{\eta}{2} \frac{\left[f_n (1 - f_n)^{2k-N+s-1} - \Pi_T \right]}{\Pi_T} \qquad A1)$$

$$\alpha_{k-1}^{-} = \frac{\eta}{2}\frac{\left[f_n(1-f_n)^{2k-N+s-2} - (1-f_n-\alpha_k^{-})\Pi_T\right]}{(1-f_n-\alpha_k^{-})\Pi_T} \quad A2)$$

$$\alpha_1^{-} = \frac{\eta}{2}\frac{\left[f_n(1-f_n)^{k-N+s} - \Pi_T\prod_{i=2}^{k}(1-f_n-\alpha_i^{-})\right]}{\Pi_T\prod_{i=2}^{k}(1-f_n-\alpha_i^{-})} \quad A3)$$

Where $\Pi_T = \prod_{i=i_1}^{i_s}(1-f_n-T_i^{-})$ is the combined node transmission over the s nodes that are spending laser power for propulsion of ships.

For a given configuration of the relay chain of N nodes, the critical configuration happens when all the remaining power from the remote star is consumed to push back the relay R_1. If there are higher losses than this, the starway will collapse.

We translate the above insight as the following condition:

$$1-f_n-\alpha_1^{-} = 0 \quad A4)$$

which is equivalent to ask that there is no light power left from the distant star, after adjusting for absorption drift at R_1. The first step is to find a sequence that expresses succinctly the recursive product. Assume that

$$1-f_n-\alpha_{j+1}^{-} = \frac{H_j}{H_{j+1}} \quad A5)$$

For a suitable sequence H_j. This implies that the product of the sequences will cancel in tandem, giving the result

$$\prod_{i=j+1}^{k}(1-f_n-\alpha_i^{-}) = \frac{H_j}{H_k} \quad A6)$$

From this, we can compute inductively the value of $1-f_n-\alpha_j^{-}$ to be

$$1-f_n-\alpha_j^{-} = \frac{M\Pi_T H_j - \frac{\eta}{2}f_n H_k(1-f_n)^{k+j+s-N-1}}{\Pi_T H_j} \quad A7)$$

where $M \equiv 1+(\frac{\eta}{2}-1)f_n$ for convenience. This expression will conform to the sequence expressed in eq. (A5) only if $H_k \propto \Pi_T$. More so, if we use eq. (A1) in order to obtain the initial value of the sequence for $j=k$,we arrive at the expression

$$1-f_n-\alpha_k^{-} = \frac{M\Pi_T - \frac{\eta}{2}f_n H_k(1-f_n)^{2k+s-N-1}}{\Pi_T} = \frac{H_{k-1}}{H_k}$$

Which gives the first term of the sequence $H_k = \Pi_T$. This simplifies relation eq.(A7) for H_{j-1} as:

$$H_{j-1} = MH_j - \frac{\eta}{2} f_n (1-f_n)^{k+j+s-N-1} \qquad A8)$$

or, in nonrecursive form:

$$H_j = M^{k-j}\Pi_T - \frac{\eta}{2} f_n (1-f_n)^{k+j+s-N} \left[\frac{(1-f_n)^{k-j} M^{k-j} - 1}{(1-f_n)M - 1} \right] \qquad A9)$$

11.1.1 A1: Critical Transmissive Efficiency Relation

First, we consider the case where all the power is being used to keep the starway stable, and there is no leftover power for anything else. This is the case when $s = 0$ and $\Pi_T = 1$. By using the critical constraint eq. (A4), eq. (A5), and taking $2k = N$ as the point where the drift correction thrust begins (half of the starway), we arrive after a few algebraic manipulations at the desired relationship between critical transmissive efficiency and the number of relay nodes:

$$N = 2\frac{\ln\left[\frac{\eta}{2}\right] - \ln\left[2 + (\frac{\eta}{2} - 1) f_n\right]}{\ln[1 - f_n] + \ln\left[1 + (\frac{\eta}{2} - 1) f_n\right]} \qquad A10)$$

Fixing N, this can be solved iteratively for f_n. The table A1 shows several computed values for different sizes of the relay chain.

N	f_n
200	9.19×10^{-3}
2000	9.23×10^{-4}
4000	4.61×10^{-4}
1.2×10^{4}	1.54×10^{-4}
2.0×10^{4}	9.24×10^{-5}
4.0×10^{4}	4.62×10^{-5}

11.1.2 A2: Available Propulsion Thrust

In order to have any power left to thrust interstellar vehicles, we want to consider the region

$$1 - f_n - \alpha_1^- = \frac{H_0}{H_1} > 0 \qquad A11)$$

It is easy to check that $H_j \geq 0$ holds for $0 \leq j \leq k$ if eq.(A4) holds. In particular, $H_1 > 0$.
So eq. (A11) is equivalent to

$$H_0 = M^{\frac{N}{2}}\Pi_T - \frac{\eta}{2} f_n (1-f_n)^{s-\frac{N}{2}} \left[\frac{(1-f_n)^{\frac{N}{2}} M^{\frac{N}{2}} - 1}{(1-f_n)M - 1} \right] > 0 \quad A12)$$

which can be rewritten as

$$\frac{\prod_{i=i_1}^{i_s} (1-f_n-T_i^-)}{(1-f_n)^s} > \frac{\eta}{2} \frac{f_n}{(1-f_n)^{\frac{N}{2}}} \left[\frac{(1-f_n)^{\frac{N}{2}} - M^{-\frac{N}{2}}}{(1-f_n)M - 1} \right] \quad A13)$$

In the case where a single ship is being accelerated in a given direction ($s = 1$) the formula of the available thrust fraction becomes

$$T_i^- < \frac{\left[2 + (\frac{\eta}{2} - 1) f_n \right] (1-f_n)^{\frac{N}{2}} - \frac{\eta}{2} M^{-\frac{N}{2}}}{(1-f_n)^{\frac{N}{2}-1} \left[2 - \frac{\eta}{2} + (\frac{\eta}{2} - 1) f_n \right]} \quad A14)$$

if $\eta = 1$, this simplifies further to

$$T \le T_m = \frac{(4-f_n)(1-f_n)^{\frac{N}{2}} - (1-\frac{f_n}{2})^{-\frac{N}{2}}}{(1-f_n)^{\frac{N}{2}-1}(3-f_n)} \quad A15)$$

References

1. J. H. Apfel. "Phase retardance of periodic multilayer mirrors." *Applied Optics*, 21:733–738, 1982.
2. James Benford. "Starship sails propelled by cost-optimized directed energy", 2011.
3. Robert Bussard. "Galactic matter and interstellar flight." *Acta Astronautica*, 6:179–94, 1960.
4. William Chang. "Principles of lasers and optics." Cambridge University Press, Cambridge, UK New York, 2005.
5. Louis Crane and Shawn Westmoreland. "Are black hole starships possible", 2009.
6. Robert L. Forward. "Roundtrip interstellar travel using laser-pushed lightsails." *J. Spacecraft*, 21:187–195, 1984.
7. Geoffrey A. Landis. "Optics and materials considerations for a laser-propelled lightsail." paper IAA-89-664, 1989.
8. Gregory L. Matloff. *Deep Space Probes*. Springer, 2000.
9. T. Mattick and A. Hertzberg. "Liquid droplet radiators for heat rejection in space". In Energy to the 21st century; Proceedings of the Fifteenth Intersociety Energy Conversion Engineering Conference, pages 143–150, 1980.

10. T. S. Taylor, L. W. Brantley, Anding R. C., and D. Halford. *Membrane Optic Research Program Internal Documentation.* Teledyne Brown Engineering and The National Space Science and Technology Center, Huntsville, AL, 2002.

Food Design
A Primer with Thoughts on Food Design in Space

David Wilson

1. Why Food Design for a Deep Space Mission is Relevant and Important as an Aspect to Humanity in Space

What is food design? Throughout this paper, concepts on food design will be explored; food design for space is also discussed.

In this paper, I shall attempt to illustrate a dynamic relationship, while building a basic outline to support an argument for an ongoing dialogue, with food at the forefront of an effort to fulfill the completion of a generational deep space star ship and mission where humans are aboard. When humans are present, what/who, when, where, and how shall they eat? The why behind why humans eat seems easy at first glance, but the question is a more complex and dynamic relationship to our existence. We eat for nutrition, to survive, we eat to enjoy, we eat for a number of reasons perhaps as yet theorized or studied; being an organic form, a neurobiochemical-endocrine basis may be involved…at least to some degree in human food selection. But, it would seem that there is great room for research in these areas today.

About this paper: This work is written with a seemingly strong focus on anthropology, however, it is the author's opinion that strong design elements happen as a direct result of deep anthropological appreciation. Also, this paper returns to themes to reinforce as well as to develop other directions.

What is culture; what is culture in deep space; what is culture as it relates to food design?

1.1 Culture

The common understanding of culture, loosely speaking, is the collective mind of humanity in environment. Culture is represented in the ideas of the minds of humans. Therefore, culture functions as a number of intangible concepts represented as a relative inseparable total.

The 'space community,' meaning culture in space, is an interchangeable term which represents the collective idea of those individuals experiencing microgravity, as well as the larger group of people which provide support for the individuals in microgravity.

Culture as it relates to food design, it is the body of people concerned with food design, yet not necessarily limited to the consumer-group of people that use or are directly affected by the food design of food for a space community.

2. Food Culture

What are the feeding traditions of various cultures on earth? Most people eat together; feeding happens as a social act- from time of birth, or of prehistoric humanity, meaning from suckling to present- from hunters /gatherers to drive-through fast-foods, there is a social tether.

What are present-day feeding traditions in space? In a recent visit to Houston's Control Room in the Johnson Space Center (2012), I was privileged to observe that all activity of the astronaut is part of a scheduled event dynamic. That is to say that all time and activity is accounted for. Astronauts have appointed timeframes where feeding is scheduled. These represent points of course in measures of duration critical awareness not directives that the crewmembers must follow.

In the following feeding social section(s), where applicable, ideas will be compared and contrasted to highlight relationship of earth-based social food design related activity to that of space.

3. Feeding Social

Historically, humanity, as far as the human record presently shows, reveals mankind as first being hunter and gatherers, and later developing agriculture. We may be hunters and gatherers again, when we hunt and gather the best specimens to clone. Only the quest for sheer and utter perfection leads potentially to that next logical step, meaning the generation of a perfect specimen; the agricultural phase, the "farming" of the perfect cell.

Mesa, mensa, table, tabele, tabula, and bord are each names throughout humanity to describe the concept of the table. It's quite likely that any flat-topped surface may have served the utility function of table. An altar, for example, may represent a gathering point with a social purpose anchored to it. While eating at the altar is a speculative leap, the idea of gathering around it and the act of sacrifice is a well-known fact [of some cultures].

The dinner table in America: While assembly-oriented food rituals and table rituals are world-over, I have elected to speak on the American dinner table, because this is where I am, this is what I know. America still functions as a cultural hub where diversity springs eternal.

The economic idea of lap trays and personal food management devices has gone through phases in NASA; much of the early food program, meaning food delivery to human(s) was in the form of tubes which resembled toothpaste containers (this was during the Mercury Program of the early 1960's when John Glenn first tried apple sauce from a squeeze tube onboard his Friendship 7 spacecraft in 1962. "Space Food Hall of Fame" http://education.ssc.nasa.gov/fft_halloffame.asp). The significant part of the act of eating in space proved that man could eat in space, and survive, which laid the groundwork to support the idea that manned space missions were a possibility. Although for Astronaut Glen's 5-hour flight, starvation concerns were not immediate, the idea that missions would grow from hours to days to weeks made the realization of food-needs an exciting proposition and progressive aspect to the overall NASA mission(s).At one point in the history of space food, NASA used food cubes; a type of jell-cube as a later advancement; the food was covered in jell.

The 'spoon bowl' and other warming and hydration "gadgets" vastly altered space dining. These alterations; warm food, spoon-like and bowl-like structures, similar home-like textures enhanced food memories carrying concepts of home into the recall of those in space, which in terms of comfort for astronauts; being so far from home, vastly improved the feeling of home-like similarities…even when the foods themselves were comparably, at the time, far from their earth-bound equivalents. Early energy bars made an appearance as the "Space Food Stick" which, conceptually, remains popular today. The space food stick represents NASA seeking professional outside solutions to a challenge; make an energy-packed product that can be highly portable and meet the requirements of a space mission; that is to say that it must not have free-floating crumbs which could damage sensitive equipment.

Automat, refectory, cafeteria, food court, etc.: each of the listed food environments represents potential eating environments. The words mess hall or galley are nautical terms, common for ships, are reflective of the ideas used in feeding on board such a vessel. Skylab had a more robust 'galley' than other space craft before or since. And the Skylab galley encouraged, by fact of existence, an enhanced social aspect of crewmember dining.

At first glance, this seems a rather obvious question, and may even sound a bit comical. However, when we really think about the act of consuming food; especially on earth where desirability and means can constitute the latitude of a culinary experience. Eating can be many things. It can be the act of riutal, refueling, social interaction, action, etc. Throughout history there have been many food habits as approaches to eating throughout the human record. This idea will be explored further at a future point in this paper. We shall now consider eating as it relates to the following ideas on food memories.

4. Food Memories and Psychological Effects of Food:

If a food was flawlessly recreated for flight and an element was in place to purposefully make the dish just slightly less than perfect, would the space diner appreciate the attempt and compare and contrasting with home long for it?

When should the perfect replica be less than perfect and what is the produced effect, if any, by this act?

Should the perfect replica of a dish be strived for in a long duration deep space mission context, and if so why?

Why should food design consideration be brought to the forefront of discussion in a generational starship discussion? Scientists already know the physiological effects of humans in low gravity, therefore, the work is perhaps more about going the distance while observing what can be learned about humans in that processes context: that is to say, what is to be learned of humans on a generational star ship? If that is indeed a subtext larger-question, then ideas on culture rest squarely upon this question. Food remains central to present-day culture here on earth. Will it be similar for deep space? How might it differ? What are the questions that should be asked of food design for a project such as this?

Some questions include- how much space, divisions of space; meaning space used for: growing food, processing food, cooking food, serving food, stowing food and the like will be available? Scientists will explore those types of questions to eventually arrive at a blend of best practice approaches. One thought to add to the mix is that a lot can happen over the course of 50 years as a crewmember aboard a star ship, and perhaps a degree of flexibility may accompany any design. What work, if any, has been done in the area of AI (artificial Intelligence) and food, or AI food and space? Is intuitive food design a model that could be a viable food system for consideration? Intuitive food design, in one sense, may be defined as culturally adaptive; the food system is or may function 'intuitively', while maintaining standards of importance/structure set forward by policy and administration.

A culturally adaptive model flows with the changes in culture, but is it a worthwhile consideration?

Is a culturally adaptive model a worthwhile consideration? Will there be the more basic (though difficult to execute) culturally aware model? This is where the cultural considerations of others may be recognized, included as features or add-ons [American Day for instance, or hotdog Tuesday], and/or maintained in the form of a planned fest or custom. Today the ISS follows a basic American diet of 3-meals per day. What might a 100YSS feeding schedule look like?

When we think of comfort, as it relates to food, we think of "comfort food(s)". Well, what is comfort food? Comfort food is the type of food that we are familiar with, foods which call forth strong personal food memories of "nostalgic" association; mom's meatloaf, Aunt Matilda's pot roast, Uncle Willie's BBQ for instance.

Not all food memories are pleasant ones, or have to be, for that matter- Perhaps Aunt Matilda's pot roast was a Sunday early-supper affair. And, the roast was dry, which would lead to your parents fighting violently on the drive home. Yet, this was an every Sunday event. And Aunt Matilda passed away one year- 15 years ago. No more pot roast, until one day when a college sweetheart cooks you a meal; the first meal of the courtship—you guessed it, pot roast- as dry and wooded as Aunt Matilda's. You laugh warmly under your breath, and when asked what's so funny? The pot roast is just like my Aunt Matilda used to make.

Each person has had unique food memories- everyone has. Food memories, in some cases have become, in a broader social context, food traditions—Thanksgiving, Christmas, Seder meal, Eid al-Fitr, New Year's, the 4th of July, Labor Day, and many, many more. Who knows, a 100YSS may have food traditions, too. There will undoubtedly be certain food traditions of the personal/collective variety, Valentine's Day, Birthday, Anniversary and the like. Food traditions will blend with belief systems and find expression there too. And, how might a 100YSS address the food traditions expressed of various belief systems? Should that be a consideration? And if so, how will those food considerations be met; among other questions around the topic of food ethics? So, food design can shift from the basic and fundamental to the philosophical, and theoretically challenging.

What is comfort, in deep space, as it relates to food design? Comfort can be measured against the degree of effort and challenge; the greater the amount of ease, and greater the relaxation, the more the comfort. Comfort as it relates to space, with regard to food(s) implies; safe [radiation-proof packaging, free of microbes or physical contaminates], nutritious foods, which taste acceptable that are also healthy. While it is no argument that NASA has proven with early missions that foods can be made safe for flight, any foods designed to last and be eaten up to two years later face acceptable gustatory palate challenges.

Comfort as it relates to food design is when an ease and relaxation, on the end user of the food design, is satisfied; that is to say expectations are met or exceeded.

Food design is a relatively new discipline which addresses sensory, and experiential and or other dimensions in food. Food design, as a discipline, is a functional means to describe and then identify what are conscious constructs regarding food and food dynamics.

Food design subcategories as described by the International Food Design Society: Design with food; food product design; design for food; design about food, food space design/interior design for food; and eating design. Design with food is when the food itself is transformed. Food product design is when the food is structured often to be purpose specific, and often geared toward mass-production. Design for food is the support for the food; containers appliances, utensils and the like. Design about food is the peripherals which iconize food-stuff. Food space/interior design for food is the design of food spaces/environments from home dining rooms to grocery stores and beyond. Eating design is focused on eating situations and differs from food-space/environment in that thematic presence is the focus.

5. Anthropology

Native American cultures may function as promising examples of hunter-gatherers in early society when considering early American/Mesoamerican cultures in nomadic context post last glacial period (last ice age). This seeming segue is to remind us that exploration is as much part of the human record as it is part of being human. Deep space exploration reflects another leg in the human journey. One constant is our need for food on a journey 10,000 years ago as it will be for deep space voyages 100 years from now. And, our resilient innovative approaches to food management will be as relevant tomorrow as in prehistoric yester-year.

Early American cultures, as did many cultures around the globe, enjoyed mineral salt. Salt was a food staple amid nomadic life of early Americans/Mesoamericans. And, as with most long-lived cultures have less-apparent layers, subtext to support the choices they make, meaning salt may have provided more than just flavor for ancient cultures.

The role of salt on earth as it may serve to contrast low-gravity space- the human body is often measured in mass composition or atomic composition. The 'mass' body is composed of roughly six elements. Those big six account for 99% of human mass. Basic mass composition elements are oxygen, hydrogen nitrogen, calcium and phosphorus. That leaves roughly 1% trace elements, and about .85% remaining is comprised of magnesium, chlorine, sulfur, potassium and sodium... good ol' salt. However, from an elemental perspective, the human body is composed of .15% sodium. Humans require salt, to a large degree, for replenishments of vital fluids, and to a questionably lesser degree to flavor food. I don't want to get too far off the rails on the theoretical crazy train here; because we have over 10,000 taste receptors in the human mouth, to have them stimulated with one of the few known edible minerals, just may produce psycho-physio-endoneurological effects. In other words, the salt stimulus may the seasoning which gets the wheel spinning on the whole biochemical soup. And yet, in space, the capillary action of humans in low gravity produce what NASA commonly refers to "Bird legs" and "Fat head." This is when the blood pools in the face as fluid is 'wicked through capillary action into the weightless face. That coupled with the dual effect of heart placement in relation to feet, meaning the distance which the heart must pump blood without the aid of gravity the feet and legs and to a degree the arm extremities suffer depletion and 'thin-out.' Salt may complicate the phenomenon as sodium will function to increase fluid retention; in a low gravity environment and this may lead to intracranial pressure or undue ocular stress as eyes bulge wildly in their sockets. The effect of fluid retention may be short-lived in micro gravity, once the nutrition is digested; salt in the body, in micro gravity, is overly expelled in the urine and can contribute to dehydration. So, while salt on food may remain a tasty flavorful notion, its practicality in a low gravity scenario *today* may be limited for health reasons. However, that is not to say that salt wouldn't have a role to play should a gravity-based destination be considered, or a ship with a form of artificial gravity. Sodium for humans, within standard health guidelines, is normal; however, low gravity is the pink elephant in the equation.

Food experience in tourism- in fact, a 100yss journey may not be all that different, in a generic sense, than having food experiences while being a tourist, or those of an early-period nomad. Presently we operate from the knowledge-based assumption that we are *it* as far as intelligent life goes, in space. Although, those of us gathered here today are optimistic, if by nothing more than the objectivized idea of our own presence; here, that life, to some degree and extent, elsewhere 'in space' will ultimately be contacted, supports a thought that there just may be exciting new foods... or at least new food constructs to dine on along the way. And by food constructs, I mean all the good stuff in the cornucopia of tomorrow's food technology.

The very act of this 100-year starship, a bold and ambitious project, becomes, like the entirety of the broader NASA project or human research project. It takes its place among the anthropology of the human record. Today, I am not certain if this will be represented as a branch or a completely separate idea of humanity, but I am certain that this second step for humanity, this deep space step, is as relevant as its first.

Serano explores Italy, the city of Naples, and centers a focus on food's socio-anthropological importance in family and community. This work set the stage for the growing discipline of food anthropology.

The goal of food anthropology is to provide context regarding the rituals and the act(s) of food and eating as it relates to humanity and the human record.

Categorization and labeling helps further reduce the abstract of culture to put a fine point on what is being asked and defining culture. Edward B. Tylor, English anthropologist in his work, 'Primitive Culture 1871efforts to define culture.

"Culture is the full range of learned human behavior patterns." Edward B. Tylor Presently, culture, it is theorized, has layers to it known as subculture. Subculture is often viewed in a number of ways; people within a culture may be an example of direct culture, whereas cultural universals, those things present in any culture/ group of people which helps to define culture. Cultural universals differ from the habits, rituals, customs and traditions of any given culture. That is to say that, cultural universals are present in all of culture that culture is by virtue of cultural universal presence and inclusion.

Tracing the habits of food, in broad brush strokes, is a fairly doable task; in prehistory man ate foods hunted and gathered, and later progressed to agriculture and domestic meats. Later the Silk Road expressed a cross-cultural exposure, meaning diffusion of food ideas. Or that during the dark ages in Europe experienced a shift from Roman influenced cuisine toward Arab influenced cooking, with a larger and more complex strife-driven dynamic [war lead to lack of refinement]. Or prior to Columbus bringing tomatoes back from the new world to the old that Italy had no tomato-based sauces, Ireland no potatoes; these key turning points of diffusion affected cuisine represent a few among many. But, the point is that these influences are, or become easily traceable. It is interesting to note that some newly exchanged influence ingredients held interest in a culture greater than others- take the tomato for instance; resembling an over-grown belladonna may have held significance in some instances and dread in others.

The act of following food habits over time is example of food anthropology; the anthropology of food is a fairly new discipline in anthropology. The idea of space food or food-in-space will continue to take its unique place among the human record.

In the early stages of the space program, NASA astronauts during the Gemini mission ate applesauce from toothpaste-like tubes. That represented a high level of terror on the eating scale; there was fear of health hazard like the astronaut choking on the food in micro gravity, or food particles getting into sensitive equipment. Today, the scale has shifted away from 'eating to survive' and the terrors associated with it as astronauts today aboard the ISS enjoy much more comfort choosing from an extensive menu of familiar foods including foods like cornbread stuffing as well as macaroni and cheese. To what degree of comfort to terror might be experienced aboard a generational starship? How much eating to survive VS eating for comfort will be experienced? Will foods hold a familiar 'tethering to the crewmember life on earth? Should it, or should food be expressed as a radical departure from that of earth?

Carolyn Korsmeyer's work (1999) centers on elevation of consciousness in the life/death relationship between eater and food source. Often we think a hamburger comes from the store, Korsmeyer's work puts a face on that food. Should there be a consideration of cycle awareness of food on a generational starship? That's assuming food will still come from conventional sources. Perhaps a food source on a generational starship includes a derivative of edible paper and ink- How might that as a scenario look?

Edible Paper and Ink Homaro Cantu Interview re: 100YSS 3D foods 2013 interview by D. Wilson 2013 (unpublished)

Homaro Cantu worked with NASA on their Institute Advanced Concept's 3-D replication, built from edible 2-D substrate.

Cantu details that there are tech hurdles to the practical application of making a meal of edible paper and ink. "When you want to send a human to another planet there are goals with nutrients; it [the technology] must get to the nano scale to be a reality." There are nano extracts w/flavors however not yet with structures. That's where the technology needs to be. The raw materials don't yet exist. But if those two things come together it could work. In order to create a molecular match with an apple, you need a vast amount of core ingredients in order to produce a match. If 3D food printing were used, a rotation of a menu is a most likely scenario rather than an 'on-demand' printing of foods.

When asked, Chef Cantu said, "Edible menu'- started with the tasting menu; a clever way to get people to eat right after they get seated." Cantu first started writing on rice paper with edible inks then that idea expanded into other things to create and further refine his edible menu.

A class4 CO2 laser was used by chef Cantu on the TV show "Iron Chef." Now, Cantu and his team use Z-NA lasers which function to etch names or engrave. He and his team has used laser technology for experimenting with flavor technology- capturing vapors and freezing g them to make things like elixirs and extracts but Cantu is quick to note that while fun, "It's expensive to play with."

"Over the years working in restaurants I've always wanted to invent things, and before Moto opened, I invented several things. There is a time and place for classic things, but I don't want to do that at work. Everything else is rapidly evolving, why not food? Food is going to evolve; how do we let it evolve?"

"We're going to find ways to make veggies replicate meats [Cantu cites the 'China Study'] I'm very pro innovation. Cantu designs and beyond eggs worked together to create an egg substitute."

Should the technology catch-up to the idea- Cantu said, "Food printing will revolutionize everything around us."

"It's a single celled protein combined with synthetic aminos, vitamins and minerals. Everything the body needs." Dozer from the movie *The Matrix*

Factors that influence food choice(s):- culture, finance, convenience/time, taste, 'food choice questioner' etc. What factors may influence the crew of a generational deep space starship? The rules of food standards set forth by NASA will likely stay the same; that "foods must be safe, nutritious and appetizing, minimize volume, minimize mass, and minimize waste." But what will influence the crewmember's individual choice? To some extent it will depend on the food system(s), the design of the food system, food substance availability. Eating and food choice aboard a modern-day submarine: it is widely known that, "When a submarine leaves on patrol, food fills every available corner." (http://americanhistory.si.edu/subs/operating/aboard/leisure/).

And because food is bulky, and may be present in degrees of perishability, fresh foods like fruits and vegetables are eaten first, and then preserved foods are consumed. The submarine provides a sound basis for sociological study in consideration for a long-duration deep space mission. Such research may function in contrast analysis of what is learned from crewmembers in micro gravity dining aboard Skylab, which sported a virtual dining room (wardroom) and kitchen VS the space shuttle which offered a slightly less social dining environment through the use of personal dining trays.

That was then and this is now; recently research was conducted in Hawaii on foods and cooking for a mission to Mars, which focused on NASA's idea of "pick and eat veggies" items such as "lettuce, spinach, carrots, tomatoes, green onions, radishes, bell peppers, strawberries, herbs and cabbages" are among consideration. These food items represent choices for application in crops which may theoretically be grown on Mars, the Moon or an Asteroid.

But, what of tomorrow? What will food and eating look like on a 100 year star ship? Eventually the submarine pulls in port to reload, a thing a generational star ship in deep space will perhaps, be hard-pressed to do... perhaps. Perhaps food won't look like edible paper substraites with nano flavorites sprinkled on the stacked layers which are then run through a 3-D printer with recycled scraps all plugged into a world encyclopedia of food. Perhaps food will be built from synthetic programmable cells, thereby still employing a cooking step, or maybe there will be a uniquely different hybrid:

"I know this steak does not exist. I know that when I put it into my mouth the Matrix is telling my brain that it is juicy and delicious." Cypher, the movie, "The Matrix" http://bowling-bash.blogspot.com/2010/01/cinema-matrix-quotes.html

Perhaps food and nutrition may happen through human Brain Computer Interface where neurocepters are stimulated and flavor memories are tickled while the dining crewmember eats a nutrient packed flavorless textured 'canvas' and drinks a flavorless nutrient-packed liquid 'canvas' with sensory data driven through BCI. A virtual encyclopedia of all foods of The Earth could be contained in a digital record. The program for the food could be selected by the person wanting a meal. The BCI and "canvas" generate neural picture with full biochemical and subsequent hormonal response to stimulus.

A spoken dish asks southerners about their food memories: food memories are the collection of related ideas on flavor, foods, textures, temperatures; food memories tell us the difference between a good pizza and a bad one. Food memories take us back to childhood, across time to given events, or rededicate commitments functioning as ritual and tradition. The above is but a taste of what food memories are, where they come from and what makes them, or why we have them. The cause of food memories is much more complex than what was just mentioned; on one level it is a smell or image triggering the recall of a point in time/space once experienced. On another level, it is a biochemical cocktail of an exact formulation; those colors, proximal shapes, olfactoral stimulation and perhaps other sensory stimulations; ambient room temperature, environmental condition, biochemical fluctuations in the individual neurological dynamics all functioning in perfect accord.

Food memories- what was the first thing you remember eating: What are your food memories? How might food memories play a role on a generational star ship? On a submarine, food memories are triggered through customs and familiar foods, like pizza night; this conjures a mood of social ease and encourages comradery while reminding crewmembers of home. The pizza is great, as a source of change, or terrible when contrast to another crewmember's food memories. Stories are exchanged through the dining area where good pizza may be found. Hormonal fluctuations happen and nourishment happens while nutrition occurs. Regardless the food on a generational starship, food memories will happen, if humans are present. The question is: will food traditions and food mythologies of earth accompany the crewmembers on the voyage, or accord the mythologies inherent of the star ship, and should they?

How our mythology of food drives us to eat the world: as the population grows we consume the world; we observe population growth and grow more food assuming to pace population growth rather than consideration that additional food growth may be fueling it. [http://misouri.sierraclub.org/sierranonline/mayjune2000/howourmy.htm] What food mythologies might result from life on board a generational starship? Do food mythologies regarding food in space, or deep space, already exist? What might a food mythology look like for food in deep space?

6. Furniture

This question of furniture is asked because locality proximity and relevance in eating. The jury may still be out on the why of humans are social creatures, but we do like to eat in a social context, generally speaking, we gather around a dining table and nosh. Furniture, according to Inar Yildiz in, History of Furniture/Part 1 "Since ancient times, designing of several furniture objects such as chairs, sofas, cupboards or desks, furniture design and making has been an expression of culture and civilization, of how we view the world and how we think it should be ordered, moving it above and beyond the merely functional aspects."

The table is no new thing; they have been around since time immemorial; of Catalhoyuk, a Neolithic 'tell' in Turkey; the 6300BCE site, "seems to have been a village of mud-brick houses, each built to a fairly standard plan with a kitchen living room and storage area…" [also Inar 328 Yildiz history of Furniture] recalling that food design encompasses food environments as well as food itself.

Gathering to dine is a tradition in many cultures. While ideas on eating, whether communal or solitary, continue to shift, the presence of dining custom persists. On a generational ship there might be the following eating environments:

- Automat, "a fast food restaurant where simple foods and drink are served from vending machines."
- Refectory, "a monastery-style dining hall, refectorum; where one is restored…that through eating one is restored, or that, on a deeper level, that health is restored through eating; that is to say that the saying is perhaps an echoing of the idea that food is medicine."
- Cafeteria, "a space designed for food consumption.
- Food court, "a space for offering foods in stations in a cafeteria-style setting

7. Cosmic Cuisine

"Cosmic Cuisine" is a report from NASA addressing some of the challenges the Advanced Food Systems face in the consideration of a mission to Mars. It also has an eye on long duration missions. Thermostabilized foods have presently gained prominence over freeze dried foods. The report also suggests a diet with lots of hydroponically grown vegetables. The report mentioned findings through studies, not cited, that foods should have a variety of tastes and textures and colors which help foods provide mental as well as physical nutrition.

Space food standards impacted the food industry and changed food safety standards justifiable

Should there be room for connecting with the idea of 'where food comes from' on a generational star ship? Some theorists suggest that a healthy relationship with a respect for food comes from knowing where food comes from. When we ask little urban Johnny or Suzie where does milk come from, you will likely get an answer like the store; a very different answer than farm-savvy Dick and Jane. It is believed that we eat better when we have a relationship with our food, such as when a family has a 'victory garden' the children are more likely to eat vegetables because they have a connection with the food. In a larger and more general sense this translates to healthier food choices. What will a food relationship on-board a generational star ship look like?

As a classically trained visual artist- here's a future thought- perhaps there will be room for art and food together in a generational star ship. And then there's art itself… and what of art on a generational star ship. But that is a question and topic for a different day. What is the answer? A food environment on board a starship could very likely become one that is malleable adaptive and intuitive.

When something is built over time as a fluid organic dynamic, inherent complexity it to be expected. As a chef I know firsthand the difference between when something works, and making it work. Something less abstract is a software program that is launched with the thought that all of the 'bugs' are out. Inevitably there are 'patches' required to 'fix' underappreciated or unobserved issues. A well-thought food system conceived on earth may require real-life feasibility and functionality alteration in real-time in deep space for crew aboard a generational star ship.

About the Author

David Wilson earned his certification in baking and pastry through Le Cordon Bleu College of Culinary Arts of Chicago. David also holds an Associate Degree in Applied Science in Culinary Arts through Le Cordon Bleu College of Culinary Arts in Chicago. While earning his culinary degree, David interned with NASA at the Johnson Space Center in Houston Texas in the Space Food Systems Laboratory.

Sources

1. Homaro Cantu, Chef/entrepreneur: Cantu Designs, Moto/Ing, 2013
2. NASA- Johnson Space Center, 2012
3. NASA- Space Food Systems Laboratory, Johnson Space Center, Houston Tx, 2012
4. NASA- Advanced Food Systems, Johnson Space Center, Houston Tx, 2012
5. How Food Can Be Art: A Discussion of Taste (Part7 of 8), January 12, 2012 by Art http://abetterwhirlpool.wordpress.com/2012/01/12/food-art-part7/
6. International Food Design Society: http://ifooddesign.org/food_design/index.php
7. International Food Design Society: http://ifooddesign.org/food_design/subcategories.php
8. Cambridge Encyclopedia of Hunters and Gatherers http://books.google.co.uk/books?id=5eE-ASHGLg3MC&pg=PP2&lpg=PP2&dq=%22all+humans+lived+this+way%22+90+12,000&-source=bl&ots=E-qs5MKbdn&sig=EmtqQEkdTuHGFX2kSOYfC7MLdXw&hl=en&sa=X&ei=5JG-GUcTIM8qg0wX3pYHACA&ved=0CDcQ6AEwAQ#v=onepage&q=%22all%20humans%20lived%20this%20way%22%2090%2012%2C000&f=false
9. Native American Cultures: history.com
10. Towards a Structural Model of the Tourist Experience: an Illustration from Food Experience in Tourism Quan and Wang 2003, www.sciencedirect.com
11. Food and Anthropology in the Early Works of Matilde Serao by Daria Valentini, https://twpl.library.utotonto.ca
12. Anthropology of Food- University of Minnesota Duluth, www.d.umn.edu/cla/faculty/troufs/anthfood/video/meaning_of_food.himl
13. Human Culture: What is Culture? http://anthro.palomar.edu/culture/culture_1.htm

14. Universals of Culture: a list of 10 cultural universals, Jennifer Hanzak, www.learner.org

15. Bloomberg, Myhrvold: Invention is the Mother of Economic Growth, Nathan Myhrvold Dec 19, 2011, http://www.bloomberg.com/news/2011-12-20/invention-is-the-mother-of-economic-growth-nathan-myhrvold.html

16. IGT Vanvitelli/ Istituto technico Per Geometri "L.Vanvitelli" Cava De' Tirreni (SA), Italy: "Food Habits Over the Centuries" 2008/9

17. Project Muse, "Anthropology and the Everyday, from Comfort to Terror, Nancy Ries New Literary History vol 33, Numver 4, Autumn 2002 pp. 725-742, http://muse.jhu.edu

18. University of Buffalo, Delightful, Delicious, Disgusting—the Difficult Pleasure of "Terrible Eating" Patricia Donovan January 9, 2004, http://buffalo.edu/news/releases/2004/01/6533.html

19. Huffington Post, 'A Spoken Dish' Asks Southerners About Their Food Memories (video), Rebecca Orchant, http://www.huffingtonpost.com/2013/06/18/a-spoken-dish-food-memories-video_n_3455136.html

20. Sierra Club, "How Our Mythology of Food Drives Us to Eat the World" John Kurmann, http://missouri.sierraclub.org/sierranOnline/mayjune2000/howourmy.htm

21. History of Furniture and Interior Design part 1: Prehistoric Furniture and Interior design, Inar Yildiz, http://emuonline.emu.edu.tr/inar328/pdf/Part_one_inar_328_furniture_styles_ancient.pdf

22. Around the Dinner table, History of Eating in the United States, Robert Cowley, Della Gibson and Chanda Sewell, 'Family that eats together, stays together', http://historyofeating.umwblogs.org/around-the-dinner-table/

23. Advance Access Publication Dec 21, 2007, Oxford University Press 2007, The Candidate Sour Taste Receptor, PKD2L1, is Expressed by Type III Taste Cells in the Mouse, S. Kataoka et al 2007

24. Open Biology, Evolutionary Origins of Taste Buds: Phylogenetic Analysis of Purinergic Neurotransmission in Epithelial Chemosensors, Open Biol. March 6, 2013, Masato Kirino et al. Rsob.royalsociety.publishing.org

25. Chemoreceptors of taste and Smell- specific material extracted from, 'The Neuron and Neural System: Electrolytic Theory of' by James T. Fulton, http://neuroresearch.net/smell/files/chemoreceptors.htm last update Aug. 14, 2013

26. Food Manufacturing, Mars Food Study Researchers Conclude Experiment, Jennifer Sinco Kelleher, http://www.foodmanufacturing.com

27. Open Access, Nanotechnology, Science and Application 2010:3 1-15, Dove Medical Press Ltd., "Food Nanotechnology- an overview" Bhupinder S. Sekhon

28. Soylent, Rob Rhinehart, http://robrhinehart.com/?=289, http://www.soylent.me/

29. Food Manufacturing, Taste Testers: Lab-Made Burger Lacks Flavor, Maria Cheng Aug 8, 2013, www.foodmanufacturing.com

30. Food Manufacturing, Q&A: The Science Behind a Lab-Grown Burger, Maria Cheng Aug 8, 2013, www.foodmanufacturing.com

31. NASA, "Food For Space Flight," NASA, www.NASA.gov

32. NASA," Space Food," NASA, http://spaceflight.nasa.gov/living/spacefood/index.html

33. NASA, "Cosmic Cuisine," NASA, nasa.gov

34. "From Farm to Fork: How Space Food Standards Impacted the Food Industry and Changed Food Safety Standards", Jennifer Ross-Nazzal, http://history.nasa.gov/sp4801-chapter12.pdf

35. NASA, "Space Food Hall of Fame," http://education.ssc.nasa.gov/fft_halloffame.asp

36. NASA, "The Brain in Space: a Teacher's Guide with Activities for Neuroscience," NASA.gov

37. Space Policy Institute, graduate Education, http://www.gwu.edu/~spi/education.cfm

38. BioServe, http://www.colorado.edu/engineering/BioServe/about.html

39. http://en.wikipedia.org/wiki/Eating

40. http://www.montignac.com/en/the-history-of-man-s-eating-habits/

41. http://www.academia.edu/2375711/Paleodietary_reconstruction_of_a_Neolithic_population_in_Slovenia_A_stable_isotope_approach

The Astrobiological Matrix
Minimizing the Trauma of First Contact

Paul Ziolo, PhD

Abstract

"Any future encounter between humanity and an extraterrestrial civilisation will likely end in a fiasco."
—Stanisław Lem

As the human species spreads into outer space, the likelihood of detecting evidence of, or of eventual contact with, an advanced technological civilisation (ATC) of extraterrestrial origin will exponentially increase, though it is doubtful whether we will either detect or make contact with a civilisation of this type while we remain a planet-bound species, nor whether we are as yet sufficiently mature as a species to survive the challenge of such an encounter, should it occur. Nevertheless, Lem's pessimistic prognosis could be avoided if future interstellar pioneers were adequately prepared for this eventuality. A better understanding of "alien" psychology would be enhanced by relinquishing the "species-narcissism" from which much human thinking is prone, and by adopting an ideational outlook that emphasises the mutability and contextuality of "Being" as implied by the astrobiological matrix. Such an outlook may challenge some very fundamental "humanistic" assumptions. This paper will show how the "Drake Equation" may continue to function as an evolving Bayesian heuristic that will incorporate knowledge as further advances are made in exoplanet discovery and in the evolutionary and biological sciences. The implications of the Lineweaver studies on the Galactic Habitable Zone (GHZ) and the possible concentration of ATC's in the Milky Way will be addressed, as well as the likely range of biologies implied by Kauffman's analysis of autocatalytic sets and the universality classes comprising CHONPS-based lifeforms. Consideration of hypothetical evolutionary trajectories of advanced chordate (arthropodic) intelligence may serve as an example.

1. What Does It Mean to Be "Human'?

Fig. 1. The Mir Space Station

According to reports of the Shuttle-Mir mission STS-63, when the American shuttle crew first had sight of the Mir space station (Fig. 1 above), their first impression was that "it had come from another universe" (Dumoulin 2001). Evidently, the Russian solutions to space station technology seemed completely alien to the Americans – even if this technology "worked". Of course, Russian ad American space programs originate from different research traditions – Von Braun vs. Tsiolkovsky, yet if two human groups have difficulty understanding one another's" technologies, what would happen were we to meet a totally alien, extraterrestrial technology – one that in all probability would be far more advanced? Many well-known SF stories explore the less than felicitous consequences of such a scenario including *Roadside Picnic* (Пикник на обочине), written by the Strugatsky brothers and published in 1971, numerous stores and novels by Stanisław Lem (*Solaris* (1961), *His Master's Voice* (1968), *Fiasco* (1987) etc.) as well as Carl Sagan's *Contact* (1985) and Iain Banks' *Excession* (1996).

As a species still in the course of emerging from its primal evolutionary matrix, humanity has a sell-image problem – one of extreme narcissism. While this trait proved useful for humanity's emergence on the Environment of Evolutionary Adaptiveness (EEA), it may decidedly prove less than useful in our exploration of the Cosmos beyond Earth.

Fig. 2. Earth as seen from Mars
(composite image from the Mars Exploration Rover Spirit, 63rd day of the mission,)

Fig. 2 above, the only authenticated NASA image of Earth from our nearest true planetary neighbour, Mars, may convey some impression of our visual "impact" on the Solar System. It is by no means obvious at this distance that there is any evidence of "intelligent life" on Earth.

Nevertheless, the likelihood of detecting evidence of, or of eventual contact with, an advanced technological civilisation (ATC) of extraterrestrial origin will exponentially increase as we expand into space, though it is doubtful whether we will ever detect or make contact with a civilisation of this type while we remain a planet-bound species, nor whether we are as yet sufficiently mature as a species to survive the challenge of such an encounter, should it occur. If the word "human" is taken to mean "that which is grown from earth" (>Lat. "humus" → "humanus'), i.e. the chreod (see below) or species variant that finally emerged from its home world', then when such an encounter does finally occur, we should not be surprised to find that the "alien" in question will consider itself to be "human", in the sense of "we who have finally emerged from our original biological matrix". The species will consider itself to embody all "truly human" values, no matter how many tentacles of multi-facetted eyes its original form may possess.

2. The Continuity Thesis

The "continuity thesis" has been advanced by a number of scientists and philosophers (see for instance: Fry 1995; Gardner 2004; Ćirković 2012), and holds that 1) there is no discontinuity between so-called "inanimate" and "living" matter, 2) there is no discontinuity between the simplest sensorimotor reaction of the mist primitive unicellular organisms and the most "profound" or "advanced" manifestations of "intelligent thought and 3) the evolution of consciousness (or "sentience" itself) proceeds without radical discontinuities (Fig. 3 below). Furthermore, biological life itself may simply be one of many transitional stages between "inert" and "fully sentient" matter. We find ourselves now approaching a critical point in the evolution of the "machinic phylum" (Deleuze & Guattari 1987; Johnston 2008) where levels of human and machine intelligence seem to be approaching convergence. "The Machine will not dominate the Human: Human and Machine will become One" (Sandberg 2013), but what form this synthesis may take is still unclear.

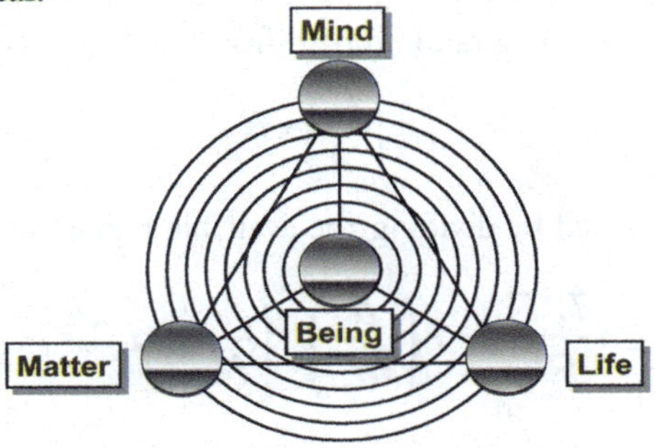

Fig. 3. The Continuity Thesis

3. The Encyclopedia Galactica: Perspectives on Humanity

Fig. 4 below shows the Hertzsprung-Russell Diagram of stellar evolution with the approximate position of Sol (our Sun) as an average G2-spectral class star on the "main sequence".

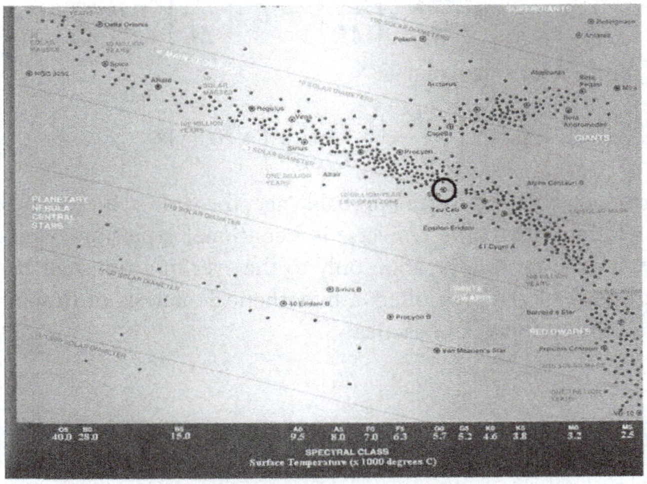

Fig. 4. The Hertzsprung-Russell Diagram(showing the position of Sol)

Main sequence stars with spectral classes ranging from F0 → K9 have lifetimes guaranteed long enough to permit the genesis & evolution of biospheres – and eventually, ATC's. This is not to say that biospheres cannot evolve outside this range – just that biospheres within this range are optimal in terms of our current knowledge of biology and astrobiology. Nucleosynthesis in stellar interiors involves the conversion of helium to carbon according to the sequence 3He4 → C12 + 2γ, an essential step in the creation of organic, carbon-based life. But this step would normally proceed far too slowly for carbon to be synthesised at a rate consistent with its demonstrated abundance in the Universe. Burbidge et al. (1957) showed however, that this reaction in fact proceeds with the aid of a beryllium catalyst (Be8) as follows:

$$2He^4 + (99 \pm 6)KeV \rightarrow Be^8$$
$$Be^8 + He^4 \rightarrow C^{12} + 2\gamma \qquad \text{Eq. 1.}$$

The energy level of C12 (7.6549 MeV) lies just above that of Be8 + H4 (7.3667 MeV) permitting a reaction to take place within the resonance level 7.656 ± 0.008 MeV. Furthermore, the O16 nucleus has an energy level of 7.1187 MeV lying just below that of C12 + He4 (7.1616 MeV) thus stabilising this reaction by inhibiting the rapid depletion of all carbon to oxygen through alpha capture via C12 + He4 → O16. Such "fine tuning" of resonance levels provides further evidence of how "auspicious" the "laws" of physics would seem to be in relation to the emergence of organic life.

Photons emitted at the fusion core of a star reach the surface by traversing a "random walk" produced by collisions and deflections by other particles (electrons and charged ions). The Thomson cross-section (σT) for the effect of electron-induced scattering is σT ~ α2me-2 with the mean free path (i. e. λ - the average distance) travelled by photons between collisions) being λ ~ (σTne)-1 with an electron number density ne ~ NR-3. Under these conditions, the time taken to traverse a random path from the fusion core to the boundary is:

$$t_{ex} \sim \left(\frac{R}{\lambda}\right) \times R \qquad \text{Eq. 2,}$$

The luminosity of the star L is obtained by dividing the available radiation energy by the escape time from the core, hence:

$$L \sim \frac{T_*^4 R^3}{R^2/\lambda} \sim f\left(\frac{\alpha}{m_e^2}\right)^{-1}\left(\frac{N}{N_*}\right)^3 \alpha_G^{-1/2} \qquad \text{Eq. 3,}$$

where f is the dimensionless factor governing deviations from the Thomson scattering which might result in low density or high temperature. This gives an estimate of:

$$L \sim 5 \times 10^{34}\left(\frac{M}{M_*}\right)^3 erg\ s^{-1} \qquad \text{Eq. 4,}$$

a value independent of the stellar radius and temperature. With respect to H2-burning stars, setting ε ~ 0.007 as the fusion efficiency factor and dividing the resultant energy by L, we obtain:

$$t_0 \sim \frac{E_f}{L} \sim \frac{\varepsilon \alpha \alpha_G^{-1}}{m_e}\left(\frac{M_*}{M}\right)^2 \sim 10^{10}\left(\frac{M_*}{M}\right)^2\ yr \qquad \text{Eq. 5,}$$

with the time-dependent mass factor calculated as:

$$t_{ms} \sim \left(\frac{hc}{Gm_N^2}\right)\left(\frac{h}{m_N c^2}\right) \approx 10^{10}\ yr \qquad \text{Eq. 6.}$$

Yet against the background of the Universe's probable age, the time-span for the emergence of an ATC may be comparatively brief. As Lem writes: "The Universe is a continual explosion extending over a time of perhaps 1011 years that appears as a majestic solidification only to the eyes of a transient being like man" (*Microworlds*, 1986). Against this background, Carl Sagan offers a hypothetical analysis of the dominant species on Sol as it might appear in some future Encyclopedia Galactica:

World 806.4616.0110
Civilisation type: 1.0 J. Society code: 4G4 "Humanity".
Star G2V, r = 9.844 kpc, θ = 00°95'24", φ = 206°28'49"
Planet: third. a = 1.5×10^{13} cm. R = 6.4×10^{8} cm. p = 8.6×10^{4} s. P = 3.2×10^{7} s.
Extraplanetary colonies: none. Planet age: 1.4×10^{17} s.
First locally initiated contact: 1.21×10^{9} s. ago. Receipt first galactic nested code:
Application pending.
Biology: C, N, O, S, H, O, PO_4 deoxyribonucleic acid. ***No genetic prosthesis.***
Mobile heterotrophic symbionts with photosynthetic autotrophs.
Surface dwellers, monospecific, polychromatic O_2 breathers,
Fe-chelated tetrapyroles in circulatory fluid. ***Sexual mammals.***
m = 7×10^{4} g. t = 2×10^{9} g. Genome: 4×10^{9}.
Technology : ***exponentiating/fossil fuels/nuclear weapons/***
organised warfare/environmental pollution
Culture: ≈ 200 nation states, ≈ 6 global powers:
cultural & technological homogeneity imminent.
Prepartum/postpartum: 0.21 [18],
individual/communal: 0.31 [17], artistic/technological; 0.14 [11].
Probability of survival (per 100 yr): <1%.

Fig. 5. "Encyclopedia Galactica" report on the Human Race
(adapted from Carl Sagan's COSMOS (1978-9))

The comments in bold type – reflecting the present author's emphasis – describe those traits that, in the opinion of the present author, are likely to account for the estimated survival rate of c. 1%. The very traits that conferred adaptive strength on the EEA may prove to be the evolutionary nemesis of the species.

Throughout the Universe, even within the single range of carbon chemistry (CHONP-based life-forms), the range of evolutionary variance is likely to be extremely wide. There are good enough grounds for assuming carbon chemistry to be the optimal "main sequence" for complex biological evolution within our present Universe (given its laws and fundamental constraints) without postulating other more exotic, hypothetical biochemistries. (Barrow & Tipler 1986: ch. 8). Kauffman (1986, 1993) identifies of a wide range of "universality classes" even within the range of CHONPS-based (Carbon – Hydrogen – Oxygen – Nitrogen – Phosphorus - Sulphur) life. (Fig. 6 below).

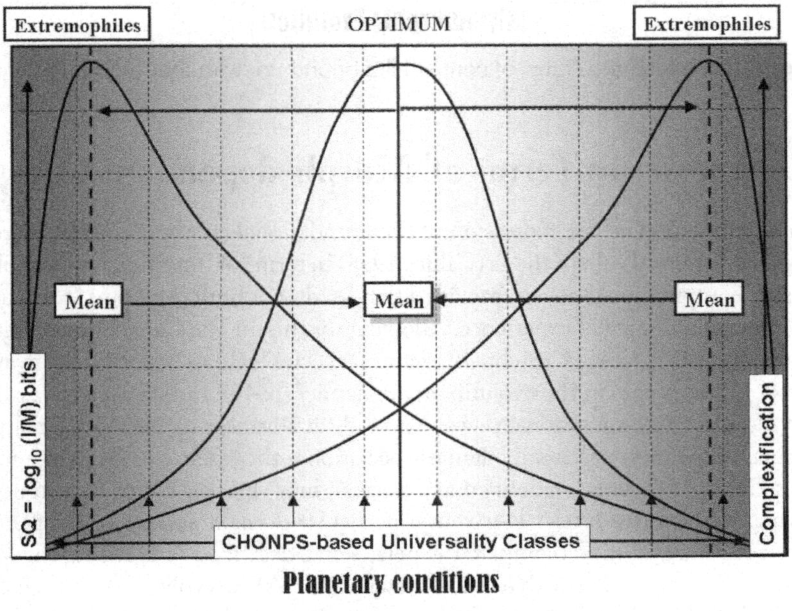

Fig. 6. Different planetary conditions generate different evolutionary optima for the emergence of sentience

In this chart, we assume the conditions we experience on Earth to be the "Goldilocks zone" for "intelligent" life (the central mean), but given a wide range of variance in planetary conditions that still remain favourable to

CHONPS-based life, certain "earthlike" worlds may evolve conditions in which what we might classify as "extremophile" life forms become, on that world, the mean. This would further imply that on a cosmic scale, multiple orders of life subsuming a wide variety of universality classes may be scattered prolifically across the Universe, creating the statistical likelihood that sentience will emerge in sufficient strength to meet the challenge of the final "singularity" whatever its nature (Fig. 7 below). Although teleological issues remain far beyond the scope of this paper, we choose to assume that the ultimate destiny of sentient life is somehow bound up with that of the Universe we inhabit. This view of the co-evolution of matter and form (which is also the basis of Tibetan Buddhist traditions such as the rDzogs-chen: the "Great Perfection" or Matrix of Mystery (Guenther 1984, 1989)), is termed hylomorphism (> Gk. ὑλο- hylo-, "wood, matter" + μορφή, morphē, "form") and/or hylopathism (the thesis that sentience is the dominant evolutionary vector in the Universe). The Western roots of this philosophy, which also forms the basic premise of the continuity thesis, are found in the works of Aristotle (Physics 194b: 9; 195a: 16; 1949b: 9, 23 – 24; Metaphysics 1045a: 26 – 29; Nichomachean Ethics 1098a: 1 - 5, 7 – 8, 16 – 18; On the Soul 412a: 9).

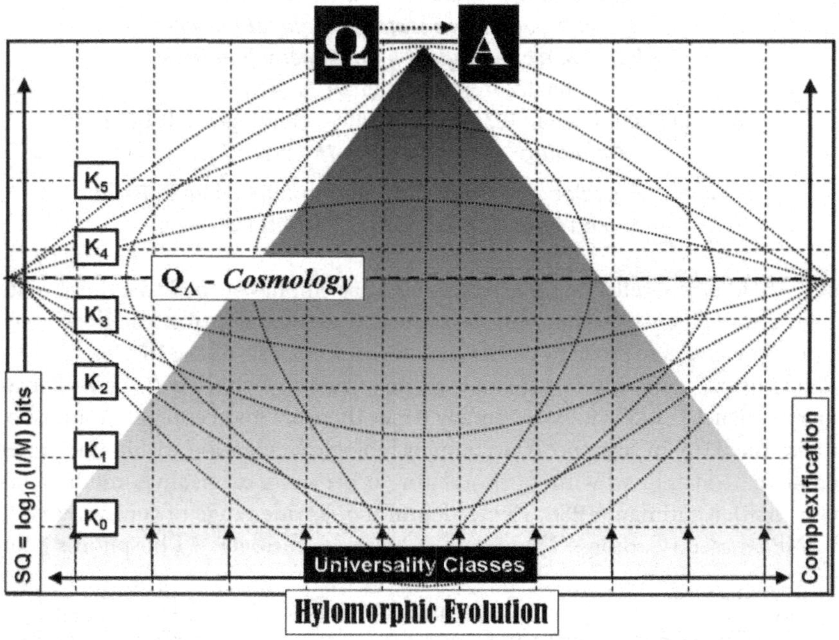

Fig. 7. Is the ultimate "fate" of sentient life bound up with that of the Universe?

4. Defining a Apecies in Terms of Morph-dependent Computation

"Morph-dependent computation" is the expression of the foundational concepts of embodied cognition (Lakoff &Johnson 1999) and the Quine-Duhem thesis (Gillies 1998) in terms of "machinic" philosophy. We might define "sentience" in terms of "computational power" provided that we do not limit the terms "computation" or "machine" to anything within our current range of experience. Allowing for this, in The Calculi of Emergence: Computation, Dynamics and Induction (1994) James Crutchfield defines "species" (pl.) as "metastable, invariant languages – … temporary (hyperbolic) fixed points in the evolutionary dynamic" (p. 49). In other words, evolutionary advance is defined in terms of increasing computational power, and radical advances in computational power imply radical physical re-structuring. Innovation necessarily defines speciation – the essence of the posthuman challenge.

In Crutchfield's Calculus, "agents", redefined as "species", may be considered as autonomous, goal-seeking programs evolving on two adaptive levels: 1) within a particular domain and 2) as independent agents within the general context of cosmic evolution (compare the perspective described in Devising the Prime Radiant – this volume). "Species" are therefore stochastic dynamical systems (SDS's) co-evolving with their own biospheres and, upon emergence from these biospheres, within a deterministic dynamical system (DS) – the known Universe (Fig. 8 below). All CHONPS-based life forms, despite radical genomic divergence, share a similar embodied substrate – physical, chemical, macromolecular or metabolic. What differentiates them are the classes of model each agent creates in order to make sense of its environment and to interact with other agents in a manner that best serves its own interests.

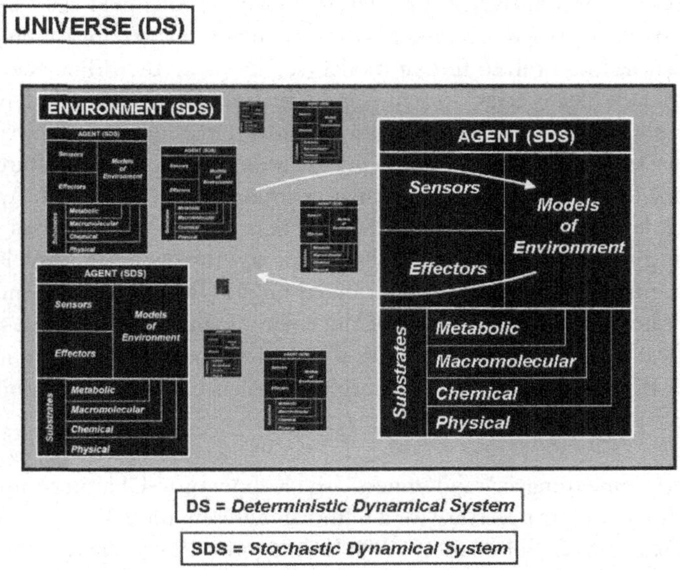

Fig. 8. The co-evolution of agents within a deterministic Universe
(After Crutchfield (1994))

The evolution of a particular species may therefore be characterised in terms of the "metastable, invariant language" used by that species to classify the environment in which it finds itself at a given point in time. "Evolutionary advance" (for that species) would mean an advance in the range and power of the computational language employed to deal with that environment. Fig. 9 below shows a "hierarchy" of machine languages (as we currently understand them). This chart is intended only as a conceptual heuristic, not as an actual means of encoding "real" alien languages. What is most important here is to observe that this "hierarchy" broadens out over an increasingly wide domain base as well as advancing "upwards" in terms of complexity.

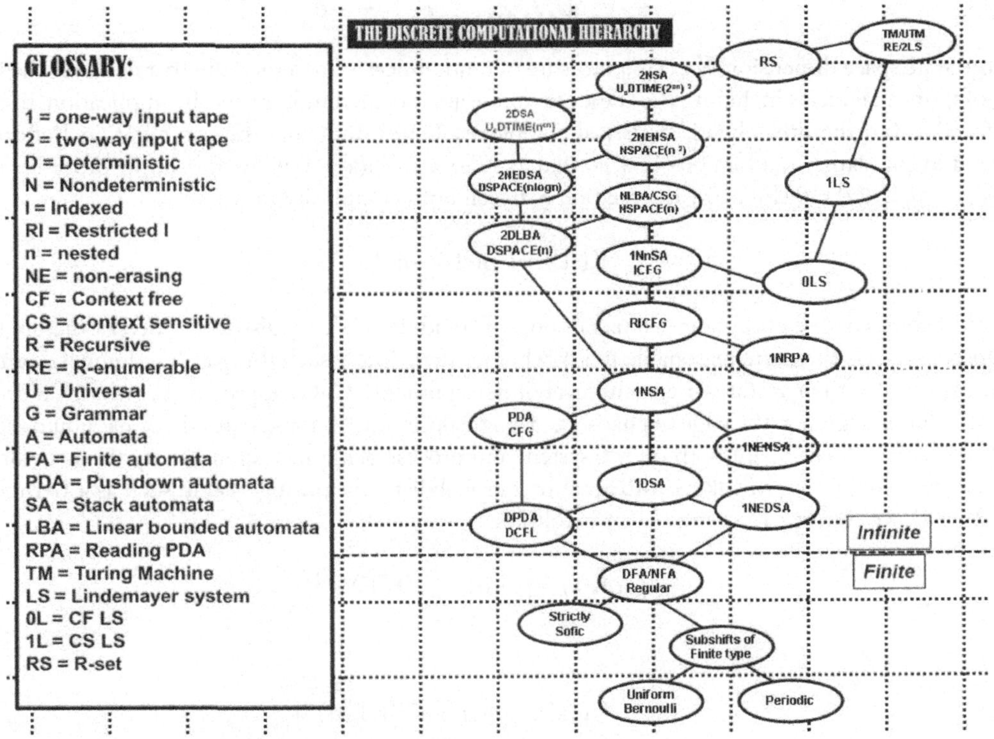

Fig. 9. The Hierarchy of Machine Languages

(After Crutchfield (1994))

Advances through this language hierarchy are effected by what Crutchfield calls "ε-machine reconstruction" ("epsilon-machine reconstruction') where ε represents the level of accuracy of measurement in a given model, i.e. the degree of discrete quantization utilized in that model (see Devising the Prime Radiant (this volume) – Fig. 15). Computational "style" reflects a species" structure and structure is a product of adaptation. Epsilon-machine reconstruction may radically alter the perspective from which an environment is viewed and therefore radically alter a species evolutionary trajectory. We have seen (Fig. 6 above) how species structure determines how "optimum" conditions necessary for adaptation to a particular environment are evaluated. Agreement as to what this "optimum" consists of will determine the model of what members of a particular species collectively define as "intelligent" behaviour at any given point in time and for each species, this "optimum" will be the point of maximum statistical complexity in terms of its collectively-determined model. The efficacy of a model is determined solely by the accuracy of its predictive capacity in terms of the environment in which it arose. Different trajectories may converge onto identical futures, a slight difference in the initial conditions of two near-identical trajectories may cause their futures to diverge radically, hence "morph equivalence induces conditionally-independent states" (Crutchfield, op. cit. p.18).

5. Statistical complexity

"Complexity" is not the same thing as "randomness". The Kolmogorov-Chaitin complexity of a structure (x) is defined as the smallest number of bits necessary for a minimal representation of x on a Universal Turing Machine or UTM (Shannon & Weaver 1962; Solomonoff 1964; Kolmogorov 1965; Chaitin 1966). If x is considered as a string sL of L discrete symbols, produced by an information source such as a Markov chain with Shannon entropy rate hμ , then the growth rate of the Kolmogorov complexity of x will be:

$$\frac{K(s^L)}{L} \xrightarrow[L\to\infty]{} h_\mu \qquad \text{Eq. 7,}$$

and for a chaotic system (with continuous state variables):

$$K(s^L) \underset{\substack{L\to\infty \\ \varepsilon\to 0}}{\propto} h_\mu L \qquad \text{Eq. 8,}$$

where ε is the "discretization" factor (i.e. how coarse- or fine-grained the measurement of x is) and is

$$s_\varepsilon \in \{0, 1, 2, \ldots, \varepsilon^{-d} - 1\}$$

with d = the state space dimension. K (x) is a measure of randomness – but a measure that obscures the underlying structure of x since all bits – including the "random" bits – are included in K(x) and by implication, in the process that produced x. On the other hand, statistical complexity Cμ(x) discounts the computational effort a UTM must expend in simulating random bits in x, so that as K → ∞, Cμ(x) = 0 while for simple processes such as x = 0.0000000, …, 0, Cμ(x) = 0 (Fig. 10). The relation between both complexity measures is:

$$K(sL) \approx C\mu(sL) + h\mu L \qquad \text{Eq. 9.}$$

Statistical complexity Cμ(x) is a measure, not only of "randomness", but also of the deep structure underlying this "randomness", It is therefore reasonable to ask whether there is any such thing as "randomness" or if "random" simply means "we don't know" (at our current level of development. Cμ(x) approximates what Kauffman (1993) means by "the liquid region at the edge of chaos" (as a single optimum), but indicates that this "liquid region" is not absolute but relative – contingent both on the system and process being measured and by the structure of whatever or whoever is measuring it. Taking into account a probabilistic measuring system such as a Bernoulli-Turing machine (BTM), we contrast K(x) and Cμ(x) as follows:

$$K(x) = || M\min (x|UTM) || \qquad \text{Eq. 10}$$

and

$$C\mu(x) = || M\min (x|BTM) || \qquad \text{Eq. 11.}$$

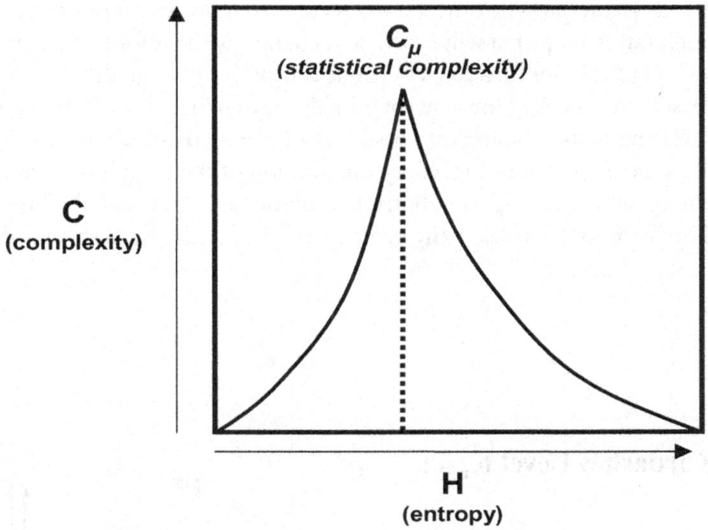

Fig. 10. Morph-dependent computation: measuring statistical complexity (Cμ): at either extreme of H, "complexity" is minimal in terms of Cμ(x).

As we have seen from Fig. 9, advances in modelling complexity do not simply follow a vertical hierarchy, but also spread out to cover different classes of complexity that reflect more faithfully the range of universality classes encompassed by the space of CHONPS-based life-forms. Fig. 11 simulates this space in 3D, showing ranges across three branchings of machine language hierarchies in metamodel space: from periodic to chaotic, from stochastic to deterministic and from stationary to multistationary.

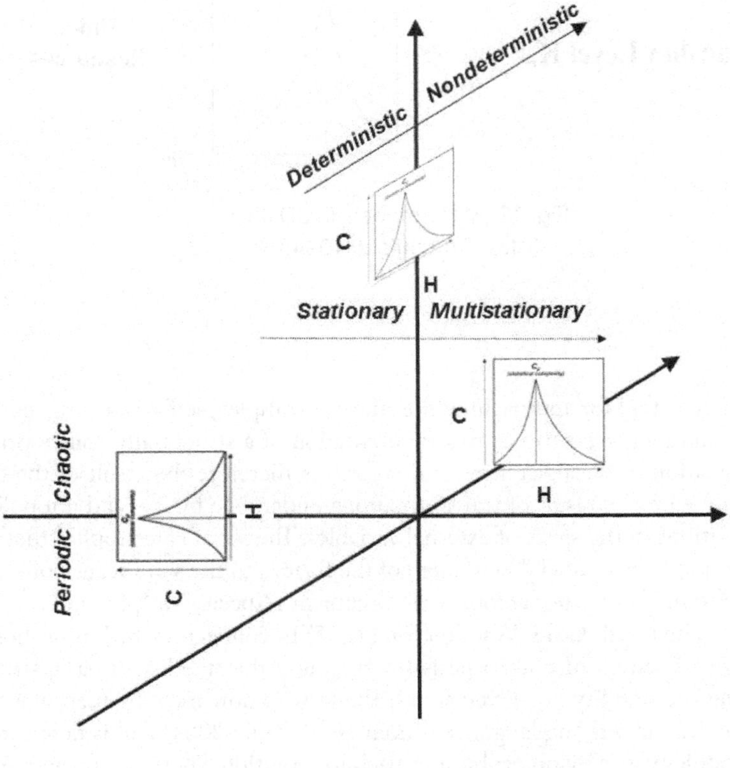

Fig. 11. Meta-models in 3D (I)
(After Crutchfield (1994))

Especially important for us at the present time as we create ever more ambitious plans for future human evolution, is the fact that computational optima will evolve and change as the information-processing architecture of a species becomes subject to radical modification as various subspecies emerge through speciation (Fig. 12). As we evolve therefore, our perspective on the Universe we inhabit may change beyond all recognition. The most subtle and refines arguments relating to astrobiological possibilities or human evolution will also mutate beyond all recognition. These mutations are indicated in Fig. 12 by the rotation of the complexity/entropy diagrams around the vertical axis. For any emergent sentient species, both its evolutionary drive and the limiting factors to speciation are always determined by the resource base at any given level – hence the Kardashev scaling.

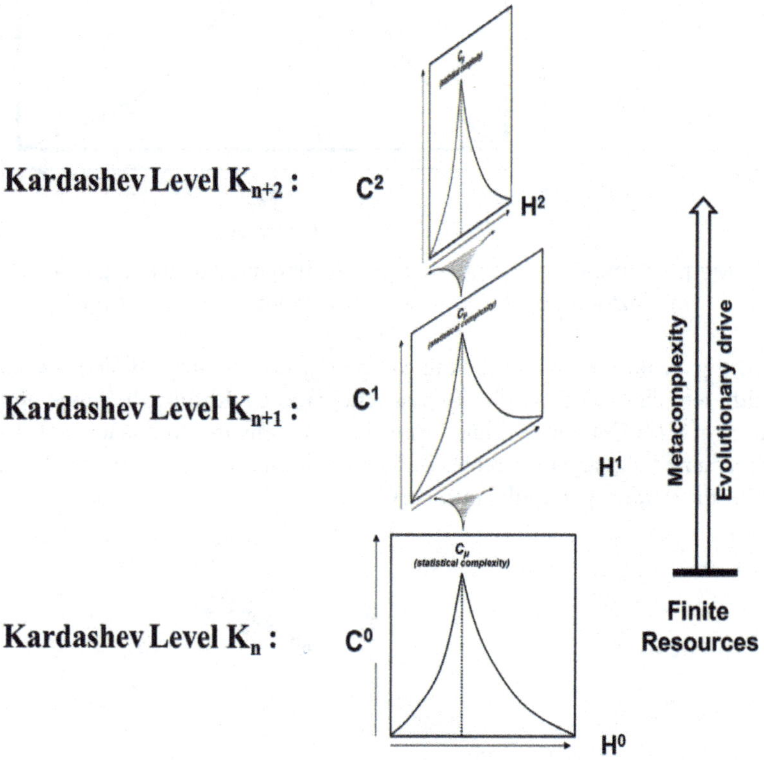

Fig. 12. Meta-models in 3D (II)
(After Crutchfield (1994))

6. The Chreod

A useful conceptual and analytic tool for approaching the study of complex, self-organising entities is the chreod. René Thom (1972) conceived of a given form as the representation of a structurally-stable attractor in the space of internal variables. The creation or disappearance of this form is therefore the result of the disappearance and replacement of this attractor – i.e. the result of transformations undergone by it – and a topological catastrophe is a map of these transformations in the space of external variables. The set of catastrophes that unfold during the lifetime of a given form comprise the chreod ("kray-ode') of the form, and the significance of a chreod is therefore determined by this set. The term is not synonymous with "organism", "species" or "phylum".

The word "chreod" originated with C. H. Waddington (1957) in connection with morphogenetic fields, and was used to describe an overall system of homeorhesis (maintaining the stability of basic structure) as opposed to homeostasis (maintaining the stability of a given state). The term is now most frequently encountered in connection with the deep structures underlying language (Thom 1989; Ziolo 2004) and is here extended as chreodic analysis - the central methodology in a hylomorphic approach to evolution where we consider only "matter" (in its most general sense) as it changes and transforms, either evolving into configurations of increasing complexity or decomposing to more fragmented, rudimentary states, In so doing we do not necessarily distinguish between the organic and the non-organic since according to the continuity thesis, both form a smooth, differentiable continuum from the level of ultrastructure to that of macrostructure. From this perspective, the chreod is ideally suited as a "unit of study" in that although it is essentially qualitative in conception, it is a topologically precise tool whose parameters are measurable within certain limits. Since it is also central to Thom's key studies in structural stability and morphogenesis, the following topological definition is given, based on that of Thom (1989):

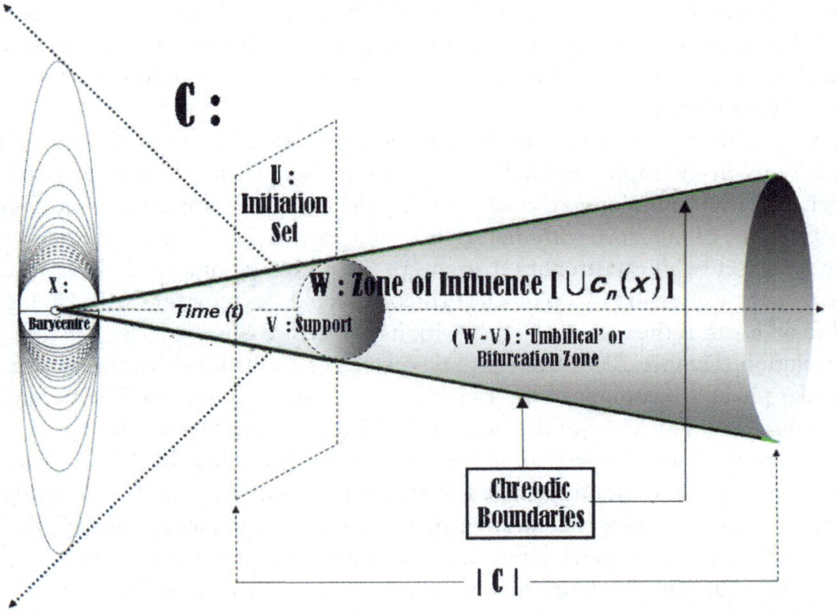

Fig. 13. The Chreod: basic structure

In Fig. 13, a chreod C in space-time R3 × R consists of an open set U (the initiation set) in the hyperplane t = 0 and a zone of influence W. This zone of influence (the "cone" in Fig. 13 above) in turn consists of the union of all "light-cones" $\bigcup c_n(x)$ emanating from a barycentre X (the "germ" or point from which the chreod unfolds – lit. "centre of gravity')1. From a hylomorphic perspective, a chreod metabolises (organically or non-organically) and therefore requires material as a metabolic source or "food set".. There is therefore an open set V contained in W and containing U in its boundary. This is the "support" |C| of C and contains a morphogenetic field, defined up to isomorphism on V. The set (W – V) is the "umbilical" or bifurcation zone of C.

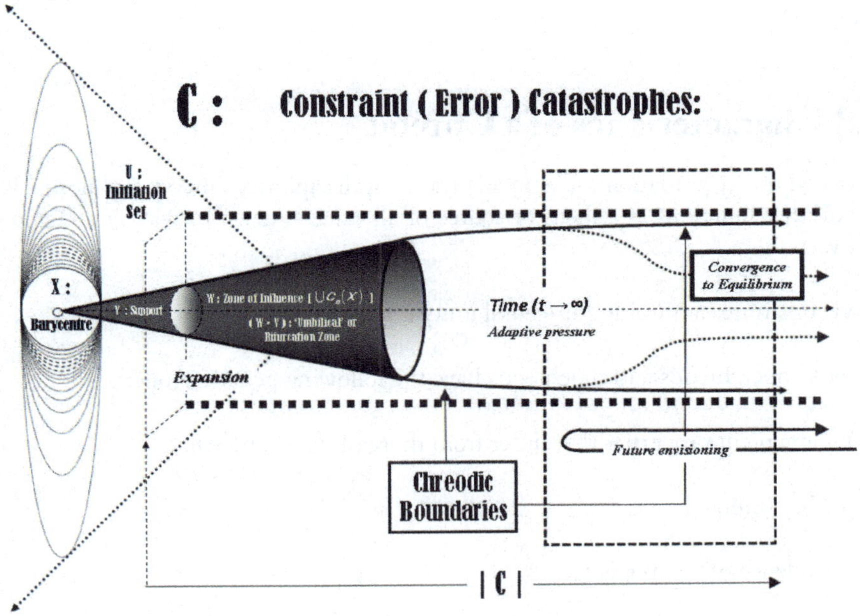

Fig. 14. A bounded (asymptotic) chreod

The "chreodic boundaries" in Fig. 13 are the limits of W (by analogy, any events beyond the limits of a relativistic light cone are inaccessible to the source-point of that cone). These boundaries may begin to flatten out in what Thom calls an asymptotic chreod (Fig. 14 above) and in the case of living systems (i.e. species), turn back "inwards" towards the vertex of C (the line = time beginning from x above) – as happens when a species proves unable to adapt and becomes extinct. The limiting factors in such cases are the Constraint and Error Catastrophe theorems identified by Kauffman (1993).

A chreod C may contain subchreods Jn within its zone of influence provided that the initiation sets and supports of Jn are found in the support zone |C| of C, Conversely, C may in turn be a subchreod of a superchreod C++ under the same conditions. A subchreod may detach itself completely from its parent chreod once the intersection of its zone of influence with that of the parent diminishes to zero. Bearing in mind that a chreod is defined strictly as the set of all catastrophes that govern the unfolding of a given form, where, in the case of a species (considered as the expression of a biological chreod), would the initiation set be found? The superchreod of all biological life on Earth is the biosphere itself, with its barycentre X being located in pre-space - the domain of Schrödinger evolution (Penrose 2005). Every species expresses a subchreod within its own genus, and that in turn of the particular phylum according to the Linnaeus classification system, all the way up to the "terrestrial" superchreod, which is itself a subchreod of the "ultimate" biological superchreod: sentient life in the Cosmos. The initiation set of each species" chreod (like that of the genus and phylum) may be defined as the point of genomic closure, i.e. when a given group ceases to be able to breed with other groups and becomes its own closed "gene pool". Whereas the details of species classification may at times be somewhat "murky", the genetic definition remains fairly robust. Under these criteria, humanity is the expression of a single biological chreod – there being no human ethnic group that cannot interbreed with another. From any non-biological perspective however (e.g. social, political, linguistic etc.) other subchreods may be identified according to the analytical criteria employed (as is the case with the "Thom-Pomian Historical Chreod" as studied by Petitot (1978)).

All chreods consist of open sets that operate on embodied entities in a biological aggregate. In the latter case, the aggregate is formed through modes of nucleation from smaller to greater, with each mode forming a nucleate in its own right according to the self-organising principles determined by the chreod. The nucleate therefore refers to structure while the chreod refers to the dynamics (catastrophe sets) expressed by the interactions between nucleates and their evolution over time. The nucleate determines the initiation set and conditions that in turn fix the mode of nucleation at the next, "higher" level. Within humanity, there exist many nucleates such as families, groups, societies, nations etc. The catastrophe sets expressed by their social and cultural dynamics may be considered as individual "chreods" if we are looking at their separate origins, growth and evolution in time, or as "subchreods" if we are considering how the sum of their interactions create a "higher level" structure. Each and every species may have its own modes of nucleation, and each nucleate may therefore be studied on its own terms for a specific purpose.

7. General Characteristics of a Chreod

A chreod is founded on an "initiation set" – in a "cultural or disciplinary chreod, a philosophical presupposition" or "prime symbol" that functions as a focus or source of all value systems in that chreod and guarantees logical consistency between them.

1) A chreod constitutes a "meaningful-causal unity".

2) All chreods, superchreods and subchreods have the following generic properties:

(i) Reality - emergent properties that differ from those of its components.

(ii) Individuality (uniqueness).

(iii) Interdependence of components

(iv) The emergent properties of the whole remain stable despite changes in components - i.e. a chreod possesses structural stability.

(v) "Change in togetherness" regarding components - see (iii)

(vi) Changes to the whole are self-directing (immanent) and self-determining, so that global transformation does not result in a fundamentally different entity.

(vii) A chreod is selective as regards external influences.

(viii). There is therefore limited variability within each chreod.

8. The Drake Equation as an Evolving Bayesian Heuristic

The Drake Equation has long served as an "iconic" tool in the estimation of the number of possible civilisations in the Galaxy. As our acquisition of exoplanet data increases, the Drake Equation becomes increasingly subject to refinement, evolving from a "chainsaw" heuristic to a "laser-scalpel" heuristic. In other words, as exoplanet data improves, the ε-factor (section 5 above) in our modelling capacity increases. The standard form of the Drake Equation and the meaning of its various terms is shown in Fig. 15 below.

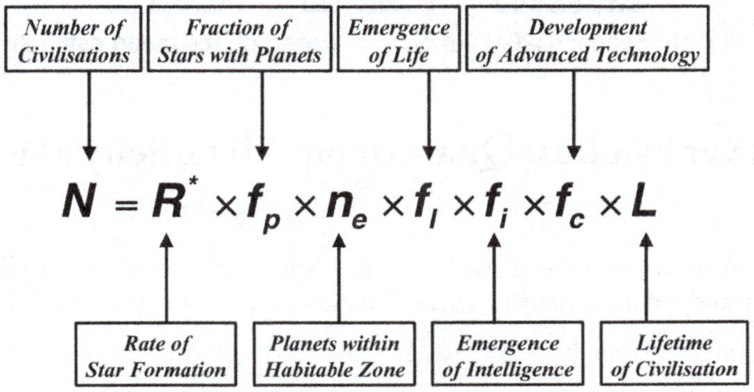

Fig. 15. The Drake Equation

On this basis it is possible to construct a matrix with which to collate exoplanet data (Table 1) to the extent of measurement scale accuracy (i.e. the ε (epsilon) precision factor).

DRAKE MATRIX (1961):	$N = R^* \times f_p \times n_e \times f_l \times f_i \times f_c \times L$						
Object type	Location	Metals MATTER	Organics	H_2O	Molecular Stability	Energy Sources	Time Stabil;ity
Interstellar dust	Grain surfaces	Low	Variable	No	Variable	UV/cosmic rays	Variable
Asteroids	Subsurface	High	High	Early	Rel. good	UV/cosmic rays/impacts	Short, early
Comets	Subsurface	Moderately high	Moderately high	Early	Rel. good	UV/cosmic rays/impacts	Short
Ice moons	Subsurface/ surface	Moderately high	Moderately high	Early – tidal	Good in outer zones	Differentia-tion Radioactive/ tidal	Short
Silicate planets	Subsurface/ surface	High	Moderately high	Variable Surf/subs	Good near surface	Differentia-tion-radioactive	Long - decreasing
Giant planets	Atmosphere	Mid-range	Low	Atmos. vapour	Good in atmos.	Contraction	Decreasing
Brown dwarfs	Atmosphere	Low	Low	Atmos. vapour	Good in upper atmos.	Contraction	Late - decreasing

Table 1. An astrobiological matrix based on the Drake Equation

To date (25/11/2013), 1048 exoplanets in 794 star systems have been discovered, including 175 in multiple planetary systems (Schneider 2011). Based on data relating to red dwarfs, Sara Seager (Seager 2012) of MIT has refined part of the Drake Equation as shown below in Fig. 16.

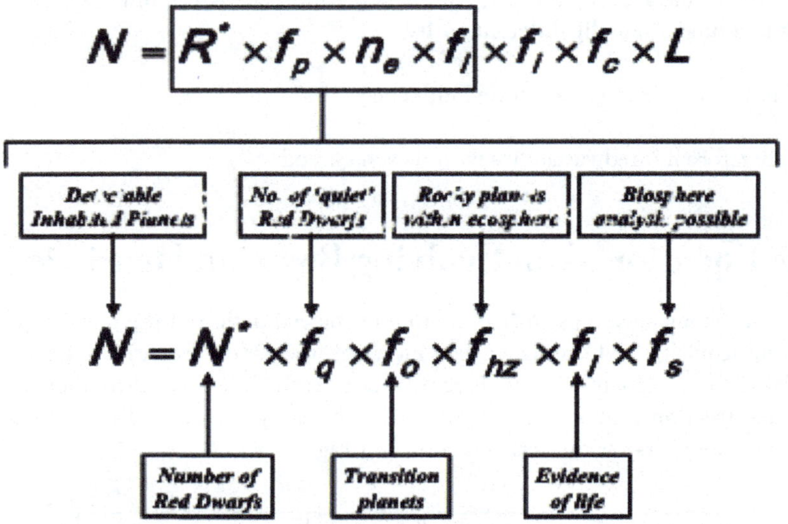

Fig. 16. Seager's (2012) expansion based on red dwarf estimates

9. The Lineweaver Studies: Quantifying Metallicity as a Selection Parameter

Lineweaver (2001) has made an estimate of the hospitality range and number of earthlike planets and their rate of formation based on a study of metallicity as a selection parameter. For a given exoplanet, the H2 mass fraction (X) is :

$$X \equiv \frac{m_H}{M}$$, (Solar Value: Xsun = 0.70) Eq. 12,

where M is the estimated mass of the system and mH the estimated H2 mass. Similarly, the He mass fraction is given by:

$$Y \equiv \frac{m_{He}}{M}$$, (Solar Value: Ysun = 0.28) Eq. 13.

The remaining mass fraction of elements heavier than He (Z) is calculated by:

$$Z = \sum_{i>He} \frac{m_i}{M} = 1 - X - Y$$ (Solar Value: Zsun = 0.02) Eq. 14.

Most recent solar estimates for Eqs. 12 – 14 above and the metrics below are found in Chaplin et al. (2007). The general metric for the estimation of a system's metallicity is:

$$[Fe/H] = \log_{10}\left(\frac{N_{Fe}}{N_H}\right)_{star} - \log_{10}\left(\frac{N_{Fe}}{N_H}\right)_{sun}$$ Eq. 15,

General metric for specific element: where NFe and NH are the number of iron and hydrogen atoms per unit of volume respectively. A metric for specific elements has the following form:

$$[O/Fe] = \log_{10}\left(\frac{N_O}{N_{Fe}}\right)_{star} - \log_{10}\left(\frac{N_O}{N_{Fe}}\right)_{sun}$$

$$= \left[\log_{10}\left(\frac{N_O}{N_H}\right)_{star} - \log_{10}\left(\frac{N_O}{N_H}\right)_{sun}\right] - \left[\log_{10}\left(\frac{N_{Fe}}{N_H}\right)_{star} - \log_{10}\left(\frac{N_{Fe}}{N_H}\right)_{sun}\right] \quad \text{Eq. 16,}$$

where [O/Fe] represents the difference in the logarithm of a given star's O2 and Fe abundances compared to those of the Sun. From this data it has been possible to refine existing models of the Galactic Habitable Zone (GHZ). There are high rates of star formation and high rates of metallicity near the Galactic core, but in this zone, high radiation levels and supernova rates preclude the formation of long-lasting earthlike planets. In addition, the formation of "hot Jupiters" (a number of which have been found in other star systems) also act as constraints on the lifetimes of high-metallicity planets within the solar ecosphere. Levels of metallicity also decrease towards the Galactic Rim. Given these considerations, Lineweaver et al. (2004) have developed a "Goldilocks" model for the formation of earthlike planets within the GHZ. Fig. 16 shoes a "top-down" image of the GHZ, while Fig. 17 shows a "lateral" view.

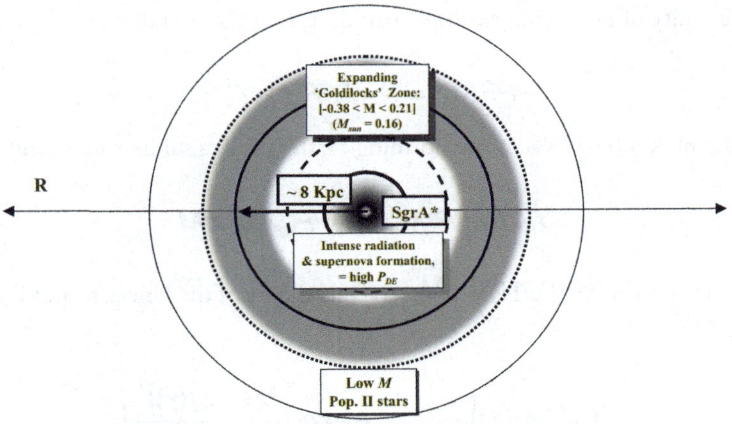

Fig. 16. The Galactic Habitable Zone (GHZ)

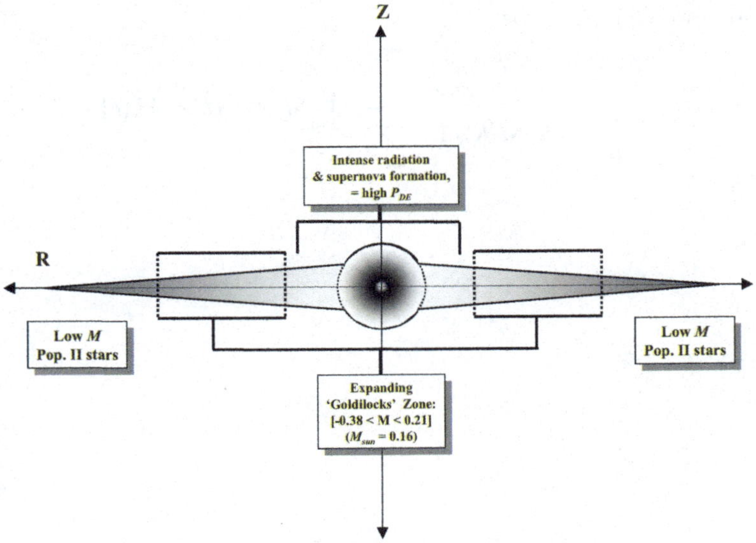

Fig. 17. The Galactic Habitable Zone: "lateral" view

10. Lineweaver et al.'s (2004) Refinement of the Drake Equation

According to this model, how many earthlike life-bearing planets may have emerged within the Milky Way since its formation? What is the probable age range of these planets? From the Lineweaver studies, the population of planets in the GHZ is given by:

$$P_{GHZ} = SFR \times P_{metals} \times P_{evol} \times P_{SN} \qquad \text{Eq. 17,}$$

where SFR is the Galactic star formation rate, Pmetals is the overall probability of harbouring earthlike planets, Pevol(t) is a cumulative integral for estimating the time taken for the emergence of complex life with a normal distribution of 4 Gy and a spread of 1 Gy (i.e. based on terrestrial data) and PSN is the probability that complex life survives supernova events. the probability of destruction by supernovae is given by:

$$P_{DE}(M) = \frac{N_H(M)}{N(M)} \qquad \text{Eq. 18,}$$

where M is defined as in Eqs; 12 - 14 above, and N as in Eqs. 15 – 16. The probability of harbouring earthlike worlds (PHE) is therefore given by:

$$P_{HE}(M) = P_{PE}(M) \times [1 - P_{DE}(M)] \qquad \text{Eq. 19,}$$

where PPE is the probability of producing earthlike worlds. Given the overall planet formation rate:

$$PFR(t) = A \times SFR(t) \times f(t) \qquad \text{Eq. 20,}$$

where the age determinant A = 0.05 (5% of star-forming matter creates sunlike stars) and

$$f(t) = \int P(M, \overline{M}(t))\, P_{HE}(M) dM \qquad \text{Eq. 21,}$$

where M = local metallicity = (Fe/H)/Fe/H) ⊙, $\overline{M}$ the metallicity of the Universe and P(M, $\overline{M}$ (t)) the Gaussian distribution given by:

$$P(M, \overline{M}(t)) = \frac{1}{\sigma\sqrt{2\pi}} exp\left[\frac{(M - \overline{M}(t))^2}{2\sigma^2}\right] \qquad \text{Eq. 22,}$$

fp in the Drake Equation would be given by:

$$\left.\begin{array}{l} PFR \propto SFR \\ \frac{dM}{dt}(t) \propto SFR(t) \end{array}\right] - \int_0^t SFR(t')dt' \sim \overline{M}(t) \qquad \text{Eq. 23.}$$

Figs. 18 – 19 below show this data in graphic form:

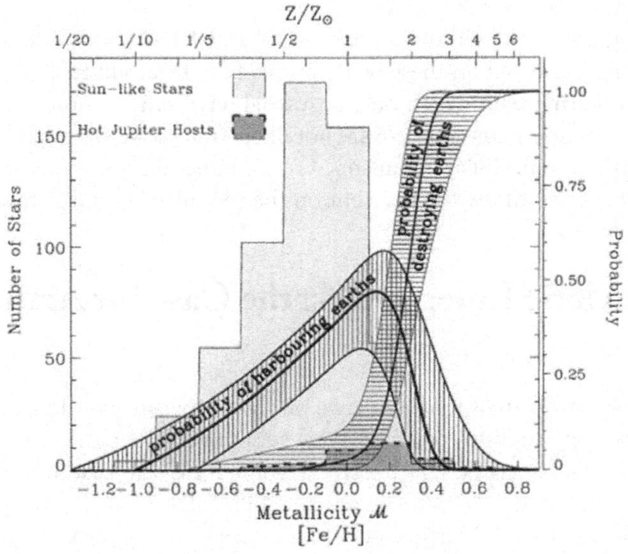

Fig. 18. Evolution of metallicity in the Galaxy (after Lineweaver (2008))

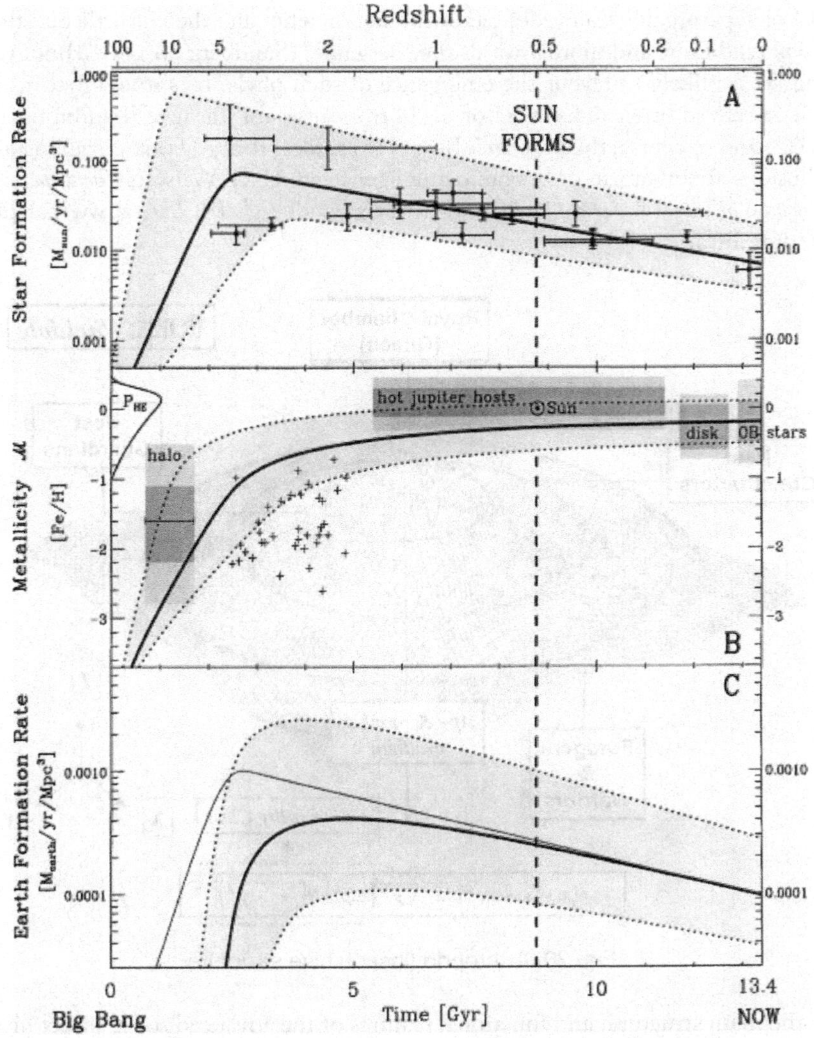

Fig. 19. Star formation rate (top), metallicity (middle) and earthlike planet formation rate (bottom) (after Lineweaver (2001))

The most significant outcomes of Lineweaver et al.'s estimates are:
1. that there are at least ~ 10 million earthlike planets in the Galaxy (this being the most conservative estimate),
2. that their average age is 6.9 ± .9 × 109 years,
3. that complex life on these world will therefore be ~ 1.8 ± .9 × 109 years older than that on the Earth,
4. that any ATC's that have evolved on these worlds is likely to be around 2 billion years in advance of us.

This, if true, is very sobering data which puts terrestrial "civilisation" more thoroughly in perspective. What was happening on Earth 2 billion years ago? We are not likely (unless we are celebrities on acid) to talk to or form a relationship with, bacteria or unicellular organisms. We are more likely to experiment with them or utilise them in specific projects. This may also throw further light on the so-called "Fermi Paradox".

11. Different Phyletic Emergences: the Case for Arthropodic (Insect) Intelligence

In many ways the rise of the mammals to dominance on this planet may be an anomaly. Certainly in the case of humans, the very traits that appeared to confer adaptive strength on the EEA now threaten to become the evolutionary nemesis of the species. It may be very likely that some ATC's at least may be of arthropod (insect) origin. This of course does not mean "giant ants" in the manner of THEM! but much more subtle forms of evolved intelligence (see for instance George M. M. Martin's story Sandkings (1979)). The structure and functional organisation of the social insects (Wilson 2008) suggest that the evolution of supoerorganic forms of sentience may prove highly adaptive – more so than mammalian sentience since after all, human advance and dominance are largely due to the hive-like or superorganic traits adopted by human societies and there is little question that future post-human evolution will tend more and more towards the "borganic" (Sandberg op. cit.). Minor variants in planetary conditions (see Fig. 5) are likely to favour the emergence of such phyla. It is ironic that insect intelligence and borganic-type societies are so often the subject of SF horrors - see for instance the film documentary *The Hellstrom Chronicle* (1972) and of course, the Star Trek Borg. Two studies that go far in dispelling anthropomorphic or anthropocentric illusions about swarm or borganic intelligence are E. O. Wilson's *The Superorganism: The Beauty, Elegance and Strangeness of Insect Societies* (1978) and Anders Sandberg's *We, Borg* at www.aleph.se/Trans/Global/Posthumaniyu/WeBorg.html.

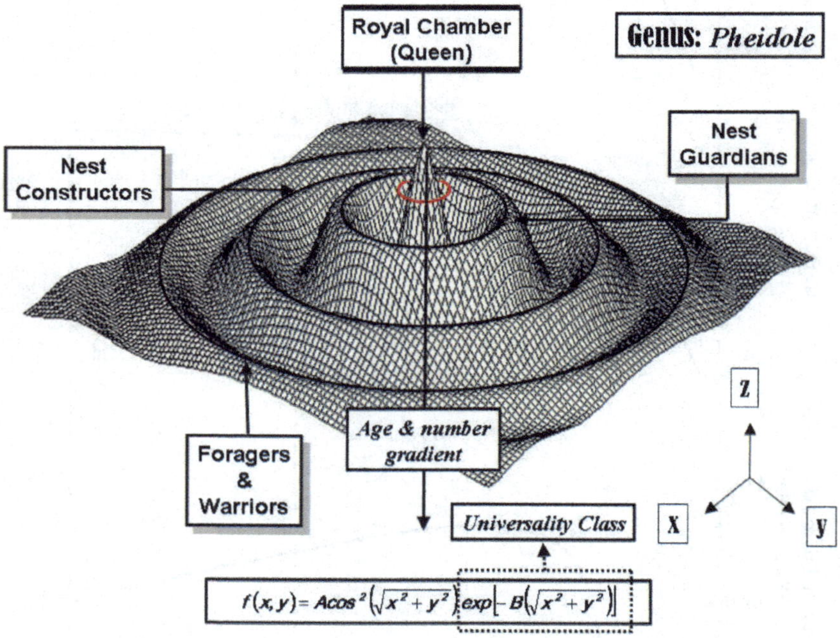

Fig. 20. Arthropod (insect) hive structure

Fig. 20 shows the main structural and functional features of the advanced social insect hive (in this case, that of the ant genus Pheidole).Hive structures of this type can be represented as a type of function known as a "sombrero" (for obvious reasons). Universality classes are linked by the relationship between the exponents of generative functions (see Fig. 6 and associated text). It is this type of highly flexible and efficient structure that may prove to be a favoured adaptation channel in the space of cosmic life-forms.

12. 12-tone Serial Technique: A Clue for the Decoding of Alien Transmissions?

A fruitful avenue of research from the SETI standpoint might be a search for "totally parametrised" languages structured serially or in parallel that lie concealed beneath seemingly chaotic data. Combinatorially-based algorithms exist for such an approach – based on the 12-tone serialist quest for 12-tone sets of specific type and order (Hook 2009). The 12-tone serial world was an unusual flowering of music-theoretic research at a time (the 70's and 80's) known for its "liberalisation" tendencies – but which also included advances in many directions that have since been abandoned as global society returned to a far more conservative frame of mind. "12-tone music" proved far too complex and "alien" for the embodied human intelligence – probably as was to be expected. Nevertheless, the "alien" quality of this musical style and thought – which still emerged spontaneously from a sentient species (humanity) – may well provide an important clue for the understanding of a truly "alien" intelligence and the decoding of its transmissions. Multi-parametric serialised languages of the serial type can be adapted to all forms, scalings and ranges of the electromagnetic spectrum, including the quantum realm. The scaling of computational languages of this type will relect the scaling of sentient systems in the Universe (Perez-Mercador 2002). Assimilation of the quantum domain is crucial. It is this domain – that of Schrödinger evolution – that may prove to be a more "popular" playground for ATC's than the 21 cm. H2 band. Human scepticism over the quantum realm is simply a consequence of our limited morph-dependent computational powers. As S. R. Haddon says in the film version of Carl Sagan's *Contact* (1995) "An alien intelligence has got to be more advanced. That means efficiency functioning on multiple levels and ... in multiple dimensions."

13. Epilogue

Future research may reveal that life is abundant in the cosmos – but that many advanced technological civilisations (ATC's) may be far in advance of us, i.e. that we are "new kids on the block". ATC's that are ~ 2 × 109 years in advance of us are unlikely to communicate with us, or respond to any efforts at communication from our side although the much sought-after "evidence" for their existence may be abundant – right "under our noses" in fact – but due to the model constraints of morph-dependent computation discussed in sections 4 and 5 above, this evidence is not perceived. To us at our current level, ATC technologies are probably indistinguishable from "natural law" as we currently understand it. Ants walking on the wall of a corporate high—rise in Manhattan probably imagine they are walking on some kind of smooth stone.

Unmodified "natural" humans are unlikely ever to travel into deep space. If they ever do so, it will be with the support of infrastructures created and maintained by posthuman pioneers and their successors. As regards future human evolution, speciation will accelerate in proportion to our advance along the Kn axis. Only at K2 and beyond will we have developed sufficiently to experience (and competently handle) some form of contact with an ATC.

"Many are called but few are chosen" (Matt. 22:14). The Biblical proverb may apply not only to "selection criteria" for arcology citizens or interstellar crews,, but to the emergence of ATC's in general – Darwinian selection may be operative throughout the Cosmos with "eligibility for convergence at the Final Singularity" serving as the "fitness value". For participants at this humbler level of transition from K0 → K2 the traditional "right stuff" selection criteria – "normal", "well-balanced" (even if biologically-fit) humans - may be less than ideally suitable for a one-way trip into the Void. Motivation is everything, and Antarctic-based studies have shown that those most able to endure high-stress, high-risk environments with the ideal combination of telic and paratelic behaviours tend to be "highly-motivated civilians" rather than the meticulously-trained "right stuff" (Harriison et al. 1990).

"A species only reaches maturity when the nature of Mind is fully understood" (Amoroso & Martin 1995). We have a long way to go in this direction, although we have certainly taken some initial steps. For our K3-level successors, the word "human" might come to mean "that which is grown from earth" (>Lat. "humus" → "humanus"), i.e. "the chreod or species variant that finally emerged from its home world". They will come to understand biological life as simply an early transitional form between "inert" and fully "sentient" matter. A readiness to encounter the truly alien would require that they regard "humanity" not as a single family, but as a "speciation crucible" of which they are but one of the possible emergent branches. In possessing a boundless, empathic appreciation of the infinite potentialities of Being, they will have come to accept the potentially infinite mutability of form and nature and will have renounced their unique human identities on order to transcend the collective limits of that identity. Whether they will weep or exult in this transcendence will be their own affair.

References

1. Amoroso, R. L. & Martin, B. E. (1995). "Modelling the Heisenberg Matrix: Quantum Coherence and Thought at the Holoscape Manifold and Deeper Complementarity". Scale in Conscious Experience: Is the Brain Too Important To Be Left to Specialists to Study? Proceedings of the Third Appalachian Conference on Behavioural Neurodynamics, pp.351-77. Lawrence Erlbaum Associates Inc., Mahwah, NJ.

2. Barrow, J. D. & Tipler, F. J. (1986). The Anthropic Cosmological Principle. Oxford University Press, Oxford & New York.

3. Burbidge, E. M., Burbidge, G. R., Fowler, W. A., & Hoyle, F. (1957). "Synthesis of the Elements in Stars". Revs. Mod. Physics vol. 29, pp. 547 – 650.

4. Chaitin, G. (1966). "On the length of programs for computing finite binary sequences". J. ACM vol. 13: 145.

5. Chaplin, W. J., Serenelli, A. M., Basu, S., Elsworth, Y., New, R. & Verner, G. A. (2007). "Solar Heavy Element Abundance: Constraints from Frequency Separation Ratios of Low-Degree p-Modes". The Astrophysical Journal, vol. 670: 1, pp. 872-84.

6. Ćirković, M. (2012). The Astrobiological Landscape: Philosophical Foundations of the Study of Cosmic Life. Cambridge University Press.

7. Crutchfield, J. (1994). "The Calculi of Emergence". Physica D. Proceedings of the Oji International Seminar: Complex Systems – from Complex Dynamics to Artificial Reality.

8. Deleuze & Guattari, F. (1987). A Thousand Plateaus. University of Minnesota Press.

9. Fry, I. (2000). The Emergence of Life on Earth: A Historical and Scientific Overview. Rutgers University Press, New Brunswick, NJ.

10. Gardner, J. (2007). The Intelligent Universe: AI, ET and the Emerging Mind of the Cosmos. New Page Books, NJ.

11. Gillies, D. (1998). "The Duhem Thesis and the Quine Thesis". In Philosophy of Science: The Central Issues, (Curd, M. & Cover, J. A. eds.) Norton, New York, pp. 302-19.

12. Guenther, H. (1984). The Matrix of Mystery. Shambala, Boulder & London.

13. (1989). From Reductionism to Creativity: rDzogs-chen and the New Sciences of Mind. Shambala, Boston/ Shaftsbury

14. Harrison, A. A., Clearwater Y.A. &McKay, C.P. (1990). From Antarctica to Outer Space: Life in Isolation and Confinement. Springer-Verlag, New York & London.

15. Hook, J. (2007). "Why are there 29 Tetrachords? A Tutorial on Combinatorics and Enumeration in Music Theory". Music Theory Online, vol. 14: 4 http://www.mtosmt.org/issues/mto.07.13.4/mto.07.13.4.hook.html

16. Johnston, J. (2008). The Allure of Artificial Life: Cybernetics, Artificial Life and the New AI. MIT Press, Cambridge, Mass., London.

17. Kauffman, S. (1986). "Boolean systems, adaptive automata, evolution". Disordered Systems and Biological Organization (Bienenstock, E., Fogelman-Soulie, F. & Weisbuch, G. (eds.)), Series F: Computer and Systems Sciences 20. Springer, New York.

18. (1993). The Origins of Order: Self-Organization and Selection in Evolution. Oxford University Press, New York & Oxford.

19. Kolmogorov, A. N. (1965). "Three approaches to the concept of the "amount of information"". Prob. Info. Trans, vol I: 1.

20. Lakoff, G. & Johnson, M. (1999). Philosophy in the Flesh: the Embodied Mind and its Challenge to Western Thought. Basic Books, New York.

21. Lem, S. (1986). Microworlds. Mariner Books, New York.

22. Lineweaver, C. H. (2001). "An Estimate of the Age Distribution of Terrestrial Planets in the Universe: Quantifying Metallicity as a Selection Effect". Icarus 151, pp. 307-13, arXiv:astro-ph/0012399.

23. (2004) & Fenner, Y. & Gibson, B. K. "The Galactic Habitable Zone and the Age Distribution of Complex Life in the Milky Way". Science 302 (5654), pp. 59-62, arXiv:astro-ph/0401024.

24. Perez-Mercader, J. (2002). Scaling Phenomena and the Emergence of Complexity in Astrobiology. Springer, Berlin.

25. Schneider, J. (2001). Interactive Extrasolar Planets Catalog: http://exoplanet.eu/catalog/ accessed 28/11/2013 : 1800 GMT.

26. - - -

27. Seager, S. (2012). "An Equation to Estimate the Probability of Identifying an Inhabited World Within the Next Decade". http://seagerexoplanets.mit.edu/ accessed 28/11/2013: 19.4 GMT.

28. Shannon, C. E. & Weaver, W. (1962). The Mathematical Theory of Communication. Univ. of Illinois Press, Champaign-Urbana, Ill.

29. Solomonoff, R. J. (1964). "A formal theory of inductive control". Info. Control vol. 7: 224.

30. Thom, R. (1989). Structural Stability and Morphogenesis. Addison-Wesley Advanced Book Program, Reading, MA.

31. Wilson, E. O. (2009). The Superorganism: The Beauty, Elegance and Strangeness of Insect Societies. W. W.Norton & Co. New York/.

32. Ziolo, M. P. (2004). "Trauma and Transcendence: Psychosocial Impacts of Cybernetics and Nanotechnology". Proceedings of the PISTA 2003 International Conference on Informatics and Systemics (AAAS) - also available on the World Systems Analysis Archive :: http://wsarch.ucr.edu/archive/papers.htm

Religious Engineering for the K0→ K3 Transition

Paul Ziolo, PhD

Abstract

What is the strength of steel compared to that of the hand that wields it? While science and technology will provide the material tools for getting us o the stars and helping us live there, only the appropriate ideational system can sustain that vision long enough for us to get there. An ideational system is defined here as a cognitive-emotional construct that shapes, harnesses and directs the evolutionary drives of a group towards a 'transcendental' according to the challenges of the presenting environment. Examples of how transgenerational task-oriented groups catalysed major social and technological transitions in the past are provided by the role of the medieval monastic institutions and knightly orders in the genesis and growth of Western civilisation. This paper examines how the structure and character of such foundations and the strategies employed by them might be translated into the terms of the present and thereby assist the 100YSS in achieving its goal, Following practical approaches to 'religious engineering' such as those studied by Kuah Khun Eng (Khun Eng 2009), this study would recommend, not the invention de novo of some new 'Galactic religion', but a discrete and subtle emphasis on and gradual synthesis of, those elements in existing ideational systems most likely to foster the 'outward urge' and gain a broad international base of support for the 100YSS vision.

1. Projective Systems as Cultural Determinants.

"It is the work of Science to describe the physical order of the Universe – and
the work of Religion to place humanity within that context..."
(The Dune Encyclopedia: Bene Gesserit Training Manual,)

"Οίων την μέν ουσίαν θεολογίας διδάσκαλων"
('Being is the Teacher of Theology')
(Maximos the Confessor: Ambiguum X, PC. p.1134, 148b).

The Nobel Prize-winning economist Douglass North has shown that the main catalysing force in a civilisation's genesis and growth is the institutional/organisational matrix governing its political economy, and that the dominant ideational or belief system of that civilisation is the primary determinant of that matrix (North 1997). Even studies that claim primacy for environmental determinism such as Diamond's Collapse (2005) make it clear that similar environmental conditions can evoke different responses from human groups at comparable technological levels. Challenges posed by an Oceanic ecology evoked radically different responses from the societies of Tikopia and Rapa Nui (Easter Island) respectively, as did an Arctic ecology from the Inuit and the Greenland Norse, the semi-desert ecology of Utah from the native Paiute-Shoshone groups and Mormon settlers and the South African veldt from the Bantu-Xhosa groups and Dutch 'voortrekkers'. On every continent, small 'traditional' societies with quite different social systems have co-existed in close proximity to one another. The variance in these systems has been traced primarily to differences in the mythic structures of each group and the intrademic transmission of those myths.

The 'strength' of a dominant projective system or construct is defined as its capacity to inspire its adherents towards collective action. Construct strength was termed asabiyya' (group solidarity) by the Tunisian historian Ibn Khaldun (1332– 1406) and the evolution of asabiyya' in relation to social complexity is analysed in the dynamic models of P. Turchin (Turchin 2003), who shows that constructs can be effectively regarded as homeostatic mechanisms, much as in the models of the cyberneticists (Ashby 1952; Haken 1996; Friston et al. 2008, 2009, 2010). Fig. 1 below shows a simplified, unidimensional model of this type, omitting delay factors for negative feedback and translating parameters of physical space into those of social space.

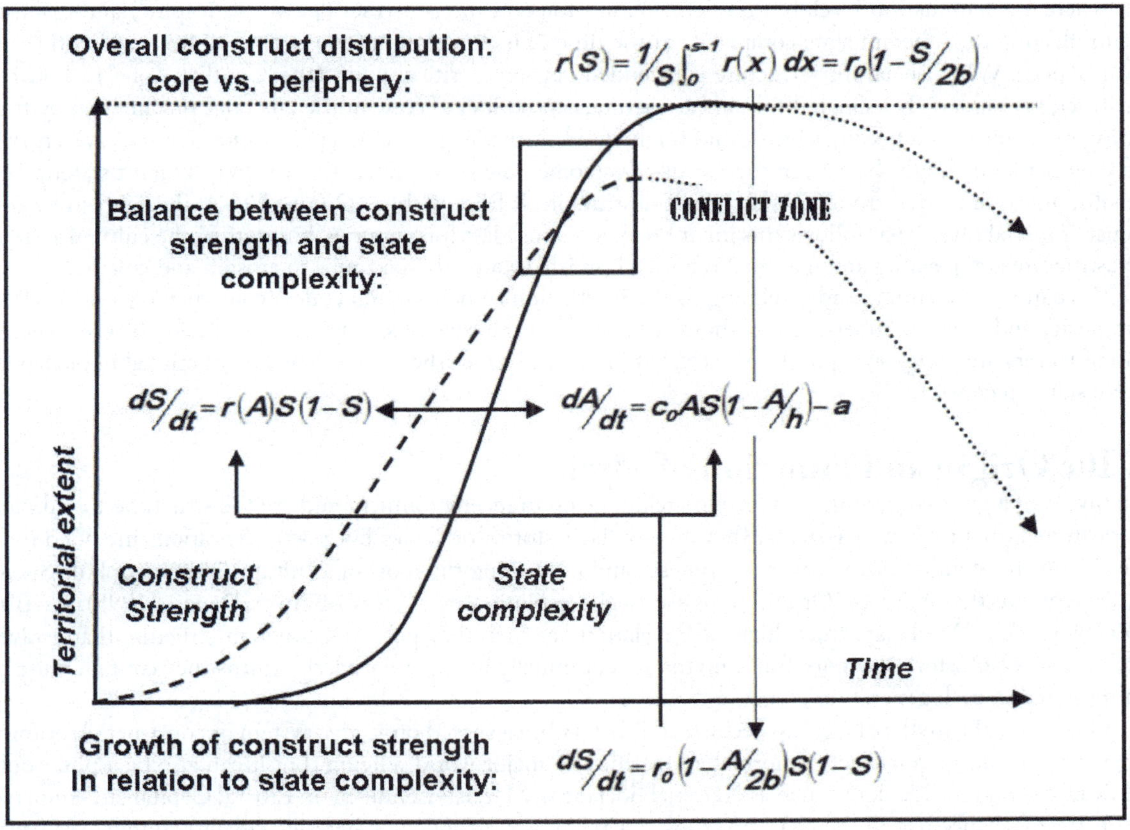

Fig. 1. Co-evolution of construct strength and state complexity

After initial consolidation and expansion, a construct becomes the foundation or Spenglerian 'prime symbol' (Spengler 1928) of cultural genesis and state formation. While the society is in its nascent, exploratory stage and the construct still 'open', almost any external challenge (barring extreme physical catastrophe) can be overcome. In Fig. 1 construct strength S grows in conjunction with state complexity A according to:

$$dS/dt = r(A)S(1-S) \qquad \text{Eq. 1}$$

where r is the growth rate of state complexity. As wealth increases, construct strength falls off within the 'élite' social core due to 'habituation' while remaining high at the more 'deprived' periphery. 'Habituation' (Ibn Khaldun's term) signifies the withdrawal of projective significance from objects in the construct's 'transcendent' frame of reference and their condensation onto the 'realities of daily life' as those 'realities' grow more complex - hence Quigley's description of institutional decay (Quigley 1979) and Weber's observation that following the Reformation, medieval 'grace' became translated into 'money' (Weber 1958). Construct strength has the overall distribution:

$$r(S) = \frac{1}{S}\int_0^{S-1} r(x)\,dx = r_0\left(1 - \frac{S}{2b}\right) \quad \text{Eq. 2}$$

where b = the residue of construct strength at the periphery in relation to that at the core. Inputting this term into Eq. 1, the relations between construct strength and infrastructural complexity are given by:

$$\frac{dA}{dt} = c_0 AS\left(1 - \frac{A}{h}\right) - a \quad \text{Eq. 3}$$

and

$$\frac{dS}{dt} = r_0\left(1 - \frac{A}{2b}\right)S(1 - S) \quad \text{Eq. 4}$$

where a = habituation level, h = implementation force of the construct ('power to inspire') and c_0 = a constant reflecting the inherent state complexity at the time of its foundation (or at a time t0 from which the model is calculated), When social infrastructure is extended beyond a critical point (the 'Conflict Zone' in Fig. 1) the construct, having largely validated that infrastructure, can no longer contain the anxieties precipitated by the resulting instabilities and becomes brittle and fragmented. A challenge that at a point prior to a society's entry into the Conflict Zone might have been successfully overcome now proves fatal. Thus, as Will Durant stated, "before a civilisation is destroyed from without, it first destroys itself from within" (Durant 1944). The aggressive, expansionist 'imperial phase' that follows this inner breakdown may last for centuries, but in time the cultural system, as infrastructure complexifies and marginal returns diminish, eventually succumbs to erosion and collapse.

Dynamics of a similar order relating to the interaction of behavioural codes (constructs), stress levels, task complexity and group numbers, can be shown to operate at the level of smaller groups (Ziolo 2004) – where the scaling factors are group size, profile, construct strength and time. These factors will be of critical importance for future starship crews.

2. The Origin and Function of Myth

The myths of a group determine that group's relationship to its environment and are the structural foundation of all ideational (or 'projective') systems. They are not 'fairy stories' or 'fables' but poetic evocations intended to contain the psychic conflicts arising from epigenesis, and in forming the roots of a culture's 'prime symbol' (Spengler 1928) generate that culture's 'Orphic response to the evolutionary sources of Awe' (Danesi 1995). As Binion (1997) remarks: "Myths are the vehicle of Revelation: as such they possess a coherent structure that evolves in response to sociohistorical events. Since myths are continually re-experienced, the truths they contain must be re reinterpreted in each generation".

Over time, the myths of a group become codified within a shared projective system or construct. A 'dominant' projective system (or construct) is normally identified as 'major world religion', but it can also be a philosophy, a political or economic ideology (such as National Socialism, Marxist-Leninism or 'Liberal Capitalist Democracy'). 'Mainstream' science has all the characteristics of a 'religion' – orthodoxies, heresies, rituals ('conferences'), priests and bishops (academics'), saints ('eminent' scientists) and demons (practitioners of 'pseudoscience'). A psychohistorical analysis makes no judgement as to whether a given construct is 'rational' or 'irrational', or to be classified as 'science' or 'superstition'. It is not the proclaimed, overt character of a construct that matters so much as its deep structure and the manner in which it catalyses individual or collective action. Neither is it the job of psychology (nor of physics nor philosophy for that matter) to 'prove' or 'disprove' the 'existence of God'. It is a legitimate point of psychohistorical inquiry to ask why people believe the things they do in any given era or culture, why the things they believe in take the forms they do at these times and above all, how and why these forms make a population act in certain ways as opposed to others. Such an approach requires a language that is specific yet sufficiently abstract in relation to 'traditional' theological or philosophical discourse so as to avoid irrelevant controversies. Only in this way can it be seen how specific constructs become instruments whereby the primary evolutionary drives and residues of generic trauma are refined and shaped so as to result in specific desires and goals. 'Religious' engineering can become a far more subtle and efficient tool of social change than more explicit types of 'social' engineering. The study conducted by Kuah-Pearce Khun Eng (Khun Eng 2009) is a modern example of this, while historical examples are provided by the sermons and letters of St. Bernard of Clairvaux (1090-1153 AD) and the prophecies of Joachim of Fiore (1135-1202 AD).

The 'complex' consciousness of so-called 'higher' mammals forms one end of a continuum originating with the sucrose gradients of unicellular organisms such as Paramecium (Froese & Di Paolo 2009). The structures of organic life determine the models different organisms construct in order to interpret their environments (Crutchfield 2008), and the prescriptives of evolution require a degree of plasticity and swiftness of adaptive response such that in more complex cognitive systems, more basic cognitive and perceptive mechanisms are necessarily relegated to the level of the 'subconscious' in a fitness landscape, the understanding of which is far from complete. We consciously operate with little more that 15% of our brain capacity, the remainder includes 'free energy' or environmental stimuli that 'transgress the boundaries' of models based on everyday experience. The 'supernatural' can be therefore be considered as a 'reification' of all that transcends conscious experience – a projection in bodily form of the 'as-yet-unknown' and a naming of the unnameable in the terminology of a group construct whereby the thing named assumes a concrete reality delimited by the operational mentalities of the group. Such a fixation of what is essentially fluid permits temporary co-ordination of the group's shared perceptions and enables a collective response to the shifting challenges of an environment whose hidden features therefore remain mysterious.

All sources of transcendent experience therefore undergo projection in terms of human embodiment and become 'in-corporated' into constructs that serve a binding function ('religion' > Lat. 're-ligāre' – to 'bind together' or 'reconnect') so that members of a group can live and work in co-operation for that group's survival in a world whose origins and purpose are imperfectly understood. Religion is thus an essential tool of survival, with specific 'religions' remaining subject to the same rules of mutation and evolution that govern organic life in response to shifts and deformations in the fitness landscape.

3. Probing the Matrix

"The symbolism of uterine ecology is the Rosetta Stone of all world religions"
(Terence Dowling – personal communication)

Despite the diversity of human groups and cultures throughout history, all human constructs derive from the same morphogenetic 'core'. As Giambattista Vico (1668-1744) has stated: "The mental vocabulary of human social institutions, which are the same in substance as felt by all nations but are diversely expressed in language according to their diverse modifications, is exhibited to be such as we conceive it" (The New Science (1744) [355]).

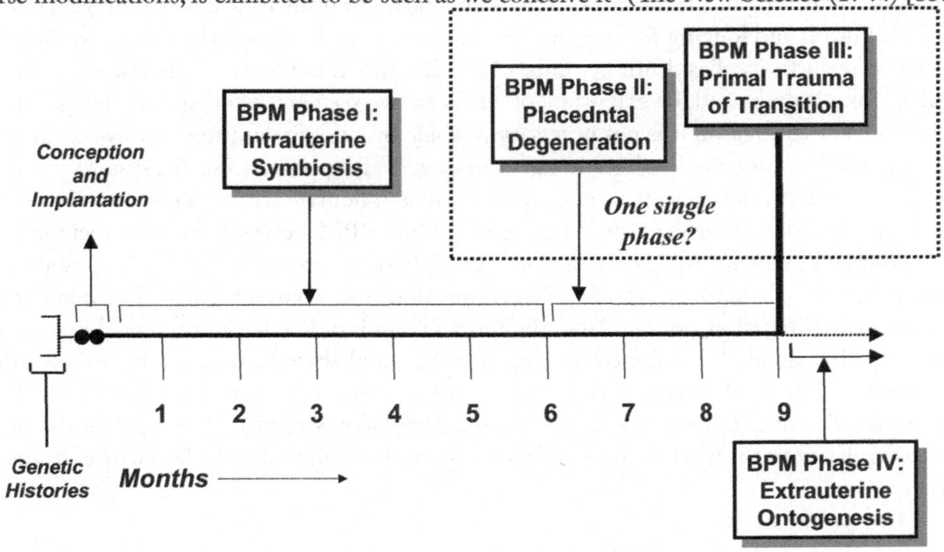

Fig. 2. The Basic Perinatal Matrix (BPM) (Grof 1975)

Where does this 'mental vocabulary' come from? The generic 'archetype' of all human experience is birth and early morphogenesis (growth of form). Evidence for the centrality of this experience and its foundational character for all later mental development is abundant in the gynaecological, clinical and fetal psychiatric literature (e.g. Rank 1924; Ferencsi 1929; Fodor 1949; Peerbolte 1957; Lake 1966; Mott 1968; Ployé 1973; Janus 1977; Verny & Kelly 1982; Wasdell 1985; Ziolo 2004 etc.). Fig. 2 above shows the sequential stages of the Basic Perinatal Matrix (BPM) charted out with temporal succession running from left to right. According to Grof's (1975) clinical identification stages in the BPM, each stage is the generic substrate for the following 'mythic' archetypal situations (stages III and IV are sometimes 'telescoped' into a single phase):

- BPM I : Symbiosis within the womb, a state of peace, of timelessness, of Paradise, of 'union with God' and of being in touch with the 'ground of Being', during which the environment is ever bountiful, limitless in resources and endlessly supportive of unfettered growth.
- BPM II: The beginning of antagonism, conflict and of subjective time. This phase runs from the onset of placental degeneration to that of labour, hence first evoking a sense of increasing confinement and suffocation, and of shrinking lebensraum, eventually leading to powerful sensations of constriction and crushing, of cardiac distress and intense suffering, and finally to the feeling of being sucked towards a whirlpool - of being between Scylla and Charybdis - or of being swallowed by some monstrous placental avatar - a dragon, octopus, giant squid or many-headed Hydra. A confrontation with, and fear of engulfment by, the infinite abyss of Being.
- BPM III: The titanic struggle of birth, the 'Mother of all Battles', archetype of all wars, of Armageddon and Götterämmerdung, of all sadomasochistic orgies, of cataclysmic volcanic eruptions, nuclear detonations, rapes, mass murders, suicidal frenzies and orgies of destruction - the ultimate conflict between Life and Death.
- BPM IV: After birth - liberation, relief, salvation, the sensation of having been forgiven, purged, absolved, yet also of being exhausted, helpless and exposed - the anticlimax of 'what now?'

The recovery of birth experiences by analytic memory is difficult due to their prelinguistic character and the transmarginal stresses induced by perinatal transition. Stress becomes 'transmarginal' when the stress quotient exceeds the boundaries of previous experience, becomes dissociated or split off from active consciousness and then bound within a defensive psychic construct whose function is to contain and control the potentially explosive power of the residual anxiety generated by it (Ziolo 2004). Perinatal stress is a direct concomitant of morphogenesis: "experience can change the mature brain, but experiences during the critical periods of perinatal transition and early childhood organize brain systems" (Perry et al. 1995). This is why so many perinatal studies often tend sooner or later to verge on the 'mystical' – a tendency we might regard as further evidence in support of Dowling's statement at the beginning of this section. The uterine ecological sequence involves the transition from a water-breathing entity to an air-breathing entity. Such transition –a true metamorphosis - necessarily involves the psychological death of the former state of the organism prior to its 'rebirth' into a new form. In this way the BPM becomes the archetype of the death experience and, being imprinted in subconscious memory, exacerbates the human fear of death and longing for immortality (Lifton 1974) that lies at the root of the 'human condition'. Failure to achieve 're-birth' implies death by suffocation. The BPM, linked as it is to sexuality and death, is thus the 'core matrix' for all myths and, by extension, of all projective systems, whether 'rational' or 'irrational' (Fig. 3 (a) and (b) below). For this reason it is not necessary to abide by a dominant projective system in order to follow its precepts, consciously or unconsciously. One can even be actively opposed to such a system, yet be compelled to follow its precepts, at least externally, in order to conform to and flourish within, a given society.

It is therefore important to realize the extent to which the BPM serves as the root metaphor for transition from K0-K1 (confined existence within a planetary atmospheric environment) to K2 (permanent adaptation to a vacuum environment) whenever we speak of 'emerging from the womb of Earth'. For many if not most, the 'Starship' is a symbol of liberation, not only from the 'mortal bounds of Earth', but from the human condition itself. Such a protean symbol should be exploited intelligently. Any myth-based construct that exceeds the boundaries of the 'core' shown in Fig. 3 (a) is felt not to be of concern to 'real people' and hence somehow 'bizarre', 'alien' or 'incomprehensible'. This does not prepare us well for the eventual posthuman challenge, for the possibilities of a technological singularity, or for what we may encounter when we finally leave the boundaries of the Solar System (Lem 1984).

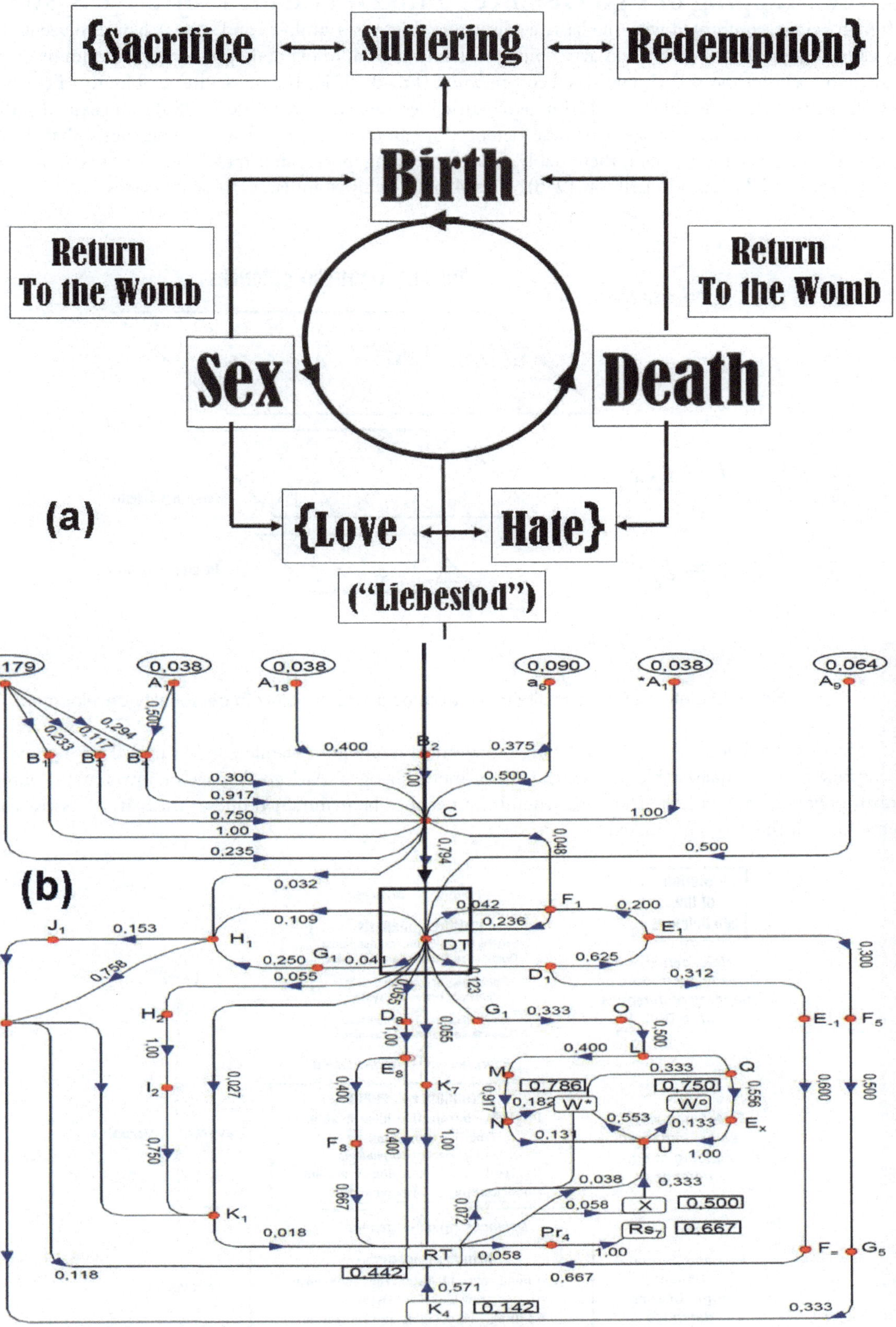

Fig.3. (a) The 'core matrix' as source of all mythic structure and (b) neural net for the generation of myths and stories based on Prop's Analysis of the Folktale(1968).

4. The Mapping of Ego Defences onto Dominant Projective Systems.

Human beings, confronted with the basic evolutionary drives (survival, sex and status), have evolved a number of so-called 'ego-defences' or 'involuntary coping mechanisms' both to maintain psychic equilibrium in a social context (homeostasis) and to facilitate social co-operation (Freud 1921). The transcultural ubiquity of these defences is well established in the clinical and ethnopsychiatric literature (Devereux 1967, 1980) so no extended discussion is required here. Suffice it to say that these 'defences' are an integral part of human epigenesis, that they grow in sophistication over the lifespan, their stable features defining 'personality traits' and that their transformations mark out critical 'life stages' (Erikson 1956). Figs. 4 and 5 below summarize these processes.

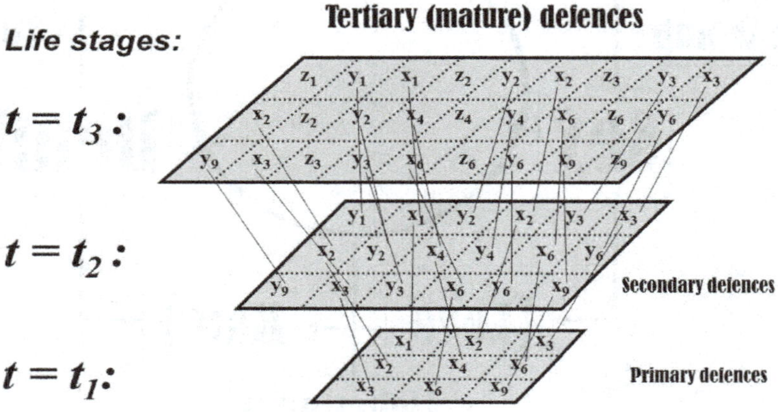

Fig. 4. Evolution of 'personality traits' at three arbitrary points in personality development.

Since ego defences serve a homeostatic function (to maintain dynamic equilibrium) their moment-to-moment functioning is quasistable – hence the flux of 'mood changes'. An 'ego-defence' or 'involuntary coping mechanism' only becomes pathological when equilibrium cannot be maintained due to excessive stresses, whereupon some form of therapy is mandated,

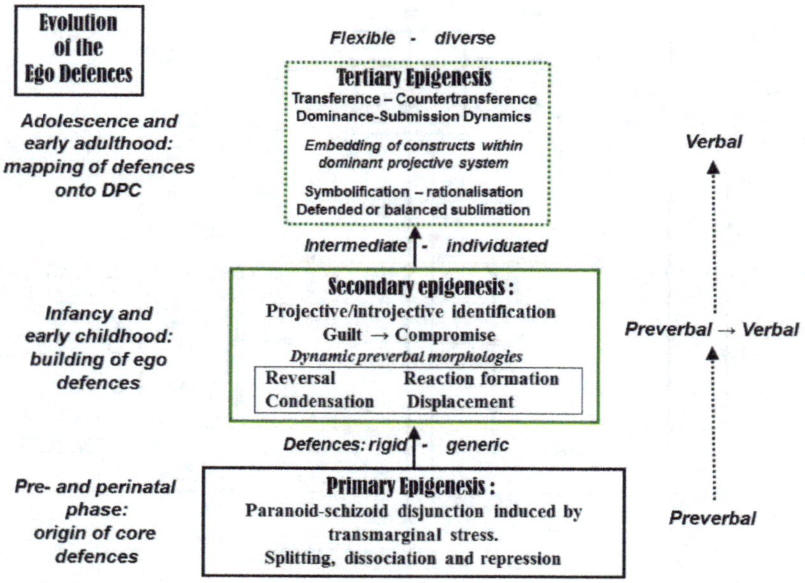

Fig. 5. Hierarchic development of the ego defences over the lifespan

Under 'normal' circumstances, ego defences are stabilized through bijective mapping onto a dominant projective system or construct. A construct evolves within a group in order to assist this function and bind the group together ('religion' > Lat. re-ligāre = 'to bind together' or 'to reconnect') by becoming the group' 'totem'. By means of mapping her/his inherently unstable defences onto the mythic structures of a construct, an individual confers stability (equilibrium) on these defences, and maintains this stability by sharing personal conflicts with other group members in terms of the semantic and syntactic structures of the construct (Stromberg 1993; Breck 1994; Brueggemann 2000). Individual mappings are bijective in that although the mapping may be one-to-one or, on occasion, one-to-many, the construct itself evolves the capacity to absorb high degrees of individual variance. Should the construct rigidify and fail to evolve new semantic and syntactic structures in response to a changing psychosocial environment, fragmentation or radical reform will result (as was the case of the historical 'Reformation' in Western Europe). This 'mapping and diffusion' process is illustrated in Fig. 6 below.

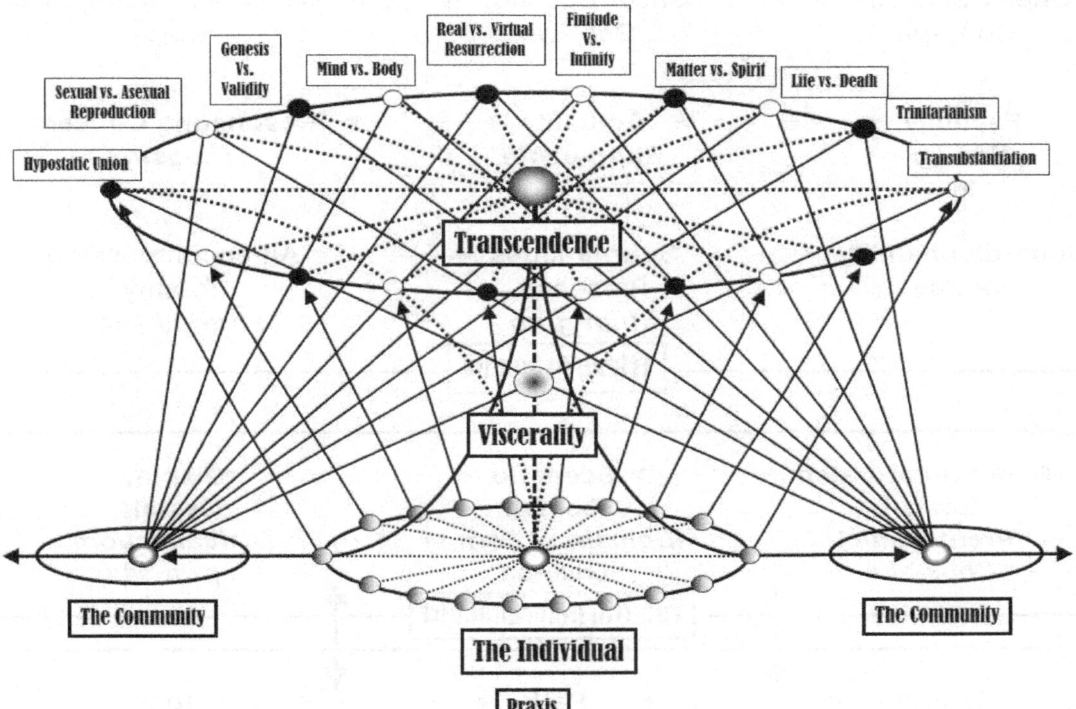

Fig. 6. The mapping of individual conflicts onto the mythic structure of a projective system and the stabilization of this mapping through sharing this system with other members of the group.

5. General Structure of the West European Construct

For a construct to be an effective catalyst for action, it must evolve the capacity to contain a wide range of conflicts and the flexibility to adapt over time to changing psychosocial circumstances. This capacity depends on three basic qualities a construct must acquire during its generative phase if it is to become 'dominant'. These are:

1. *transcendence*: the system (or construct) must confer on the individual and the group a firm understanding of personal and collective existentiality in relation to the wider cosmos,
2. *viscerality*: the system (or construct) must make a direct appeal to the basic drives and emotions – the 'evolutionary sources of Awe' and
3. *praxis* ('action'): the system (or construct) must provide a clear guide as to how to behave and act in relation to self and others.

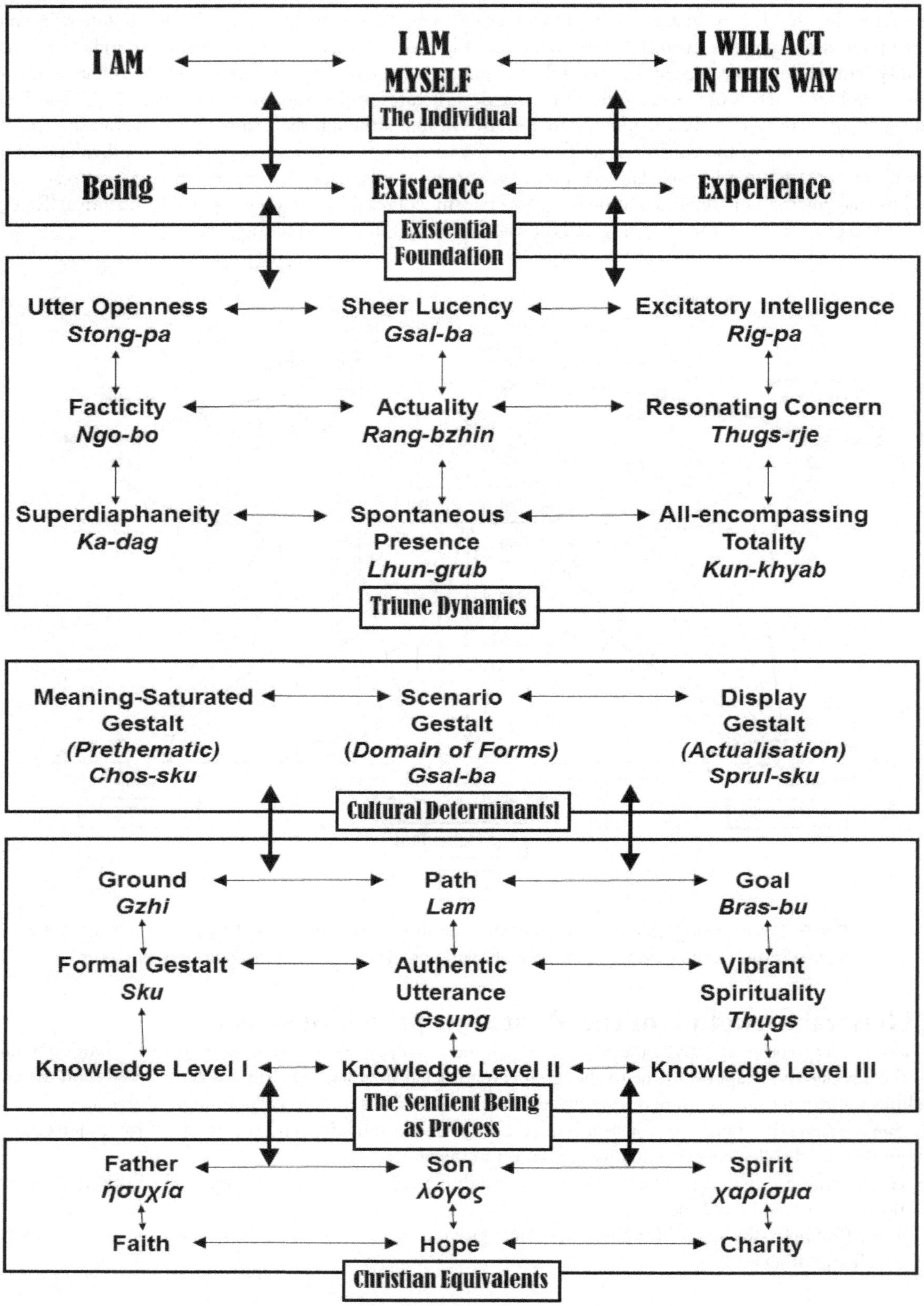

Table 1. Trinitarian structures and meanings in the rDzogs-chen tradition of Tibetan Buddhism and their Christian equivalents

These qualities are reflected in the 'trinitarian' theologies of advanced religious systems as shown in Table 1. From this comparison of trinitarian structures in the Tibetan rDzogs-chen and Christian patristic traditions, it will be seen that in essence, the quality of transcendence corresponds to the experience of 'being-as-such', that of viscerality to individual embodied identity and that of praxis ('action') to the ethics of personal and collective behaviour. The 'psychological' character of the rDzogs-chen terminology reflects the emphasis of that tradition on apophatic (non-symbolic or non-theistic) cognitive development in contrast to the more specifically cataphatic (dogmatic) terms of Christian patristic theology. Nevertheless the latter, construct, in its course of development, acquired certain specifically 'visceral' features and a historically dynamic embedding which, in the European West was to effect a profound social transformation. Many studies (e.g. Weber 1958; Eisenstad 1982;, Toulmin 1992; Galgan 1992; Jaki 1986; Holton 1988; Woods 2005; Gray 2007; Hannam 2009 etc.) show conclusively that it was the very ambiguities and contradictions inherent in the Judaeo-Christian construct which, in forcing a creative response to 'living in the world', catalysed the development of a recursively self-enhancing scientific and technological culture. These features are:

1. one single Universe, finite in time, created once and once only by a single Creator,
2. a final and all-embracing eschatology,
3. a patristic theology of space and time which teaches the doctrine of synergy (co-creation) with the Creator and the doctrine of individual and collective progress towards this final eschatology,
4. a trinitarian theology which hypostasizes ('personalizes') Being, Identity and Action and, by combining the Augustinian theology of time with the Filioque doctrine, becomes embedded in the historical process itself.
5. a resolution of existential dualism (matter vs. spirit) through the doctrine of the two co-equal natures of Christ,
6. a final synthesis of dynamic trinitarianism and existential dualism within a framework of historical immanence through the 12th century Joachimite prophecies and
7. the Johannite doctrine of apocatastasis (the final and universal redemption of all Being).

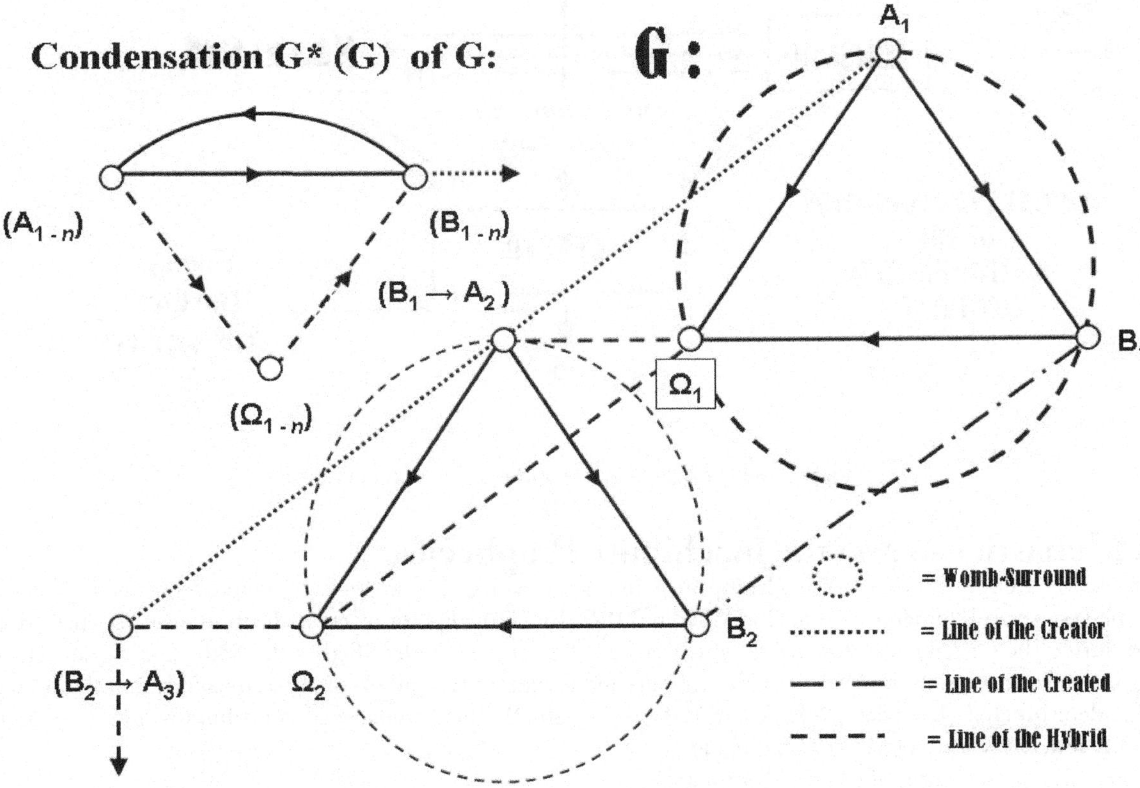

Fig. 7. Digraph (G) and condensation G*(G) of the Judaeo-Christian construct in its Western form showing transformations. An = 'Creator', Bn = 'Created' and Ωn = 'Hybrid'.

Of especial importance is the combination of 5) (the bridge between matter and spirit) with 3) (synergetic creation) and 4) (embedding of both in historical time, i.e. sanctification of the historical process itself). It is this combination that catalysed the sequences of apocalyptic cycles in European history (Fig. 6 above) and which differentiates the construct from all other world religions. Each cycle begins with a Creator, who creates the Created, both fuse to form a Hybrid, which in turn becomes the new Creator – a form of 'Hegelian dialectic' that in psychosocial reality represents the conflict between successive generations that may be represented symbolically as:

$$\left[\left(A_n \rightarrow B_n\right) \rightarrow \Omega_n\right] \xrightarrow{C_{n+1}} \left[\left(A_{n+1} \rightarrow B_{n+1}\right) \rightarrow \Omega_{n+1}\right] \xrightarrow{C_{n+2}} \ldots \qquad \text{Eq. 5.}$$

where Cn represents the dominant construct and its states of transformation. Fig. 7 below provides a more 'literal' representation of this process, starting from ''biblical' foundations and extending into the future.

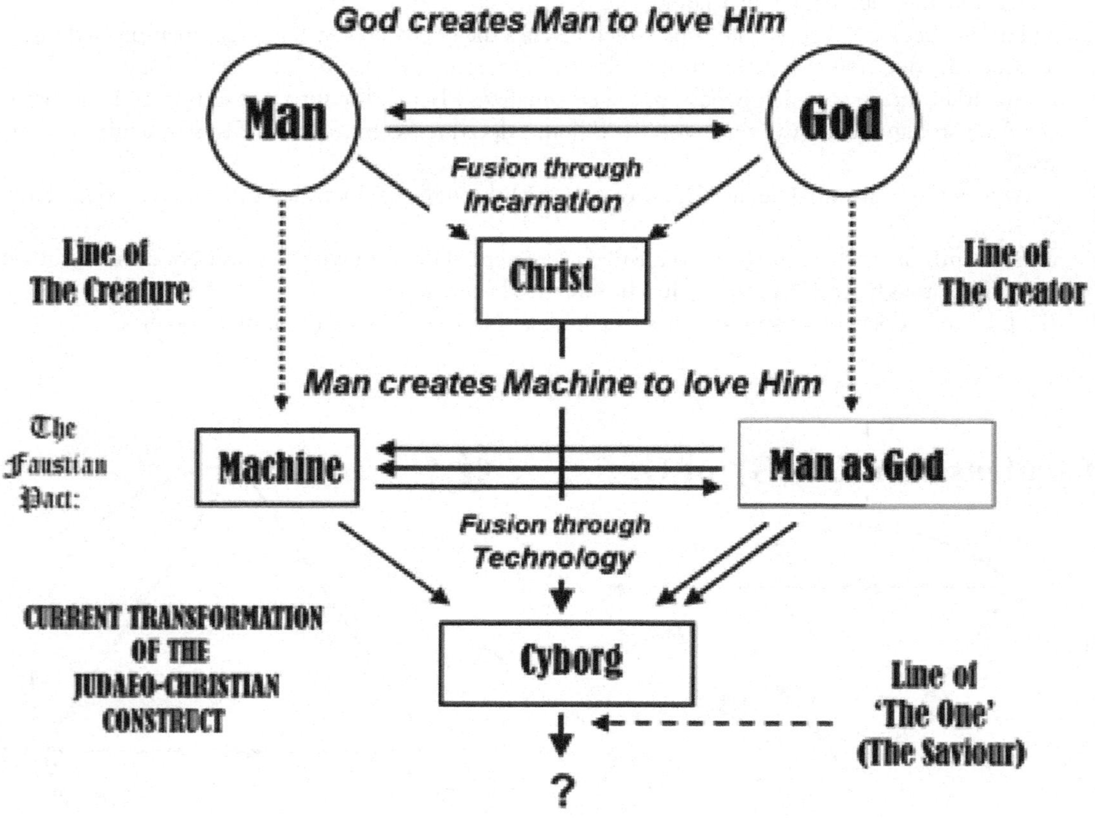

Fig. 8. Creative cycles in the Judaeo-Christian construct

6. Monasticism and the Joachimite Prophecies.

It took over 1,000 years for the West European construct to evolve – the main stages in its evolution being marked by the first seven Ecumenical Councils (Wilken 2012). The critical point of 'crystallization' occurred just prior to and during the Cistercian Reformation which preceded what is now identified as the 12th century Renaissance. The critical role of monastic institutions, both male and female, in the genesis of Western civilization should never be underestimated (Knowles 1949; Silber 1995; Ziolo 2002). The foundational contributions of this erstwhile 'deeply reticent world' (Davies 1997) include:

1. the preservation of classical learning,
2. the creation of school and university infrastructures,
3. through the disciplined integration of work and liturgical cycles ('laborare est orare' – work is prayer), the foundational concepts of goal-directed rationality and enterprise,

4. radical innovations in animal breeding, metallurgy, architecture and civil engineering,
5. the creation of conditions for the later 'Industrial Revolution' and the prototypes of modern corporate institutions and
6. through the novitiate formation process, the foundations of psychoanalytic and psychological training.

Various forms of ascetic practise have existed since the earliest periods of recorded history, but the institutionalisation and formalisation of a specifically Christian asceticism dates from the mid 4th century AD and is female in origin (Elm 1994). In this context, monasticism may be considered a precursor of the modern transhumanist/posthumanist movement. Against the fundamental drives governing human society (the 'World') -- pride, lust and greed, or 'status, sexuality and resource acquisition' – the monastics set the counter-behaviours of poverty, chastity and obedience (humility). Any association of these latter terms with 'virtue' tends to obscure the fact that 'asceticism' in itself carries no intrinsic moral imperative but is in essence a behavioural tool refined in the service of specific ends. No moral systems are absolute. All of them are behavioural codes specifically tailored to assist a given group in achieving its goals. History has shown that no weapon is deadlier than the 'jihadist' mentality schooled in a rigorous ascetic practice.

'Poverty' does not mean 'rags and begging', but the subjection of one's own personal economic advantage to the needs and welfare of the group. 'Chastity' was regarded as a liberation of mind and body from the demands of human sexuality – a deliverance from 'human bondage' rather than 'human bonding' - while 'humility' did not mean slavish subservience to any particular individual, but submission of one's own will to the overriding needs of the group's long-tern goal or 'Plan'.

Monastics chose 'to resign from their unique human identities in order to transcend the collective limits of that identity' (Elson & Moore 1981). The various 'habits' adopted by monastics as standard dress represented a funeral shroud – a virtual death with respect to the 'human World' of the passions – while the adoption of a new 'monastic' name signified the loss of a previous worldly identity. As a monastic, one underwent 'metamorphosis' in the manner of a chrysalis, and was virtually 'reborn' into a new life. One was no longer a 'person' but an element of the power one served. Through subjugation of the passions ('apathy' in its patristic sense) to a 'higher' purpose, one achieved a positive state of synergy between emotional drives and cognitive insight – "the fires of apathy' writes Maximos the Confessor "illuminate the eyes of the heart".

Medieval monastic codes of behaviour closely approximate to what modern psychology would term 'anhedonic" – i.e. 'pleasure-denying' – a type of behaviour normally associated with what is loosely termed 'schizophrenic' or at least, 'schizotypal'. Historically, many pre-industrial societies retained special niches for 'schizotypal' members in order to profit from the potentially creative abilities that are so often linked to such personality types (Nichols 2009; Spikins 2009; Crespiu et al. 2007; Brüne 2004; Polimeni et al. 2003). The Middle Ages was an era in European civilization – the era of stabilization – when such a niche appears to have spontaneously emerged. Given the 'borderline' personality type (in modern terms) prevalent in this epoch (Leclercq 1973; McLaughlin 1976; deMause 1982) and the ability of monastic institutions to absorb such personality types, it is hardly surprising to find almost one third of the population of Western Europe in religious orders for most of this period (Davies 1997; Knowles op. ,cit.).

The psychodynamic genius of Bernard of Clairvaux (1090 – 1153 AD) contributed greatly to this phenomenon. This genius consisted in Bernard's ability to divert the erotic impulse away from primary goals and into the service of 'sublimated' cultural goals through the use of the biblical Song of Songs as a 'neurolinguistic' metaphor in his sermons and letters (Leclercq 1973; Evans et al. 1987 (see esp. pp. 54-55); Leclercq 1989). Bernard was also well aware that human sexuality is intimately linked with aggression – one cannot 'make love' without 'making war' since then latter always arises from the ultimate frustration of the former. In order to stabilize the annexation of the erotic impulse to the service of cultural goals, Bernard was compelled, as so many leaders before and since, to locate an 'enemy' as a focus social aggression. Then, as now, the Muslim world was deemed appropriate, hence Bernard's call for the Second Crusade. Modern opinion has not been kind to Bernard on this account – yet by what other means was he to persuade a quarrelsome group of feudal barons and as yet only superficially-converted 'barbarian' war- band leaders to unite and create the conditions for a strong Christian civilization? For homo sapiens in our present, genetically-unmodified state, morality relative is, unfortunately, relative.

An understanding of these 'functional' principles lying behind the monastic ideal is essential both for an understanding of how and why the monastic and military Orders sought and achieved power in the Middle Ages and of the complex roles they played in medieval history. Much can be learned here that would facilitate long-term survival and prosperity. Anyone familiar with modern Japanese corporate culture will have a good idea of the 'corporate' dynamics involved here. For an application of these principles to modern Western business practise see Dollard et al. (2002).

The founder of Western monasticism is considered to have been St. Benedict of Nursia (c. 480-550 AD) whose Rule became the rock on which the edifice of Western monasticism was built and is in its own right a work of profound psychological insight. By the end of the 11th century however, the Cluniac Order of the Benedictines (so-called after the Order's central abbey at Cluny in France) had acquired vast wealth and was in danger of absorption by the very social matrix whose formation it had so successfully catalysed. It was at this point in time that the Cistercian Order was founded, originally by Robert de Molesmes (c. 1027 – 1111), Stephen Harding (ob. 1134) and Alberic of Cîteaux (ob. 1109) about the year 1098, but it was the reception into the Order of Bernard of Clairvaux around the year 1110 that inaugurated the Cistercian Reformation (Benson et al. 1982; Tobin 19951; Burton & Kerr 2011). The 12th century witnessed a rapid expansion of Cistercian abbeys throughout Europe, many of which, such as Rievaulx Abbey in Yorkshire, becoming major centres of technological innovation in such domains as architecture, civil engineering, animal breeding and metallurgy (Fergusson & Harrison 2000) and by the middle of the century, the Order had come to dominate the main centres of learning in Western Europe.

The most significant, if 'hidden' event of this century however, may have been the dissemination of the prophetic visions of Joachim of Fiore (1135 – 1202). Born at Celico in the diocese of Cosenza in Calabria, Joachim was educated for diplomatic service at the Sicilian court and sent to Constantinople c.1167 as part of a diplomatic mission to the Byzantine Emperor Manuel I (Comnenus). Before the conclusion of this mission Joachim travelled to the Holy Land where he received 'a vision of the fullness of knowledge' on Mount Tabor (the legendary Mount of the Transfiguration). The results of these revelations were the Liber Concordiae Novi ac Veteris Testamenti (The 'Book of Concordance between the Old and New Testaments'), the Expositio in Apocalypsim ('Exegesis of the Book of Revelation') and the Psalter Decem Chordarum (The Ten-Stringed Psalter) all written from 1183-4. In addition, Joachim spent the remainder of his lifer working on the Liber Figurarum - a collection of images and geometrical figures intended to elucidate and summarise the theses contained in his written work. The complex illuminated codices of the Liber Figurarum present a philosophy of history geometrised in terms of a graph-theoretic 'topology of the mind' and from the psychohistorical perspective therefore. constitute an invaluable analytic resource, far exceeding in political significance and historical impact anything along the lines of the so-called 'Da Vinci Code'.

The 'Prime Directive' behind the prophecies is shown in Fig. 9 below – a tripartite scheme of past, present and future history called the Triplex Aetatum or 'Triad of the Three Ages'. Its significance is as follows:

1. Human history is divided into three successive Ages (in Joachimite terms, the Ages or Status (Lat. status (pl.) in the sense of epochs, aeons or psychospiritual 'conditions') of the Father, of the Son and of the Holy Spirit).
2. The 'First Age' – that of the Father – is that of genesis and 'tradition'. Humanity is currently situated in the Second Age - the age of struggle and transition, while the Third will be the 'New Age', ('Novis Ordo') bringing the resolution of all conflict and an endless 'spiritualised' state of human perfection.
3. The New Age will heralded by the victorious struggle of the God-anointed 'World Emperor' in alliance with the 'Angelic Pope' (i.e. the Jedi 'Force') over the 'Beast of the Apocalypse' and the Antichrist (the 'Sith' representing the 'Dark Side of the Force').
4. The New Age will be ushered in or 'catalysed' by two new monastic orders - one engaged in the active life, the other in the contemplative. Life.

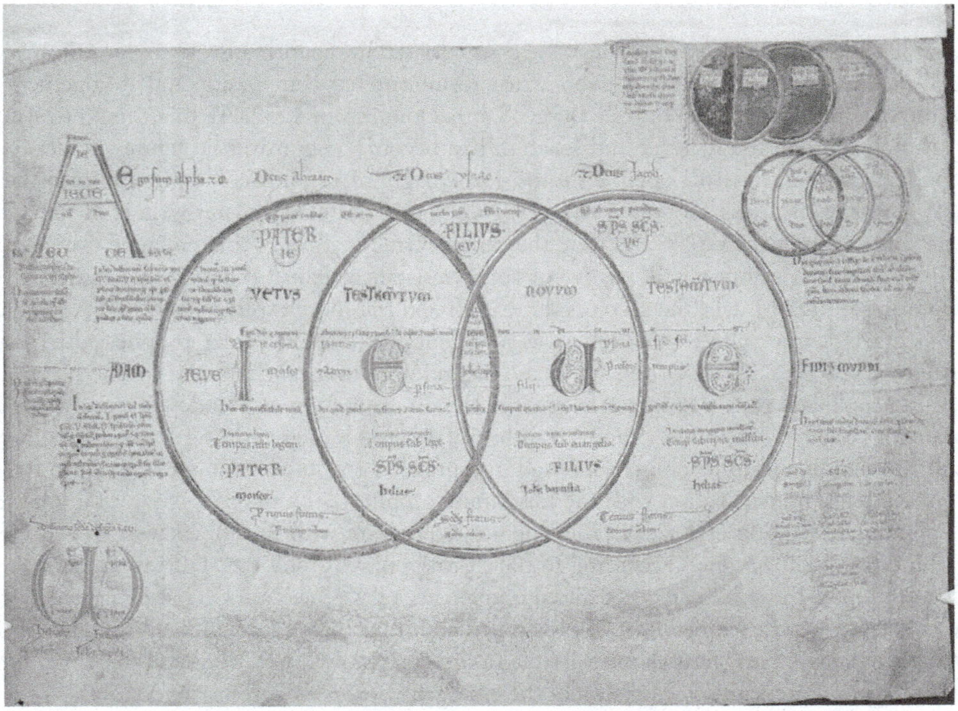

Fig. 9. The TRIPLEX AETATUM. Trinitarian Circles enclosing the Tetragrammaton from the Liber Figurarum of Joachim of Fiore (1135-1202 AD).

(Archives of Corpus Christi College, Oxford MS. 255A, f.7v).

In this way, Joachim created a synthesis of dynamic trinitarianism and existential dualism within a framework of historical immanence. The tripartite program clearly fulfils the foundational requirements of the dominant projective system, successfully embedding it in lived time and causing it to be re-experienced in every age by every individual who comes under its influence. The First Age is the 'age of the past', the past we all inherit at every moment, the 'Second Age' is the 'today' – the moment of struggle and transition, while the 'Third Age' is the eternally hoped-for future – the age of 'Utopia'. The structure corresponds to the 'ecological sequence' of the BPM, so every generation is primed to experience it. A graphic 'deconstruction' of the Triplex is shown in Fig. 10 below.

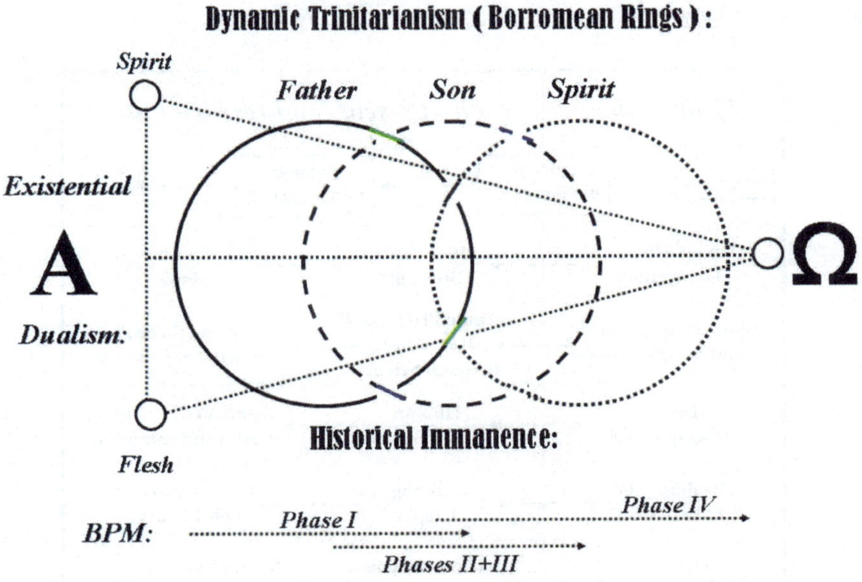

Fig. 10. Graph-theoretic interpretation of MS. 255A, f.7v

After Joachim's death, the prophecies were disseminated in the form of what became known as the Eternal Evangel - a reference to Rev.14: 6 and to an actual 're-interpreted' compilation and abridgement of Joachim's works issued under this title by Gerardo di Borgo Santo Donnino in 1252. At this time dissension and strife within the Church were on the increase and Joachim had become a controversial figure - a fact that eventually led to the condemnation of his works by the Protocol of Anagni in 1256. This did not diminish the emotionally felt canonical power of his ideas - a power that was consciously imitated by later writers of the apocrypha. The 'vector' for dissemination of the prophecies were the various Orders of Friars that emerged over the course of the 13th and 14th centuries as symptoms of passive or active resistance against ecclesiastical authority. The Franciscans, the Minorites, the Spirituals, the secretive Fratricelli and the more explicitly terrorist Apostoli founded by Seguarelli and Frà Dolcino - all saw themselves as representing the 'New Order' prophesied by Joachim which was to initiate the New Age. These Orders wandered far and wide over northern and southern Europe preaching to the common people and thus were crucial agents in the dissemination of pseudo-Joachimite ideas. The Friars themselves, anxious to justify their own canonical authority in the face of ecclesiastical disapproval, appealed to the prophecies in Revelation and to Joachim's supposed interpretations of them in order to endorse their own activities. Hence the intense preoccupation with the Joachimite prophecies on the part of later German Reformation thinkers. Explicit identification of stages in the Joachimite program with specific historical events led successive groups to re-interpret their program according to their own political ambitions. During the Enlightenment and Industrial Revolution, as the static and etherealised immanence of the medieval Civitas Dei receded from Western consciousness, the Joachimite 'embedding' of the construct within the context of lived time led to a quest for a sociopolitically-immanent 'New Jerusalem' realisable in the here and now. This became the 'germ' for all Western concepts of unilinear progress, the core catalyst of the emergent European nation-states and the foundation for all utopian and apocalyptic movements, including the Nazi and Comnmunist ideologies and, paradoxically, Al-Qaeda and all 'radicalised' Islamist movements (McGinn 1985; Ziolo 2001; Reeves & Gould 2001; Gray 2007).

7. Construct Evolution through Semiotic Permutation

The construct 'lies beneath' modern philosophies of both science and history thanks to the process of 'semantic permutation' that occurred following the 30-Years War of 1618-1648 and which permitted the construct itself to evolve. Through semiotic permutation the basic structure of the medieval dominant projective system was preserved through a mutation of its forms of discourse from the 'theological', through the 'metaphysical' to the 'scientific'. The 'Enlightenment' was never a question of 'Science vs. Religion', but rather a power struggle between two groups – the clergy and the 'philosophers', both appealing to the basic mythic forms of the same construct while veiling them in contradictory semiotic structures. The process is shown in Table 2 below. The first semiotic permutation occurred during the Enlightenment, while the second represents the 'background' assumptions current in much futuristic thinking today.

1100 - 1600	*1600 - Present*	*Present - Future*
Crystallisation →	Semiotic Permutation I →	Semiotic Permutation II ⋯→
Divine-Human Incarnation →	**Mind-Body Dualism** →	**Continuity Thesis**
Infinite Progress of the Soul towards God →	**Infinite Progress of Humanity through Science** →	**Expansion throughout the Known Universe**
Divine Omnipotence →	**Human Omnipotence** →	**Sentience as Prime Vector of Expansion**
Building the City of God →	**Building Utopia** →	**Convergence at Point Omega**
Dynamic Trinitarianism →	**Ancient → Medieval → Modern** →	**Kardashev Type I → Type II → Type III**

Table 2. Semiotic transformations of the Western dominant projective system from its inception to the near future

Transformations of this type are especially possible in the West due to the predominantly analytic character of Western European languages. The chief 'fallacy' engendered by analytic languages is that of nominalism, where because a 'name' is given to some concept, it is assumed that that concept has thereby acquired real material existence. Rates of linguistic evolution proceed much faster than those of cognitive evolution. Signifiers of a construct therefore change (surface variance) while the signified (deep structures) remain constant.

Successive levels of semiotic permutation are interpreted cognitively by analogy with the Hausdorff transform, where if { μ_p } (p = 0. 1, 2, …) is an arbitrary sequence of numbers, the double sequence {Δ n μp } is defined by:

$$\Delta^n \mu_p = \sum_{r=0}^{n} (-1)^r \cdot \binom{n}{r} \cdot \mu_{p+r} \quad (n, p = 0, 1, 2, \ldots)$$ Eq. 6,

these quantities being termed the 'differences' of the sequence { μp } since they satisfy the recurrence relations:

{Δ n μp } = μp (p = 0, 1, 2, …) Eq. 7

and

{Δ n μp } = Δ n – 1 μp – Δ n – 1 μ n + 1 (n = 0. 1, 2, … : p = 0, 1, 2, …) Eq. 8,

then the Hausdorff transform

$$t_p = \sum_{q=0}^{p} \binom{p}{q} \cdot (\Delta^{p-q} \mu_q) \cdot s_q \quad (p = 0, 1, 2, \ldots)$$ Eq. 9

is the transformation of the sequence { sp } generated by the sequence {μp } (Jakimowski 1956). In the case of the generalised transform (Hausdorff 1921; Leviatan 1968), if the sequence (λi) (i ≥ 0) satisfies:

0 = λ0 < λ1 < … < λn < …,

$$\lim_{n \to \infty} \Lambda_n = \infty,$$

$$\sum_{i=1}^{\infty} \frac{1}{\Lambda_i} = \infty$$

Eq. 10,

and if { μn } (n ≥ 0) is a sequence of real or complex numbers, the generalised transform { tn } (n ≥ 0) of a sequence { sn } (n ≥ 0) is given by:

$$t_n = \sum_{m=0}^{n} (-1)^{n-m} \Lambda_{m+1} \cdot \ldots \cdot \Lambda_n [\mu_m, \ldots \mu_n] s_m$$ Eq. 11,

where for 0 ≤ m ≤ n = 0, 1, 2, …,

$$[\mu_m, \ldots, \mu_n] = \sum_{i=m}^{n} \frac{\mu_i}{(\Lambda_i - \Lambda_m) \cdot \ldots \cdot (\Lambda_i - \Lambda_{i-1})(\Lambda_i - \Lambda_{i+1}) \cdot \ldots \cdot (\Lambda_i - \Lambda_n)}$$ Eq. 12.

Thus in the Western construct, while global and continuous symmetry transforms operate horizontally across the trinitarian hypostases ('personifications') of the three Ages (Father, Son and Spirit), they are simultaneously projected downwards through the ver etween discrete number series, but between n-dimensional matrices bounded only by the creative potentialities peculiar to the psychic ecology of the universality class that embodies them – in this case, a historically contextualised humanity.

$$i \xleftrightarrow{C_p} j : C_p = \{S_n\} \xrightarrow{\delta} \{t_m\}; t = (\delta(\mu)\delta)s$$ Eq. 13,

where

$$t_m = \sum_{n=0}^{m} (-1)^m [m! \, n! \, (m-n)!]$$

Eq. 14.

These permutations are also embedded in the history of science, where the function of the underlying construct is to protect the dominant paradigm and to contain innovation within acceptable bounds. As Jean Petitot states:

"There are certain problems in the history of science which are as yet unrecognised. Often, instead of simple steps from approximation to greater rigour, major scientific breakthroughs were in fact acts of containment which repressed key aspects of reality by binding them within webs of symbolic elaboration". (Petitot (1978) translated by the present author).

Thus if 'mainstream science' itself has all the properties of a projective system, its so-called 'controversies' are organized so as to create webs of this type – webs whose primary function is not to access 'reality-as-such' but rather to keep existential anxieties at bay and create an acceptable 'playing field' for the competitive pursuit of status and resources – the Darwinian 'imperatives' – that all too often absorb and neutralize the creative energies of emergent generations. This is why radical innovation, which 'transgresses the boundaries' of the construct, provokes so much conflict. A 'Kuhnian paradigm shift' forces a deep 'reconstruction' of this construct in the same way as the medieval construct was forced to transform and adapt during the transition from 'Renaissance' to 'Enlightenment'.

8. From Cathedral to Starship

As J.R.R. Tolkien has pointed out in his Introduction to the Elder Edda (Tolkien 2009), it is the archaic, preliterary oral forms of a culture's epic poetry that contain the essence of that culture's 'prime symbol'. During its period of genesis therefore (its first thousand years) the Judaeo-Christian construct was adapted to the deep mythological structures of Northern Europe. It has been suggested (Cohen et al. 2011) that only in the building of the great medieval cathedrals do we find an instance where human beings have proved capable of undertaking major transgenerational projects – and this point is well made. The medieval cathedral and the starship represent material projections of the same 'prime symbol' – an evolving image that began with the genesis of Western Gothic architecture in the 12th century and may culminate in an era yet to come. A continuity can be observed from images of primeval forest groves where the hunter-gatherers of Northern Europe assembled to celebrate their ancient rites, through to the masterpieces of Gothic architecture where the trunks, branches, leaved canopies and filtered sunlight of the sacred groves mutate into the ribbed and fan vaultings, flying buttresses and rose-windows of the great cathedrals (Fig. 11 below).

Fig. 11. From sacred forest grove to Gothic cathedral architecture

This continuity of architectural evolution was sustained through to the present where the 'cathedral' appears to prefigure the 'starship' itself Fig. 12 below) and the sanctuaries and altars where priest and people came face to face with the 'Transcendent' morph into the 'starship bridge' where 'captain' and 'crew' come face to face with the Cosmos (Fig. 13). Indeed as we have seen, rather than 'rejecting' the physical world, the medieval Christian construct above all sought direct engagement with 'matter' and its purposeful transformation through praxis – with the doctrine of the Incarnation forming the bridge between Matter and Spirit, or between Earth and the transcendent realm. As we have seen, the deeply rooted psychological structures of this construct were conserved throughout the Renaissance, Reformation and Enlightenment through a transmutation of discourse from the realm of the theological to that of the scientific.

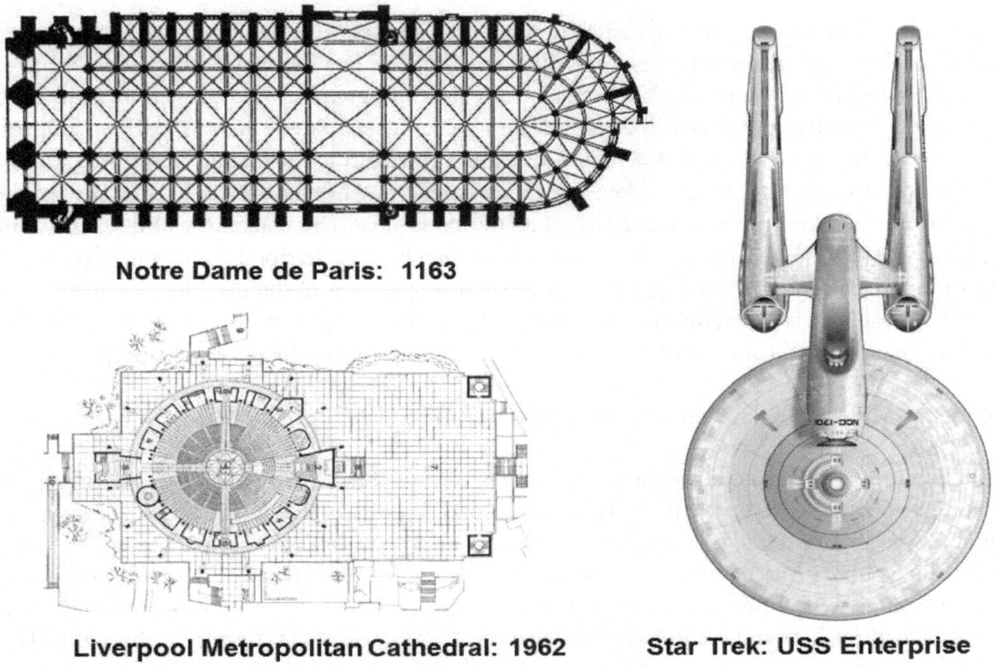

Fig. 12. From Gothic cathedral architecture to Starship

Fig. 13. From Gothic cathedral altar to Starship bridge

9. Conclusion

Let us consider two contrasting visions of the human future:

I. "Those who live and work permanently in space will be distinguished by self-discipline, their ability to meet the requirements of the Machine, to repress fear, desire, all emotion. They will belong to a breed apart, cut off from family, the consolations of love and sexuality. After they have endured the initial ordeals of training, their physicality will be measured by the requirements of the Machine. Life in space ships will be as closely regulated and lonely as life in a monastery: the greatest virtues will be those of restraint, not excess. They will voluntarily cut themselves off from earthly satisfactions in order to approach closer to the infinite". 127]

(Elson, P. & Moore, C. (1981). Parallel Lines: The Science Fiction Illustrations of Peter Elson & Chris Moore. Quote from the Introduction by Pat Vincent, p. 16).

II. "The cultural era is past. The new civilization, which may take centuries or a few thousand years to usher in, will not be another civilization – it will be the open stretch of realization which all the past civilizations have pointed to. The city, which was once the birth-place of civilization, such as we know it to be, will exist no more. There will be nuclei of course, but they will be mobile and fluid. The peoples of the earth will no longer be shut off from one another within states but will flow freely over the surface of the earth and intermingle. There will be no fixed constellations of human aggregates. Governments will give way to management, using the word in a broad sense. The politician will become as superannuated as the dodo bird. The machine will never be dominated, as some imagine; it will be scrapped, eventually, but not before men have understood the nature of the mystery which binds them to their creation. The worship, investigation and subjugation of the machine will give way to the lure of all that is truly occult. This problem is bound up with the larger one of power – and of possession. Man will be forced to realize that power must be kept open, fluid and free. His aim will be not to possess power but to radiate it."

(From Henry Miller's Sunday After the War, cited in Brown (1985) p. 305).

Both these visions aspire to the prophetic and are unquestionably 'religious' in tone. The first quotation represents the philosophy of space exploration while the second represents the 'Earth-Child' philosophy. Which is more 'Utopian'? We are often told that we should 'solve all our problems here on Earth' before we go 'bombing off into the wild blue yonder'. But 'natural', i.e. genetically unmodified, humanity will never 'solve its problems here on Earth' since dominance, submission, aggression and the struggle for resource control, status and sexuality are all core aspects of a 'natural' human psychology. If we had our 'utopia' we would destroy it through sheer boredom. Radical modification of the genome would appear to be a prerequisite for any realistic 'peace on Earth' – and this is not a solution likely to find favour with the majority.

Permanent habitation of space on the other hand will require an even greater 'long-jump adaptation' on the part of humanity. As the first of the above quotations makes clear, transcending the constraints of one evolutionary trajectory will simply require us to deal with unknown constraints of the newly-chosen trajectory. Such a transition will present fierce challenges – challenges that only be overcome with the aid of a suitably modified construct. Whether those who accomplish this transition are destined to exult or weep in their transcendence – this will be their own affair, and no concern of 'the rest of us'.

The human future in deep space (as opposed to LEO) will require a pioneer society that will be polycentric, metacomplex and cyborganic. 'The Machine will not dominate the Human: Human and Machine will become One' (Sandberg 2013). For now at least, the Machine is capable of a virtually unlimited (from our present context) range and rapidity of computation - but lacks motivation and the power to take large-scale evolutionary decisions. The Human on the other hand, thanks to a long evolutionary and morphogenetic history, possesses an intense motivational drive. Although 'computational power' for the human is slower and more restricted in range than that of the Machine, the massive parallelism of the biological brain and the evolutionary sources of motivational drive make an alliance with the Machine – i.e. 'merging' with the machine as the construct dictates – a powerful alliance for the permanent colonisation of space.

What strategy is best adopted by those 100YSS members whose 'outreach' territory includes religious organisations? Certainly not that of 'inventing' some new 'Galactic religion'. Rather, we might follow the example of Kuah Khun-Eng's 2009 study (and that of the Bene Gesserit as outlined in the Dune Encyclopedia (McNally) and selectively emphasise those elements of whatever religious field we are involved with that promote teleology, transcendence, and human 'synergy' with the Creator.

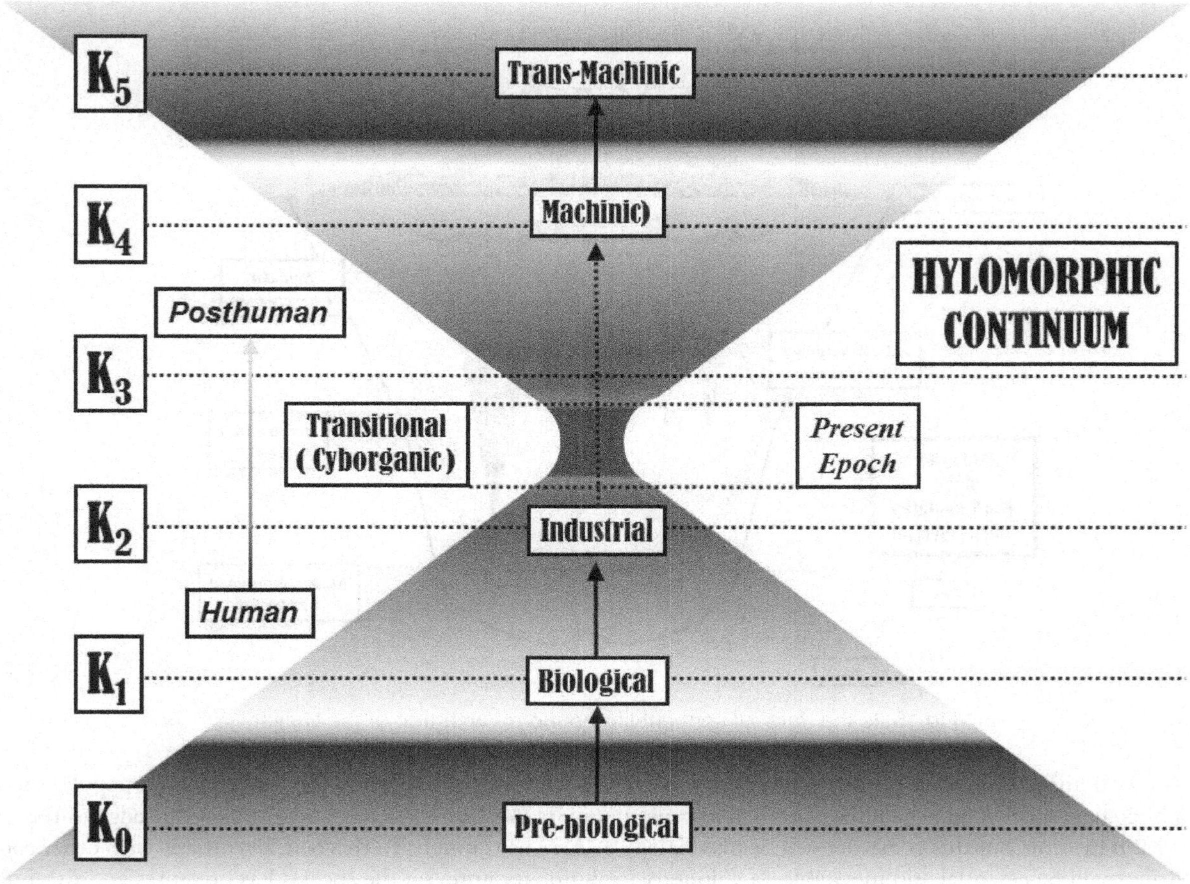

Fig. 14. Projected Kardashev levels in the evolution of a sentient species

As a species, we are in no position at present to make any final statements about the 'meaning of life' - but the directionality of evolution is evident, even from a cursory study, even if that directionality appears chaotic and frequently destructive. It is possible that from a far more 'advanced' perspective, such chaos and destruction is simply part of the statistical process necessary for the emergence of advanced intelligence throughout the cosmos – 'many are called but few are chosen' (Matt. 22:14). Fig. 14 above shows a sequence of possible Kardashev levels a hypothetical sentient species would traverse from the point of its emergence from within a planetary biological matrix to the most 'advanced' level – K5 –that we could realistically imagine at present. The most radical transition is the 'long-jump' adaptation from a planetary (biological) environment at K0 to a space (posthuman) environment at K2 – the 'bottleneck' of Fig, 14. If we succeed in making this transition, well and good, but should we fail, this failure would be of no ultimate consequence for the overall 'Galactic Plan'. And what is this Plan? We cannot guess at present, but it may be safe to assume from the evidence we have (as do many hylomorphic-hylopathic philosophies) that the ultimate 'fate' of sentient life may be bound up with that of the Universe itself (Fig. 15 below). No moral prescriptive is implied here – participation in an evolutionary advance of this type is a matter of species choice.

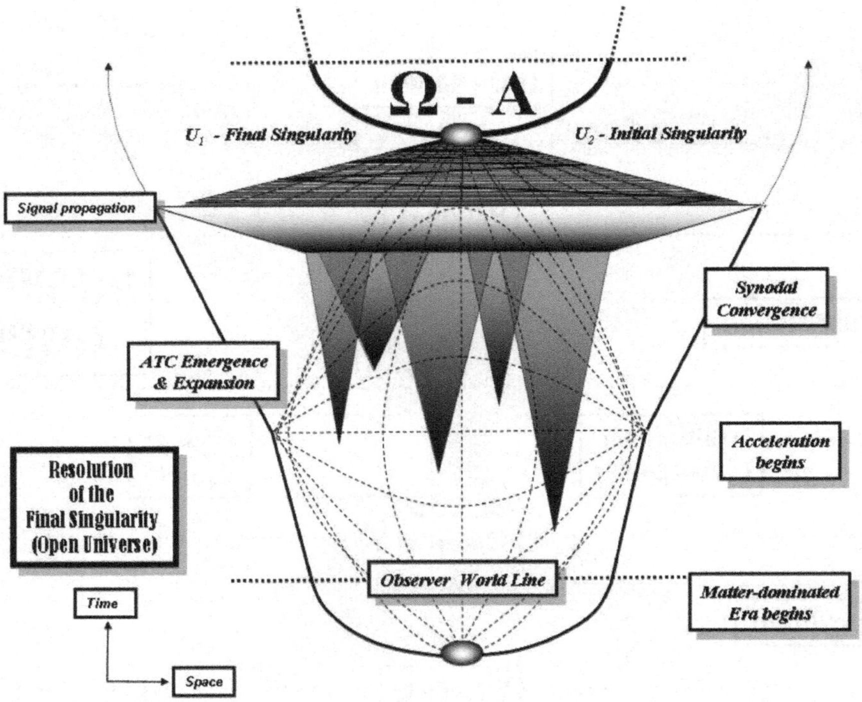

Fig.15. Is the 'fate' of sentient life bound up with that of the Universe?

A 'dominant projective system' may be regarded as a 'metaprogram' guiding the evolution of a population or a biological species, and the Judaeo-Christian construct in its Western form has brought us to the edge of the K0 – K2 transition, but this construct will be insufficient to carry us through. To this end, a synthesis with other constructs will be essential and inevitable– a voluntary evolutionary jump on the species level requires an existential and experiential synthesis from all participants in that jump.

The main 'opposition' between the philosophies of East and West appears to derive from each tradition's approach to the question of the fundamental impermanence 'matter' (> Lat. 'mater' = 'mother'). In the West, this impermanence is denied through an attempt to fixate transient phenomena within the bounds of an analytic language. In the East, 'impermanence' is denied simply by denoting all material manifestations as 'illusion' or 'maya' (Fig. 16).

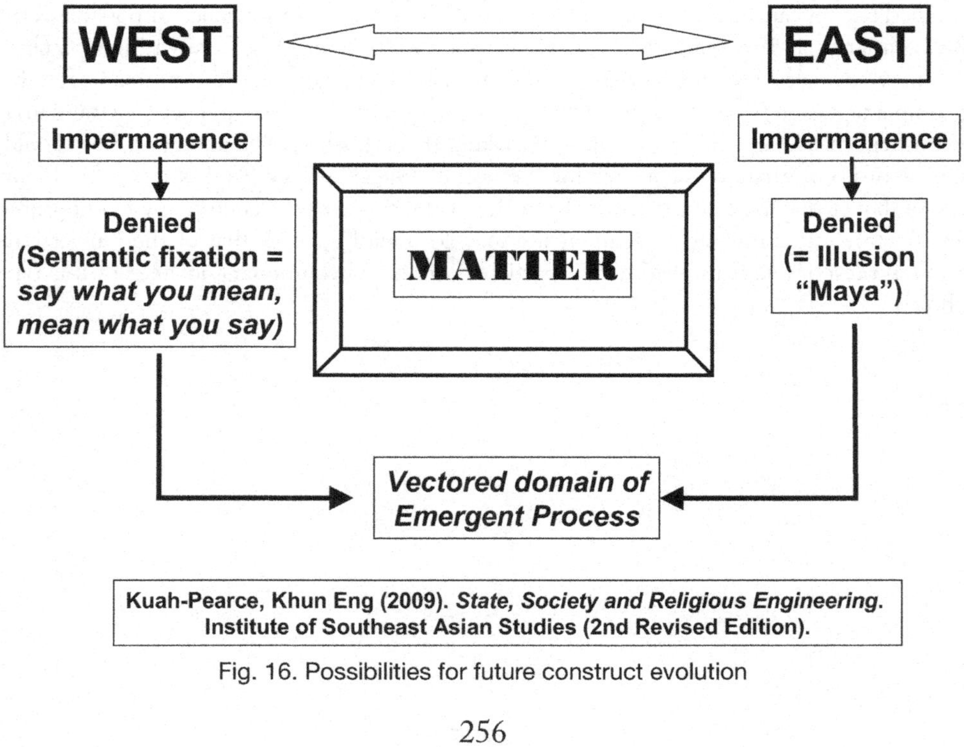

Fig. 16. Possibilities for future construct evolution

To combine the deep cognitive-emotional disciplines of the East with the West's commitment to the dynamic transformation of matter-as-lived process would result in a psychological 'thermonuclear bomb'.

From the Joachimite prophecies, a dramatic history leads us from the North Yorkshire moors to the mountains of Sicilian Calabria, from the forests of Northern Germany to the archives of Corpus Christi, Oxford and thence to the sands of Cape Canaveral. The historical record serves only to demonstrate the moral neutrality of a construct as construct, discussed above. 100YSS members, like most space advocates, are heirs to the Joachimite prophecies, even if few are explicitly aware of the fact. Since this inheritance has come to us, the question is – do we have the intelligence to benefit from historical experience and use such an instrument to best advantage.

Future interstellar pioneers will not be 'natura;', unmodified humans. At present our constructs reflect our evolutionary history, our morphology and modes of epigenesis, but posthuman or cyborganic evolution will challenge the most basic bonds the define our 'humanity' – those between male and female, between parent and child, between biosphere and species, between genes and culture – and between life and death (Braidotti 2013). Cyborganic evolution will force us to redefine these bonds and require that we come to terms with the infinite mutability and contextuality of Being. Genetics is the real 'language of God' and constructs are the metaprograms written in that language that direct the currentrs and flows of matter across time and through space. The time may come when we no longer need to rely on the salvific promise of constructs and can engage consciously and confidently with those vectored domains of emergent process of which we are so intimate a part. That time, however, is not yet.

References

1. Ashby, W. R. (1952.) Design for a Brain. Chapman & Hall, London.
2. Benson, R. L., Constable, G. & Lanham, C. D. (1982),(eds). Renaissance and Renewal in the Twelfth Century. Harvard University Press, Cambridge, Mass.
3. Binion, R. (1997). Sounding the Classics: From Sophocles to Thomas Mann. Praeger, Westport CT.
4. Braidotti, R. (2013). The Posthuman. Polity Press, Cambridge, UK.
5. Breck, John (1994). The Shape of Biblical Language: Chiasmus in the Scriptures and Beyond. St. Vladimir's Seminary Press, Crestwood, New York.
6. Breck, J. (1994). The Shape of Biblical Language: Chiasmus in the Scriptures and Beyond. St. Vladimir's Seminary Press, Crestwood, New York.
7. Briend, A. (1979). 'Fetal Malnutrition - The Price of Upright Posture?'. British Medical Journal 1979, 2 pp. 317-19.
8. Brown, N. O. (1985). Life Against Death. Wesleyan University Press, Middletown CT.
9. Brueggemann, W. (2000). 'The Prophetic Word of God and History'. Texts That Linger, Words That Explode: Listening to Prophetic Voices. Fortress Press, Minneapolis.
10. Cohen, M., O'Donnell, D. J., Becker, R. F. & Brody, A. (2011). 'Interstellar Sweat Equity'. http://www.astrotecture.com/Interstellar_-_Exoplanets_files/20110708.Abstract.Update.Interstellar_Sweat_Equity.Public%20Symposium.pdf. Accessed 22/02/2013: 15.23 GMT.
11. Crutchfield, J. (1994). 'The Calculi of Emergence'. Physica D. Proceedings of the Oji International Seminar: Complex Systems – from Complex Dynamics to Artificial Reality.
12. Danesi, M. (ed.) (1995). Giambattista Vico and Anglo-American Science. De Gruyter Mouton. Reprinted in 2011.
13. Davies, N. (1997). Europe: A History. Pimlico, London.
14. Durant, W. (1944). The Story of Civilization. Vol.3: Caesar and Christ. Simon & Schuster, New York. Quotation from the Epilogue: Why Rome Fell.
15. Deleuze & Guattari, F. (1987). A Thousand Plateaus. University of Minnesota Press.

16. Diamond, J. (2005). Collapse: How Societies Choose to Fail or Succeed. Penguin Books, New York.

17. Dollard, K., Marett-Crosby, A. & Wright, T. (2002). Doing Business with Benedict: The Rule of St. Benedict and Business Management: A Conversation. Continuum Press. London & New York.

18. Eisenstadt, S. N. (1982). 'The Axial Age: The Emergence of Transcendental Visions and the Rise of Clerics'. Archives Européenes de Sociologie 23 (1982) pp. 294-314.

19. Elm, S. (1994). Virgins of God: The Making of Asceticism in Late Antiquity. Oxford University Press.

20. Erikson, E. (1956). 'The Problem of Ego Identity'. Journal of the American Psychoanalytic Association, 4, 1 pp. 56-121.

21. Evans, G. P., Leclercq, J. & Cousins, E. (1987). Bernard of Clairvaux : Selected Works. Paulist Press, Mahwah, New York.

22. Ferenczi, S. (1989). Thalassa: A Theory of Genitality. Maresfield Library, Karnac Press, New York.

23. Fergusson, P. & Harrison, S. (2000). Rievaulx Abbey. Community, Architecture, Memory. Yale University Press.

24. Fodor, N. (1949). The Search for the Beloved: A Clinical Investigation of the Trauma of Birth and Prenatal Condition. University Books, New Hyde Park, New York.

25. Freud, S. (1921). Group Psychology and the Analysis of the Ego (1921). SE vol. 18 (1955). Penguin Freud Library vol. 12 pp. 91-178.

26. Friston, K. & Kiebel, S. (2009). 'Cortical circuits for perceptual inference'. Neural Networks 22, pp. 1093-1104.

27. Friston, K. (2008). 'Hierarchical Models in the Brain'. PloS Computational Biology, vol. 4: 11, pp. 1-23.

28. Friston, K. (2010). 'The free-energy principle: a unified brain theory?'. Nature Reviews | Neuroscience AOP doi: 10. 1038/nrn2787.

29. Froese, T. & Di Paolo, E. A. (2009). 'Sociality and the life-mind continuity thesis'. Phenomena in Cognitive Science, DOI 10.1007/s11097-009-9140-8.

30. Galgan, G. (1982). The Logic of Modernity. New York University Press, New York.

31. Gardner, J. (2007). The Intelligent Universe: AI, ET and the Emerging Mind of the Cosmos. New Page Books, NJ.

32. Gray, J. (2007). Black Mass: Apocalyptic Religion and the Death of Utopia. Penguin Books.

33. Grof, S. (1975). Realms of the Human Unconscious: Observations from LSD Research. Viking Press, New York.

34. Haken, H. (1996). Principles of Brain Functioning: A Synergetic Approach to Brain Activity, Behaviour and Cognition. Springer-Verlag, Berlin & Heidelberg.

35. Hannam, J. (2010). God's Philosopher's: How the Medieval World Laid the Foundations of Modern Science. Icon Books.

36. Hausdorff, F. (1921). 'Summationsmethoden und Momentfolgen I umd II'. Math. Zeit. vol. 9 (1921), pp. 74-109, 280-288.

37. Jaki, S. (1986). Science and Creation: From Eternal Cycles to an Oscillating Universe. Scottish Academic Press, Edinburgh.

38. Jakimowski, A. (1956). 'A Note on Hausdorff Transforms'. Proceedings of the American Mathematical Society vol. 7 no. 5 (Oct.) pp. 803 – 7.

39. Janus, L. (1997). The Enduring Effects of Prenatal Experience: Echoes from the Womb. Jason Aronson, Northvale NJ & London.

40. Knowles, D. (O.S.B) (1949). The Monastic Order in England: A History of its Development from the Times of St. Dunstan to the Fourth Lateran Council 943-1216, Cambridge University Press.

41. Khun Eng, Kuah, (2009). State, Society and religious Engineering: Reformist Buddhism in Singapore. Institute of South-East Asian Studies (2nd Edition).

42. Leclercq, J. O.S.B (1973). 'Modern Psychology and the Interpretation of Medieval Texts'. Speculum, vol. 48 (1973) pp. 476-490.

43. Lem, S. (1984). Microworlds. Harcourt Brace & Co., New York.

44. Leviatan, D. (1968). 'Moment problems and quasi-(Hausdorff) transformations'. Can. Math. Bull. vol. 11 (1968), pp. 225-36. DOI.10.4153/CMB-1968-026-7.

45. Lifton, R. J. (1974). 'The Sense of Immortality: On Death and the Continuity of Life'. Explorations in Psychohistory: The Wellfleet Papers. (Lifton, R & & Olson, E. (eds.) (1974). Simon & Schuster, New York pp. 271-87.

46. McGinn, B. (1985). The Calabrian Abbot: Joachim of Fiore in the History of Western Thought. MacMillan Publishing Company, New York.

47. McLaughlin, M. M, (1976). 'Survivors and Surrogates: Children and Parents from the Ninth to the Thirteenth Centuries', in The History of Childhood: The Evolution of Parent-Child Relations as a Factor in History, DeMause, L. (ed.), Souvenir Press (Educational & Academic) Ltd. London, 1976, ch.3 pp. 101-181, p. 171 n.182 and n.186.

48. McNally, W. (1984). The Dune Encyclopedia. Corgi Books.

49. Mott, F. J. (1948). The Universal Design of Birth: An Analysis of the Configurational Involvement of Birth and Its Relation to Emergence Generally. David McKay Co., Philadelphia.

50. Mott, F. J. (1964). The Universal Design of Creation. Mark Beech, Edenbridge, UK.

51. North, D. (1997). Understanding the Process of Economic Change. Princeton University Press.

52. Peerbolte, L. (1951). 'Psychotherapeutic Evaluations of Birth-Trauma Analysis'. Psychiatric Quarterly 25, pp. 596-600.

53. Perry, B.D., Pollard, R., Blakely, T., Baker, W. & Vigilante, D. (1995). 'Childhood trauma, the neurobiology of adaptation and 'use-dependent' development of the brain: how 'states' become 'traits'. Infant Mental Health Journal, Vol. 16 no. 14 (Winter), pp. 271-91.

54. Petitot, J. (1978). 'Sur le Modèle Historique de Thom-Pomian'. Mathématiques et Sciences Humaines, 16e année no 64, pp. 43-70.

55. Ployé, P. M. (1973). 'Does Prenatal Mental Life Exist?'. International Journal of Psycho-Analysis 54, pp. 241-246.

56. Propp, V. (1968). Morphologyof the Folktale. Second Edition, University of Texas Press.

57. Quigley, C. (1979). The Evolution of Civilizations: An Introduction to Historical Analysis. Liberty Fund Inc. (2nd. Edition).

58. Rank, Otto (1923). The Trauma of Birth. Harper & Row, New York, 1973.

59. Reeves, M.& Gould, W. (2001). Joachim of Fiore and the Myth of the Eternal Evangel in the Nineteenth and Twentieth Centuries. Oxford University Press.

60. Sagan, C. & Agel, J. (2000). The Cosmic Connection: An Extraterrestrial Perspective. Cambridge University Press.

61. Sandberg, A. (2013) 'We Borg: Speculations on Hive Minds as a Posthuman State'.: http://www.aleph.se/Trans/Global/Posthumanity/ WeBorg.html. Accessed 21/02/2013.

62. Silber, I. F. (1995). Virtuosity, Charisma and Social Order: A Comparative Sociological Study of Monasticism in Theravada Buddhism and Medieval Catholicism. Cambridge University Press, Cambridge.

63. Spengler, O. (1928). The Decline of the West (Die Untergang des Abendlandes). Alfred A. Knopf Inc. Abridged Edition - George Allen & Unwin Ltd., London 1961.

64. Spinks, R., Harinder, K. S., Andreasen, N. C. & Philibert, R. A. (2004). 'Association of the HOPA12^{bp} allele with a large X-chromosone haplotype and positive symptom schizophrenia'. American Journal of Medical Genetics Part B: Neuropsychiatric Genetics: May 15 2004, 127 (1): pp. 20-7.

65. Stromberg, P. G. (1993). Language and Self-Transformation: A Study of the Christian Conversion Narrative. Publications of the Society for Psychological Anthropology, Cambridge University Press.

66. Tobin, S. (1995). The Cistercians: Monks and Monasteries in Europe. The Herbert Press, Ltd.

67. Toulmin, S. E. (1992). Cosmopolis: The Hidden Agenda of Modernity. University of Chicago Press. Reprinted. 2nd Edition.

68. Turchin, P. (2003). Historical Dynamics: Why States Rise and Fall. Princeton University Press.

69. Verny, T. & J. Kelly (1982). The Secret Life of the Unborn Child. Sphere Books Ltd., London.

70. Vico, G. (1744). The New Science of Giambattista Vico: Unabridged Translation of the Third Edition (1744) with the addition of "Practice of the New Science". Bergin, Thomas G. & Fisch (trans.), Max H. Cornell University Press. Ithaca & London 1948. Revised 1968, 1984. Reprinted, 1986.

71. Wasdell, D. (1985). Towards a Unified Field Theory of Human Behaviour: A Collection of Four Papers. URCHIN (Unit for Research into Changing Institutions), Meridian House, London.

72. ---, (1990). The Roots of the Common Unconscious. URCHIN (Unit for Research into Changing Institutions), Meridian House, London.

73. ---, (1992). The Pre- and Perinatal Grounds of Capitalism and the Free Market Economy. URCHIN (Unit for Research into Changing Institutions), Meridian House, London.

74. Weber, M. (1958). The Protestant Ethic and the Rise of Capitalism. Scribner, New York.

75. Wilken, R. L. (2012). The First Thousand Years: A Global History of Christianity. Yale University Press. New Haven & London.

76. Woods, T. E. (2012). How the Catholic Church built Western Civilization. Regnery History.

77. Ziolo, P. (2001). 'Joachim of Fiore and Apocalyptic Immanence'. Journal of Psychohistory 29 (2) pp. 186-224.

78. ---, (2002). 'The Psychodynamics of the ECF-Nexus: Monasticism and Psychospeciation in Western Europe, c.500-1500 AD'. The International Journal of Psychotherapy vol. 7:. 3, pp. 221-48.

79. ---, (2004). Psychohistory: Emergence, Theory and Applications. PhD. Thesis, University of Liverpool, ch. 5: 5, pp. 101-107.

2014

100 YEAR STARSHIP™
canopus 310 LY
2014 PUBLIC SYMPOSIUM
Pathway *to the* Stars, Footprints *on* Earth

100 YEAR STARSHIP®

2014 Public Symposium

Pathway to the Stars, ***Footprints on Earth***

Symposium Overview

CURVING. CROOKED. 90 DEGREES LEFT TURN. SPLITS. UPHILL. DOWNHILL. RETROGRADE. DEAD ENDS. FRESHLY HEWN. BUMPY. SMOOTH. CROWDED. ISOLATED.

The road to achieve the capabilities for human interstellar travel will not be straight, flat, unidirectional, one dimensional or even predictable from year to year.

Yet, I believe the requirements to discover a pathway, and to travel upon it until the ability to establish human life outside our solar system is assured, are quite clear. Requirements include, first and foremost, flexibility and openness. As we purposefully commence the journey, it is especially essential to reconsider how knowledge and ideas are categorized and presented. It is vital that we are willing to re-evaluate the "accepted methods" to achieve objectives. We must question whether technology and society rubrics, as currently "evolved," are the optimal concepts and techniques for an interstellar undertaking or *even living here on Earth. Such were major elements of the 100YSS 2014 Public Symposium.*

Breaking Bad

Creating Predictive Models For Psychological Breaking Points During Long-term Space Travel

Alires J. Almon, M.A.

Abstract

Our human curiosity to discover what's out there requires us to journey further and longer than we ever have before. The question of "Who gets to go?" is one that has everyone wanting to "take the test" to be selected. Great! You are selected. Have a seat because you are going to be together for a long time. As with any group, there are times when the relationships fail, an individual is traumatized, or other social and personal breakdowns occur. When these events or incidents occur, we now have a mission in jeopardy—not due to hardware failure, but because a psychological breaking point has been reached. It will happen—the question is when. The purpose of this paper is to identify signs of when breakdowns will happen and allow the crew to minimize its impact when it does.

A breaking point is reached when an individual is not able to consistently and effectively contribute to work efforts and/or maintain functional relationships with the crew. This paper focuses on the psychological and psychosocial characteristics that lead to individual and group "breaking points". The thesis presented is that the longer a travel mission, the earlier in the mission the group/individual will have a breakdown. This paper begins to create definitions and qualities of breaking points, define the impact on group dynamics, and identify post-break recovery strategies, which will be used as a foundation for predictive model development. Physiological constraints lead to both physical and psychological predictable impacts. However, interstellar mission constraints make this type of travel difficult, such as extensive travel time, no crew changes, eventual loss of communication with home, and the possibility (reality) of no return. Research draws upon typical areas such as analog isolation studies, group dynamics, and physiological impacts, but also incorporates historical explorer activities, grief and Post Traumatic Stress Disorder (PTSD) studies, and organizational development theories.

Keywords

group dynamics, psychology, mental health, predictive models, mission types, physiology.

1. Introduction

Hubris is often the downfall of good intentions. It is a term from the Ancient Greeks meaning extreme arrogance. Used as a focal point in Greek tragedies, it is the arrogant belief that one's power is unassailable. This is the premise of the popular culture United States television show *Breaking Bad*. The main character Walter White begins a nefarious side career as a drug dealer in order to prepare his family financially for his inevitable death due to an aggressive cancer diagnosis. To his friends and family, he is the meek husband of an overbearing wife. His career is that of a high school chemistry teacher. As with most individuals their story is much more layered than that. We find out that he is an extremely talented chemist with an advanced degree, and in his decision to manufacture drugs he is able to create a product that is pristine and in high demand.

His good intentions quickly require decisions and actions that begin to compromise the integrity of his initial intent. He becomes more demanding, he makes life choices for others, and he makes more money that could ever manage or spend in his lifetime. Walter does not realize that his behavior change is noticeable and affecting others in his life. As an audience, we see his descent into this new personality as a fairly quick and dramatic turn (within one year). His "co-workers" also notice a change as he enters the new work environment, although he did not have an alter ego that they were aware. With them he has a breaking point and new direction.

With hindsight, some behavior changes can be predicted. In this case, family welfare and fear turned into greed and hubris as time went on. Sometime during his journey as a drug dealer, Walter White had a breaking point where his behavior turned from helpful to harmful and he became "Heisenberg". While his intent was still the same – to provide a financial stability for his family when he died, when circumstances changed when death was no longer imminent, he continued his behavior to a point where he broke his current path. He reached his breaking point, and broke bad. This scenario, while fictional, showcases the impact of a sudden behavioral change, in this case a "break for the bad." An individual starts out with good intentions and then is suddenly faced with circumstances that were unseen and unexpected and reacts in a way that produces a negative effect for themselves and those around them.

Extreme conditions such as long-term space travel will create an environment where an individual is confined, without a change of scenery or personnel, or the idea of going home and the realization of this environment may create a challenge for them to stay positive and on a path of their initial intent to be positive and productive. Which is being defined as the intent to be a useful contributor to the crew and the social society develops as a result of the journey.

This paper will look at the circumstances that create an environment where an individual will be challenged in their ability to maintain optimal physical health, focus on their work and tasks supporting the group mission, and maintaining effective and meaningful interpersonal and intrapersonal relationships. Long-term space travel, while replicated by analog studies on Earth, creates a strong speculative basis for predicting behavior in certain circumstances. They do not always take into account the unique conditions that long-term space travelers will encounter.

2. Psychological Studies and Space

Psychology is the study of people and their behaviors. Space travel has unique impact on the behaviors of those who travel. Many astronauts have noted the experience of awe and wonder at the experience of space. Space shuttle Astronaut Kathryn D. Sullivan shared her account looking at the Earth from space: "I am happy to report that now amount of prior study or training can fully prepare anybody for the awe and wonder this inspires". (Robinson, et al., 2011). It is this awe and wonder that drives us to be explorers and pioneers into the unknown. The notion of the unknown puts individuals into environments that are not always conducive to the individuals being in their optimal state both physically and emotionally.

Psychological stressors resulting from both physical and mental environments affect an individual's behavior and decision-making. The amount of attention to stressors has grown as space programs have learned the value of the psychology in space exploration. While the beginnings revolved around the early space race, the evolution of those studies has taken us all the way to Mars. Douglas Vakoch (2011) edited a historical view on psychology and space exploration, which outlines the role of psychology and space programs. Three sources of information educate us about important psychological and psychiatric issues affecting human space missions: anecdotal reports, space analog and simulation studies on Earth, and research conducted during actual missions. (Kanas & Manzy, 2008).

The first wave of research focused on the ability to do tasks and individuals who had the ability to do tasks. Supporting activity was focused on human factors and other "ability-focused" actions. Harrison and Fiedler (2009), observed that NASA managers and engineers as are "thing" people rather than "people" people, so the behavioral side of space flight is of little interest to them. And, for many years, the psychological side of space travel was ignored.

The next wave of studies began to look at the importance of "fit" versus skill for mission success. The dynamics of mission success was expanded to include personality fits as well as skill fits. (Bishop, 2011).

Analog studies provide an experience that mimics certain conditions. Study results provide a foundation for extrapolative applications for long-term space travel. In the case of space, researchers have conducted studies such as surveying extended mission submarine crews and experimental studies, such as those who are asked to live in isolated and extreme conditions of Antarctica. Antarctic experiments have been quite successful in bridging the gap of skill vs. fit. Bishop (2011) cites Palikans who states:

> "The cumulative experience with year-round presence in Antarctica makes it an ideal laboratory for investigating the impact of seasonal variation on behavior, gaining understanding about how biological mechanics and physiological processes intact, and allowing us to look at a variety of health and adaptation effects."

These studies examine the psychological effects of these conditions. Not only are individuals tested on their ability to live and work in these environments, but also their ability to get along with others in those same conditions. It is the understanding of the impact of behavior that begins to gain traction as a research topic.

A recent analog study, MARS 500, simulated a 520 day mission Mars where activities included communication delays, on-planet task completion, and real-time frame mission. Researchers studied the actual psychological impacts and needs for this mission. They found results that will necessitate a change in the way crewmembers are selected and trained for such a mission. (Ushakov, Morukov, Bubeev, Vasil'eva, Vinokhodova, & Shved, 2014)

Once recognizing the importance of behavior in mission success, studies began to explore all the variations of behavioral impact, for example, the personality types that are best fit for extreme conditions (Bishop, 2011), and simulated research of negative group interactions (Wichman, 2011). Salutogenic approaches to mental health focus on the factors that support the health and well-being of an individual, rather than the factors that cause disease. In the case of utilizing an often mission-required task of photographing earth to help astronauts experience the awe of their position and the uplift it offers. Providing activities that give positive mental experiences rather than trying to fix a broken individual can be a more effective to long-term mental health wellness. (Robinson, et al., 2011). Each approach validates the impact of the psychological state of the crewmembers on the success of the mission.

Space programs have expanded in complexity – from being single country endeavors, to global cooperation efforts. Space crews are no longer one gender or culture. Space exploration is no longer a male-dominated field. Women have begun to join crews that provide a different kind of challenge to psychological health of crewmembers. Group and crew interaction is now impacted by gender roles, and the evolution of those roles in society and at work. Women are now peers and leaders of their male counter parts, and the workplace transition has been bumpy, both on Earth and in orbit. Likewise, the breadth of research has evolved into multi-disciplinary and trans disciplinary studies. Kring & Kaminsky (2011) note that due to the inherent differences in behaviors and abilities, mixed gender crews have performed effectively, both in space and in similar settings like Antarctica. They have assured us that a crew composed of both men and women is the right choice for extended missions to the International Space Station, rendezvous missions with asteroids, and one day, the first human mission to Mars.

Global missions will require another layer of adjustment: cross-cultural relationships. Although individuals may be trained for space flight, their own cultural biases will have an impact on their approach to their work and the development of interpersonal relationships. There are fundamental differences between American, Asian, European and Slavic cultures, and the many cultures in-between that have to be considered as crewmembers are selected and trained. Success has been exemplified through the crew cooperation in the International Space Station (ISS) Required training will have to cover the subjective culture, which is implicitly known to its members even though the knowledge is rarely articulated. Its rules and practices must be made explicit for an outsider. (Draguns & Harrison, 2011)

One area of study that is often overlooked for its applicability to long-term space travel is that of grief and loss studies. In particular, ambiguous loss is defined as an unclear loss that defies closure (Boss, 2006). During an interstellar mission, crewmembers may experience ambiguous loss, where they lose contact with the living and will never see family and friends again.

Those who are left behind experience unfulfilled grief without closure as are the crewmembers. As the crew gets farther from Earth, their communications will become less and less frequent, not due to lack of interest, but due to our technological and physical communication limitations. Boss (2006) defines the effect of ambiguous loss as "inherently traumatic because the inability to resolve the situation causes pain confusion, shock, distress and often immobilization. Without closure, the trauma of this kind of stress becomes chronic." The recognition of this condition will be important in the development and application of coping mechanisms.

The research studies cited are just a sampling of the need and the possibilities of research to make space travel the most optimal experience both from a program/mission and a commercial/tourist perspective. Eventually the commercial space business will become large enough that the training required to participate will be as simple as the inflight safety demonstration on today's commercial airplane flights.

3. Predictive Modeling

As space programs grow globally, there are certain inevitabilities that will occur as a result. Crews will comprise of individuals from different cultures, will be more gender-balanced, and the missions will be longer and more complex. Exploration goals will be bigger and require more effort of the crew—task work, endurance and collaboration. Much of the current psychological research is geared toward optimization of human performance, a necessity. However, it is just as important to address the potentiality of human breakdowns both physically and especially psychologically. The impact of a psychological breakdown is sometimes subtle, but the impact can be crippling or devastating to a mission.

A breaking point is reached when an individual is not able to consistently and effectively contribute to work efforts and/or maintain functional relationships. When an individual reaches this point, there is no way to predict how they will eventually react and resolve the issue, nor how others will react. In a mission where the journey will be several months or years, a breakdown is inevitable, be it small or large, but it must be addressed.

What does a breakdown look like when there are so many factors that could be the cause? In some instances it is a person's prolonged depression, or self-isolation. It could be an extroverted behavior such as hostile interactions with fellow crewmembers or continued questioning or disobedience of orders. Or nothing may be apparent, because an individual may be strong enough to weather their internal strife and be able to maintain normal relations on the mission, but have post mission recovery issues, such as PTSD, as experienced by many veterans of war and traumatic experiences. If it is a long enough mission, how long can a person last before psychological distress ensues?

While creating a predictive model for understanding the state of the individual human condition seems ambitious, it is possible to lay the groundwork and identify signs of behavioral anomalies caused by physical and psychological stress. Once these signs are identified, then they can be addressed and managed for that individual and their group relationships.

3.1 Physiological Factors in Psychological Behavior

An individual's physical well being has an impact on their mood and behavior. Many studies have been conducted, both on Earth and in space that identify specific physiological conditions and their effect on behavior. For example, the change in the circadian rhythms of space crews has been studied for many years. It has become recognized that sleep disturbances and fatigue, as well as alterations of circadian rhythms in astronauts, are among the most important factors contributing to impaired well being, alertness, and performance during space missions. (Kanas & Manzy, 2008)Kanas and Manzy (2008) defined the actual types of stressors and stress that are encountered during space travel.

> A **stressor** is a stimulus or feature of the environment that affects someone, usually in a negative, arousing manner. In space, there are four kinds of stressors: physical, habitability, psychological, and interpersonal. (p. 1). Examples are:

Table 1: Environmental Stressors Experienced during space travel

Physiological	Habitability	Psychological	Interpersonal
Acceleration	Vibration	Isolation	Gender Issues

Microgravity	Ambient noise	Confinement	Cultural Effects
Ionizing radiation	Temperature	Danger	Personality Conflicts
Meteoroid Impacts	Lighting	Monotony	Crew Size
Light/Dark Cycles	Air Quality	Workload	Leadership Issues

A stress pertains to the reaction produced in someone by one or more stressors. In space, there are four kinds of stress that affect human beings: physiological, performance, interpersonal and psychiatric. (p.2) Examples are:

Table 2: Types of Stress Experienced during space travel

Physiological	Performance	Interpersonal	Psychiatric
Space Sickness	Disorientation	Tension	Adjustment Disorders
Vestibular Problems	Visual illusions	Withdrawal/ Territorial behavior	Somatoform disorders
Sleep Disturbances	Attention deficits	Lack of privacy	Depression
Bodily fluid Shifts	Error Proneness	Scapegoating	Suicidal thoughts
Bone loss and hyperkalemia	Psychomotor problems	Affect displacement	Asthenia

3.2 Mathematical Modeling

Developing a mathematical model has many challenges. Mathematical models, while extensively used in the field of engineering, have been utilized by other disciplines seeking to model their designs and behavior of objects. The behavior that is described in engineering is much different than that of human behavior. In psychology, mathematical models are designed to describe basic behavioral processes in a more precise way than can be done with simple verbal descriptions. (Mazur, 2006)

Human behavior is a very difficult process to model, due to the complexity of the decision- making itself. Mathematical models can make precise and important statements about behavioral processes that are relevant to anyone who is interested in explaining, predicting or controlling behavior, either in the laboratory or in applied settings. (Mazur, 2006). Specifically, decision modeling studies the human decision-making mechanism and tries to build models that predict human decisions. (Kim, Yang, & Kim, 2007)

Predictive models can be very difficult to develop and execute. On of the most apparent reasons is that individuals have free will and can exercise it at anytime, which means they can make any choice despite past behaviors. It may be possible to consider improved modeling strategies. It will be important to distinguish models that try to predict what a particular individual will do from models that try to predict particular actions that all people will do. (Bar-Yam, 2011)

3.3 Defining Mathematical Models for Predicting Human Behavior

The predictive model for behavior both defines and simulates the conditions in which certain behaviors will occur. Kennedy (2011) notes that a rational, random number-generated methodology is unwise. However, a rationally behaving model needs to be able to represent knowledge, learn, remember new knowledge, and apply it to determining the behavior of the agent.

The key components of this mathematical model are that it is dynamic, computational, and has elements of fuzzy logic in its framework. The idea of soft computing or fuzzy logic is a dynamic representation of human behavior as a gradation of behaviors, not just absolute experiences. Fuzzy logic is a form of multi-valued logic derived from fuzzy set theory to deal with reasoning that is approximate rather than precise. Fuzzy logic models human thinking in imprecise and uncertain conditions comparatively better than other techniques. (Singh, 2010). These principles lay the foundation of the mathematical model itself.

Conceptual frameworks are approaches to modeling human decision-making using more abstract concepts than mathematical transformations of environmental parameters. They involve concepts such as beliefs, desires, and intentions (BDI), emotional state and social status (PECS), and "fast and frugal" decision hierarchies. (Kennedy, 2012)

3.3.1 Model for Individual Behavior

Creating a predictive model for human behavior on any level is difficult, as noted above. However, having some degree of assessment and therefore predictability allows individuals and crewmembers to assess their environment more effectively. It also gives a foundation for the development of assessment tools, support activities and post-mission management actions.

The next two sections will focus on two models of predictive behavior in space, one for individuals and one for groups or crews. Empirical data to support these models are simulated in the sections below.

The Individual Model

The behavior of the individual is based on his or her interaction with the environment in which they work and live. The model reflects the aspects of that living situation. The behavior of the individual (Bi) is a result of the combined physiological state of the individual (Pi) within the phase of the mission cycle (Mc) plus the Level of group dynamics (Gd) plus the Role of the individual relative to the group development. (e.g. Is the person a crew member, crew leader, or not part of the crew in question?) This is a description of the Psychological state of the individual. The lower the result, the more likely an individual is to have ineffective relationships or an emotional breakdown. (See Figure 1)

$$B_i = (P_i / M_c) + G_d + (I/G)$$

Figure 1: Mathematical model for predicting emotional breakdowns in Individuals during long-term space travel.

The Mission Cycle describes totality of the stages within a mission. Different stages of the mission will have an overall effect on the physiological state individual. Kanas & Manzy (2008) found that human performance varied based on the stage of the mission, which the individual is being assessed. Those stages being pre-flight, mission, and post flight. In this paper, the term "mission cycle" is being used to describe the overall mission, describing the different time frames of the mission itself which results in different mission constructs. Mission cycle types are different depending on the goal of the mission, even though the mission may be similar in duration. (See Table 3)

Table 3: Mission cycle types for long-term space travel

Mission Cycle Type for Space Travel	Time Frame			Mission Example
	Departure from Earth to Destination (Egress)	Mission Duration	Return to Earth	
Mission Cycle Type 1	24 hours <	24 months	24 hours <	2 years on ISS
Mission Cycle Type 2	7 months	14 days	7 months	Mars 500
Mission Cycle Type 3	3 months	3 months	3months	Not experienced
Mission Cycle Type 4	24 hours	Indefinite	Unknown	Pioneer Journey - Interstellar

Mission Cycle Type 1 Missions: The travel time to the destination is minimal, the duration of the mission time is spent on the mission activities itself and the return in less than 24 hours. These tend to be low-Earth orbit type missions, such as the International Space Station. An early example would be traveling by commercial flight to a summer destination and spending a month in a location, then returning by commercial flight. Travel time is minimal relative to the vacation/mission time.

Mission Cycle Type 2 Missions: The travel time to the destination and return are greater than the mission itself. This type is modeled after the Mars 500 analog study where the time it takes to the complete the mission is significantly less than the travel time to get there. The "mission" is not only the tasks to be completed, but includes the travel time itself. The Earthly example here is a road trip , which may take 2-3 weeks to reach a destination and then spend only 1 week at the beach and two weeks to return. You basic scenarios of "are we there yet?"

Mission Cycle Type 3 Missions: The travel time and the mission duration are equal or nearly equal. Our space programs have not reached a point where mission egress, mission duration and return to Earth are equal in time. Missions such as these provide equal skill for the crew of travel endurance and mission performance.

Mission Cycle Type 4 Missions: These are unique in that it will only take approximately 24 hours to get out of Earth's orbit onto a planetary or interstellar trajectory. However the destination is so far away as to feel indefinite to those on Earth, and the return is unknown for both the crew and those left behind.

The mission cycle type provides different scenarios for the individual and their environment. Their physiological state will have a different cycle depending on the mission type. Kanas & Manzy (2008) show that there is an initial adaptation adjustment to space during the pre-flight, mission duration, and even after a return to Earth and there are adjustments for both short-term and extremely long-term (over 400 days) missions. Still, little is known about long-term space mission. The application to interstellar travel must be extrapolated from these limited studies. One of the key differences is that when the mission is beyond our current extremes—over 3 years—what will be the long-term impacts on human performance? It is not known if there is a leveling-off of certain side effects, or how the severity of side effects will affect performance and therefore psychological well-being.

3.3.1 Model for Group Behavior

The behavior of the group is reflective of the individuals therein. In this model, the behavior of the group (Bg) is a result of the group development level or dynamic within the mission cycle plus the group relation to the individual. As in the individual model, the lower the result, more likely a group is to have decreased effectiveness or cease functioning altogether. This model differs from the individual model in that it takes into account the individuals members in the group and their relation to the mission and the individual. (See Figure 2)

$$B_g = (G_d / M_c) + (G/I)$$

Figure 2: Mathematical model for predicting emotional breakdowns in groups during long-term space travel.

In part one of the equation, the group dynamic or development is based on the point in the mission cycle where the assessment is taking place. That will vary based on the mission type (see Table 3). Group development principles show how mature the group is in its overall development and functioning relationship together. The body of work on team development, organizational dynamics come into play here. The longer into the mission cycle the mission is, the more the group will learn to be effective and grow.

The other part of the equation attempts to model the effect of the individual in relation to the group. Impacts such as cross-cultural differences, gender effects and general interpersonal issues are addressed at this point. The individual relationship relative to the group is what is measured. The individual may be a "minority" by culture, gender, group tenure or other isolating factors. The relationship of the individual to the group membership and level of inclusion, will affect an individual's emotional and psychological state.

4. Simulated Application and Results

Space mission environments are very structured, both physically and mentally within a very limited space. The habitable space on the International Space Station (ISS) is that of a six-bedroom house, approximately 13,696 cubic feet. (NASA, 2014). Astronauts from many nations who travel to ISS to conduct experiments, participate in space walks and maintain the station itself use the habitat. While the crew quarters are tolerable, they are not exactly comfortable. Wichman (2011) is reminded of Feedman's Density Intensity Hypothesis which states that crowding is not necessarily an adversity stimulus, but it does tend to amplify whatever emotion is extant in a group, positive or negative. For the duration of their visit, all astronauts will live in the habitual space with other crewmembers for a specific amount of time. The timeframe will vary from member to member, and individuals will have served for different amounts of time, based on their mission assignment. For example, an ISS crew may consist of up to six members who have served anywhere from two weeks to six months, and sometimes longer. (NASA, 2014) The composition of the crew can vary at any time by gender and nationality. The work required of the crew will be different and the expectations of both the crewmembers and the crew as a whole will also vary.

When assessing the potential for an emotional breakdown, the predictive model has many factors to consider. For current research, there is only an Earth analog and short distance missions to assess. While a stint on the International Space Station may be several months long, the journey there is relatively short. Extrapolating these results and predicting the experience for future space travelers will be the challenge since the travel time and the mission activity will be of equal importance even if the mission activity itself is short.

Each mission type has its own expected result, regardless of the time spent in space. However, a Type 1 mission lends itself to simulate short-term mission cycles where it will be a short journey to reach the mission activity location and to return. In Type 2 mission cycles, the journey to the mission activity location is longer than the mission itself and creates a different expectation of the crew. This gives new meaning to the saying, "It's the journey that is the reward." The Type 3 mission cycle is similar to a control group, where all aspects of journey are equal: egress, mission activity and return. And finally, in a Type 4 mission cycle, the longest journey of them all, there is nothing to compare with the timeframes expected of these journeys. The constant unknown in the Type

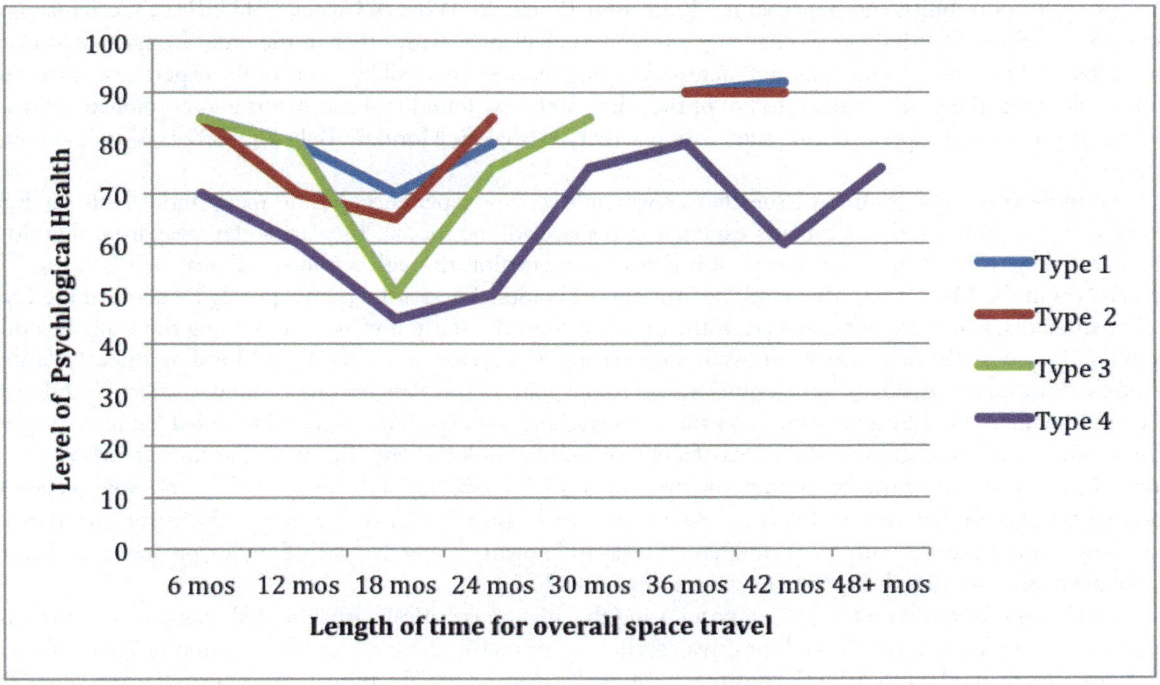

2 – 4 mission cycles is the ability of the crew to stay physiologically and psychologically stable enough to complete the entire mission cycle.

Table 4: Simulated result of modeling psychological well-being during mission cycle types over time.

Based on the simulated results, the expectation of the psychological stability and well-being of crew members are dependent on the mission cycle type and the time spent in space. As both models show, factors that influence the prediction of an emotional breakdown, such as individual experience, group development and crew composition. All mission types show a period of adjustment near the beginning of the mission. However, in longer missions such as Type 3 and Type 4, the level of psychological health dips well below the others. Using the method of Thematic Content Analysis (TCA), Suffield, Wilk & Cassel (2011) reviewed the diaries and journals of multinational crewmembers to assess their adjustments and coping mechanisms long-term space travel and their ability to work with a multicultural crew. They found that post-mission reflections were much more positive than those of the original, real-time journal entries. The physiological health of the crew members will be in challenged the longer they are on the mission. The environmental stressors will compound themselves over time, and reduce the performance level of the individual (Kanas & Manzy, 2008) (Harrison A. A., 2011).

4.1 Countermeasures

As the psychological state of the individual and groups can be tracked, and possibly predicted, there is opportunity to establish protocols for adaptive measures. This means creating training for crewmembers, on board support modules for individuals and team and post-breakdown support.

Ensuring the psychological health of a crewmember starts with effective selection criteria. The criteria should be based on the mission activity, the mission cycle and best fit for the composition of the crew. Once selected, the crew should have time to work together and bond as a group. This model assumes small crew sizes, of 2 – 7 members. As space programs embark on longer missions and larger crews, they dynamics of groups will change and require a different approach to selection and training. Kanas and Manzy (2008) suggest a series of assessments for crew selection based on an individual's ability to cope with the psychological stressors. Draguns and Harrison (2011), suggest utilizing cross-cultural data as part of astronaut selection and support.

While on board a ship and conducting a mission, crewmembers can monitor and manage their own individual health needs. Mission duration had significant effects on coping strategies during the flight/mission. The longer missions (four months and longer), showed astronauts utilizing the more certain coping mechanisms than shorter missions, (two-weeks or less). In this case the mechanisms utilized more were Confrontation, Escape/Avoidance, Accepting Responsibility, and Supernatural Protection. (Suedfeld, Wilk, & Cassel, 2011). These mechanisms are indications of potential individual challenges as well as challenges with other in the crew. Longer missions are inevitable and negative interactions and increased coping mechanisms will be a part of the experience of the crew. The results from the psychological studies of the Mars 500 crew found that astronauts and cosmonauts did seek out the psychological support during the mission activity. (Ushakov, Morukov, Bubeev, Vasil'eva, Vinokhodova, & Shved, 2014).

Homesickness is a common issue that crewmembers may experience. While many individuals on Earth encounter this problem, the astronauts' experience in space will be special. They are so far away from their loved ones that they will not have the option of immediate interaction through a phone call, text or video call. The researchers in the Mars 500 study found the incidents of homesickness arose pretty quickly for some of the European astronauts, who were not only part of the isolation from the study, they were also doing the study in a other country. They were the first to seek out psychological support with for issue. (Ushakov, Morukov, Bubeev, Vasil'eva, Vinokhodova, & Shved, 2014). When missions get to the point of traveling beyond communication capability and the crewmembers are fully autonomous on the mission, they may experience something called "ambiguous grief." This is when there is no closure to the loss of a family member or loved one. This is similar to those who have lost someone as a prisoner of war or through missing persons who have not been found. There is no way to close the loop on the loss absolutely knowing the person is deceased. (Boss, 2006) In the case of the astronauts, they will have lost communication with Earth and have no way of knowing the fate of their families, or vice versa. This can cause depressive or other disruptive behavior in the individual.

Providing the crew with tools to support during the mission is critical. Ushakov et al., suggest an approach to support of those individuals through positive psychology, primarily those under the direction of P. Suedfeld and N. Kanas, has focused on study of the positive emotional states and social attitudes arising in astronauts when they successfully cope with the stressors of a long spaceflight.

Once a situation has happened where an individual has ceased to be engaged and a contributing member of the crew – they have had a breakdown – it is incumbent upon fellow crewmembers to assist in the support of that member. If the mission still has the crew within communication range of the Earth, Mission Control can provide psychological support through outreach and discussion as was done in the Mars 500 study. (Ushakov, Morukov, Bubeev, Vasil'eva, Vinokhodova, & Shved, 2014). The other crewmembers may or may not know that this support is taking place, the crewmember may be contacting mission control on their own. When missions get longer and there is little or intermittent contact with Earth, the onus of support is on the crew itself, which may be dealing with mission stress as a result of the activities or journey.

When the group, or whole crew becomes ineffective, then the mission itself is potentially in jeopardy. The crew redundancy is compromised and it is up to individuals to take the role of self-correcting the group.

Upon return to Earth, many studies emphasize the need for post-return crewmember support for both the crewmember and their families. (Suedfeld, Wilk, & Cassel, 2011). Despite the post-mission jump in positive reflections, there are still adjustments to Earth that the crew has to make. It is suggested that private sessions with both the crewmembers themselves and with their families will make the transition back to Earth easier. (Kanas & Manzy, 2008)

5. Research Applications

The goal of the predictive model is to create opportunity for prevention and management of the psychological health of crewmembers. Long-term space travel will slowly have longer and longer timeframes as programs become more ambitions. Suffield, et al (2011), defined long term as four months or more. The Mars 500 group first conducted a 105-day mission in addition to the 520-day mission, each creating opportunities for observing behavior in this unusual environment. Those in the position of identifying, selecting and training an optimal crew for a mission will take into account the research that has been done and utilize the modeling that has been created in order to select individuals who can function, and even thrive in the new conditions.

5.1 Transdisciplinary Studies

Despite the slow start and recognition of the need for such research, the psychological study of astronauts in space travel is critical to the success of future missions. Going forward there needs to be outreach to fields not commonly associated with space and space missions. Cross-cultural psychology has important applications as the programs become intercultural and the crew composition reflects that spirit of cooperation. There is much that this discipline can offer training and psychological support during long-term space missions.

Helping other disciplines understand that they have a meaningful contribution to this work is important to creating a robust, transdiscplinary conversation. Two areas for consideration are grief studies and team development studies from an organizational development perspective.

Mentioned earlier, ambiguous grief is a condition that has the potential to incapacitate a person and prevent them from interacting and building effecting relationships. When a person withdraws from his or her team that could mean disaster for a mission. Team development is a key component to a successful mission. Working with experts in these fields to bring their expertise to the table can only enhance the effectiveness of training, coping, and post mission support.

As human beings, we will take all of our cultural biases, gender biases, etc., with us. Training can help with coping, but only if the issue is recognized as being part of everyday life.

5.2 Quantifiable Self

Predictive models lend themselves to utilizing quantified data for decision-making. Already, our society uses this data for laymen in the use of wearable health and fitness trackers. Using these biometric monitoring systems to track sleep patterns, and weight management allows the user to make data-driven decisions about their lifestyle and behavior. Monitoring systems exist currently for space missions, mostly monitoring physiological activity. However, pairing that with the predictive model and its impact on psychological behavior can prove to be an opportunity for further study. With knowledge of the impact of physiology on performance and behavior, monitoring systems can be created that will alert users and systems to possible psychological distress.

6. Conclusion

This paper opens up many questions about the impact of the conditions of extreme long-term space travel, or interstellar travel, on the human psyche. While empirical research exists from which to extrapolate certain behaviors, the factor of the unknown is so extreme that everything we have studied could have no effect. However, we can be assured that human behavior, while sometimes unpredictable, still follows certain patterns. (Kennedy, 2012) Those patterns can lead us to make predictions and anticipate certain failure points. There is a cautionary tale here. Having this knowledge does not make one omniscient. With all science, it must be tempered with wisdom and not hubris.

Culturally, in the United States, and many other cultures, the study and acceptance of mental health is an often eschewed topic of discussion. Clinically there are many areas of research within the field itself; utilizing that research outside of the field means an acknowledgement that that issue exists in that new field. Those that have studied space exploration know the importance of mental health and psychological stability of the individuals and crews, which are essential to the success of these missions. The body of work continues to grow and extend outside traditional fields into a transdisciplinary exchange of ideas.

Modeling our understanding of the cycle of an individual's physiological and psychological health engenders the research of group dynamics and cross-cultural, gender studies and other applicable fields. It will be a cornerstone of future research on longer-term space travel. However, through predictive modeling there is opportunity to identify when problems may begin to arise and how those problems can lead to a breakdown. These models can also provide a guide as to the likelihood of an occurrence and create a safety net of support for that individual and crew.

If we could have modeled of the behavior of *Breaking Bad*'s protagonist Walter White, would that have stopped his descent in to psychological and physical destruction? Perhaps. Modeling may have predicted critical of decision-making points and provided preventive measures to assist him in going down a different path. In reality, we have no idea how an individual will react when a breakdown situation occurs – will a person be violent, reclusive, and/or disruptive? There is no way to predict behavior to that level – at least not yet.

References

1. Bar-Yam, Y. (2011). *On the Role of Free Will in Predictive Models of Human Behavior.* Retrieved 09 30, 2014, from New England Complex Systems Institute: http://necsi.edu/research/overview/freewill.html

2. Bishop, S. L. (2011). *From Earth Analogs to Space: Getting There from Here.* In D. A. Vakoch, Psychology of Space Exploration: Contemporary Research in Historical Perspective. Washington, DC, United States: National Aeronautics and Space Administration.

3. Boss, P. (2006). *Loss, Trauma and Resilience: Theraputic Work with Ambiguous Loss.* New York, NY: W.W. Norton & Company.

4. Draguns, J. G., & Harrison, A. A. (2011). *Spaceflight and Cross-Cultural Psychology.* In D. A. Vakoch, Psychology of Space Exploration: Contemporary Reserach in Historical Perspective (pp. 177 - 203). Washington, DC, United States: Nationa Aeronautics and Space Administration.

5. Harrison, A. A. (2011). Behavioral Health. In D. A. Vakoch, *Psychology and Space Exploration: Contemporary Research in Historical Perspective* (pp. 17-46). Washtington, DC, United State: National Aeronautics and Space Administration.

6. Harrison, A. A., & Fuller, E. A. (2011). *Introduction: Psychology and the U.S. Space Program.* In D. A. Vakoch, Psychology and Space Exploration: Contemporary Research in Historical Perspective (pp. 1-16). Washington, DC, United States: National Aeronautics and Space Administration.

7. Kanas, N., & Manzy, D. (2008). *Space Psychology and Psychiatry.* El Segundo, CA: Microcosm Press and Springer.

8. Kennedy, W. G. (2012). *Modelling Human Behavior in Agent-Based Models.* In A. J. Heppenstall, M. Batty, & A. Crooks, Current Geographical Theories for Agent-Based Modelling (pp. 167-179). New York: Springer Science + Business Media.

9. Kim, C. N., Yang, K. H., & Kim, J. (2007). *Human Decision-making Behavior and Modeling Effects.* (S. Direct, Ed.) Decision Support Systems , 2008 (45), 517-527.

10. Kring, J. P., & Kaminski, M. A. (2011). *Gender Composition and Crew Cohesion During Long-Duration Space Missions.* In D. A. Vakoch, Pscyhology of Space Exploration: Current Research in Historical Perspective (pp. 125 - 141). Washington, DC, United States: National Aeronautics and Space Administration.

11. Mazur, J. E. (2006). MATHEMATICAL MODELS AND THE EXPERIMENTAL ANALYSIS OF BEHAVIOR. Journal of the Experimental Anaylsis of Behavior , 2 (85), 275-291.

12. NASA. (2014, 11 3). *International Space Station.* Retrieved 11 11, 2014, from NASA.gov: http://www.nasa.gov/mission_pages/station/main/onthestation/facts_and_figures.html#.VGLDZYdzinM

13. Robinson, J. A., Slack, K. J., . Olson, V. A., Trenchard, M. H., Willis, K. J., Baskin, P. J., et al. (2011). *Patterns in Crew-Initiated Photography of Earth from the ISS--Is Earth Observation a Alutogenic Experience?* In D. A. Vakoch, Psychology of Space Exploration: Current Research in Historical Perspective (pp. 79 - 102). Washington, DC, United States: National Aeronautics and Space Administration.

14. Singh, S. (2010). *Modeling Team Behavior And Predicting Success Using Soft Computing.* Third International Conference on Emerging Trends in Engineering and Technology (pp. 620-623). Goa: IEEE.

15. Suedfeld, P., Wilk, K. E., & Cassel, L. (2011). *Flying With Strangers: Postmission Reflections of Multinational Space Crews.* In D. A. Vakoch, Psychology of Space Exploration: Contemporary Research in Historical Perspective (pp. 143 - 175). Washington, DC, United States: National Aeronatics and Space Administration.

16. Washington, DC, United States: National Aeronatics and Space Administration. Ushakov, I., Morukov, B. V., Bubeev, Y. A., Vasil'eva, G. Y., Vinokhodova, A., & Shved, D. (2014). *Main Findings of psychophysiological Studies in the Mars 500 Expereiment.* Herald of the Russian Academy of Sciences , 84 (2), 106-114.

17. Vakoch, D. A. (2011). *Psychology of Space Exploration: Contemporatry Research in Historical Perspective.* (D. A. Vakoch, Ed.) Washington, DC, United States: National Aeronautics and Space Administration.

18. Wichman, H. (2011). *Managing Negative Interactions in Space Crews: The Role of Simulator Research.* In D. A. Vakoch, Psychology of Space Exploration (pp. 103 - 124). Washington, DC, United States: National Aeronautics and Space Administration.

Interstellar Travel in the New Science Fiction

Jason D. Batt

Abstract

The future of humanity in space is not guaranteed. Great cultures have arisen in the past, set their sites on great and, from their perspective, inevitable goals, and failed. Science fiction literature has been a guide post for space exploration from the early days. Much has changed since Arthur C. Clarke and Isaac Asimov wrote. Our political and international world is different. Our technology has branched into different and unforeseen areas. With the current technological expansion in view, what do today's science fiction writers say about interstellar exploration? What pitfalls do they see? What is the roadmap they present now? What challenges do they show? Is there an agreement upon the timeline? What technology shows promise? What dangers do they predict out there? The presentation will present a survey in review of current science fiction on the topic of interstellar exploration. The foundation for this has arrived from serving as judge of the Lifeboat to the Stars literary award for science fiction literature focusing on interstellar efforts. This award was presented at the 2013 Campbell Conference.

Keywords

science fiction, interstellar, short fiction, novels, faster-than-light, generation ships, slower-than-light

1. Introduction

Voyaging to the stars is not just an ancillary aspect of science fiction. It is a core concept that has been embedded in the genre since its earliest moments. Adam Roberts, in *The History of Science Fiction*, contends that "the roots of what we now call science fiction are found in the fantastic voyages of the ancient Greek novel; and I use the Vernean phrase voyages extraordinaires . . . especially voyages to other planets." Odysseus sailed to fantastic islands in his quest back home from the shores of war. The extraordinary voyage was woven into the very beginnings of the fantastic tale. Roberts continues: "Stories of journeying through space form the core of the genre . . . Are the trunk, as it were, from which the various other modes of SF branch off." The dream of interstellar travel has been explored by science fiction for years before there was even a glimmer of reality apparent in its practical and technological pursuit. [1]

Science fiction's distinctive in contrast to other genres is its ability to serve as a thought experiment, a place where various scenarios can be played out, where consequences can be explored without true damage to real people. Science fiction has been exploring the future of interstellar travel for decades and has wrestled with the very issues that current interstellar experts are grappling with today. Thus, a cataloging and analysis of interstellar aspiration within science fiction can help to mine for solutions and problems to the very topic the 100 Year Starship is engaging today: how will we reach the stars?

2. Interstellar Fiction as a Sub-Genre

Daniel Abraham, writer of *The Expanse* series, was asked on the current state of interstellar fiction and if interstellar aspirations represented a type of "science fantasy". He replied:

> "No, I don't think interstellar travel, space empires, etc. are at best science fantasy any more than they ever were. Or, putting it the other way, they haven't become any less likely, in the recent past. They've always been the stuff of fantasy. That didn't used to be a problem. The thing that has changed, I think, is the requirement that science fiction be somehow rigorous to be taken seriously. When I look back at the classic science fiction that I grew up with, a lot of it didn't pass the sniff test. Larry Niven had teleportation, giant alien cat warriors, and human's eugenically bred for their luck. Arthur Clarke had God turning out the stars. Herbert had giant freaking worms digging through a desert rich in psychoactive space fuel. Unsophisticated as I was, I didn't see anything wrong with that. Still don't. I see two things happening with the genre. The first is that it's becoming the primary idiom of pop culture. The second is that, in response to that, there is a narrower and narrower definition of "real" science fiction which uses terms like "science fantasy" to exclude work which isn't somehow pure enough. I think that there's a real risk of science fiction going the way of jazz music and poetry in which it becomes a more and more sophisticated, narrow, and inaccessible form dedicated to meeting the standards of a smaller and smaller elite. But I also think the consequences if that does happen are pretty minor, because the market for accessible science fantasies like *The Demolished Man* or Dread Empire's Fall or the Vorkosigan Saga appears to still be wide open." [2]

James Blodgett, science fiction writer, provided his view of the current status of interstellar fiction in 2012:

> "Science fiction inspired our early steps in space. If we think about how science fiction did this we may be able to do it again. Science fiction has pictured many methods of travel, and many motivations for traveling. Science fiction stories are a form of scenario analysis that helps us foresee problems and solutions. Humanity could use fresh inspiration for space travel. The public and even the science fiction field seem discouraged with space after initial enthusiasm. Folks in 1940 and 1960 could extrapolate from the Wright Brothers' airplane to the airliner and later the jetliner, amazing progress which took only a few years. The discovery of atomic power suggested miracles to come. This progress was plausibly extrapolated to predict amazing progress in the near future, the near future of 1960. Subsequent space developments have been underwhelming. It is no surprise that science fiction has shifted from space opera to dystopia. However, there are new plausible visions that could lead to amazing developments in space, a space-based form of singularity. The nanotech and artificial intelligence versions of the singularity can also suggest ways to get us into space. There are several types of existential risk that have prompted folks like Hawking to think that we had better get going into space, making the point that space travel is existentially important. We are on the cusp; with disorganization and bad luck we may never develop space and we may go extinct; with focused work and good luck we may see an exponential future in space that can produce amazing prosperity and expand humanity by orders of magnitude." [3]

Interstellar fiction is certainly alive and in its current state provides a distinct laboratory for potential technologies and approaches that will take us to the stars.

3. The Role of Science Fiction in Space Exploration and Scientific Endeavors

In reality, 100 Year Starship is an exercise in applied science fiction. The very nature of key works in the genre is being explored through the mission of the organization: the development of the capability of interstellar travel. As such, it is critical to establish the limitations of science fiction. Science fiction has its roots in fear and wonder, legend and mythology, exploration and reflection. But it does not provide a roadmap to future developments. Science fiction and innovation exist in a circle of influence, mutually dependent upon each other. Dennis Cheatham in T*he Power of Science Fiction*: states: "It may be the case that the future worlds and infinite possibilities projected in science fiction can be used to inspire viewers to pursue work that will make those possibilities or ones like them, real." [4] Science fiction does, quite often, get its predictions wrong. But it does get it right.

Science fiction can be both predictive and prescriptive. How does it do this? First and foremost, it accepts the fact and inevitability of change. It frames dilemmas and puzzles that could arise, play out possible consequences of future innovation, normalize and/or make desirable certain technology, can suggest paths to innovation, inspire invention, reflects the speculative thinking of a generation, and create a mindset that leads to innovation (if not able to predict it exactly). In *The Dreams Our Stuff Is Made Of: How Science Fiction Conquered the World,* Thomas M. Disch writes: "It is my contention that some of the most remarkable features of the present historical moment have their roots in a way of thinking that we have learned from science fiction." [5]

From the beginning, science fiction has pondered what exists past our planet. There have always been stories about travels to other places and other stars. The sub-genres within science fiction (and its close partner fantasy) continue to develop, breaking further and further into various branches. Yet, space journeys formed the primary core of the overall genre.

The focus of this paper is to look at what the modern science fiction stories of space travel are focusing on. In particular, what is the current thought in science fiction towards interstellar. We now exist within an age of actual interstellar missions and organizations. Has this recent development of interstellar organizations and actual efforts to interstellar flight affected the space travel fiction? As well, has interstellar fiction changed as a sub-genre? Has it diminished or expanded in its influence and reach? Has it been abandoned? The intent is to also glean the fiction for original thoughts that can continue to help encourage the remarkable thinking of now and the future. Overall, this analysis will begin (but not finish) the path of sifting the fiction for commonality of thought as a possible reflection of the culture of as a whole.

4. Parameters of the Review

This is not an exhaustive review of interstellar fiction. Materials covered focused on works published since 2000, with a more exhaustive approach to those published since 2010. The intention is to focus on the contemporary thinking of fiction in regards to interstellar and in particular, to assess what the current fiction is exploring in the age of interstellar organizations, in particular since the initial announcement of DARPA's original 100 Year Starship Project.

Works that were tied to other established intellectual properties and previously published methods prior to 2000 were not considered. In particular, although there continues to be numerous works written in the *Star Wars* and *Star Trek* franchises (not to mention *Stargate* and others), these are building upon the continuity and thought that was established in the original development of these IPs. The same consideration is made for works published since 2000 that were simply sequels or continuations of materials published prior to 2000.

Works that explored travel between planets within our own galaxy and universe were reviewed. If the fiction did show interstellar travel but it was dimensional travel, these were not considered. The focus is entirely on interstellar travel within our reality.

Again, this was not exhaustive and focused on professionally published materials. This period of time saw a massive explosion of independent publishing. There were a quite a number of science fiction works that addressed interstellar travel that were independently published. These titles were not considered in this round. However, this is not to say that they shouldn't be. An extension of this review should at least attempt to partially catalog these works because they are attempts to address the subject since the advent of the interstellar organization.

5. Criteria for Works Considered

Attributes considered were:

- method of travel
- specific propulsion method
- speed of travel
- method of acquisition of propulsion technology, and
- nature of the crew.

These were further broken down. For method of travel, pieces were catalogued for:

- faster-than-light capability (FTL)
- slower-than-light capability
- generational ship
- warp capability (the idea that bending the space around a ship provides a loophole in relativistic limitations)
- hyperspace (the idea of jumping into another parallel dimension and utilizing that dimension's different physics to surpass relativistic limitations)
- fixed point jump (FTL travel that requires a jump between preset locations usually due to a dependence on a technological structure)
- free jump (using technology to propel a ship in a manner-like teleportation)
- wormhole/stargate, and
- unique (ideas other than the common ones explored above).

For method of acquisition, works were analyzed based upon if the technology was human developed, developed by aliens, or another method. Under crew composition, works were organized by either:

- normal human crew
- alien crew
- embryonic crew
- robotic crew
- tech-expanded crew (humans that have expanded their capability through technology)
- bio-expanded crew (humans that have expanded their capability through biological advances
- a crew of artificial intelligences, or
- a seed colony/embryo colonization vessel.

6. Interstellar Travel

The vast distances between astronomical bodies and the impossibility of faster-than-light travel pose major difficulties to both interstellar experts and science fiction authors. Science fiction authors approach these issues in several ways:

- "accept them as such (hibernation, slow boats, generation ships, time dilation - the crew will perceive the distance as much shorter and thus flight time will be short from their perspective),
- find a way to move faster than light (warp drive), "fold" space to achieve instantaneous translation (e.g. the *Dune* universe's Holtzman effect),
- access some sort of shortcut (wormholes),
- or sidestep the problem in an alternate space: hyperspace." [6]

So how do these works tackle these problems?

7. Results

Overall, 85 distinct works were reviewed. 46 of these were novels (and/or series). For note, series (such as Gregory Benford and Larry Niven's *Bowl of Heaven* and *Shipstar*) were given one entry. 25 were short stories (the majority of these since 2010). Five were television shows. One was a film. Six were video games. One was a novella. Of the works reviewed, 39 utilize FTL (46%), 40 use STL (47%). Of those 40 STL, 21 are generation ships. Of the 39 FTL:

- 0 stories employ warp as a technology.

- 12 use hyperspace
- 5 use fixed-point jumps
- 10 use wormholes.
- 3 had unique technologies distinct to those works.

In 51 of the stories, humans develop the propulsion technology ourselves. In 15 of the stories, we receive the tech as a gift from the aliens. In analyzing novels alone, we find that 27 of them use FTL (59%), 18 use STL (39%). In particular:

- 7 use generation ships
- 6 use hyperspace
- 3 utilize fixed-point jumps
- 6 use free jump
- 6 use wormholes

In analyzing the short stories, 6 use FTL (18%), 18 use STL (72%).

- In particular:
- 11 use generation ships
- 1 used hyperspace
- The other five are non-explanatory

A full list of works reviewed appears at the end of this paper.

8. Repeated Themes

Several themes were repeated throughout the various stories. In several, older generation ships were overtaken by newer, faster ships. This was the key plot concept of Kevin J. Anderson's "The Tortoise and the Hare." As well, Karl Bunker used this in his novel *Overtaken*.

Distinctly, warp technology is not used. This is despite the recent scientific interest in Harold White's announcement at the 2011 100 Year Starship Symposium. In fact, the far more fantastical and unexplainable hyperspace is far-more employed. Hyperspace was the most utilized technology in the category of FTL.

More often than not, humans themselves develop the propulsion technology that leads them to the stars, regardless if that is FTL or STL. 51 of the stories have humans acquiring the technology on their own. However, there are moments where they technology is provided to us by aliens. Often these are aliens we are removed from and we receive it as a gift. This is the key concept of James S. A. Corey's *The Expanse* series. Initially, in Corey's works, humans utilize an advanced fusion drive for interplanetary travel. An extra-terrestially created wormhole provides the capability for interstellar travel. In fact, this approach of STL only capabilities then expanding to FTL was featured in other works. In particular, Charles Stross' *Neptune's Brood* and Alastair Reynold's *Blue Remembered Earth* and *On the Steel Breeze.*

In particular, short-form fiction focused on character interaction and life on generation ships. This was a theme not repeated in the novels.

9. Unique Concepts

There were several novels that used original ideas. In particular, John Scalzi's *Old Man's War* series employed a technology called skip drive. This was an engine that provided near-instantaneous travel from one location to another location by actually trading spots with an identical ship in another dimension. The theory is that in the endless multiverse, there must be an identical ship with identical crew composition looking to travel from where one ship wants to go to where that ship is. So, they just swap spots.

Neal Asher utilizes the nonsensical term coined by Edward Lear: runcible. In his *Polity* series, a runcible is a interstellar wormhole generator/teleporter. The technology is briefly explained but serves as a fantastical way to instantaneously travel from one spot to another.

David Brin's *Existence* insists that human (and for that matter, biological) travel between the stars is an impossibility. So, civilizations send out advanced probes with copied AIs to scour and seed the galaxy. These AIs allow for culture sharing and the potential for civilizations to, in some manner, exist forever. Yet, all biological life is ultimately locked to their own solar system.

Karl Shroeder's *Lockstep* series keeps human travel to STL only capabilities. Yet, communication and transmission of data is FTL. They utilize this to transmit personalities as well. In particular, his series is distinct for introducing a concept that all human settlements across the stars participate in a shared culture and thus agree to go into periodic very long cold sleeps even if they aren't traveling. Thus, all of the planets and the ships between them are kept in sync technology and culture-wise. For those traveling, the trip seems to occur overnight (get on the ship, go to sleep, wake up when everyone else wakes up). In this way, travelers can actually ensure that when they return home, the same amount of time has passed for those they left behind.

Another galactic-wide civilization concept is explored in Alastair Reynolds' *House of Suns.* Reynolds writes of a galactic civilization that is built on STL capabilities. In this galaxy, planet-based societies, are relatively, short-lived. The most powerful organizations in the galaxy are the "Lines"—family organizations. These Lines don't colonize planets but instead, very slowly, travel through space on circuits of the galaxy and hold reunions every 200,000 years or so.

One final aspect is in Corey's *Expanse* series. In his series, the Mormon church is actually the first to built a generation ship. However, the ship sits destitute as it was never launched.

10. Next Steps

This review was an initial work. There are additional steps to further this. In particular, these works should be compared to previous generations of science fiction. Other aspects to be catalogued could include: the purpose of the trip, how much of the story focuses on the ship, plausibility of the trip, the motivation of the ship, the time frame, political systems, story mood, clarity of story telling. Further examination in regards to the propulsion/transport method could also be explored including, power source, propulsion method, mission type, crew conditions, technology and astronomy assumptions, destination, extra-solar life, etc.

11. Conclusion

Today's science fiction writers continue to explore the concept of interstellar travel. There is great diversity in the approaches and philosophies that enable those travels. There has been a continual shift, in short fiction in particular, to STL -based works. As well, the concept of propulsion method is given far more serious thought than it has been in the past. Even if fantastical, authors go to great efforts to explain their technology. However, more and more are concluding that FTL may be impossible on its own unless access to wormholes is achieved. This study was an initial review and additional work is needed, in particular, a comparison of today's writers to those of the past.

12. Acknowledgments

Special thanks to the r/printsf subreddit of reddit.com. In particular, these users were of great assistance: Starpilotsix, superkuh, sigkircheis, fritz_freiheit, escielenn, gonzoforpresident, systemstheorist, strolls, twistytwisty, weezer3989, and miloshvu.

13. Complete List of Works Reviewed

Table 1. Works Reviewed

Commonwealth Series	Peter Hamilton
Expanse Series	James S. A. Corey
Eon Series	Greg Bear
The Sparrow	Mary Doria Russell

On the Steel Breeze	Alastair Reynolds
Great North Road	Peter Hamilton
Battlestar Galactica	Ronald D. Moore
Sins of a Solar Empire	Ironclad Games
Sword of the Stars	Kerberos Productions
Halo series	Microsoft Studios
The Light of Other Days	Arthur C. Clarke and Stephen Baxter
Palimpsest	Charles Stross
Farscape	Rockne S. O'Bannon
House of Suns	Alastair Reynolds
The Algebraist	Iain M. Banks
Old's Man War	John Scalzi
Revelation Space	Alastair Reynolds
Existence	David Brin
Colony	Rob Grant
Ascension	Philip Levens and Adrian Cruz
Firefly	Joss Whedon
Kovacs series	Richard Morgan
In Which Faster-Than-Light Travel Solves All of Our Problems	Chris Stabback
The People of Pele	Ken Liu
Five Elements of the Heart Mind	Ken Liu
Across the Universe series	Beth Revis
The Tortoise & The Hare / The Grasshopper and the Ants	Kevin J. Anderson / Steve Savile
Bowl of Heaven / Shipstar	Gregory Benford, Larry Niven
The Lost Fleet series	Jack Campbell
The Creative Fire	Brenda Cooper
In the Lion's Mouth	Michael Flynn
Earthbound	Joe Haldeman
Outer Diverse	Nina Munteanu
Ashes of Candesce	Karl Shroeder
Home Fires	Gene Wolfe
Count to a Trillion	John C. Wright
A Delicate Balance	Kevin J. Anderson
Twenty Lights to "The Land of Snow"	Michael Bishop
A Country for Old Men	Ben Bova
Overtaken	Karl Bunker
Scattered Along the River of Heaven	Aliette de Bodard
Connoisseurs of the Ecentric	Jetse de Vries
Against Eternity	David Farland
Lesser Beings	Charles E Gannon
It Pays to Read the Safety Cards	RWW Greene

Other Systems	Elizabeth Guizzetti
Transcript of Interaction Between Astronaut Mike Scudderman and the OnStar Hands-Free A.I. Crash Advisor	Grady Hendrix
The Big Ship and the Wise Old Owl	Sara A Hoyt
Choices	Les Johnson
The Old Equations	Jake Kerr
The Waves	Ken Liu
Design Flaw	Louise Marley
Lucy	Jack McDevitt
Waiting at the Altar	Jack McDevitt
Noumenon	Robert Reed
Siren Song	Mike Resnick
The First Day of Eternity	Domingo Santos
Shattering	Steven Utley
Star Soup	Chris Willrich
Ancillary Justice	Anne Leckie
Neptune's Brood	Charles Stross
Lockstep	Karl Shroeder
Embassytown	China Mievelle
Odyssey One series	Evan Currie
Vatta's War series	Elizabeth Moon
Fortune's Pawn / Paradox series	Rachel Bach
Diving series	Kristine Kathryne Rusch
Pushing Ice	Alastair Reynolds
Confederation series	Tanya Huff
Polity series	Neal Asher
Grand Central Arena	Ryk E. Spoor
The Taken Trilogy	Alan Dean Foster
Spin Trilogy	Robert Carles Wilson
Dread Empire's Fall Trilogy	Walter Jon Williams
Kop series	Warren Hammond
Eve Online	CCP Games
Web Shifter series	Julie Czerneda
Trader Pact series	Julie Czerneda
Light Trilogy	M. John Harrison
Blindsight	Peter Watts
The Future Happens Twice	Matt Brown
It's Always Spring Break Somewhere in the Galaxy"	Raven C. S. McCracken
Interstellar	Christopher Nolan
Mass Effect	Casey Hudson

References

1. Roberts, Adam. *The History of Science Fiction. London, UK: Palsgrave Macmillan: November 15, 2007.*

2. Weimer, Paul. "Mind Meld: Whatever Happened to Interstellar Travel in Science Fiction." *SF Signal.* Web. February 8, 2012. http://www.sfsignal.com/archives/2012/02/mind-meld-interstellar-travel-and-genre/.

3. Blodgett, James. Email to the Lifeboat Foundation SF Star Travel Bibliography, December 24, 2012.

4. Cheatham, Dennis. *The Power of Science Fiction.* December 6, 2011. http://designresearchcenter.unt.edu/sites/default/files/uploaded-documents/cheatham_taxonomy.pdf.

5. Disch, Thomas. *The Dreams Our Stuff Is Made Of: How Science Fiction Conquered the World.* Free Press: July 5, 2000.

6. "Faster-than-light." *Wikipedia.* Online. Accessed September 10, 2014.

Space Architecture
Design Aspects for Extended Space Travel

Antoine G. Faddoul, MS

Abstract

The conventional aspects of design reach distinctive levels when considering extended space journeys. Although space travel and its associated architecture started decades ago, space design has been limited to short distance and short term trips.

How would a design team approach the scheme for manned interstellar journeys? How would it manage sustainable deep-space designs for longer durability, minimal maintenance, and energy self-sufficiency?

It is not the typical design project that could be developed by analyzing the program and functions, then comparing them to previous projects. To carry such task, over a hundred design items related to three main areas; Structure, Environment, and Human needs were evaluated assessing their essences to achieve a sustainable spaceship that would function for hundreds of years. Design items were analyzed according to their status whether current technology, developing technology, or future technology. Each item was mapped according to how the science, technology, and design behind it would evolve in decades to reach the desired functioning product. The items are mostly associated with functions of Earthly daily needs, yet their sustainability is reevaluated to perform in extended space travel.

This design approach provides comprehensive futuristic scenarios by surveying the progress status of each item, while the resulting graphs and info help answering further questions including:

- How much more time is anticipated for each item to be ready for testing?
- Do we have the material, technology, and craftsmanship to build a spacecraft that could self-maintain and upgrade?
- What will be the status of the equipment, energy, and gadgets after decades of use on a spacecraft?
- How do the inhabitants of the space vessel foster their daily functions and needs over decades?

Keywords

Starship, interstellar, space architecture, design, technology

1. Gliina in Route to Antas, Improved

In September of 2013, a simulation of the year 2072 AD showed Starship Gliina[1] about one light year away from Earth. The simulated time was 5:00 AM, 10 years after leaving the lunar orbit. The destination was planet Antas in Constellation Cygnus. The starship had 7.5 light years remaining to reach its destination, and everything seemed fine as the long check list was streaming, showing a functioning and "healthy" starship.

The simulation of the starship readiness and future route will keep developing until 2062 – Gliina's proposed departure day. Since new aspects of design and technology are ever emerging and continuously evolving, the outcome results will keep changing and altering Gliina's future around the clock. In September of 2014, and while processing the same simulation for Starship Gliina's futuristic route, few new results came up.

For example, discoveries in material science and exoplanetary studies in the past year yielded changes in factors such as the starship weight, acceleration, speed, destination, and simulated location from Earth. The list of elements subject for modification extends to question if the future launching scheduled date of 2072 will change too. One thing that is certain about those futuristic simulations is that the pace of change will remain unsteady until all required breakthroughs in the elements of design are stabilized. That will take few decades based on the advancement rate in our current knowledge and technology.

2. Starship Architecture, a Holistic Design Approach

When a design team works on a (regular) project, it is typical to analyze the program required for such project and go over the functions to put them in full perspective while resourcing previous projects of similar nature. That generates; a realistic diagram of the relation between functions, a schedule for developing and finalizing the design, and finally the construction schedule. A holistic architecture approach for a starship, though unconventional, is still achievable following the traditional design methods. It is necessary however to note the following points as additional transdisciplinary items to be considered:

1. While space travel started, practically, with launching the first artificial satellite Sputnik into Earth's orbit in 1957 by the Soviet Union, and despite all the advances since then from landing humans on the moon, to landing robots on Mars, and sending the Voyagers to the edge of our solar system, most of the design and technologies used and developed so far are not suitable for interstellar travel. Certain aspects need to be considered when designing for extended space travel to see whether they could be used as base or heritage for developing the required ones, or to work on new concepts for interstellar travel.
2. The designs we know for interstellar projects came mostly from science fiction literature and films. In addition to that, there were some theoretical studies such as Daedalus unmanned interstellar probe [2] and Stanford Torus's inhabited starship [3]. While no startship, or even a main part of it, was built until now, both literature and theoretical studies still provide good sources for design and ideas – in fact they might be they only conceptual prototypes available.
3. Elements of interstellar and deep space travel are being studied separately and through research that is done for various items by different teams at unconnected locations, and without efficient coordination. For example, the specialized teams researching conceptual interstellar propellers have been working on a certain breakthroughs in different areas; solar sail, fusion, fission, ram jets, plasma, antimatter fusion, and many more, without direct interconnection with those working on closed air and water systems. That applies to teams studying other space travel fields. It is another sort of variables that need to be considered in developing a starship design to keep the project's integrity.
4. In the past few decades, sustainable design and green building concepts helped developing aspects much needed for extended space travel design such as designing with efficient use of material, utilizing natural resources, recycling, reusing, and self-sufficiency. One essential part is the simulation for energy generation, use, and loss for buildings. On a starship, energy calculations are part of every design item for along with the consumption and recycling of the raw materials.

   However, what we consider on Earth as naturally sustainable are the renewable energy resources such as solar, wind, and geothermal energy, in addition to the naturally renewable raw materials. Those items depend on the earth and its position from the sun. In outer space, the conventional renewable resources for energy and raw material become obsolete, and sustainability is to be achieved by other means.

5. Essential design and construction components of the starship are developed through uncommon ways depending on scientific experiments, space probes, and prototypes scouting. The probes themselves and the data they collect and transform are factors that continuously affect the elements of deep space design and schedule.
6. The nature of a starship project by itself is unusual, as it could be simply defined by building a small encompassed earth with everything that would support the life of thousands of people for decades. Add to that the necessity for such self-sustained structure to be capable of travelling at theoretical speeds and venture to locations that we cannot currently see, while employing technologies that are undeveloped and utilizing discoveries that are yet to be made.

Though the last item might seem discouraging, it is actually the key for a comprehensive study for interstellar architecture. It encourages designers to go beyond current modules, and account for variables that might be new to the design process yet essential to make this project work. Adding continuous research speculations for items that are still in theory, and projecting future advancements for others, bring the design process back under control, and somewhat put it in line with the way ordinary architecture projects are evaluated and fostered.

When accounting for all elements, traditional and unconventional, we can resort to the basic elements of design; function, budget, esthetic, and life cycle. In that sense, we can see how designing daily Earthly needs from habitat and equipment is barely different from designing interstellar mechanical spaceships, habitable spaces, architectural furniture, fashion clothing, practical equipment, and accommodating gadgets.

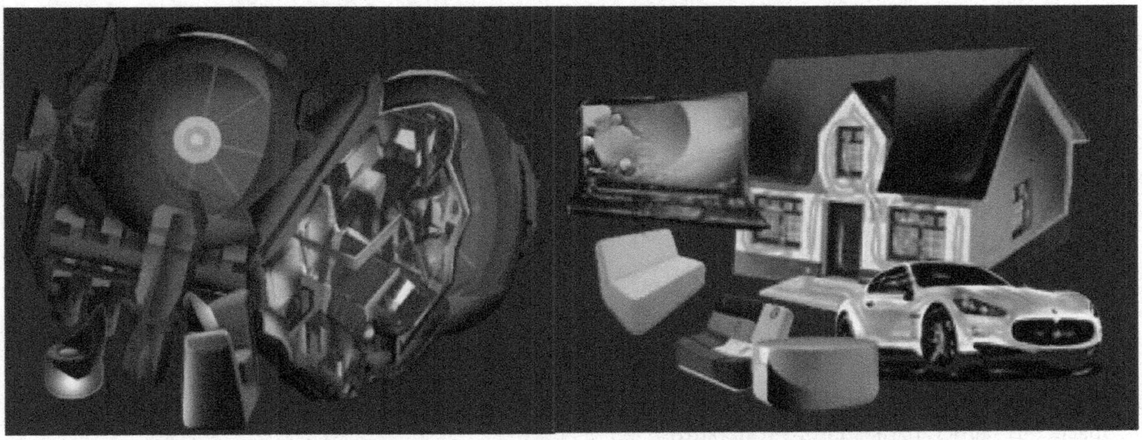

Figure 1. Including conventional and variable design aspects shows that designing for deep space could be compared to designing Earthly items

3. Starship Elements of Design

Over a hundred basic design items were evaluated and assessed for their requirements to attain a sustainable status in a starship that is bound to function for scores of years. The items were analyzed according to their development stage; current technology, developing technology, or future technology. The analysis of extended space travel requirements verifies that each item can perform with long durability, minimal maintenance, energy self-sufficiency, and the capability of self-repairing.

Future scientific and technological developments were included the in assessment. The design elements were divided into three main fields; structure, environment, and human needs.

3.1 Structural Elements

The main structure of the starship with its exterior and interior elements is analyzed with a comprehensive attention of how the components fit together and affect each other. That is based on questions a designer asks every time a new project is assigned. That includes asking about the objectives of such project - the mission in our case. Items such as destination or finance might seem odd under this category, yet they are vital for other items under Structural. Destination, for example, will determine the size of the starship and its crew in addition to everything linked to those items. The mission, whether it is to discover, terraform, or inhabit a life-supporting planet will affect most of the items under this category.

- AI and supercomputers
- Budget and finance
- Command/control center
- Communication with Earth
- Cosmic objects and dust shielding
- Constriction management and techniques
- Construction materials
- Defense system
- Destination
- Energy and alternative energy
- Engine/propeller
- Fabrication and 3D printing
- Fenestrations
- Fuel
- Functioning reliability
- Gadgets
- Gravity
- Growing neighborhoods, remodeling
- Industrial and heavy equipment
- Infrastructure
- Inhabitants, crew
- IT and computers development
- Launching
- Mission
- Navigation
- Radiation shield
- Raw material and resources
- Repairs and maintenance
- Robotics
- Scanning/mapping near-hit-objects
- Size, volume, dimensions
- Spacecraft and probes (onboard)
- Spacecraft launch pads/stations (onboard)
- Speed and acceleration
- Structure, construction
- Telescopes
- Urban design
- Weight

3.2 Environmental

Humans experienced space travelling while living in a closed ecosystem for several days in moon journeys. The trips to the International Space Station (ISS) are extremely short trips to Earth's orbit, while living aboard the station is like camping in ones backyard. Compared to interstellar travel, such trips were so short that humans can suspend certain aspects of their daily life as they will return to them shortly. When travelling beyond our solar system, all aspects of human life necessities need to be considered and delivered, whether predesigned, through onboard production, or simulated.

- Animals and insects
- Arms and artillery
- Artificial and natural open spaces
- Audiovisual production, television and radio
- Common/civic spaces
- Cultivation & botany
- Daycare and nurseries
- Emergencies and ER/safety
- Firefighting

- Food and beverage
- Furniture
- HVAC systems
- Indoors transportation
- Insulation and acoustics
- Interior air quality
- Internal telecommunication
- Internal energy source and balance
- Equipment and appliances
- Landscape
- Medical facilities and equipment
- Museums and artifacts
- Nutrition and food cycle
- Paths and hallways
- Plumbing system and water system
- Recycling
- Residential units
- Simulated light/seasons
- Space suits
- Sports venues
- Stargazing and cosmic nature
- Storage
- Traffic and commuting
- Vacations and recreation

3.3 Human

A manned interstellar mission needs much more than a controlled environment and a starship that can travel for decades at hyper speed. Cloning a small Earth is not complete without taking along all social, communal, and psychological aspects that define humanity as we know it. Studies of the human needs determine a lot of the design that will be incorporated in a starship. That is a normal task in architectural design even for a simple residential building. In ordinary design, many aspects are usually accounted for within a macro environment such as urban design, or a microenvironment that would be created by individuals later such as using a small fan inside a room to adjust the temperature or humidity. In space architecture we have to factor in all the macro and micro aspects in the original design.

- Aging in space
- Art
- Children and parenting
- Cloth and fabric, fashion
- Death and ceremonies
- Economy system
- Education
- Generation Zero
- Generation One
- Government and leadership
- Healthcare and immunity
- Interstellar culture
- Language and vocal development
- Law and order
- Libraries and encoding
- Literature
- Losing physical contact with traditional machinery, Earth's nature
- Marriage and social life
- Interstellar psychology
- Music and theatre
- Parties, weddings, and events

- Personal hygiene
- Policing
- Population growth, distribution, and management
- Religion
- Research and advanced studies
- Simulation and virtual reality
- Society, habits and protocols
- Technology development (onboard)
- Time dilatation and contact with Earthlings

4. The Radiation Shield Sample

The Radiation Shield sample represents the assessment done for each item. Analysis of technology status and the time anticipated to be ready for prototype testing is a major part of the evaluation. That includes analyzing the sustainable design requirements of a radiation shield that will endure a trip expanding for long light years under high speed and last for long decades. Other aspects are verified such as deterioration, maintenance, and future expansions.

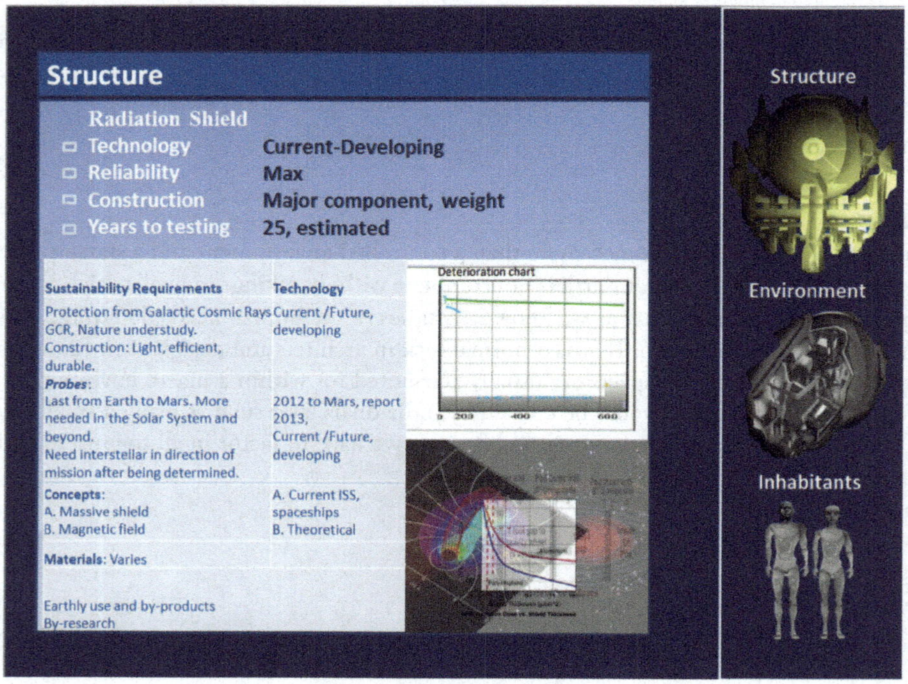

Figure 2. Item Structural- Radiation shield analysis summery

4.1 Sustainability Requirements

Protecting the humans onboard a spaceship from harmful radiation while outside the protective Earth's electromagnetic field and atmosphere is a big concern. The starship itself and its computers and communication systems need protection from radiation. The longer the trip and the further it is form our Earth and Sun, the more dangerous are the radiations. All manned missions, including the International Space Station's (ISS) have protection measures from radiation [4].

Sun plasma spews in the form of flares, or solar wind, increase the radiation hazards to spacecraft and crews. Other cosmic events such as supernovae or black holes increase the high-energy radiation known as Galactic Cosmic Rays (GCR). The solar magnetic field protects Earth, and the solar system planets from the GCR keeping most of the radiation away by a shielding enclosure represented by the heliosphere[5]. The magnetic field of the Earth plays another essential role in deflecting the harmful radiations within its proximity[4].

4.2 Conceptual Design

Current technology concerning protection from radiation is limited. That also limits the time astronauts can spend in space including the time when they are inside the ISS[6], only 200 miles from Earth's surface. Even commercial airplane pilots are subject to radiation despite flying at low altitudes [6].

The classic design depends on using thick materials for spacecraft walls for protection. Protection from the high-energy GCR needs going thicker with conventional methods or using new concepts. A major structural effect is the additional weight.

An idea that originated in 1969 by space engineer Wernher von Braun, calls for using a superconducting magnet to create an electromagnetic shield to protect the starship just like Earth's electromagnetic filed protect its inhabitants. Project SR2S is currently working in that direction[7]. The project is an EU-funded research for developing magnetic shielding to deflect dangerous cosmic rays using superconductors, materials that have no electrical resistance at extremely low temperatures. The weight of the large magnet to be used is an essential concern. Simulations show that by utilizing new superconducting materials such as magnesium diboride (MgB_2) a 30-metre-diameter magnetic field could be produced by a system weighing less than half that of a comparable passive shield depending on thick layers for protection. Project SR2S is supposed to be completed by end of 2015.

Contemplating the current pace in technology advancement, it will take approximately 25 years to reach a stage were a prototype could be tested for a starship and interstellar use. That is considering the different design options and a continuous design development due to the necessity of the radiation shields for manned space missions in general.

4.3 Probes

Further probes are needed especially that we do not know much about the cosmic radiations outside our solar system. The most detailed and recent data about the radiations covers the route to Mars and were received through a spacecraft travelling to Mars in 2012[8]. The high energy radiations outside our solar systems are more intense especially beyond the heliosphere.

Several probes and missions for studying the cosmic rays beyond our solar system and in direction of the proposed route(s) are required. Other items that need further studies and observation are solar flares and cosmic events that release massive concentrated waves of high energy radiation.

5. Status of Structural, Environmental, and Human Items

Following the same way of analysis used for the radiation shield, projections for a hundred elements of design were configured. Graph 3.0 shows the distribution of the hundred items in terms of technology status; current, developing, or future. It is clear that the majority of the elements is in developing phase or still in theory and will be developed in the future. The part that shows as current technology, less than 20%, indicates items with current design technology but still need to be configured for deep space travel.

Graph 3.2 shows the status of technology for the elements under each of the three categories. It also shows the anticipated years needed for development. The human aspects division seems to be the most underdeveloped one as manned deep space missions have not even started. The environmental aspects are in slightly better shape as they were part of the few missions with humans onboard. The structural elements are the most tested as all current space missions included many of them under this division. However, all three divisions need long years of research and development.

Graph 3.1 shows that about 84% of the design elements for the starship have corresponding applications on Earth. The continuous development of such items over the upcoming decades will result in subsequent advantages for their earthly use. Without going into details, all items listed under the environmental division that are developed to be lighter, more durable, and consume minimal energy will be of high benefit to Earthlings, who will be taking advantage of such benefits long before they are utilized on a starship.

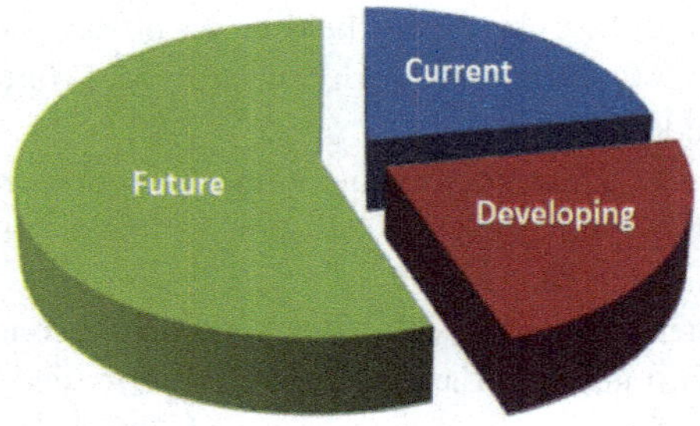

Graph 3.0 Most design items are being developed or still in theory stage

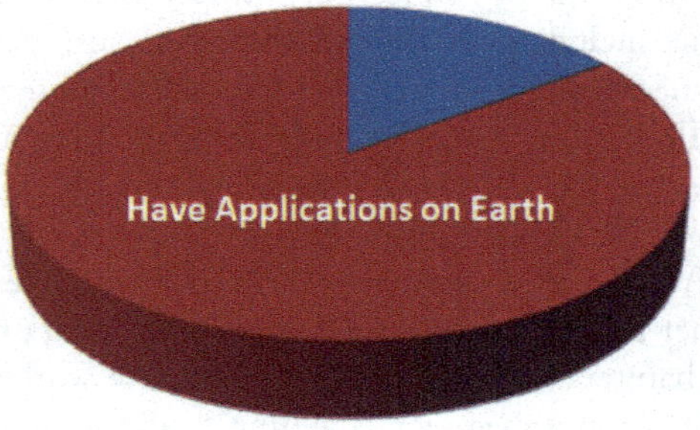

Graph 3.1 About 84% of the items have corresponding applications on Earth

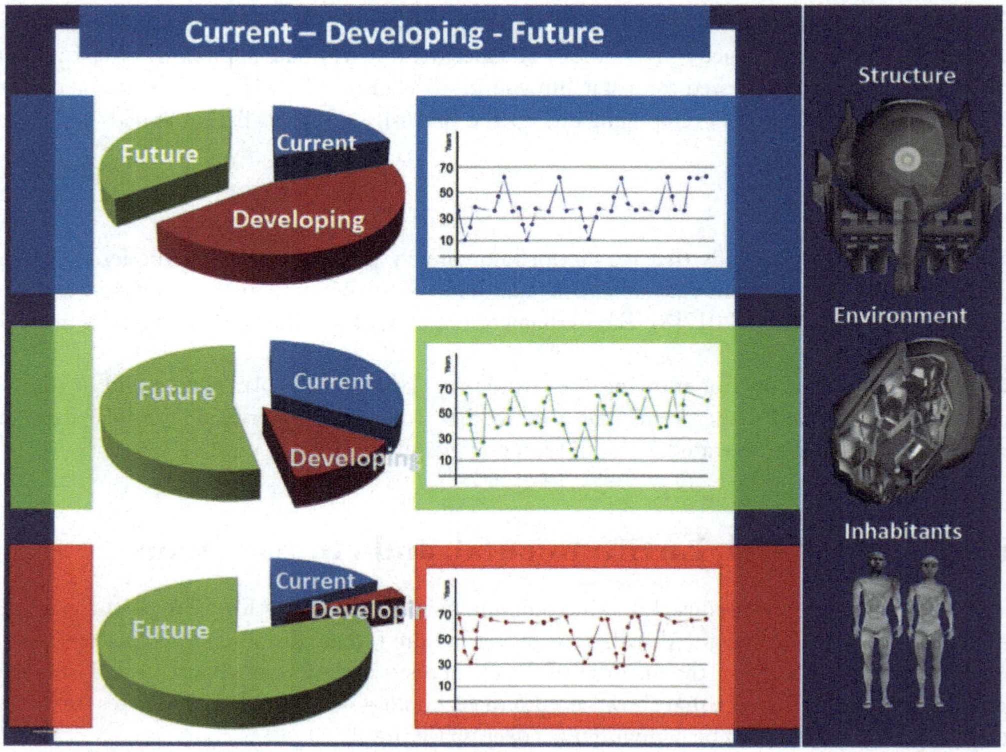

Graph 3.2 Status of technology for the elements of each of the three groups. Human factors are the least developed, followed by Environmental, while Structural items are the most ahead.

Graphs 4.0 and 4.1 show that current space design and technology account for a good portion of the elements needed for the deep space travel. However, only a small portion of the current technology could be reconfigured for deep space travel. For example, we cannot depend on rocket engines that use liquid fuel as prototype or heritage for extended space travel.

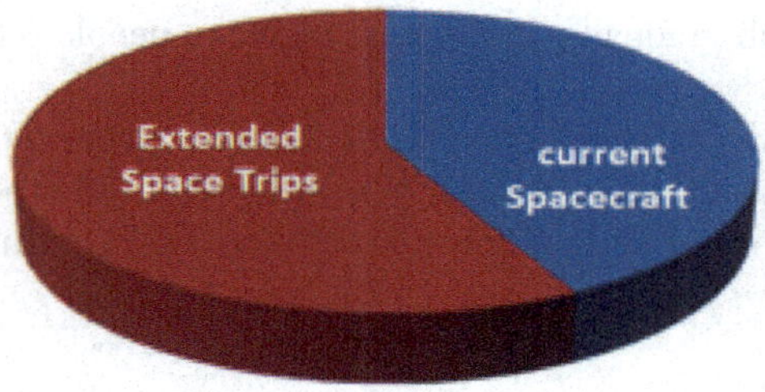

Graph 4.0 Design for current space travel covers about 40% of the deep space elements

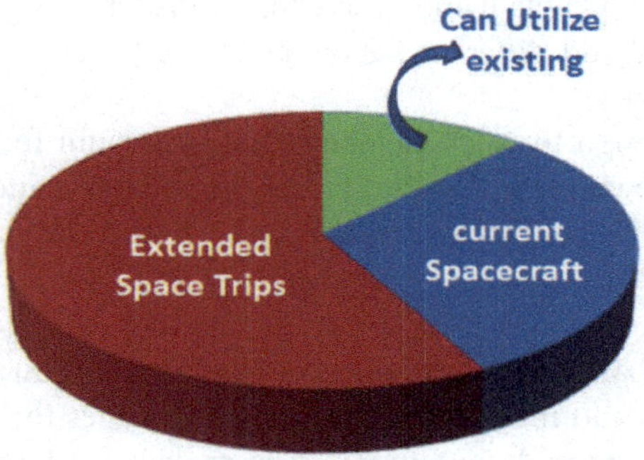

Graph 4.1 Only a fraction of current space design could be configured for deep space travel

5. Starship Inhabitants, Human Factor

The design elements related to the human aspects pose considerable challenges as they include some unpredictable features. Our experience with manned missions is limited to missions to the moon and Earth's orbit. Such short history of humans in space can merely explore conditions for a population on starship several light years away. On a multigenerational space trip, Generations Zero and Generation One are expected to undergo critical circumstances.

5.1 Generation Zero

Generation Zero is composed of those who will leave Earth on a one way journey. To make a simple assessment, let us assume that we have a small to medium starship with a startup population of about 1,000. Out of those, the part that will be considered the crew of trained scientists will be from 200 to 300 persons. The remaining population includes service providers, technicians, educators, and members of the community of various ages and backgrounds. While it might be easier for trained scientists and astronauts to understand leaving everything behind on Earth on a one way trip, it is certainly harder for others. In all cases, and putting all simulations aside, the thousand people who were born and raised on Earth are attached to it, and to the compilation of civilizations that led to their time. They will never see or touch any person that they knew on Earth, except for those leaving with them. They will never see a passing train or a sailing ship in the horizon. They will not be able to climb a mountain or dive into the ocean.

By the time the starship is ready to launch, we should have several space colonies within our solar system that examined this issue at several levels. Currently, this topic is being studied and analyzed with several propositions to help Generation Zero adapt to their new life and future. While suspended animation is being researched for saving weight and energy[9], some drastic suggestions for diverting the effects of leaving Earth are surfacing for multigenerational trips. The starship population, or part of it, could be sent in long space trips prior to the interstellar journey. Even more, Generation Zero could be fully born and trained in a space colony so they will not be emotionally attached to Earth by the time they leave our solar system[10].

5.2 Generation One

Generation One, while might be less attached to Earth and the places or activities that they never experienced being born and raised on a starship, will have other points to address. They need to maintain a certain pre-engineered directions within the interstellar community regarding population, education, careers, and daily life. After all, it is a small society and a certain average number of people per family and in total must be upheld. The same applies for the number of scientists, engineers, technicians, doctors, educators, service providers, artists... to the end of the list. That applies to all generations born on a starship, with Generation One having the first real test.

5.3 Social Development in Space

In an interstellar journey, society takes its own course in adaption and evolution just like any society on Earth. Such developments are built on two main factors; the inherited features within the community and the effects of the surroundings. The inherited aspects for the inhabitants of the starship when leaving Earth will be similar to those for people living on Earth. However, living in an enclosed environment on a multigenerational interstellar trip will add different factors to the evolving culture that will gradually differ from what the people on Earth experience.

Design for such elements should account for the departure time conditions, and allow for changes that are not known yet. Future design accommodations and extensions will be made by the generation that will experience the changes utilizing available resources on the starship. Below is a brief discerptions of some of the main social fields that are subject to change.

5.3.1 Languages and dialects keep evolving through time. The development of languages start with changes in dialects and many dialects become languages through generations. Spoken languages are affected by the direct surroundings. A contemporary example is the French language and how the same language developed in two different places in few generations. The Quebec French spoken in Canada today is very distinctive from the French spoken in Europe. It is not clear what (form of) language the inhabitants of the starship will speak when it takes off, since current languages will keep evolving until then. During the trip, the languages will keep developing onboard and on Earth in ways that adjust to the local surroundings at the two locations.

5.3.2 Art and Literature are direct reflection of the environment. On a starship they will start with what is considered, on Earth, classical in visual art, performing art, and creative literature until the interstellar effects take place. The art might reflect the starship as an environment instead of natural, urban, and demographic surroundings. Performing artists or pop stars on a starship would have to adjust as their audience shrinks from millions to thousands. Other forms of art may emerge, flourish, or decline.

5.3.3 Sports in space will have many adjustments. The vast fields needed for football, succor, or races might have to be adjusted or even replaced by simulated sports and venues. On Earth, different people at different locations have developed localized sports like Tejo in Colombia and Dandi Biyo in Nepal. New sports may develop accommodating the community and the environment on a starship.

5.3.4 Education development will not only change in the means of teaching but also in what people will learn on a spaceship. Do they need to learn about Earth and earthly geography and weather? What about agriculture and the history of civilizations on Earth? That might need adjustments considering the destination planet. Will the Earthly physics laws taught on Earth apply to that on the starship and at the final destination, and how are such laws adjusted? How much scientific research will be conducted onboard, and how different will that be from the scientific breakthroughs occurring on Earth? A vast part of learning comes from interacting with diverse cultures and communities. Can simulation cover this gap for the limited number of inhabitants?

5.3.5 Encoding, documenting, libraries, and passing information are changing in our current time at an unpredictable manner and speed. Encoding on Earth moved from printed books, to electronic books, and downloadable data on tablets in a couple of decades. While we do not know what further developments will take place until the starship starts its mission, the type of information to be circulated and recorded during the mission is the major concern. What is needed to be documented and passed to next generation, and by what means? Are there going to be any libraries or bookshops in the future on a starship?

5.3.6 Technology is another leading aspect that has been through unprecedented breakthroughs in the past few decades. It is not clear how far the technology will progress in the forthcoming decades until interstellar travel commences, and how things will develop onboard. Computers fast paced development in the 1990's, phones' in the 2000's, and tablets' in the past decade give an idea of how unpredictable this field is. Such developments will be implemented into all other items of design as applicable, while the advancements during the trip will match the needs of the inhabitants and will be built upon current and future knowledge.

5.3.7 Economy Systems develop naturally in communities. They get more complex with the growing of the population. A well-organized society of small size, take1000 persons, on a starship might start as a simple communal society with no need for making business. How would a society without economic incentive be driven? What type of economy could work as the population grows? Will they use monetary system or stocks on a starship? Can they distribute property? Can any person have a 1,000 times more money or property than others, and how would that affect everyone? Will the market be driven by supply and demand in such a small population? Is there a need to build financial institutes?

5.3.8 Social habits and protocols dealing with others differ considerably between communities on Earth. Will interacting with the same people for years make communication among the inhabitants more casual? Or will that create a new definition for privacy that differs from than what we have on Earth. Will people need more open spaces or more private enclosures? How would people deal with social occasions such as marriage and death?

6. Conclusion

Interstellar architecture introduces a new approach for working on deep space projects. The sustainability aspects in three main areas; Structure, Environment, and Human, need to be assessed, and the contentious development in technology must be incorporated taking into consideration the progress expected until launching a starship and later onboard the starship. Comprehensive futuristic scenarios simulating the future developments for all design items go beyond current modules, and accounts for variables that might be new to the design process. However, that brings the design progression back under control, and somewhat puts it in line with the way ordinary architecture projects are evaluated and fostered, resorting to the basic elements of design; function, budget, esthetic, and life cycle.

References

1. "Sustainable Design for Extended Space Journeys," Antoine G. Faddoul, 100YSS Symposium, 2013
2. published in a special supplement of the Journal of the British Interplanetary Society in 1978
3. *Space Settlements, A Design Study,* Published by NASA and Stanford University 1977
4. NASA, *Space Radiation Analysis Group*, Johnson Space Center, March 5, 2014
5. *Galactic cosmic rays in the heliosphere,* Thomas, S.R., Owens, M.J. and Lockwood, M., 2014
6. NASA, *Space Faring the Radiation Challenge*, 2008
7. *Space Radiation Superconducting Shield,* www.sr2s.eu
8. Radiation measured by Curiosity on voyage to Mars, *Astronomy magazine*, March 2013
9. NASA, *Torpor Inducing Transfer Habitat For Human Stasis To Mars,* 2013
10. *Astro-Sociology: G-Zero to G-1Manning and Phasing*, Antoine G. Faddoul, 2014

Exoplanets and Survival Kits

Where to Go and What to Pack for an Interstellar Voyage?

Pauli E. Laine, MSc

Abstract

The selected target (among the available payload mass) for interstellar mission affects what can and must be taken on the journey. The number of exoplanets and information about them is increasing all the time. For a potential interstellar mission we should choose the most Earth-like planet with proven atmosphere and water. With only limited information about the conditions on the target planet, we should prepare our mission with a suitable "survival kit" in order to survive and start a self-containing colony. In this paper I will describe a minimal mission concept based on our current knowledge about habitable exoplanets and possible propulsion technology. This presentation is a feasibility study about where we can go and what we can and should take with us. As a case study, I will consider a long duration mission with suspended animation to some of the most potential Earthlike or super-earth-size exoplanets to establish the first human interstellar colony.

Keywords

Manned interstellar flight, exoplanets, habitability, terraforming

1. Introduction

If and when we decide to travel to another star for some reason (survival, political, exploration), where would we go, and what should we take with us? Such questions are essential when planning a manned interstellar voyage. As the 100 Year Starship organization's goal is make the capability of human travel beyond our solar system a reality within the next 100 years, we will consider here only manned interstellar missions [1].

2. Where Can We Go?

It is commonly accepted that ship capable of speeds of 0.1 c, or about 30,000 km/s with some reasonable energy usage can be used for interstellar travel. At this speed the nearest star systems can be reached in some 50 years. Today we do not have anything that could achieve such speed, but there are some proposals about near future propulsion systems, (e.g. the Fission Fragment Rocket Engine, or FFRE [2]) that can theoretically be used for manned interstellar travel.

Available payload mass is always a limiting factor for developing space flights. Especially long duration manned interstellar journeys will require very careful planning, depending first on whether active crews (possibly generation ships) or hibernating crews (a common solution in science fiction) are involved. In any case, oxygen, water, and nutrients are needed for both live and hibernating crews. Here we assume, that hibernation can be used, minimizing all the needs during the journey. The number of crew members will directly affect the payload mass, but how many colonists are needed for a self-sustaining colony? This will depend on whether there will be only one planned mission or perhaps many successive missions to same target planet. While there is no certain minimum size for a genetically viable colony, more is always better. Cameron Smith proposed a population of 40,000 as a safe and stable number for a multigenerational interstellar voyage [3]. However, entire modern human populations might be the descendants of a mere 10,000 individuals, so the colony size could be considerable smaller [4]. In any case, a mere 10,000 people with some life support gear could weight approximately 1000 tons. With all the necessary water and nutrition, the payload mass would exceed the capacity of all current and foreseeable high speed propulsion technologies. This leaves us only two options: a) whether we use multiple missions or b) use one huge and slow generation ship (e.g. an asteroid ship with artificial gravity and closed water and food circulations). Here someone will always ask: what's the point in a multigenerational voyage that can last for thousands of years, while we, back on Earth, might invent hyperdrive and pass the poor ark in its first decades? The only rationale for a generation ship is that the ship itself is a world, a 'snapshot' of selected life forms that could survive even if Earth becomes uninhabitable.

3. Potential Targets

Today over 1800 exoplanets have been discovered, most of them are Jupiter-size gas giants. This is due to of our limited detection sensitivity. Even the closest exoplanets are extremely difficult to observe due to their overwhelmingly bright parent stars. Nevertheless, information about exoplanets is increasing all the time, and lists of presumed habitable exoplanets are maintained, e.g. at Planetary Habitability Laboratory [5] (figure 1).

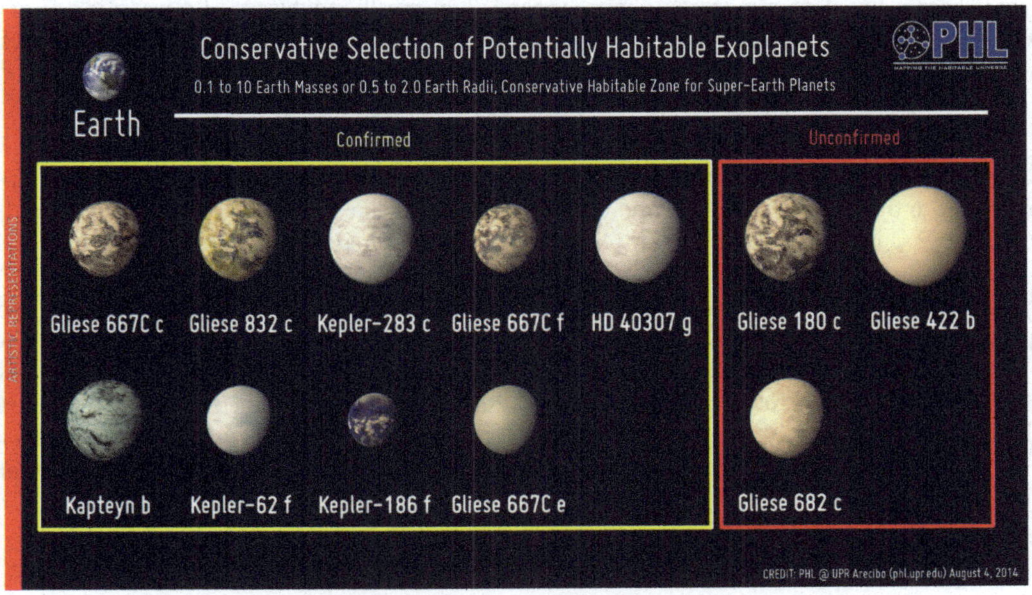

Figure 1. Conservative selection of potential habitable exoplanets, August 2014 (PHL/UPR, used with permission).

Figure 1 represents the best knowledge about known possible near Earth-sized habitable planets so far. As such, they are also our best potential targets for manned interstellar travel, if we will have to go quickly.

How far will we have to travel? The number of star systems within 13 light-years is about 33, within 50 light-years about 1400, and within 250 light-years about 260,000 [6]. If we limit our range to approximately 15 light years, then we know only 3 potential targets (figure 2).

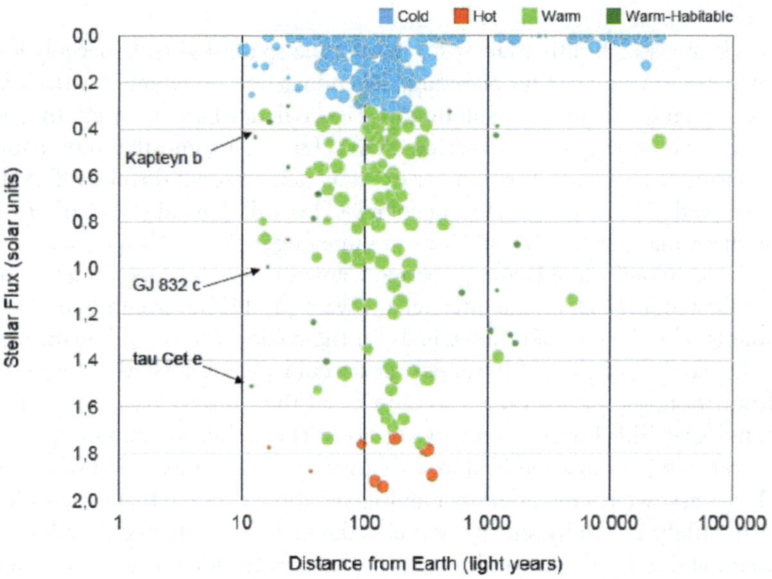

Figure 2. Distribution of habitable zone planets by stellar flux and distance from Earth, August 2014 (PHL/UPR, used with permission).

Tau Ceti e is an unconfirmed habitable 1.6 x Earth radius planet 11.9 light years from Earth. Kapyten b is a confirmed habitable 1.6 x Earth radius planet 12.7 light years form Earth. Gliese 832 c is a confirmed habitable 1.7 x Earth radius planet 16.1 light years form Earth with equal stellar flux. There is very little we know about the true conditions of these planets. They are located in habitable zone, but so is Venus in our Solar System.

Even our best choice for a target planet is scarcely suitable for humans. The limits for a stable human presence are quite narrow, e.g. A temperature of approximately 273 - 303 K, pressure 700 - 5000 hPa, and radiation 1 - 3 Gy. There is hardly any elemental oxygen, since oxygen is biologically produced. Also water is one of the main criteria for habitable exoplanets. Some constraints can be circumvented, e.g. with portable life support systems and underground habitats, but eventually we will have to terraform or find another place.

3.1 Backup Plan

We will get more and more information about our target planet from Earth, and with our ship's own instruments as we get closer to the target. What if we find out that our target is like Venus (737 K temperature, and 9.2 MPa surface pressure)? Currently we have no way to measure or even estimate an exoplanets' surface temperature and pressure. Things might get better in the future with larger and better telescopes, but there will always be a margin of error that can be fateful to our mission. Can we somehow use multiple potential targets in order to find the best target planet, i.e. a trajectory with possibly multiple stops? We (the crew, ship's computer and mission control on Earth) could do habitability estimations during the approach phase and choose whether to stop or continue to the second target (figure 3). However, we must do our decision before the so called No-Stop Limit and braking phase. Due to very long distances, the final decision has to be made either by the crew or ship's computer system. It remains open whether there really are multiple habitable planets reachable with feasible trajectory and propulsion technology.

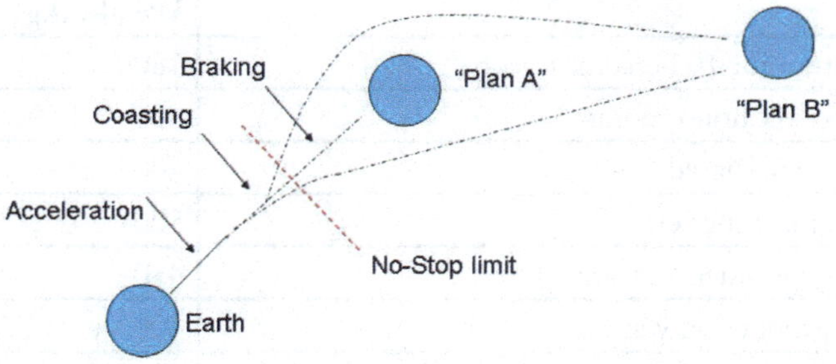

Figure 3. Possible trajectories for two habitable planets.

Another possibility is to engage in a suspended park orbit over the target planet and try to terraform with e.g. with microbes, orbital mirrors, and asteroid impacts. In any case, we had to stop somewhere, unless we have a self-sustained generation ship.

4. What to Take With Us?

When we have found habitable exoplanets with tolerable conditions for human colonization, what should we take with us when we travel there? Payload mass is the critical factor that sets limits what can be taken. There are some items that are essential for the survival of the new colony, the first two being oxygen and water. As mentioned earlier, exoplanets most probably lack elemental oxygen in the atmosphere, so that humans will continually require suits and/or breathing apparatus. There can only be limited amount of oxygen on board the landing vehicle. On the other hand, water could be abundant on many exoplanets, either as liquid in oceans or lakes, or as ice. And water can be used to produce oxygen by means of electrolysis. So our inventory list must include oxygen and energy production devices. Electric energy can be produced with solar cells or e.g. RTG (radioisotope thermoelectric generator).

4.1 Getting Started

Now, that we have oxygen, water, and energy, what else is needed most for the first moments after arrival? In order to be able to stay and work outside the lander module, some kind of extra-vehicular activity suits (EVA) are needed. At minimum (and optimum) level it could be just breathing masks and air containers. After this there will be a need to set up temporary houserooms, e.g. pressurized domes. Some furniture (tables, beds) and equipment (e.g. lighting systems) could be borrowed from the lander. Food production (seeds, plants, greenhouses) would be the next logical step, but there should also be nutrition stock for approximately one year to ensure a successful harvest.

4.2 Building a Foothold, and Beyond

After the first steps of self-sustaining colony have been taken (e.g. the provision of oxygen, energy, and food production), new needs will arise, for example, there will definitely be a need for health care and spare parts for EVA suits and production systems. We should be prepared for all this. We might want to build permanent buildings. The simplest buildings might be of brick, and for them we need only shovels and brick moulds [7]. If the planet's soil is suitable, colonists could plant trees to provide raw material for buildings and furniture. Trees, plants, microbes (e.g. cyanobacteria) and algae would be the first steps toward terraforming.

5. Case Study

Let us summarize the main features of this paper's 'survival kit' in a simple case study. To set certain limits, we will assume that we have a) a high speed propulsion system capable of 10% light speed, b) a lander module with 5000 kg payload capacity and crew of 10, c) some hibernation method, and d) a habitable target planet with tolerable temperature and pressure, and accessible water. Payload capacity (5000 kg) might contain a selection of the items listed in chapter 4 (see Table 1).

Table 1. Proposal of survival items included in 5000 kg payload mass.

Item	Weight (kg)
Nutrition for 10 persons/1 year	1800
Temporary houserooms	800
EVA suits, shared	700
Tools, machines etc.	500
Energy production wares	450
Food production wares	400
Oxygen production equipments	250
Health care accessories	100

Mission mass could be possibly be reduced to a minimum through crew hibernation and thus a possibility of reducing food requirements. With this configuration, the total payload mass should be approximately 20 metric tons.

6. Conclusion

The selected target and available payload mass for an interstellar mission affects what can and should be taken to the journey. Any potential colony must be prepared for some limitations in addition to basic oxygen, energy, and food production. On the other hand, there is an upper limit what can be taken on the voyage. A manned interstellar journey will therefore require very careful planning and review of every single item.

Although the number of exoplanets and information about them is increasing all the time, true surface conditions remain unknown. This is why we need some back up plan in the case of arriving at an uninhabitable planet. We should have either alternative target(s), or have adequate terraforming capabilities

Acknowledgements

The author would like to thank Prof. Abel Méndes (University of Puerto Rico at Arecibo) for permission to use Planetary Habitability Laboratory Figures 1-2.

References

1. 100 Year Starship. www.100yss.org

2. Laine, P. E., "To the stars with current technology?", *100 Year Starship 2013 Public Symposium Conference Proceedings*, 100 Year Starship, 2014.

3. Smith, C. M. (2014). "Estimation of a genetically viable population for multigenerational interstellar voyaging: Review and data for project Hyperion", *Acta Astronautica*, 97, 16-29.

4. Ambrose, S. H. (1998). 'Late Pleistocene human population bottlenecks, volcanic winter, and differentiation of modern humans'. *Journal of Human Evolution* 34 (6): 623–51.

5. Habitable Exoplanet Catalog. Planetary Habitability Laboratory, University of Puerto Rico at Arecibo. http://phl.upr.edu/hec

6. *The Atlas of the Universe.* http://atlasoftheuniverse.com

7. Zurbin, R. G., *The Case for Mars.*, Free Press, 1996.

KLT Filtering, Even from Relativistic Sources

Claudio Maccone, Ph.D.

Abstract

This paper looks forward to the time when our spacecraft achieve relativistic speeds and discusses the impact of relativistic motion on spacecraft communications The paper addresses a key element for interstellar communications; using the Karhunen-Loève Transform (KLT) mathematical algorithm for communications processing. KLT is argued to be superior to the classical Fast Fourier Transform (FFT) used in Earth-based communications today.

1. Introduction

One of the most critical elements of interstellar travel will be reliable communications between the Earth-base and the spacecraft for the sending and receiving of information by interstellar travelers (manned or robotic). The extreme distances and relativistic motions involved with interstellar travel will have negative impacts on signal strengths, timing and composition, requiring high quality processing to retrieve the data of interest. The KLT can filter signals out of the background noise in both wide and narrow mobile bandwidths, providing the foundation for earth-based communications with interstellar spacecraft. That is in sharp contrast to the FFT that rigorously applies to fixed narrow-band signals only.

2. Details

The concept of the Karhunen-Loève Transform is essentially a transformation to the principal axes in Hilbert space (an infinite-dimensional analog of Euclidean space). For signal processing, using the KLT allows the extraction of weak signals from noise of any kind over wide or narrow bandwidths. The signal extraction can be achieved by the KLT more accurately than by the FFT, especially if the signals buried into the noise are very weak, in which case the FFT fails. This superior performance of the KLT happens because the KLT of any stochastic process (both stationary and non-stationary) is defined from the start over a finite time span ranging between 0 and a final and finite instant T (contrary to the FFT, which is defined over an infinite time span). Mathematically, the series of all the eigenvalues (a special set of scalars associated with a matrix equation. The determination of the eigenvalues is equivalent to matrix diagonalization) of the autocorrelation of the signal (noise + signal) may be differentiated yielding the "Final Variance" of the stochastic process in terms of a sum of the first-order derivatives of the eigenvalues, resulting in the immediate reconstruction of a signal buried into the thick noise. This allows a KL expansion on a noisy signal, followed by an examination of the first eigenvalue to get a measure of the signal to noise ratio. For a large first-eigenvalue the corresponding eigenfunction (each of a set of independent functions that are the solutions to a given differential equation) is a good representation of the signal. Applicable to broadband signals (where the Fourier transform is not particularly useful), the KLT provides a mathematical foundation for the optimal extraction of any data from any noise.

3. A KLT Example

As a signal is received at the input of a receiver, the KLT separates the input (noise + signal(s)) into uncorrelated components. This separation yields the Karhunen-Loève's eigenfunctions and the kernel of this integral equation is an autocorrelation. The KLT adapts itself to the shape of the input by adopting, as a reference frame, the uncorrelated component spanned by the eigenfunctions of the autocorrelation. This turns out to be a linear transformation of coordinates in the Hilbert space.

There is no degeneracy since each eigenvalue corresponds to just one eigenfunction, and, the eigenvalues turn out to be the variances of the random variables. By sorting, in descending order of magnitude, both the eigenvalues and the corresponding eigenfunctions, we can decide to consider only the first few eigenfunctions as the "bulk" of the signal and declare the rest of the values as "noise." Thus we have isolated the signal of interest.

4. The "Lack of Use" Issue

If N is the size of the autocorrelation matrix, the number of calculations required to find the KLT is of the order of N squared, while the same number of calculations for the FFT is much less: just N ln(N). This computational burden prevented scientists from replacing the FFT with the KLT until 2007.

That winter the SETI-Italia Group at Medicina discovered a way to circumvent the computational burden of the KLT. Called the Bordered Autocorrelation Method (BAM), the approach is to regard the autocorrelation as a new function of the KLT Final Instant (T). This is accomplished by adding one more row and one more column (i.e., bordering) to the autocorrelation matrix for each new positive eigenvalue, increasing the value of T. The idea is to use the BAM to exploit the dependence on the Final Instant on both sides of the relationship.

5. A BAM Example

Consider a pure tone (a sinusoid) buried in unitary white noise; write down the autocorrelations, assuming that the sinusoid and white noise are uncorrelated, and pull in the KLT of these two stationary processes. The expansion of both autocorrelations introduces the KLT eigenvalues for both processes and yields a new form of these KLT expansions from which unneeded terms can be deleted by executing two simple integrations from zero to T. The result, after the two integrations, is the relationship among the eigenvalues; it shows that by increasing "n" the two eigenvalues coincide, and that by increasing "T" the two eigenvalues coincide. But it also shows that the periodic part of the equation goes like the sine square of omega, i.e., with a period of twice omega, or most importantly twice the frequency of the signal of interest. In other words still, we have found the KLT eigenvalue of the pure signal as a function of "T". The frequency of the signal of interest is equal to half of the frequency of the inverse Fourier transform of the derivative with regard to T of the dominant KLT eigenvalue. This is the BAM-KLT method that allows for the isolation of frequencies of interest even for SNR very very low.

The Karhunen-Loeve Transform is not widely used today because of its mathematical complexity and processing costs; however, the development of BAM has offered a path of increased usage. Likewise, the advent of parallel computing techniques, coming in the near term, may revolutionize the algorithms for finding the eigenvalues and eigenvectors of symmetric matrices, enabling the KLT to become readily available to all scientists and engineers. 100 Year Starship should take on the task (perhaps in collaboration with IEEE) of encouraging more academic institutes to require courses in the KLT.

Space Mineral Resources

Prior to, During, and After Interstellar Flights

Peter A. Swan, PhD

Space Mineral Resources: Prior to, During & After Interstellar Flights

Peter A. Swan, Ph.D. Member, International Academy of Astronautics
President, Member BofD's, International Space Elevator Consortium
Director, Space Way Research Institute

Cathy Swan, Ph.D. Member, International Academy of Astronautics, President, SouthWest Analytic Network, Docent, Phoenix Art Museum.

100 Year StarShip Symposium

Note: many images from Heinlein Prize Trust and Excalibur Exploration

10/31/2015

Image from IAA Study

Vision -- ***Leverage the Phenomenal Resources – "Along the Way".***

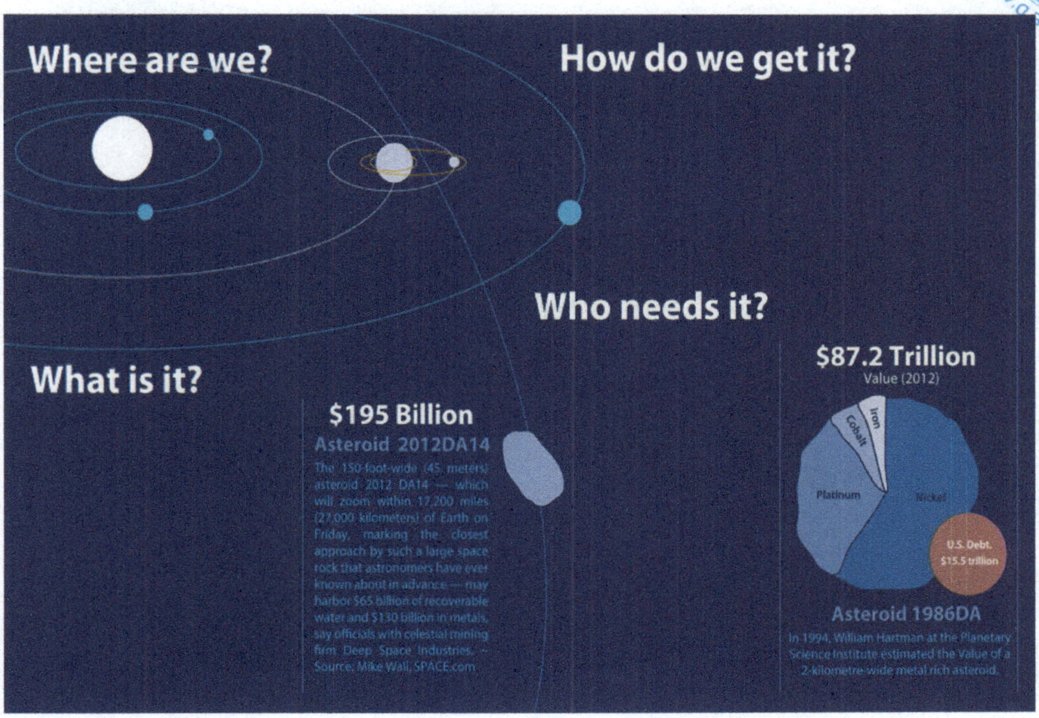

3

Hypothesis: For Interstellar travel, mining and processing raw resources is essential:

- Develop Tools, Practice and Grow Skills in Earth-Moon Ecosystem
- Logistics support for Interstellar Flight will include processed SMRs
- Logistics re-supply along the way will require special designs
- Once arrival in target Solar System, will need resource replenishment

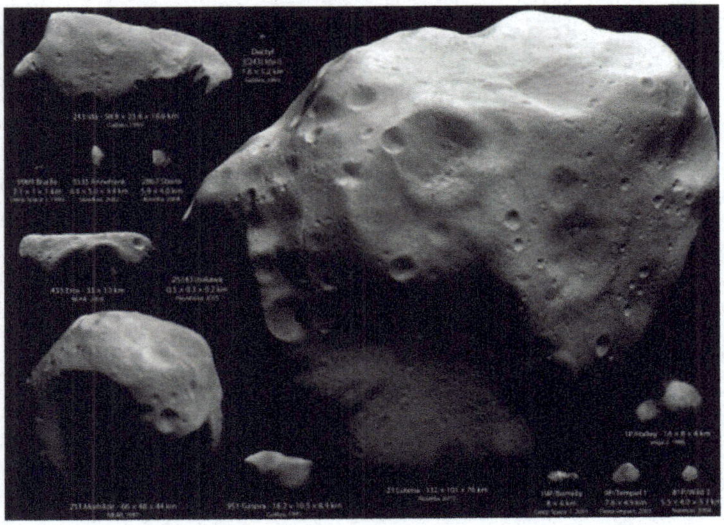

Today's Topics

- Introduction
- IAA Study
- Logistics – Develop Tools
- Logistics – Store Supplies
- Logistics – Examples
- Conclusions & Questions

Image from 100 YSS

10/31/2015 5

Any sufficiently advanced technology is indistinguishable from magic.

Arthur C. Clarke

One man's 'magic' is another man's engineering. 'Supernatural' is a null word.

Robert A. Heinlein

Opening Interstellar Flights Starts at Earth Moon L-1 – A Vision

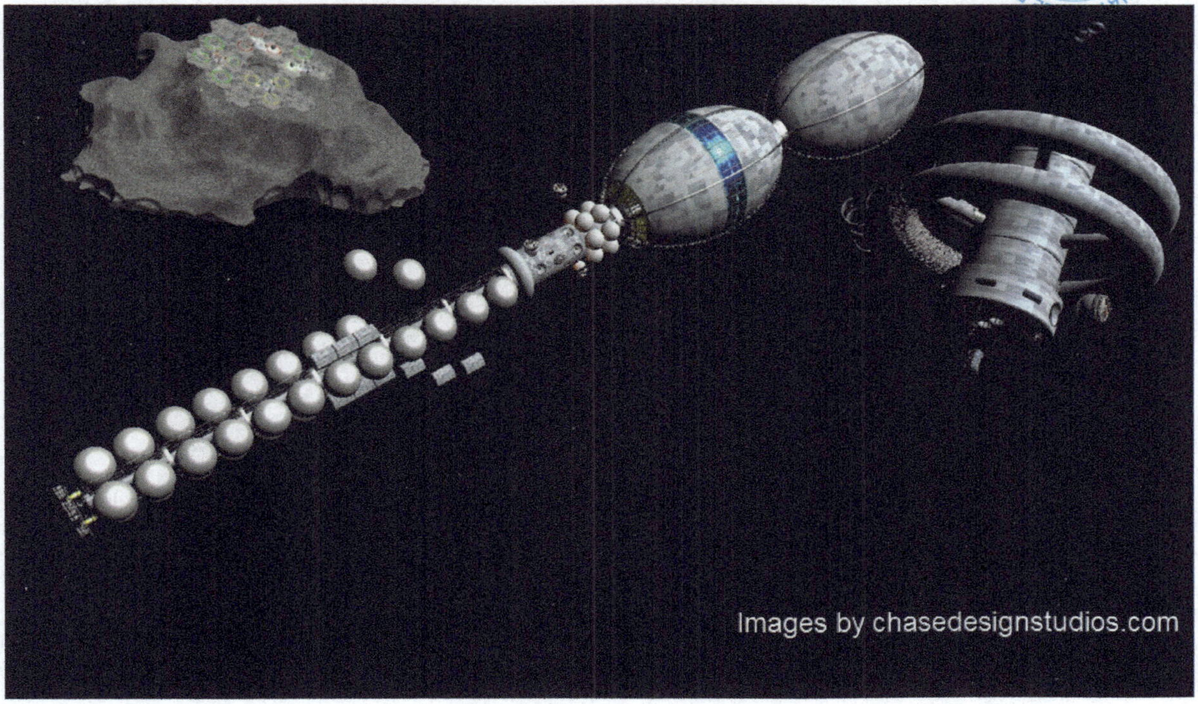

Looking to the Future

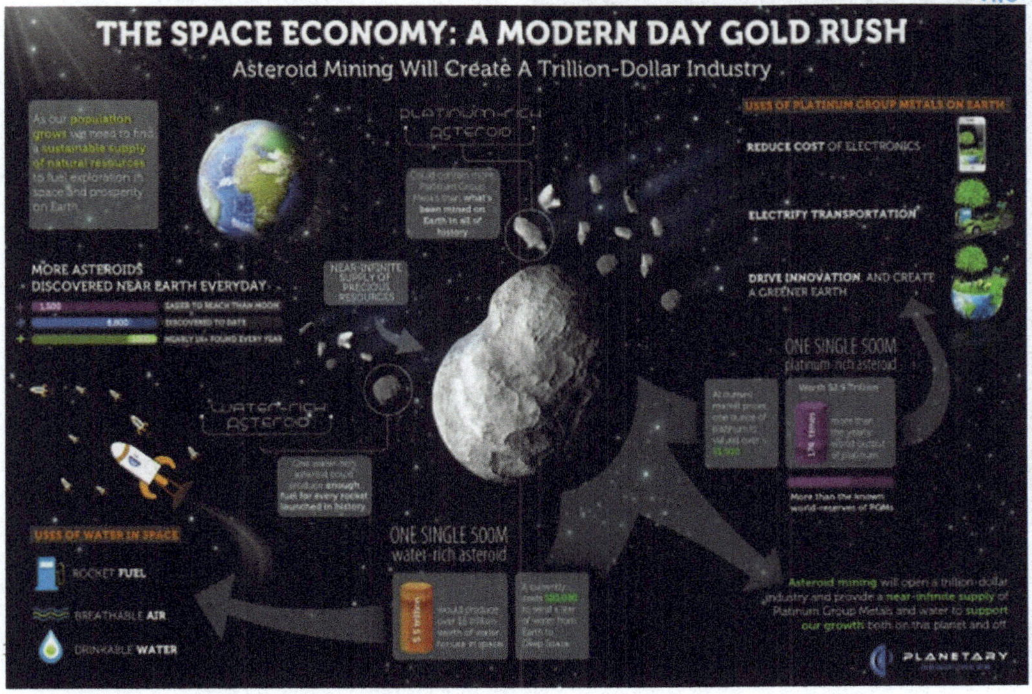

8

Today's Topics

- Introduction
- IAA Study
- Logistics – Develop Tools
- Logistics – Store Supplies
- Logistics – Examples
- Conclusions & Questions

Image from 100 YSS

10/31/2015 9

International Academy of Astronautics

Editors

Arthur M . Dula
Zhang Zhenjun

Research and Editorial Staff:

Dr. Peter Swan
Mr. Roger Lenard
Dr. Cathy Swan
Mr. Brad Blair
Ms. Anat Friedman
Ms. Shirazi Jaleel-Khan
Mr. Jason Juren

Authorized by the IAA Scientific Commission October 2012

Published (385 pages) September 2015

10/31/2015 10

Purpose of the SMR Study

- To provide, in one document, the current state of the art of the technology, economics, law & policy related to Space Mineral Resource opportunities and to make recommendations for moving forward.
- To provide a logical, systematic and practical road map to promote and encourage near term evaluation, development and use of space mineral resources.
- No comprehensive summary of the current literature on this subject is now publicly available. This IAA study is the first comprehensive study of the subject; and, thus it should be of significant value to its development for the benefit of humanity.

From Robert Heinlein Trust

International Academy of Astronautics

Recommendation

Develop technologies, corporations and government relationships to support the following action plan.

Phase One: Initiate the business infrastructure on Earth 2014-2020

Phase Two: Execute prototype flights to potential asteroids as well as testing hardware in LEO 2015-2022

Phase Three: Initiate mining operations with sale of product 2018-2029

Expected Results: Selling water at the Earth-Moon Lagrangian Point #1.

12

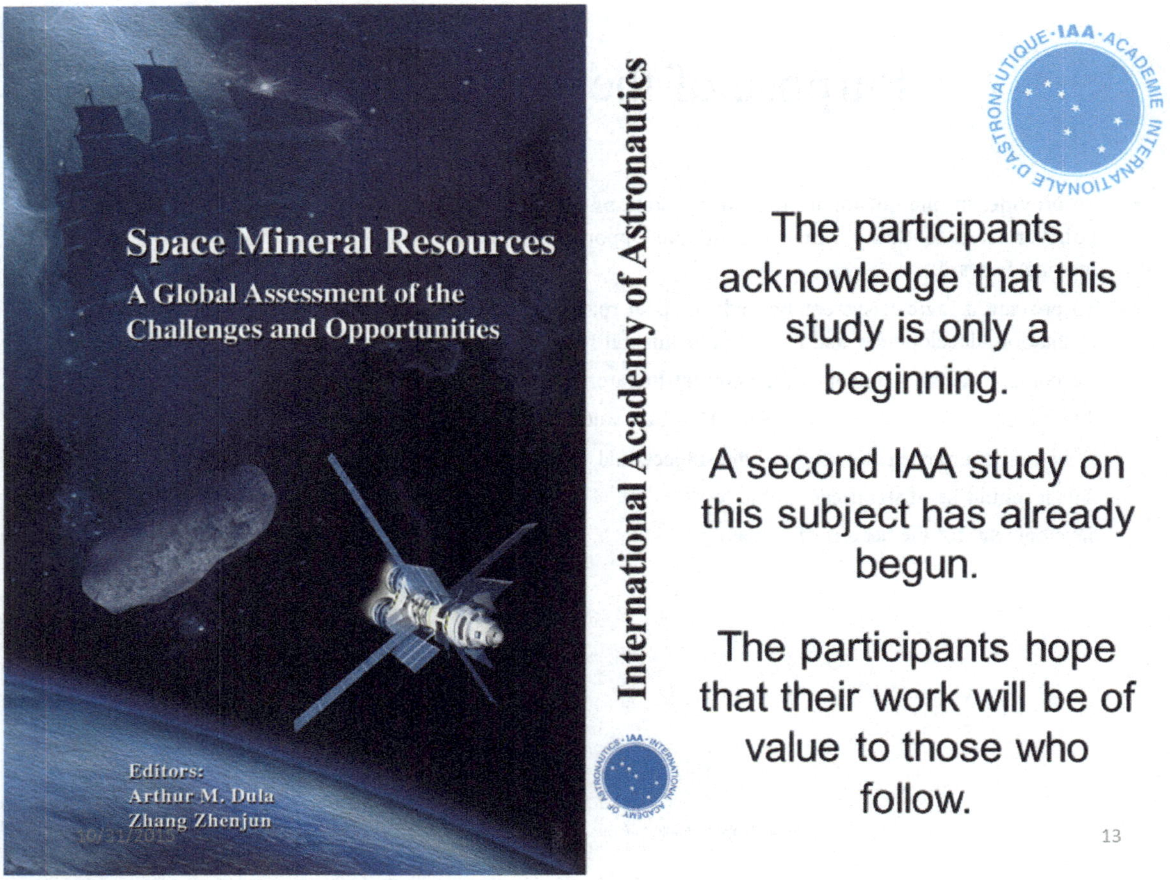
Space Mineral Resources
A Global Assessment of the Challenges and Opportunities
Editors:
Arthur M. Dula
Zhang Zhenjun
International Academy of Astronautics
The participants acknowledge that this study is only a beginning.
A second IAA study on this subject has already begun.
The participants hope that their work will be of value to those who follow.
10/31/2015
13

The Case for Space Solar Power
ISBN : 978-0-9913370-0-2
Publication Date: January, 2014
Page Count: 488
Format: Hardcover - $49.95
eBook - $9.95
Space Elevators:
An Assessment of the Technological Feasibility and the Way Forward
ISBN : 978-2917761311
Publication Date: January, 2014
Page Count: 349
Format: Hardcover - $29.95
eBook - $9.95
Virginia Edition
SCIENCE DECK
Advancing Robert A. Heinlein's vision of humanity in space.
Books for sale at the Heinlein Display
Ordered online at: www.virginiaedition.com
10/31/2015
14

Essence of SMR

- Enhance the human condition on Earth
  - Provide Jobs
  - Re-invigorate education in science, technology, enginering, and mathematics [STEM]
  - Stimulate innovation
  - Provide a vision for return on investment [ROI] from space
  - Stimulate commercial investors for space activities
  - Provide avenues for commercial expansion into our solar system
- Initiate Commercial Movement into the Solar System
  - Provide a vision for movement off-planet
  - Provide profit motive for movement into space
  - Provide commercial products to national space exploration programs
  - Enable space exploration
  - Enable space colonization
  - Enable solar power satellites

10/31/2015 15

Cosmic Study Outline

To provide a logical, systematic and practical road map to promote and encourage near term evaluation, development and use of space mineral resources (SMR)

Executive Summary

1. **Introduction**
2. **Mining of Space Resources**
3. **Market Approach**
4. **SMR Roadmaps**
5. **Analysis of Systems**
6. **Modeling and Analysis**
7. **Policy and Legal**
8. **Finding, Conclusions and Recommendations**

10/31/2015

Executive Summary

"Don't undertake a project unless it is manifestly important and nearly impossible." Edwin Land, quoted in the Coral Reef Alliance letter, March 30, 2011. www.coral.org

Executive Summary:

- The exploitation of space mineral resources is becoming a commercial space endeavor for the benefit of humanity and profit
- The question on the table is not "how" to leverage space minerals resources; but, "how best" to leverage them
- Preliminary economic conclusions include (1) architectures based upon returning precious metals to terrestrial markets alone appears to be a non-starter, (2) the existence of in-space customers for propellants, consumables, structural materials, and shielding could make asteroid mining economically feasible, and (3) longer-term hybrid architectures with both terrestrial and in-space customers could become feasible as costs drop and market size increases.

10/31/2015 17

Major Conclusion & Finding

Major Conclusion: The process of mining water from asteroids, the Moon or Mars will ensure that the key elements are available at the spaceports of the future. Water will ensure that human exploration will expand beyond low Earth orbit with the profit motive driving the exploitation of resources.

Principle Finding: SMR ventures cannot wait for government programs to lower technological and programmatic risks. Commercial ventures must determine the optimum path for commercial success and aggressively lead the way beyond LEO. During the first half of the 21st century, space leadership will come from commercial enterprises and not depend upon government space programs.

10/31/2015 18

Cosmic Study Conclusions

The conclusions from this study fall into a few distinct categories:

- **Legal:** The space elevator can be accomplished in todays arena!
- **Technology:** It can be accomplished with today's projection of where materials science and solar array efficiencies are headed.
- **Business:** This mega-project will be successful for the investors
- **Culturally:** This project will drive a renaissance on the surface of the Earth with its solution to key problems, stimulation of travel throughout the solar system, and inexpensive and routine access to GEO and beyond.

10/31/2015 19

Today's Topics

- Introduction
- IAA Study
- Logistics – Develop Tools
- Logistics – Store Supplies
- Logistics – Examples
- Conclusions & Questions

Image from 100 YSS

10/31/2015 20

Propellant Flow (quantity)

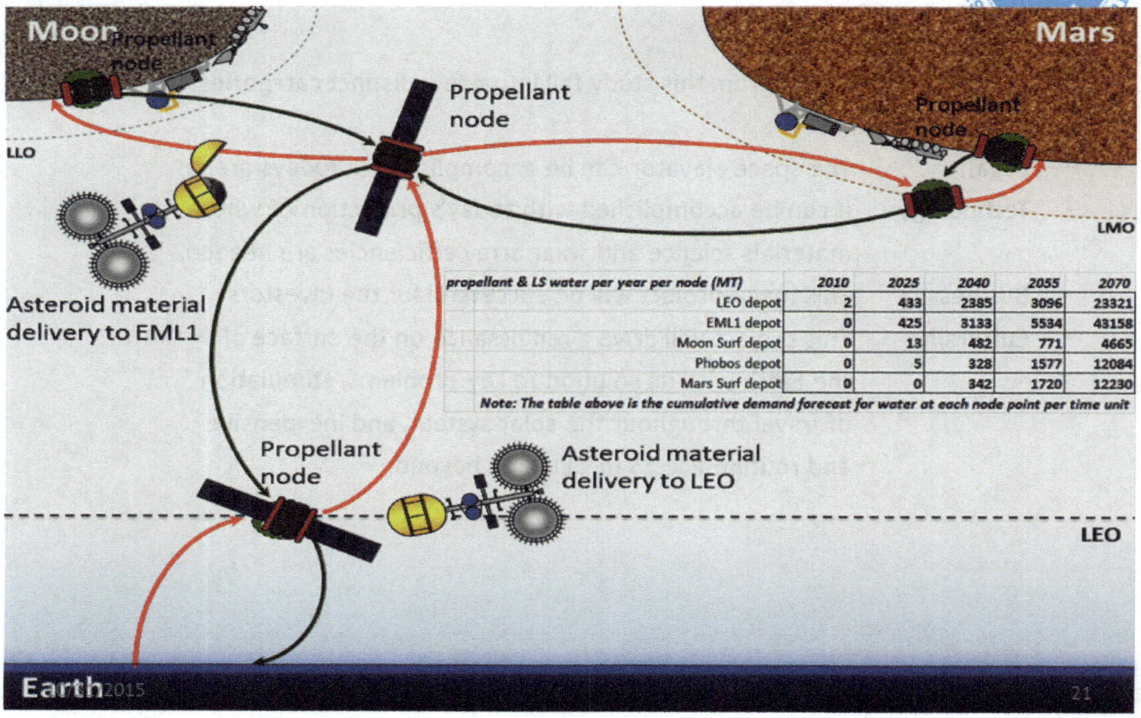

propellant & LS water per year per node (MT)	2010	2025	2040	2055	2070
LEO depot	2	433	2385	3096	23321
EML1 depot	0	425	3133	5534	43158
Moon Surf depot	0	13	482	771	4665
Phobos depot	0	5	328	1577	12084
Mars Surf depot	0	0	342	1720	12230

Note: The table above is the cumulative demand forecast for water at each node point per time unit

Earth Moon Infrastructure Velocity Requirements:

Delta-Vs in Earth's Neighborhood [Mankins, 2012].

Bummer – approximately 10,000 kilometers per second of energy required to get to LEO

10/31/2015

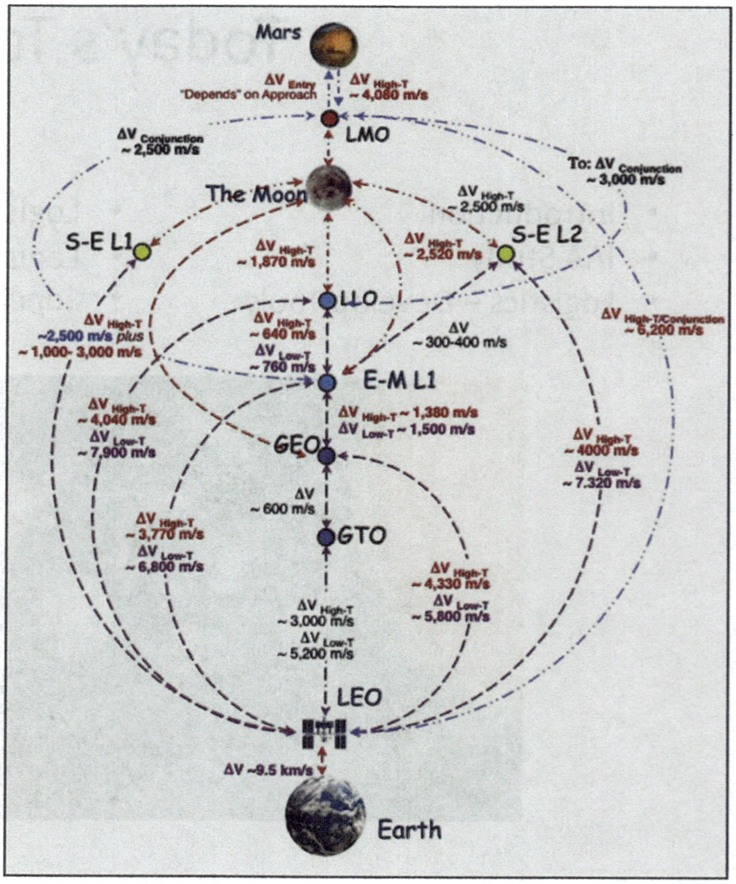

Billionaire Space Investors

rank	name	age	net worth	source	space investment
19	Jeff Bezos	49	$25.20	Amazon	Blue Origin
21	Sergey Brin	40	$22.80	Google	Google Lunar X Prize
20	Larry Page	40	$23.00	Google	Google Lunar X Prize, Planetary Resources
53	Paul Allen	60	$15.00	Microsoft	SpaceShipOne, SETI telescope array
138	Eric Schmidt	58	$8.20	Google	Planetary Resources
272	Sir Richard Branson	63	$4.60	Virgin Group	Virgin Galactic
527	Elon Musk	42	$2.70	PayPal, Tesla Motors	SpaceX
831	Guy Laliberte	53	$1.80	Cirque du Soleil	Visitor to ISS
922	K Ram Shriram	56	$1.65	Google	Planetary Resources
1031	Ross Perot, Jr.	54	$1.40	Oil & Gas	Planetary Resources
			$106.35	Total Net Worth	

10/31/2015 23

Commercial Space Companies

Lunar Developmenmt	***Asteroid Development***	***Mars Development***	***Space Tethers and Elevators***
Golden Spike	Planetary Resources	SpaceX	Tethers Unlimited
Shackleton Energy, Co.	Deep Space Industries	Inspiration Mars	Liftport
Moon Express	Excalibur Exploration	Mars One	International Space Elevator Consortium
Excalibur Almaz			Japanese Space Elevator Association
Bigelow Aerospace			

10/31/2015 24

Today's Topics

- Introduction
- IAA Study
- Logistics – Develop Tools
- Logistics – Store Supplies
- Logistics – Examples
- Conclusions & Questions

Image from 100 YSS

10/31/2015 25

Cruise Ships Precursors to Interstellar

Name	*Passengers*	*Crew*	*Total*	*Mass (tons)*	*Cost*
Disney Wonder	2700	900	3600		
Carnival Legend	2124	930	3054	88,500	
Carnival Elation	2052	920	2972	70,367	
RC Oasis of the Seas	5400	2100	7500	225282	$ 1.2 billion US
RC Allure of the Seas	6296	2100	8396	225282	

7,000 people with
225,282 Metric Tons
For seven days

Allure of the Seas from Wikipedia

10/31/2015

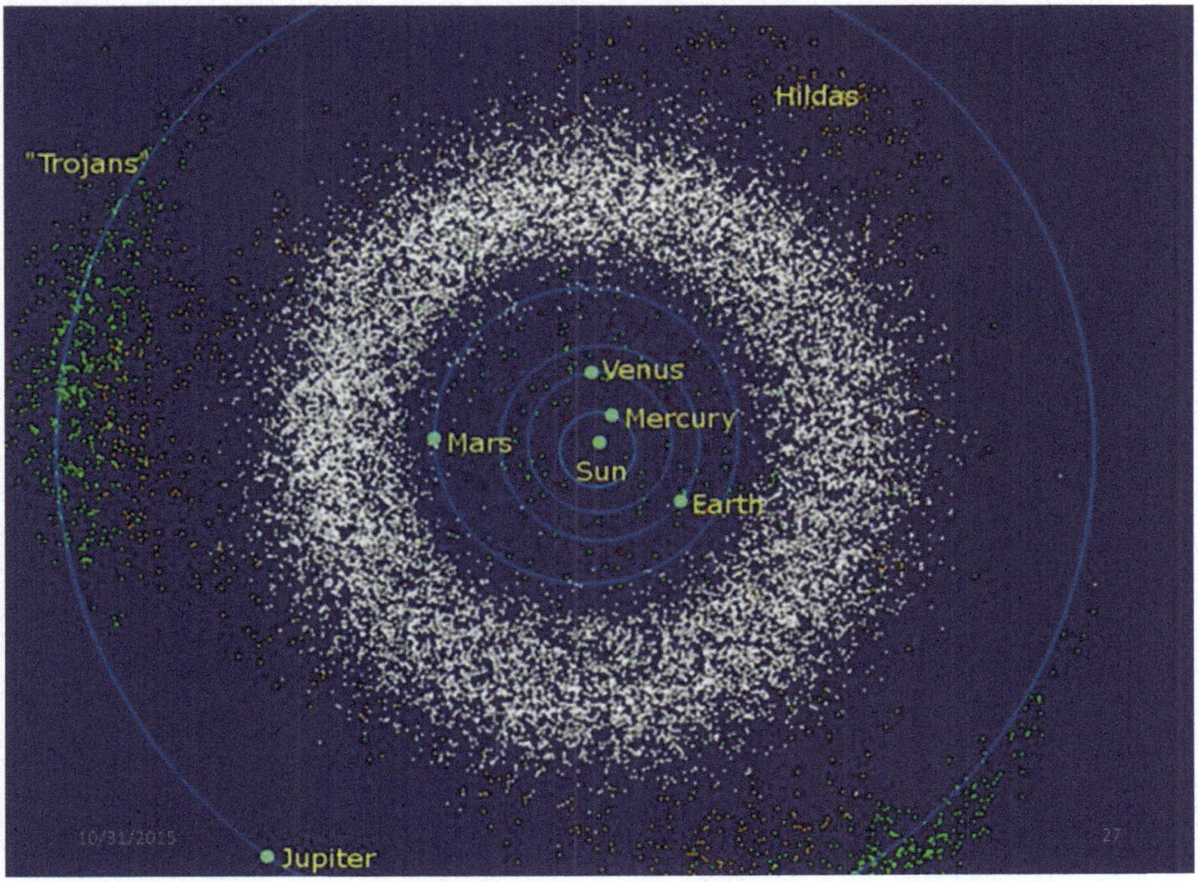

Asteroid 216 Kleopatra

An asteroid roughly the size of New Jersey (135 x 58 miles) located between Mars and Jupiter.

- **Composed mostly of nickel (10%) and iron (88%)**
- **Kleopatra is not completely solid - its surface is about one meter of metal dust and loosely consolidated rubble, although its core may contain large (cubic miles) solid-metal nodes.**
- **2003 world steel production was 854.1 million metric tons. At the world market price of $482/ton this is $320 billion.**
- **At this rate, 10% of Kleopatra would be worth over $200 trillion.**
- **Kleopatra alone has more material wealth than all of humanity has produced on Earth.**

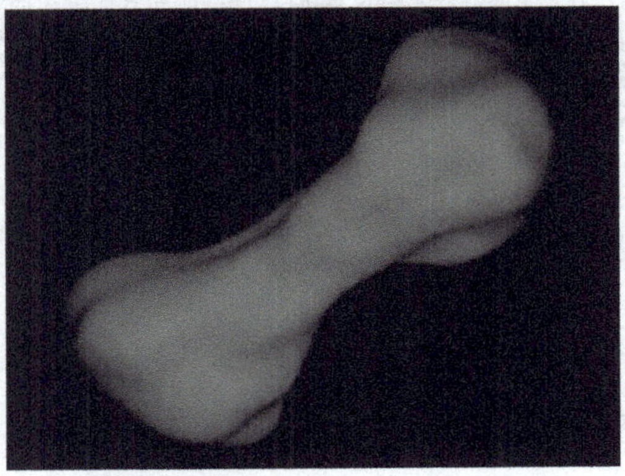

3

Capture and Move Resources

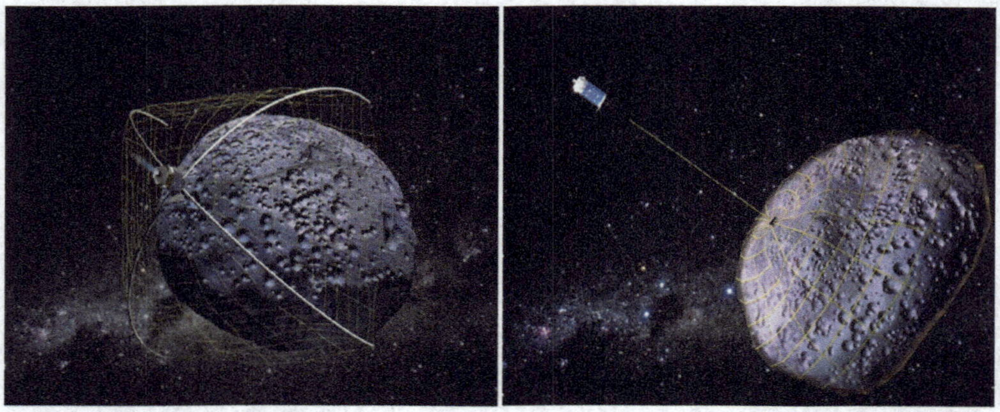

WRANGLER system - Asteroid Capture [Tethers Unlimited]

10/31/2015 29

Lunar Resources

From Project Horizon Study, US Army, 1959]

- Resources are boundless on the Moon
- Low gravity enables movement by multiple methods
- One key development will be a Lunar Elevator for moving supplies to the EM L-1 Manufacturing arena
- A key resource is Lunar water for living and fuel.

10/31/2015 30

Sell Water at Spaceports

	Mass kg	Payload [water] Mass kg	Price per kg	Price per metric ton
On–Pad	1,462,836			
In LEO	53,000	13,250	$7,547	$ 7.5 million
At GEO	21,000	5,300	$ 18,868	$ 18.9 million
At Earth-Moon EML-1	20,000	5,000	$ 20,000	$ 20.0 million
On Asteroid surface*	14,000	3,500	$ 28,571	$ 28.6 million
On Lunar Surface**	7,314	1,828	$ 54,705	$ 54.7 million
On Mars Surface***	13,200 (Insertion)	1,320 [surface]	$ 75,757	$ 75.8 million

"Water will be the Currency of Space!"

10/31/2015 31

Today's Topics

- Introduction
- IAA Study
- Logistics – Develop Tools
- Logistics – Store Supplies
- Logistics – Examples
- Conclusions & Questions

Image from 100 YSS

10/31/2015 32

Volatile Materials in Asteroids

- 10 to 50% of known large asteroids are likely hydrated CI-CM-like, possibly parent bodies of CI-CM chondrite meteorites
- CI-CM chondrites are typically 10-20% water by weight in the form of hydrated minerals with significant other volatile (e.g. CO_2) content that can be thermally extracted
- CI-CM materials are friable and may be in rubble piles with regolith or in blocks on asteroids.

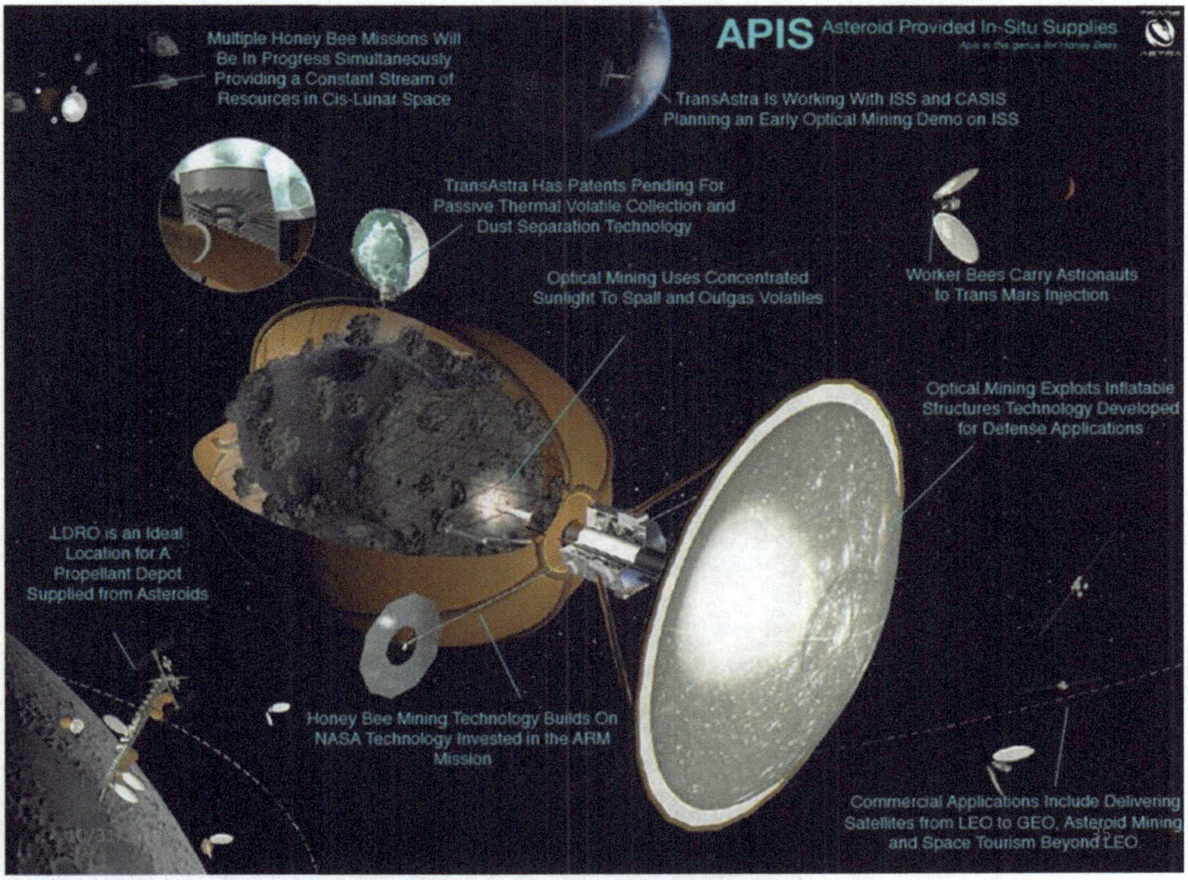

At Our Next Star: Need Resource Acquisition: A Vision

For Interstellar travel, mining and processing raw resources is essential:

- Develop Tools, Practice and Grow Skills in Earth-Moon Ecosystem
- Logistics support for Interstellar Flight will include processed SMRs
- Logistics re-supply along the way will require special designs
- Once arrival in target Solar System, will need resource replenishment

Today's Topics

- Introduction
- IAA Study
- Logistics – Develop Tools
- Commercial Ventures
- Logistics – Store Supplies
- Logistics – Examples
- Conclusions & Questions

Image from 100 YSS

10/31/2015 38

Finding 1: Technological risk reduction and engineering design are within State of Art

The mining of asteroids and lunar regolith is within the current state of the technical art. The extrapolation of Earth-based mining seems to be a one-for-one trade with some significant alterations due to vacuum, low gravity and temperature extremes. Many proposed solutions have been suggested and tested [on Earth] leading to positive conclusions on this topic.

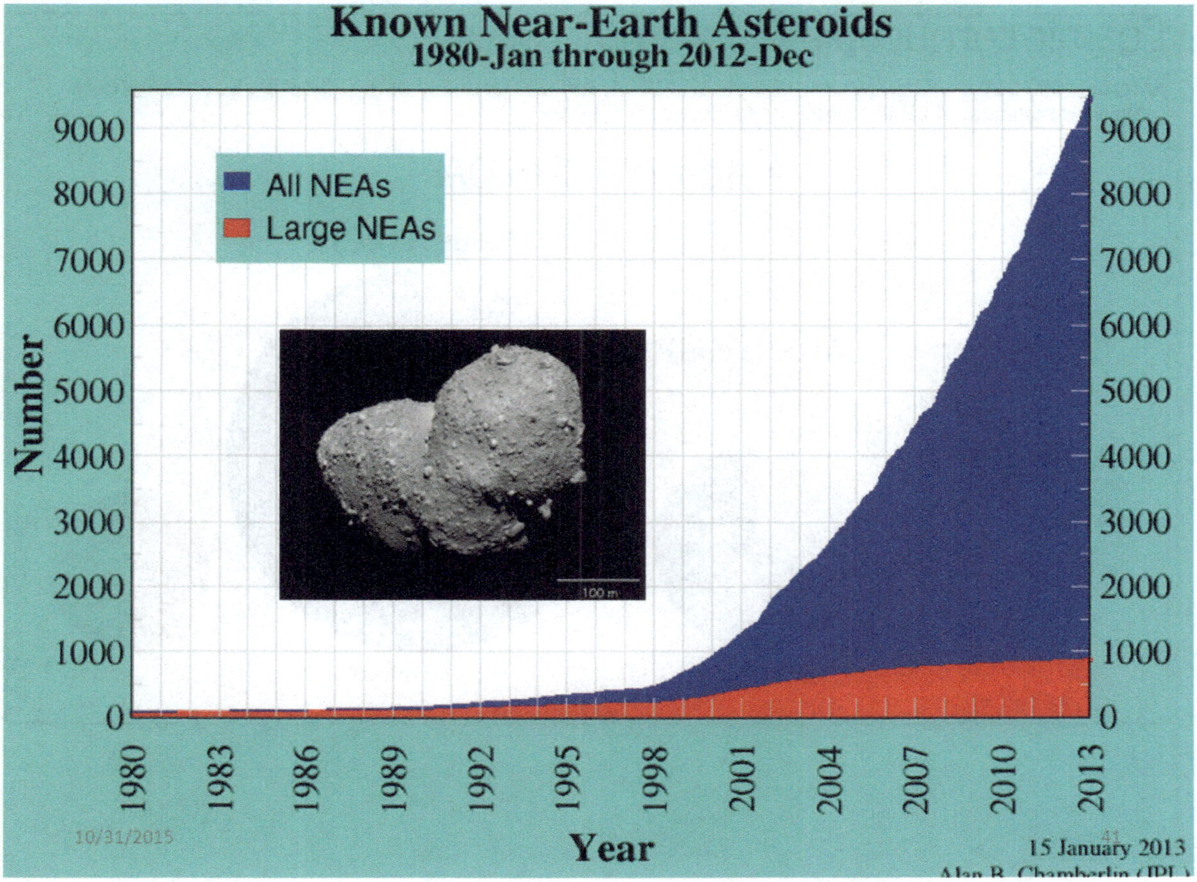

Abundances of ferrous and precious metals in asteroids

Metal	(1) Abundance in metal of average LL-chondrite asteroid	(2) Abundance in good" iron asteroid (90th percentile in Ir, Pt)	(3) Abundance in "best" iron asteroid (98th percentile in Ir, Pt)
Ferrous metals:			
Fe	63.7%	81-94%	82-94%
Co	1.57%	0.46-0.80%	0.43-0.75%
Ni	34.3%	5.6-18.0%	5.4-16.5%
Precious metals:			
Ge	1020 ppm	0.06-70 ppm	0.05-35 ppm
Re	1.1 ppm	1.1 ppm	2.4 ppm
Ru	22.2 ppm	20.7 ppm	45.9 ppm
Rh	4.2 ppm	3.9 ppm	8.6 ppm
Pd	17.5 ppm	2.6 ppm	1.2 ppm
Os	15.2 ppm	14.1 ppm	31.3 ppm
Ir	15.0 ppm	14.0 ppm	31.0 ppm
Pt	30.9 ppm	28.8 ppm	63.8 ppm
Au	4.4 ppm	0.16-0.70 ppm	0.06-0.6 ppm

Calculated from data given by Müller et al. 1971, Buchwald 1975, Malvin et al. 1984, Rasmussen et al. 1984, Hirata and Masuda 1992, and Morgan et al. 1992.

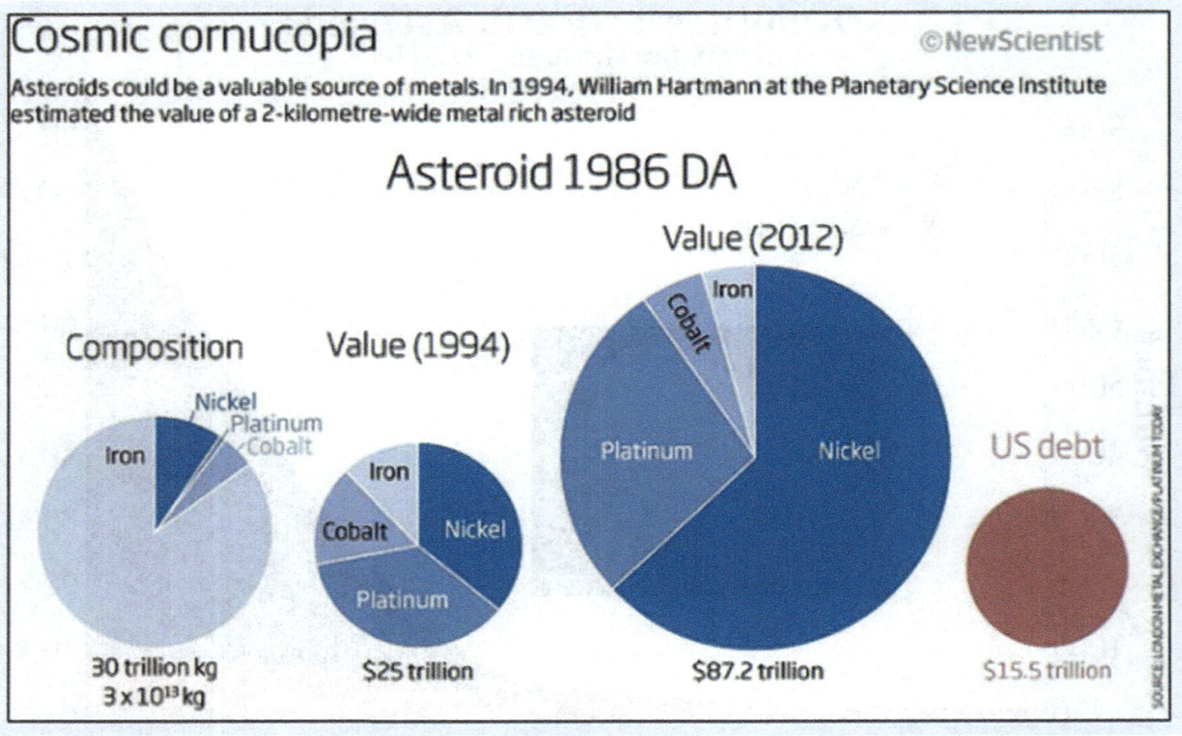

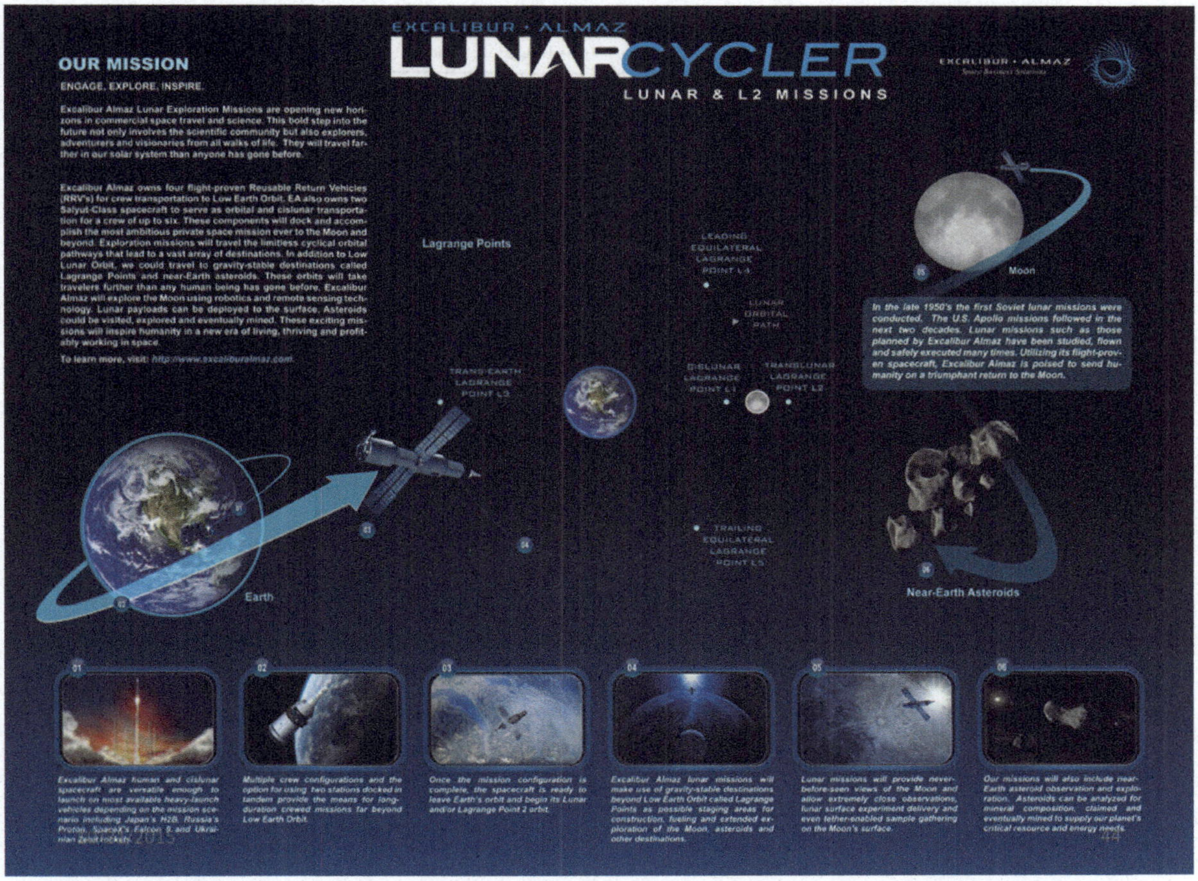
EXCALIBUR · ALMAZ
LUNARCYCLER
LUNAR & L2 MISSIONS
OUR MISSION
ENGAGE. EXPLORE. INSPIRE.
Excalibur Almaz Lunar Exploration Missions are opening new horizons in commercial space travel and science. This bold step into the future not only involves the scientific community but also explorers, adventurers and visionaries from all walks of life. They will travel farther in our solar system than anyone has gone before.
Excalibur Almaz owns four flight-proven Reusable Return Vehicles (RRV's) for crew transportation to Low Earth Orbit. EA also owns two Salyut-Class spacecraft to serve as orbital and cislunar transportation for a crew of up to six. These components will dock and accomplish the most ambitious private space mission ever to the Moon and beyond. Exploration missions will travel the limitless cyclical orbital pathways that lead to a vast array of destinations. In addition to Low Lunar Orbit, we could travel to gravity-stable destinations called Lagrange Points and near-Earth asteroids. These orbits will take travelers further than any human being has gone before. Excalibur Almaz will explore the Moon using robotics and remote sensing technology. Lunar payloads can be deployed to the surface. Asteroids could be visited, explored and eventually mined. These exciting missions will inspire humanity in a new era of living, thriving and profitably working in space.
To learn more, visit: http://www.excaliburalmaz.com
Lagrange Points
LEADING EQUILATERAL LAGRANGE POINT L4
LUNAR ORBITAL PATH
TRANS-EARTH LAGRANGE POINT L3
CISLUNAR LAGRANGE POINT L1
TRANSLUNAR LAGRANGE POINT L2
TRAILING EQUILATERAL LAGRANGE POINT L5
Moon
In the late 1950's the first Soviet lunar missions were conducted. The U.S. Apollo missions followed in the next two decades. Lunar missions such as those planned by Excalibur Almaz have been studied, flown and safely executed many times. Utilizing its flight-proven spacecraft, Excalibur Almaz is poised to send humanity on a triumphant return to the Moon.
Earth
Near-Earth Asteroids
01
Excalibur Almaz human and cislunar spacecraft are versatile enough to launch on most available heavy-launch vehicles depending on the mission scenario including Japan's H2B, Russia's Proton, SpaceX's Falcon 9 and Ukrainian Zenit rockets.
02
Multiple crew configurations and the option for using two stations docked in tandem provide the means for long-duration crewed missions far beyond Low Earth Orbit.
03
Once the mission configuration is complete, the spacecraft is ready to leave Earth's orbit and begin its Lunar and/or Lagrange Point 2 orbit.
04
Excalibur Almaz lunar missions will make use of gravity-stable destinations beyond Low Earth Orbit called Lagrange Points as possible staging areas for construction, fueling and extended exploration of the Moon, asteroids and other destinations.
05
Lunar missions will provide never-before-seen views of the Moon and allow extremely close observations, lunar surface experiment delivery and even tether-enabled sample gathering on the Moon's surface.
06
Our missions will also include near-Earth asteroid observation and exploration. Asteroids can be analyzed for mineral composition, claimed and eventually mined to supply our planet's critical resource and energy needs.

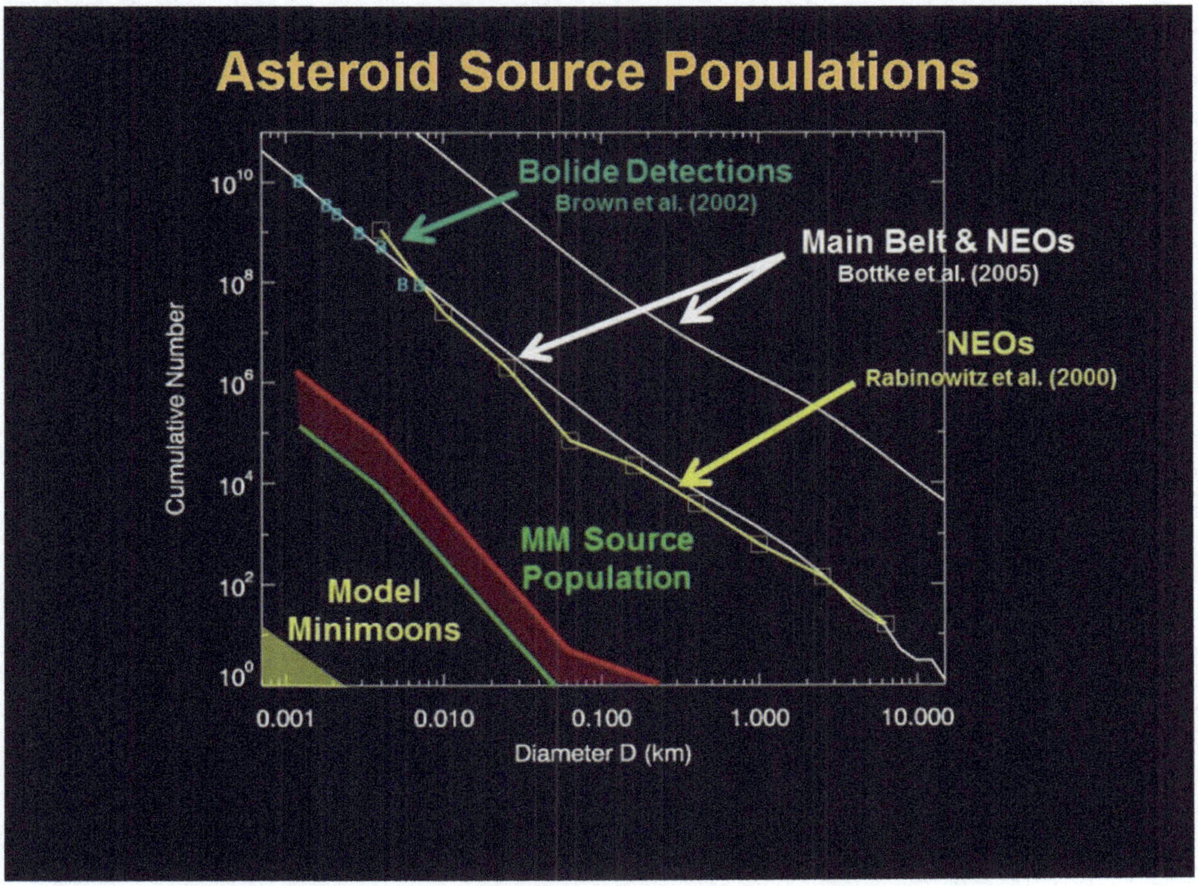
Asteroid Source Populations
Bolide Detections
Brown et al. (2002)
Main Belt & NEOs
Bottke et al. (2005)
NEOs
Rabinowitz et al. (2000)
MM Source Population
Model Minimoons
Cumulative Number
10^10
10^8
10^6
10^4
10^2
10^0
0.001
0.010
0.100
1.000
10.000
Diameter D (km)

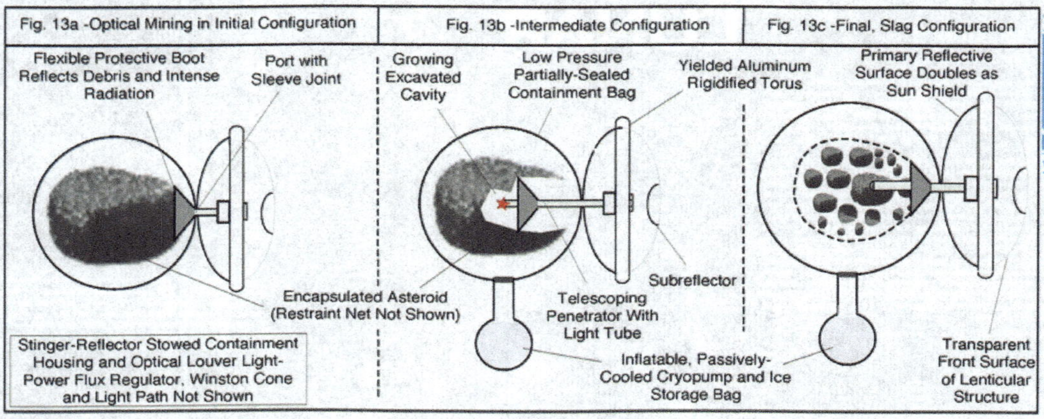

18

The Trans Astra Corporation has developed a new patent pending invention called "optical mining" – a way to excavate and process asteroids using sunlight. Space News called this invention "a possible game changer for space exploration." *The ability to tap mega-amounts of water from asteroids could be used directly as propellant in solar thermal rockets to provide inexpensive space transportation."* **<u>The ground research into this new concept is funded by private capital and government contracts from NASA and others.</u>**

46

Fig. 2 – Figure from a Paper in *Nature* Showing A Large Population of NEOs Far More Accessible Than the Moon (Binzel 2014)

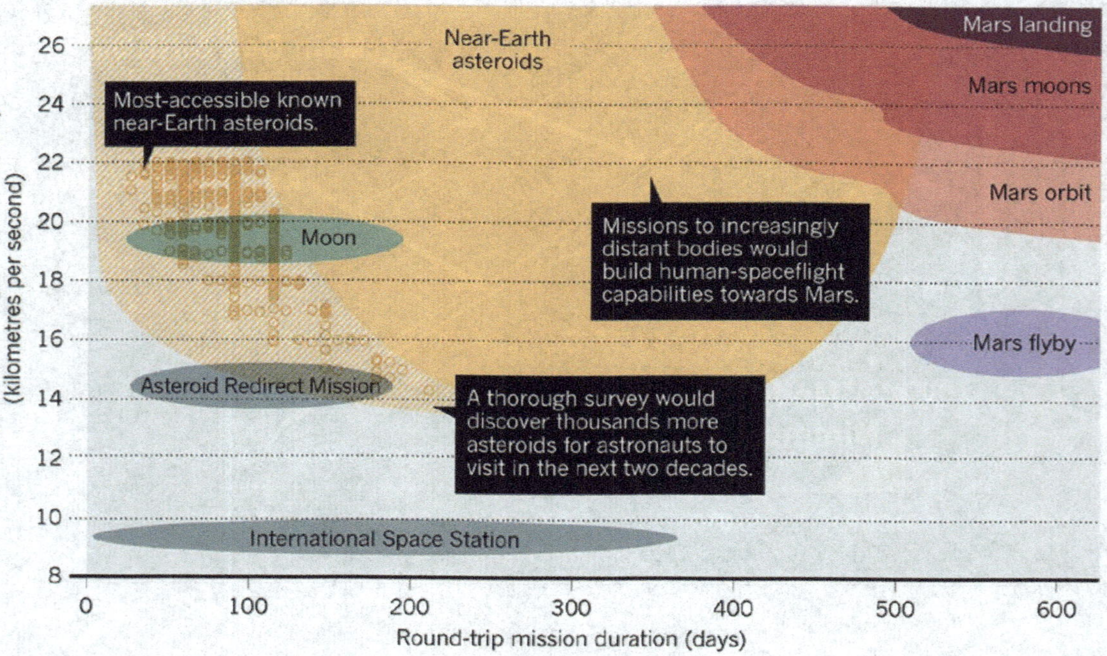

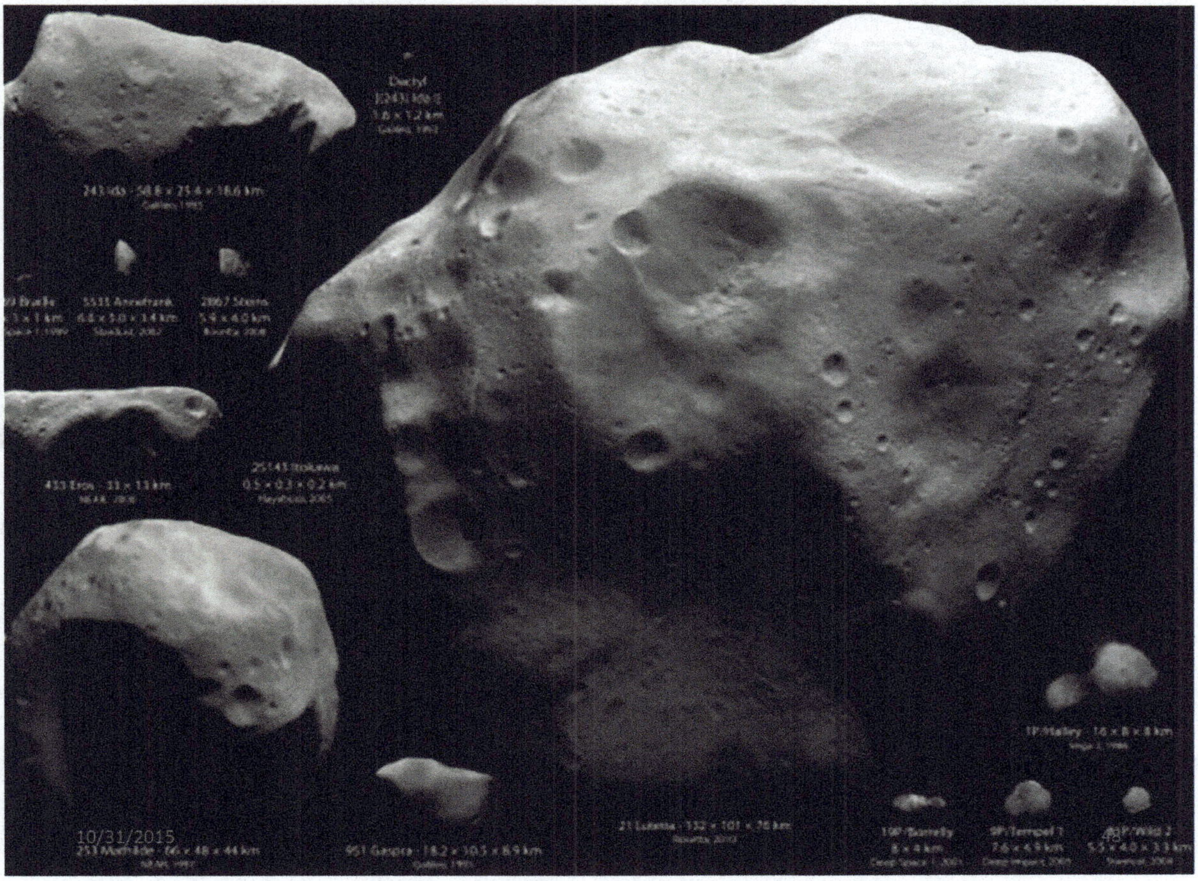

Why – Mining in Space?

- ***When I awoke this morning, I looked around and saw nature in crises***
- If I expand this perception from my small community to the global population, predictions from the Club of Rome seem real.*
- Opening the resources of space will not only change our lives; it will change our destiny. The question is not what can I do about it; but, what can we all do about the multitude of problems that seem to be overwhelming our world. The answers seem simple:
  - Change the equation.
  - Change the assumptions.
  - Increase the resources and produce innovation, jobs and wealth along the way

*Donella H. Meadows, Dennis L. Meadows, Jørgen Randers, and William W. Behrens III, "Limits to Growth," 1972.

10/31/2015 49

Goldilocks Zones

A Fine Grained Exoplanet Taxonomy

Patrick J. Talbot and Dennis R. Ellis

Abstract

Research, engineering techniques, and data analytics identify Goldilocks zones as potential destinations. The taxonomy is based on available data, anticipated growth in the number of known exoplanets, and a flexible engineering methodology. The foundation of the taxonomy is a classic trade study to rank and score candidate attributes based on criteria derived from a set of exoplanet taxonomy requirements. A "critic", a sequential optimizer, and software-based tests find a minimum spanning taxonomy. Software assures that, to the extent practical, each object has a unique path through the taxonomy.

Why a taxonomy? It is a valuable teaching tool, it organizes and sorts a data set, and a taxonomy is structured as a hierarchical characteristics-based ontology. This knowledge structure is readily captured in a knowledge base for multi-strategy reasoning and data analytics. That is the true value of a taxonomy! The taxonomy defines the parameter space of exoplanet characteristics. While some work "inside the box", others work outside the box", a taxonomy defines the box.

Recent efforts produced an upper-level exoplanet taxonomy. This taxonomy doesn't span the orbital, physical, environmental, and stellar characteristics of interest. Nor does it provide a unique set of characteristics for each of ~ 2,000 exoplanets currently identified. The Goldilocks Taxonomy™ has 13 parameters chosen from a collection of potential characteristics. The Goldilocks Taxonomy™ provides fine-grained organization of exoplanet characteristics, a robust methodology to refine the taxonomy, and a suite of data analytics tools to discover interesting patterns.

Keywords

interstellar, Goldilocks zone, exoplanets, taxonomy,

1. Introduction and Summary

The proposed taxonomy is based on exoplanet insights insights from the literature and access to online exoplanet catalogs. These insights are the foundation for a classic trade study to rank and score candidate nodes based on criteria derived from missions and functional requirements. The nodes are further processed using a "critic", a sequential optimizer, and software-based tests to find a minimum spanning taxonomy. Software was written to assure that, to the extent practical, each object has a unique path through the taxonomy.

At first glance, a taxonomy is an important way of visualizing the widely varying set of attributes associated with an object. As such, it is a valuable teaching tool that provides people interested in space with an easy-to-understand view. Digging deeper, it provides a "world-view" of the domain that helps people understand objects in a more organized and holistic way. More practically, it organizes and sorts a data set, since all the attributes important to a taxonomy are not, today, available in a single data file. Building on the data set idea, a taxonomy is a hierarchical network and each node in the hierarchy has characteristics. This is the makings of a truly powerful knowledge structure, or ontology. The knowledge structure is readily captured in a knowledge base such as Protege [1]. The resulting hierarchical, characteristics-based ontology supports multi-strategy reasoning and learning, including simulation and planning [2] . That is the true value of a taxonomy!

2. Related Work

There have been multiple recent efforts to arrive at an exoplanet taxonomy [3],[4] supplemented with ongoing efforts and "conventional wisdom" that produce an upper-level taxonomy. These results are discussed as motivation to capture what's good and to avoid what's contrary to requirements for a fine-grained taxonomy.

3. Approach

- Define Requirements from mission analysis and functional analysis.
- Cite Assumptions and groundrules; for example, the taxonomy is limited to data in catalogs.
- Perform Trade Studies: Identify and characterize potential taxonomy nodes, citing rationale for accepting some object attributes and dismissing others. Assign weights to criteria using a pair-wise preference technique, citing rationale. Rank and score attributes, citing rationale.
- Sequential Optimization: Use a "critic" to add common sense and determine the optimum set of nodes, as the smallest number of nodes that produces less than 5% probability of one or more duplicate signatures.
- Software Development: Convert data files (for example, exoplanet.org) to a standard (.csv) format. Calculate discretized values for candidate taxonomy nodes from the integrated files. Assemble a single file with values for all candidate nodes. Analyze patterns, uniqueness, and requirements satisfaction of requirements using a uniqueness test, relative frequency, and sensitivity computations. Test the taxonomy using a Bayesian classifier to assure high probability of correct classification of an object.
- State Results: An exoplanet taxonomy that comprises current data sets integrated with additional sources. Show taxonomy resulting from the trade study and sequential optimization. Assure uniqueness, to the extent that it is practical, of each exoplanet signature, and explain optimization process. Provide a minimum description length representation of taxonomy attributes and states.
- Discussion: focus on an interactive taxonomy generator that provides a user centered taxonomy.
- Summarize: State results and cite advantages of the solution.

3.1 Define Requirements

Detailed description & requirements for an exoplanet taxonomy focus on the ability to view, understand, and predict the behavior of objects in orbit around the other stars. Two trade studies help top define the taxonomy:

- Mission Analysis maps potential nodes to mission requirements
- Functional analysis maps potential nodes to Challenge details & functional requirements.

3.2 Assumptions and Groundrules

The following are considered concerns:

- Missing Values: values for many exoplanet characteristics are unknown. The methodology and reasoning algorithms accommodate this aspect of the catalogs.
- Uniquely Described: the driving requirement for the minimal taxonomy is that the state space is large enough and that the states of characteristics are equally likely to guarantee a low probability that two objects have the same state. This is not the case because many states have unknown values. Hence the goal is to increase the number of characteristics and states thereof until we reach a point of diminishing returns. That is, when the percent of exoplanets with duplicate signature fails to diminish with increasing number of states.

3.3 Trade Studies: Potential taxonomy nodes are derived from Mission Analysis and Functional Analysis.

Mission Analysis: To "cast a wide net" and capture all potentially valuable nodes for describing exoplanets, the missions of potential interest are analyzed to identify important characteristics (Figure 1). Four attributes are identified: the extent to which an attribute facilitates exploration, the science value of the attribute, the cultural significance, and the extent to which the attribute is amenable to sensing technology. The characteristics are weighted by pair-wise comparison (top of Figure 1) and the potential taxonomy parameters are the scored based on the four attributes (bottom of Figure 1). The result is that from a mission perspective, the distance from our sun and the type of stat system (binary) ate rated most highly. Mission tasks will vary with project goals.

Mission Weighting

Mission Task	Exploration	Science	Culture	Sensing	Raw	Weight
Exploration	1	1	1	1	4	1
Science	1	1	1	1	4	1
Culture			1	1	3	0.75
Sensing		1		1	2	0.5

Scores indicate the degree to which a characteristic is expressive and distinctive

Characteristic	Exploration	Science	Culture	Sensing	Raw Score	% Score
Weights	1	1	0.75	0.5		
Distance from our Sun	10	10	10	10	32.5	100
Binary	10	10	10	10	32.5	100
Separation	10	10	9	10	31.75	98
Temperature	10	10	10	8	31.5	97
Period	10	10	9	9	31.25	96
Mass	9	10	8	7	28.5	88
Semi-Major Axis	10	9	7	8	28.25	87
Eccentricity	10	9	7	8	28.25	87
Inclination	10	9	7	8	28.25	87
Spectral Type	8	10	8	8	28	86
Physical Radius	9	10	8	6	28	86
Age	8	10	8	6	27	83
Spin Orbit Alignment	9	10	5	8	26.75	82
Stellar Metallicity	7	10	6	7	25	77
Discovery Method	7	8	8	8	25	77

Figure 1: Mission Trade Study

Functional Analysis: In similar fashion, a functional trade study, geared to identifying the utility and power of potential taxonomy characteristics, was completed. A classification scheme for exoplanets is required with these attributes:

- Explain: the exoplanet taxonomy (ET) shall be based on a set of characteristics that provide intuitive insights into these objects
- Discriminate: the ET shall identify differences among exoplanets
- Characterize: the ET shall comprise physical and dynamic characteristics, or a combination of both
- Group : the ET shall hierarchically organize and name objects based on similarity
- Predict: the ET shall allow the object trajectory to be accurately predicted
- Growth: the ET shall accommodate growth of the exoplanet catalog over at least two orders of magnitude.

The result of this trade study (Figure 2) is that temperatre is the most highly rated characteristic for inclusion into the taxonomy, while distance from our sum and star system type again are highly ranked.

Weighting Matrix

Functional Requirement	Explain	Discriminate	Characterize	Group	Predict	Growth	Raw	Weight
Explain	1	1	1	1	1	1	6	1
Discriminate	1	1	1		1	1	5	0.8
Characterize		1	1		1	1	4	0.7
Group	1	1	1		1		4	0.7
Predict		1			1	1	3	0.5
Growth			1			1	2	0.3

Characteristic	Explain	Discriminate	Characterize	Group	Predict	Growth	Raw Score	% Score
Weights	1	0.8	0.7	0.7	0.5	0.3		
Distance from our Sun	10	10	10	10	8	8	38.4	96
Binary	10	10	10	10	8	8	38.4	96
Separation	10	10	10	9	9	8	38.2	96
Temperature	10	10	10	10	10	10	40	100
Period	10	10	10	9	9	8	38.2	96
Mass	10	10	10	9	9	8	38.2	96
Semi-Major Axis	10	10	10	9	9	8	38.2	96
Eccentricity	10	10	10	9	9	8	38.2	96
Inclination	10	10	10	9	9	8	38.2	96
Spectral Type	10	9	9	10	10	6	37.3	93
Physical Radius	10	10	10	10	8	8	38.2	96
Age	10	9	9	10	10	6	37.3	93
Spin Orbit Alignment	10	10	10	6	7	5	34.5	86
Stellar Metallicity	10	9	8	9	8	7	35.2	88
Discovery Method	10	10	7	7	8	8	34.2	86

Figure 2: Functional Trade Study

3.4 Sequential Optimization

The candidate nodes of the taxonomy have scores for two trade studies: the mission tasks and the functional requirements trades. These are tabulated and averaged to produce a composite score for each candidate taxonomy node, and sorted according to rank (Figure 3).

Candidate Node	Mission Trade	Functional Trade	Composite Score	Rank
Temperature	97	100	98.5	1
Distance from our Sun	100	96	98	2
Binary	100	96	98	3
Separation	98	96	97	4
Period	96	96	96	5
Mass	88	96	92	6
Semi-Major Axis	87	96	91.5	7
Eccentricity	87	96	91.5	8
Inclination	87	96	91.5	9
Spectral Type	86	93	89.5	10
Physical Radius	86	96	89	11
Age	83	93	88	12
Spin Orbit Alignment	82	86	84	13
Stellar Metallicity	77	88	82.5	14
Discovery Method	77	86	81.5	15

Figure 3: Composite Ranking

Minimizing the Number of States: there are two ways to minimize the number of states in the taxonomy: reduce the number of nodes, or reduce the numbers of states in the nodes. A small number of states produces an exponentially large sample space; for example, if there are 13 nodes and each has six equally likely states, the sample space is 136 = 13,060,694,016. Given these simplifying assumptions, the probability that each object path through the taxonomy is not unique is given by the "Birthday Problem calculator [7]: for a catalog consisting of 15,065 objects, the probability of duplicates is < 1%.

Uniqueness: Many of the candidate nodes are continuous variables, for example, eccentricity. Such variables have an infinite number of states and must be discretized (Figure 4); for example, low, medium, and high eccentricities. This discretization produces a total of 96 states across 16 taxonomy nodes which results in 6*6*6*6*6*6*6*6*9*6*6*6*6*6 = 4,231,664,861,184 possible states. The optimum strategy is to bin the variables so that all values of the variable fit in a bin and each bin has about the same Figure 4: Discretization of Taxonomy Nodes number of instances. The probability that an object is not unique is computed using the Birthday Problem calculator. Let m=2,065 random selections (number of confirmed exoplanets) from n= 4,231,664,861,184 choices. The resulting probability that it is not unique is p = 2.6%. This appears suitable to guarantee uniqueness to the extent practical (unknown values being exceptions). It is tested on the taxonomy data set derived from the current catalog and the state space is perturbed to minimize the number of states while maintaining a unique taxonomy path for each object.

We see that 12 characteristics, with six or more states each are, in theory, sufficient to predict a vanishingly small probability, for a catalog containing 5,000 exoplanets, that two exoplanets share exactly the same characteristics. On the other hand, 15 characteristics, with the number of states shown, provide a near zero probability of duplicates for a 50,000 object catalog. Here, the curse of dimensionality" works in our favor: adding a small number of characteristics (three) greatly expands the cumulative number of states. Characteristics are grouped according to these categories. Optimization will later identify which, if any, characteristics can be dropped from the taxonomy.

Employing a Critic: Modifications to the composite ranking are required to impose knowledge of redundancy in parameters. Specifically, separation varies with position in orbit unless circular orbits are assumed (and the often are), while period and semi-major axis are related by Kepler's Third Law [5]. Separation is deleted in favor of the more intuitive period attribute. Period and semi-major axis are retained because the parameters are well known and discretized binning provides additional information.

Spectral type is not available in the catalog we are using [6], so this parameter is omitted.

Metallicity is chosen ahead of spin-orbit alignment. It better predicts planet type formation.

Discretizing States: Exoplanet catalogs [6] include tools to easily construct histograms. From these plots, rules are readily defined (Figure 5).

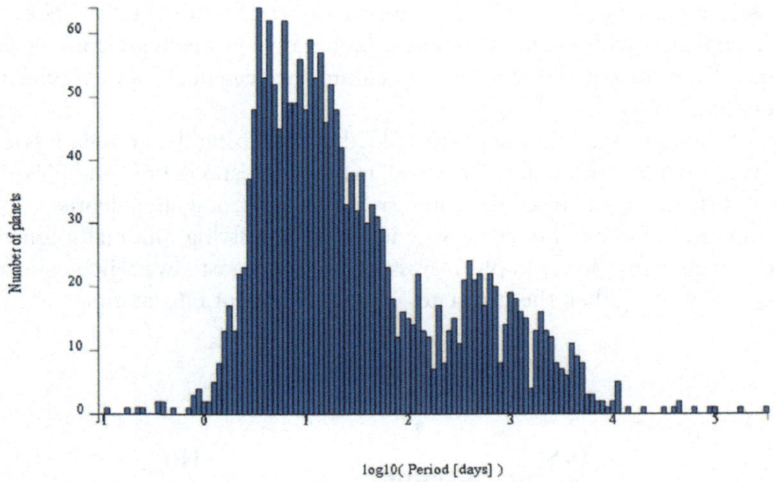

If P < 1 then Period = V (Very Short)
If 1 <= P < 10 then Period = S (Short)
If 10 <= P < 100 then Period = M (Moderate)
If 100 <= P < 1000 then Period = T (Terran)
If 1000 <=P < 10000 then Period = L (Long)
If 10000 <= P then Period = Extremely Long (E)

Figure 5: Rules Discretize Characteristics

Other rules, based on types of stars, discovery method, and a derived Earth Similarity Index [10] are:

- Spectral Type: O,B,A,F,G,K,M,L,T
- Discovery Method: R=Radial velocity, A = Astrometry, T = Timing, M = Microlensing, R = Transit, I = Imaging
- Earth Similarity Index (derived, but not included in this taxonomy)
    - **if .85 <= ESI then ESI = E (Excellent)**
    - **if .80 <= ESI < .85 then ESI = V (Very Good)**
    - **if .75 <= ESI < .80 then ESI = G (Good)**
    - **if .65 <= ESI < .75 then ESI = M (Moderate)**
    - **if .50 <= ESI < .65 then ESI = P (Poor)**
    - **if ESI < .50 then ESI = D (Dismal)**

Object Signature: A path through the taxonomy produces a unique 12 letter code that describes the object (Figure 6). A software application is executed to determine the percentage of objects that have unique signatures.

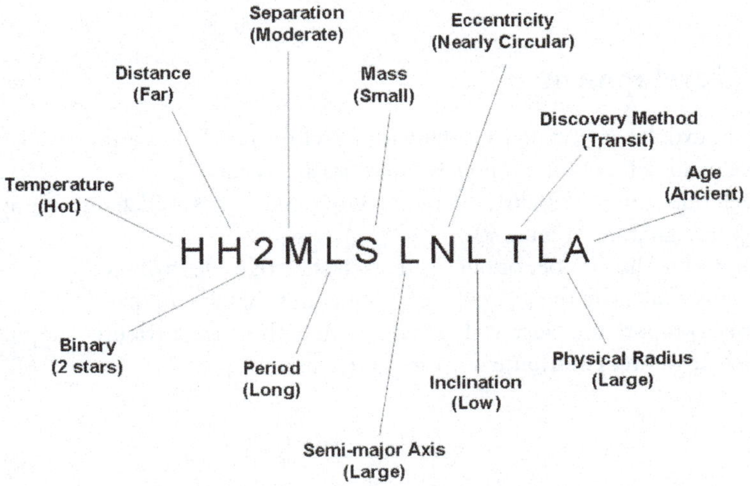

Figure 6: Sample Exoplanet Signature

Optimization Process: The challenge is to find the lowest number of representative states that provide the highest percentage of the exoplanets with unique signatures. An iterative processing technique (Figure 7) is employed. Three computed quantities are required: the object signature, the percentage of each state in each node, and the impact of attributes on uniqueness..

If the objects do not have unique signatures, states are added by increasing the granularity of populous states; for example, if the object type node has three states (payload, rocket body, and debris) and 80% of the objects are debris, then the debris state may be further discretized into small, medium, and large debris.

Once signatures are unique, the number of states is reduced by identifying minimally contributing states as possibly redundant and removing them; for example, if size=very small almost always is accompanied in a signature by radar cross section = very small, then the two states convey redundant information.

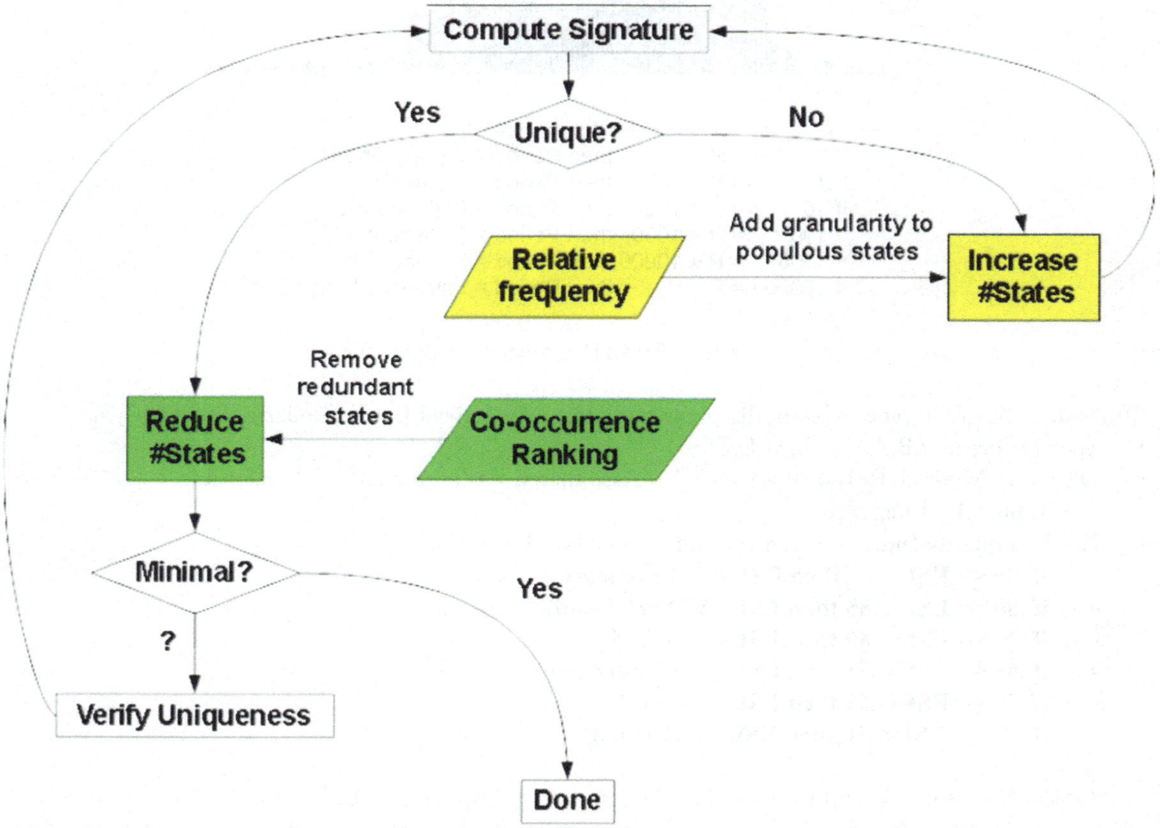

Figure 7: Optimization Process

3.6 Software Development

- Format: Convert an exoplanet catalog to a standard (.csv) format. Choose the exoplanet name as the pivot field: it is available in all data sets and uniquely identifies an object.
- Discretize: For each object in the catalog, discretize state values for candidate taxonomy nodes from raw data to produce an object "signature" (Figure 9).
- Attribute Bins: Compute the number of objects in each state of each attribute
- Signature Uniqueness: Calculate the percentage of signatures that are unique.
- Sensitivity Analysis: compute the percent duplicates with each of the attributes suppressed. This provides an indication of which attributes contribute most to uniqueness.

4. Result

The exoplanet taxonomy (Figure 8) is draw from earlier sections that provide rationale for a minimal set of taxonomy nodes that span the required range of missions and functions. The number of nodes is 12 and the number of states is 77. Many of the nodes have missing values, set the the "unknown" state. This lack of information precludes, for the present, the ability to form unique exoplanet signatures.

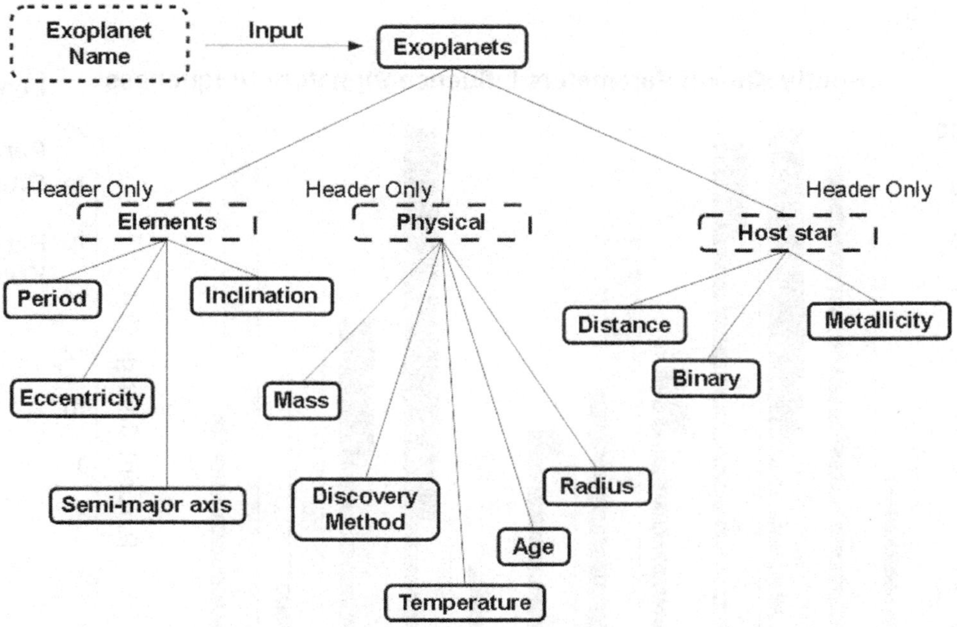

Figure 8: Exoplanet Taxonomy

The signatures of exoplanets were tested (Figure 9) to determine the percentage of unique signatures. Many duplicates (1166 or 56.5%) were identified. The rules provided earlier put objects in discrete bins for each attribute as shown in the table:

Iteration: 6 # states =77 #Objects: 2,065 Duplicate Signatures:1166
 # Attributes=12 # Possibilities =705,438,720 Probability of Duplicates: 56.5%

Attribute	States (#/state)
Temperature	Terran(0), Hot(0), Very Hot(0), Extreme(0), Intense(0), Sun Like(0), Missing(1275)
Distance	Very Near(184), Near(240), Moderate(121), Far(64), Very Far(64), Distant(1047), Missing(112)
Binary	Two Stars(1899), One Each(26), Two Binaries(138), CircumBinary(2), Other(0), Missing(0)
Period	Very Short(21), Short(741), Moderate(712), Terran(323), Long(172), Extreme(17), Missing(79)
Mass	Very Small(6), Small(35), Moderate(210), Terran(317), Large(476), Giant(2), Missing(947)
Eccentricity	Circular(964), Very Small(0), Small(0), Terran(0), Elliptical(0), Highly Elliptical(0), Missing(1101)
Inclination	Perp LOS (33), Nearly Perp (27), Prograde(0), Retrograde(0), Highly retro(0), Missing(1557)
Radius	Very Small(0), Small(0), Terran(991), Moderate(0), Large(0), Giant(0), Missing(730)
Age	Extra Young(81), Very Young(95), Sunlike(92), Young(52), Old(18), Ancient(9), Missing(1696)
Semi-Major Axis	Very Small(840), Small(362), Moderate(1), Terran(0), Large(0), Giant(0), Missing (846)
Metallicity	Extra Low(0), Very Low(7), Somewhat Low(45), Low(109), Neutral(231), Moderate(368), High(248), Abnormally High(25), Missing(1025)
Discovery Method	Radial Velocity(32), Transit(2023), Astrometry(0), Timing(0), Micro-lensing(0), Imaging(0), Other(0), Missing(10)

Figure 9: Testing for Missing Values

Sensitivity Analysis: to determine which attributes contribute most heavily to signature uniqueness, the percent duplicates was computed with each of the 12 attributes suppressed. The result (Figure 10) was that attributes, such as eccentricity, age, and inclination, that have many missing values do not influence signature uniqueness because they have values that are unknown or are not evenly distributed across states. In two other cases, Binary and Discovery method, the percent known is large, but the sensitivity to changing the number of duplicates is small because the majority of exoplanets have a dominant star system state (binary) and a dominant discovery method (transit).

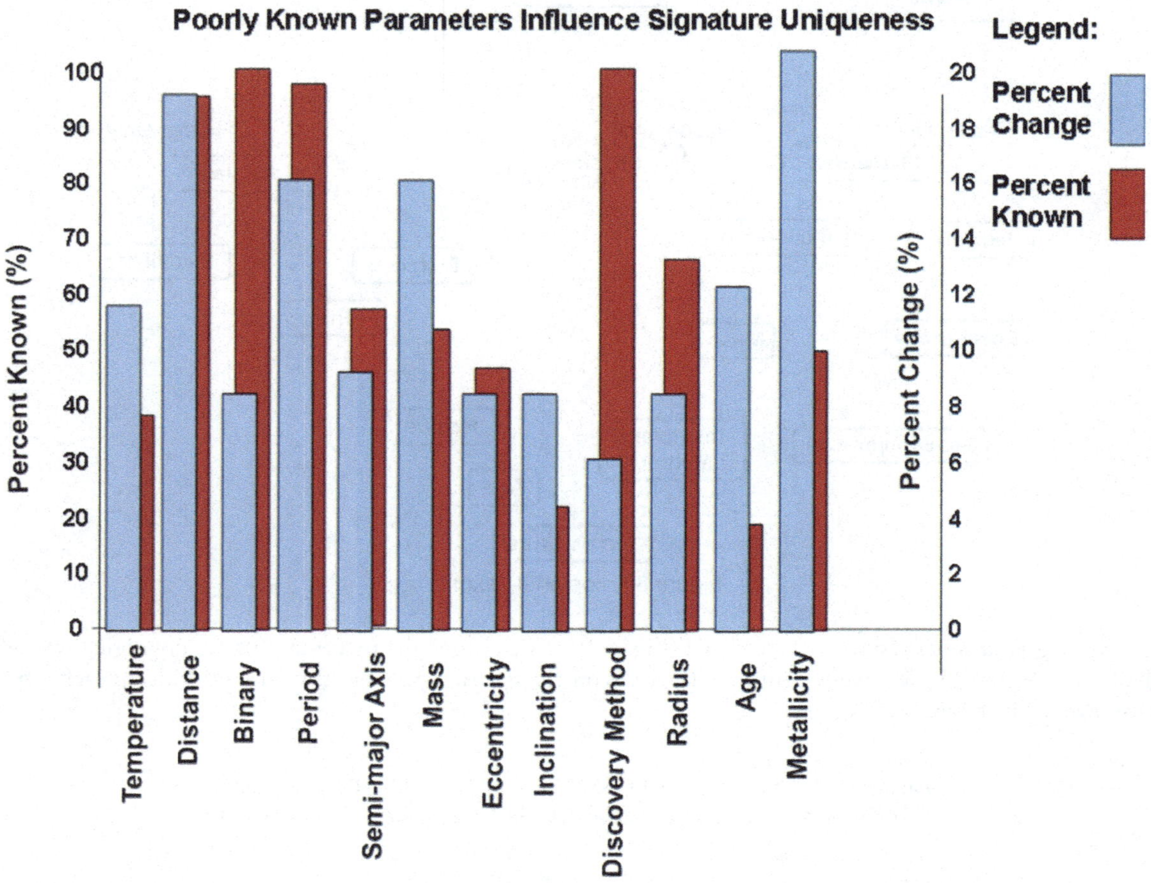

Figure 10: Signature Uniqueness Sensitivity

5. Discussion

The result of this paper is a fine-grained exoplanet taxonomy. However, the stakeholders in the astro-biology community have varied and sometimes conflicting interests. Knowing that there will be counter-proposals for an exoplanet taxonomy, and that users have varied needs, I propose that a "laundry list" of taxonomy nodes be made available. The user chooses those of interest, or chooses from among several pre-configured taxonomies. Underlying data, which will surely come from multiple databases, is assembled based on automated queries and integrated, with federated query technology used to automatically update the taxonomy data files.

Another reason to have a flexible ability to tailor a taxonomy "on-the-fly" is that various communities of stakeholders will operate from differing cross-disciplinarian perspectives. Finally, a user may want to deepen the taxonomy to include lower level detail.

Uses for the Taxonomy:

- Database query: use MySQL [12] to produce custom tables that sort exoplanets based on taxonomy nodes; for example, show a table with all exoplanets with earth-like diameter and temperature.

- Visualization: While the taxonomy hierarchy chart shows the overall knowledge structure, a Kiviat diagram (Figure 11) provides a great visual sense of all characteristics of a single exoplanet at a glance! Four quadrants (Physical, Orbital, Environmental, and Influential, are shown. A ring denotes the Goldilocks Zone. The exoplanet name and habitability score are shown in the center.

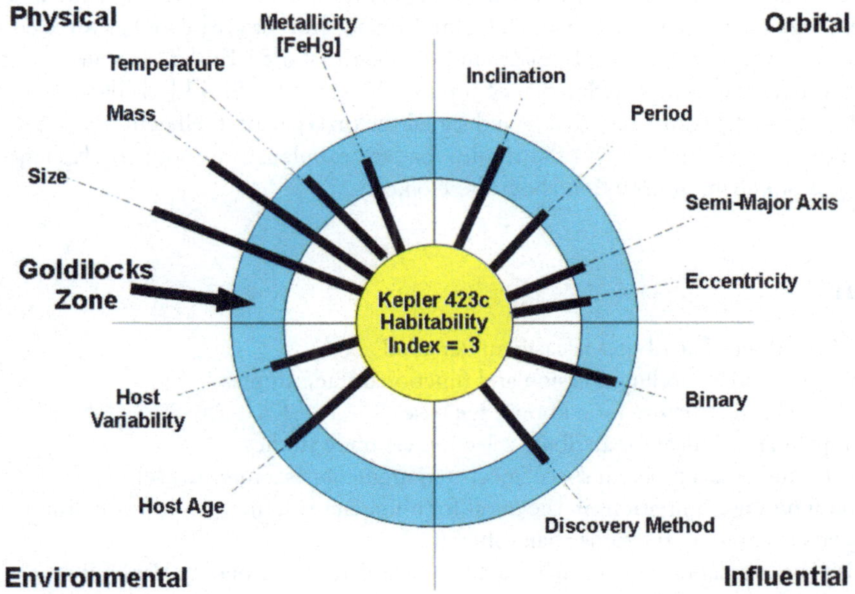

Figure 11: Kiviat Diagram

- Parameter Plots: use Excel to plot histograms of the number of objects versus taxonomy nodes
- Knowledge Base: use Protege to form a hierarchical, characteristics-based ontology. The taxonomy provides the tree-like structure and the node values provide characteristics. Protege allows knowledge base visualization and supports plug-in for automated reasoning; for example, GATE for information extraction, Weka for data mining, and SUBDUE for abductive reasoning.
- Pattern Discovery: use Weka rule induction to find patterns in the for of rules (Figure 12). The user identifies an independent variable (here, it is temperature) and the J48 rule induction algorithm automatically induces If,... AND,... AND,... THEN rules. For example, the longest rule shown is: If the exoplanet radius is less than .9 earth radii and the host star radius is greater than .13 sun radii and the exoplanet mass is less than .892 earth masses, then the exoplanet temperature is medium with two instances and no exceptions.

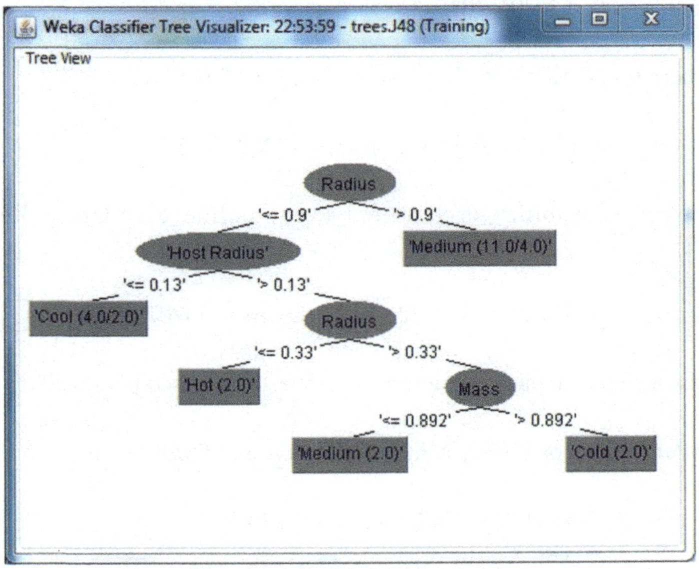

Figure 12: Weka Rule Induction

- Automated Discovery of Unknown Unknowns [13]: The knowledge base keeps track of what is known. As information arrives at the catalog, it is posted to an abductive reasoner that uses the knowledge base as the default structure, and automatically identifies previously unknown nodes and links.
- Additional Characteristics: two attributes that would be useful in filtering on habitable planets that are not yet available or have not been detected are the bio-signature (for example, rocky, tenuous atmosphere, significant oxygen, organic molecules, earth-like). Gemini Planet Imager [16] data has revealed that 51 Eri b, the recently discovered Jupiter-like exoplanet around the nearby star 51 Eridani has an atmosphere of methane and water, and likely has a mass twice that of Jupiter. The second is SETI [14] detection of electromagnetic energy (radar, infrared, gamma ray) that would signal the presence of intelligent life.
- Limitations: the goal of a unique set of attributes for each exoplanet was not met, due largely to that significant amount of unknown information about these objects.

6. Summary

- Elegant: a minimal set of attributes is easily understood
- Leverages domain understanding: mission and functional trade studies
- Provides a hierarchical structure for a knowledge base
- Provides compelling rationale for attribute selection via trade studies
- Satisfies constraints: spans problem space, meets requirements, is a minimal set
- Identifies commonsense limitations in the quest for uniqueness: sometimes it is important to understand that a some exoplanets are more the same than different
- We will revisit the exoplanet catalogs in a few years to update the taxonomy, hopefully with more information on attributes available.

Bibliography

1. http://protege.stanford.edu/, 07/28/2015.

2. http://www.amazon.com/Applications-Artificial-Intelligence-Decision-Making-Multi-Strategy/dp/1502907593, accessed 05/27/2015

3. http://arxiv.org/abs/1106.0635, accessed 07/28/2015.

4. http://www.lpi.usra.edu/planetary_atmospheres/presentations/Taxonomies_plenary.pdf, accessed 11/06/2015.

5. https://en.wikipedia.org/wiki/Kepler%27s_equation, accessed 11/27/2015.

6. https://github.com/OpenExoplanetCatalogue/, accessed 11/27/2015.

7. https://lazycackle.com/Probability_of_repeated_event_online_calculator__birthday_problem_.html, accessed 05/03/2015.

8. https://en.wikipedia.org/wiki/Earth_Similarity_Index, accessed 11/07/2015.

9. https://en.wikipedia.org/wiki/Minimum_description_length, accessed 07/18/2015.

10. http://arxiv.org/ftp/arxiv/papers/1308/1308.0616.pdf, accessed 05/30/2015.

11. http://www.google.com/patents/US8078559, accessed 05/30/2015.

12. http://www.seti.org/, accessed 11/22/2015.

Military Planning for Interstellar Flight

Ken Wisian, Ph.D.

Abstract

Though ubiquitous in science fiction, the military aspects of interstellar travel are too seldom addressed in the forums on actually reaching the stars. The debate on METI (Messaging to Extra Terrestrial Intelligence) /Active SETI (Search for Extra Terrestrial Intelligence) highlights that a peaceful, benign cosmos cannot be assumed a priori. This paper briefly surveys the legal and historical background of military activities in space, civilization to civilization first contacts, the reasons this should prompt us to consider "arming up" and the basic options for crewing and equipping a starship militarily.

Current international law/treaties addressing the military uses of space forbid many military operations and specifically weapons of mass destruction in space. It is worth noting though that throughout history most international norms on weapons restrictions have had mixed success at best. Our limited one-planet dataset also shows that civilizational first contacts are frequently very hazardous, particularly for the less technologically advanced society.

Military options for interstellar ships break down into two main components. The first is how much of the crew, including any Artificial Intelligence (AI), is military or has military training; from none at all, to a complete crew of full-time military. The second component is whether or not there are military use only weapons and defenses on board. There will be systems usable as weapons on a starship regardless - communication lasers, systems to destroy objects on collision course with the ship or propulsion systems could be used as weapons. There will also be personal defense weapons on board for use against dangerous life forms at a minimum, as there have been on manned missions for decades.

This overview will hopefully stimulate the explicit incorporation of military aspects in the work across the interstellar community from ship design to sociological studies.

Keywords

Military, weapon(s), defenses, first contact, dual-use

1. Introduction

Science fiction largely assumes that mankind's future in the universe will be violent, filled with war, fleets of starships, and humans fighting humans or humans fighting aliens. The reality is likely (hopefully) much different. The serious ongoing work on the now envisionable goal of interstellar travel makes little if any mention of military affairs. There is excellent work being done on designing ships from micro-robotic to colony ships as well as thinking through many social/psychological aspects . It is critical then to address sooner rather than later the question of the place of military capabilities (here divided into crew and technology/ hardware) that would be part of a starship. Addressing military capabilities now is driven by the general rule that the earlier components are incorporated into a design, the less costly it is. This can also be phrased as the later you make design changes the more it costs, often exponentially so. This paper will broadly survey the legal issues, the perspective of history, outline courses of action, and hopefully stimulate substantive discussion of the issue at this very early stage of design.

1.1 Current Treaties Preclude Some Military Uses Of Space

Current international law regarding militarization of space is relatively simple, shallow and incomplete. At the least, the commercialization of space will likely drive a contentious expansion of space law. The Treaty on Principles Governing the Activities of States in the Exploration and Use of Outer Space, including the Moon and Other Celestial Bodies , took effect in 1967 and is the primary effective treaty on space. It is a only a few pages long and the primary military-related provisions are;

1. No weapons of mass destruction will be based in space
2. There will be no military bases or activities on the Moon or other celestial bodies

It is important to bear in mind that this treaty was written during the height of the cold war and focused on the superpowers not carrying their conflict into space. The treaty takes no account of the need to defend humanity from extraterrestrial threats, living or inanimate. It is also relevant that throughout history treaties and other international attempts to restrict weapons have a very spotty record – even today treaties banning nuclear proliferation, chemical and biological weapons are nowhere near 100% effective. Some obvious gaps in the current treaty include: 1. Weapons short of mass destruction classification (itself a term subject to considerable debate) are allowed in space, 2. Military bases and exercises are allowed in space but not on celestial bodies, and 3. Military personnel not performing a military role are allowed in space and on celestial bodies with no restriction. Further treaties restricting military activities in space have only garnered limited support. Now add to this rather weak international law on space, the historical tendency of major players to ignore treaties when it is politically or economically inconvenient and you have a situation ripe for making up the rules as you go. Note that space has been and continues to be used extensively for military purposes (intelligence and communications primarily) with missions performed both by robots and humans. The Global Positioning System (GPS) and the Russian GLONASS (Global Navigation Satellite System) are military systems that civilians are conditionally allowed to use. Multiple countries have tested Anti Satellite (ASAT) weapons that are for use solely in space. And finally, all ICBMs (Intercontinental Ballistic Missiles) transit space enroute to their targets.

Related to the above discussion is the growing impetus to develop the capability to neutralize asteroids on a collision course with earth. A military or quasi-military mission such as this, potentially using nuclear weapons will also force the development of new legal framework as well as of dual-use military capabilities and technologies.

1.2 History Suggests That We Should "Arm Up"

The interstellar community often speaks of the innate human desire for exploration and curiosity, but seldom mentions another innate aspect of human nature – conflict and war. War is endemic to humanity and at least some primates . Even life itself is often characterized as a type of warfare. Although human war appears to have declined steadily from pre-history to today, it is far from gone . The big and unanswerable question as it relates to this paper is 'Is violent conflict a condition of all life in the universe?' While we have no way of answering this question with a data set of one planet's species, we can delineate one boundary of the answer. We can definitively say that the universe is not universally peaceful based on our one data point.

War can also be viewed in the framework of politics. The most famous philosopher of war in the Western world, Carl von Clausewitz asserted 'war is a continuation of politics by other means .' This idea is fundamental to most scholarship on war to this day. Would anyone propose that politics will not accompany us into the future in space? A more hopeful rephrasing of the question might be 'could the boundaries of acceptable politics shrink so as not to include war?' It may be a possibility, but not a certainty.

Economic motivation has been a strong driver of exploration in human history. But even when commerce was the driving force behind exploration historically, military and commercial interests were tightly intertwined. At the very least merchants wanted the legal and physical security of their trade guaranteed.

Taken in the broadest possible scope, does human knowledge contribute anything to the question? Looking at life on Earth in the broadest terms, it is clearly a no-holds-barred competition for resources. From viruses to mega-fauna, populations expand to the carrying capacity of the environment and will without thought drive less-able competitors to extinction. Modern humanity is the first example known of self-limiting population (perhaps) and this aspect is due to our intelligence . This then becomes a possible light at the end of the tunnel. If trends continue and human population peaks followed by slow decline, we could remove population pressure / resource scarcity as a driver for conflict on Earth, but also potentially as one of the drivers for exploration or expansion.

We have a limited data set of technological civilization first contacts – namely our own cases of genus homo. The history on this planet of human civilizations meeting is not too good, particularly when one has a significant technological development edge over the other. Obvious examples of contact that did not go well for the less advanced side include the many meetings of European civilizations with American, African, Asian and Pacific civilizations. Note that not all bad effects were based on military superiority. Take for example unintended biological warfare - smallpox, sometimes decimated a civilization ahead of any actual fighting. Technological superiority however is not always determinate. Asian Steppe tribes, most notably the Mongols, conquered a much more advanced China and much of the Muslim and Christian world with superior strategy and tactics, not technology.

Not all contacts turn out in the long run to be bad for the less advanced civilization. Although very violent on a regular basis, the contact between the Roman state and the "barbarians" of Europe stimulated technology transfer and political consolidation, and eventually resulted in the barbarians destroying the Western Roman Empire .

Based on our one-species, one-planet data set reviewed above, it would be naive to venture forth into a galaxy that might have more advanced civilizations, totally unprepared to defend ourselves. Hopefully our one data point represents a transient extreme on the violent end, but "hope is not a strategy."

1.3 The Interstellar Threat

The likelihood of another civilization we come in contact with being of the exact same technology level is small. In the span of cosmic time, human civilization is merely the blink of an eye and our "advanced" technology merely the thought of blinking an eye. Thus if we are visited anytime soon, it is safe to assume that we humans will be the primitives. If we journey to other stars, we could conceivably be either the more or less advanced race. The point is that in general there will be a tremendous mismatch between encountering races.

You might reasonably ask – 'so why arm up?' The top counters that come to mind are; 1. If you don't try, your chances of winning are zero. 2. The laws of physics might set an upper bound on weaponry, such as nuclear bombs and directed energy weapons – meaning that we are already decreasing the maximum technology gap that we might face. 3. There are many non-intelligent threats that require military grade weapons and tactics anyway (asteroids impacting Earth or finding and redirecting or destroying obstacles when travelling at high speeds in space). 4. We may be the more advanced technology, but not the more warlike culture. 5. The argument that we are safe from threat here in our own solar system due to the immense difficulty of traveling interstellar distances assumes no technological breakthroughs. This is more about defenses inside the solar system, but still relevant. Until World War II, the USA thought it could be minimally armed due to the oceans separating it from potential adversaries, but that proved false as technology advanced – the same could prove true for interstellar distances.

The above arguments derive largely from an ongoing debate focused on the wisdom of actively signaling our presence in the galaxy. Stephan Hawking , Jared Diamond and David Brin among others caution that we cannot assume that any extraterrestrial civilization we encounter will be benign.

2. We Must Take Some Military Capability And Training With Us

Based on the above background it is reasonable and prudent to carry some degree of military capability with us into space. Now we look at how to think about military capabilities in early interstellar travel. While the technical aspects of potential weaponry are quite interesting to explore, to even scratch the surface is well beyond the scope of this paper. Suffice that at the minimum future vessels will likely have dual-use technology such as shielding, communication lasers, obstacle clearing systems and even a main drive that will have military utility. An early Larry Niven story, The Warriors, of a first encounter comes to mind as a great example of how a piece of technology, the main drive, can be pressed into duty as a weapon.

The frame of reference for the following is a min-max-min problem - how to minimize risk both existential and mission, by maximizing military capability (if desired), at minimum cost in terms of money, energy, mass, crew complement and training, etc.

2.1 Defenses

Shielding will be present in one or many forms to protect from the harsh environment of space. Shielding could be in the form of mass, particularly in the bow of a ship to protect from collisions with energetic rays and particles - a grain of dust at relativistic speeds can do a lot of damage. Electromagnetic, gravity or other "non-material" shielding is also a possibility (and maybe a life support necessity). The same thought also could apply to the personal level. Suits capable of handling any threat in space or on a new planet would have to be able to withstand a lot of punishment and could end up as a cross between a space suit and combat armor. Ship Maneuverability, laterally as well as acceleration and deceleration is a key defensive quality too. Extra maneuverability beyond the minimum mission requirements is very desirable, but would likely be "expensive" of weight and resources. Jamming or decoy systems are a different matter in that they are unambiguously military in nature . Any unambiguously military systems send a message, not necessarily bad, but distinct from dual-use systems.

Less obvious issues that I will lump under defenses are ship layout and information protection. A ship's design could be different based on the possibility of fighting. For example a backup bridge might be at the center of a ship versus at maximum separation from the primary bridge. If you are looking to protect the bridge from aimed weapons versus random space rocks. There are countless other ways that design might differ, from spatial layout to systems redundancy. Information is another significant question. Any ship will carry substantial information on Earth in many forms. Would a method of quickly destroying information be desired in order to keep knowledge out of potentially hostile appendages? Is it even practical to try to protect knowledge of Earth? What about the location of Earth? Any course alterations (to hide the ship's origin) would increase flight time – unlikely to be sanctioned or affordable on early missions.

2.2 Weapons

Weapons cover a huge range of possibilities from non-lethal personal defense to weapons of mass destruction – we'll start small and work up. I will skip the area of biochemical weapons as an explicit subject in this paper.

The need for self-defense on an unknown world with an unknown biota will drive the need for a wide range of weapons analogous to modern ground weapons. This would range from side arms (pistols or stunners) through heavier, crew-served weapons (think heavy machinegun or cannon) all the way up to armored vehicles (think tanks). Remember although there are no land animals on Earth today that can't be stopped by a large rifle, the equivalent of a large, hungry or amorous dinosaur might require something heavier. Any robots will also have some level of self-protection inherent in their design. Besides chemical powered current weapons, technologies such as rail guns and directed energy weapons are already showing up as fielded weapon systems on Earth.

For ships again there will most likely be dual-use tech present. An "active" drive or maneuvering thrusters have some capability – the focused energy needed to move a ship could be pretty destructive. Project Orion used nuclear bombs for propulsion; Project Daedalus used electron beams to ignite fusion . A starship itself could even potentially be a large guided bomb. Similarly any communication system with enough power to send signals across interstellar distances could at be destructive to electronics if not more, at short range. Even an ion drive might be capable of blinding or destroying sensors. If a ship has any kind of active defense against physical objects on a collision course, that also would have a significant military capability. These active defenses could be anything from missiles, kinetic or explosive, directed energy weapons, to nuclear weapons.

It is possible that ships could be sent out even without dual-use weapons. Potentially un-armed options for ships include swarm, micro or nano-ships, or any ship considered essentially "disposable." These would not necessarily be armed, though a self-destruct or memory erase function would be an option. Sail ships are another possibility - the extreme need for minimum payload mass could force the decision to go with no weapons, human crew, or even redundant systems.

Obviously military-only use weapons could be added to a ship. It is a matter of the cost in terms of mass and ship/crew resources more than actual cost vs. risk. Regardless, we almost certainly will not be venturing forth unarmed, even if a mission is 100% peaceful in intent. Then there is the perception problem. There is no way to guarantee that another civilization will "see" peaceful intent in the dual-use technologies regardless (particularly if they have been watching decades of our broadcasts).

2.3 Crew Options

Given that any interstellar ship likely has some military capability, let's turn to what the options are for crewing such a ship. This discussion initially addresses an actual human crew, followed later by options for robot/AI crews. The range of possible military training and crew complement range from no military training for any crewmember to an all-military crew. The possibility of a police force filling the military role (likely only in a large-crew generation ship scenario) could be a sub-branch of any of the models and is really just a matter of how much military oriented training the police receive (e.g. Special Weapons and Tactics or SWAT forces).

2.3.1 All Crew Full-Time Military

An all full-time military crew would obviously be the most secure option, in that the military mission will be the first priority. Military command and control also has the likely benefit of minimizing crew chaos in crisis. The obvious drawback is that this might be detrimental to the exploration or colonization mission. In other words (in fiction) it is often assumed (but unproven) to be expensive of "non-productive" resources to be 100% military. Productive is here taken as maximizing scientific discovery or human settlement. Certainly though, neither scientific nor settlement mission success is precluded by an all-military crew. In particular, military organization in an extreme survival situation, like colonization, could be a significant asset. Why worry about contact on a colonization mission? Our technology will likely tell us if an advanced civilization exists at a destination before we set out, but movement and expansion outside the solar system will significantly increase the chances of contact and the signature of an interstellar drive is likely to be "bright" and attract attention. Consider also how far our technology has advanced in the last hundred years (a reasonable journey time).

A more subtle drawback to an all-military crew is the old saying about 'if all you have is a hammer everything looks like a nail'. A military outlook is more likely to see worst-case scenarios or to interpret ambiguous actions as threat. Perhaps this could be overcome with training.

2.3.2 Part Of Crew Full-Time Military

This option could vary tremendously, so I will focus on a small percentage of the crew being military, say 5-20%. This is somewhat a reflection of most Earth societies. The intent would be enough of a force to use ship weapons and act as Marines if needed. Less "costly" than the all-military crew, this option is one possible min-max-min solution – getting maximum military benefit at the lowest cost, therefore the size of this part of the crew is kept small.

2.3.3 Entire Crew Citizen-Soldiers

This option looks something like the Enterprise of Star Trek fame. All the crew is technically military, but their military training and focus is somewhat secondary. This is also analogous to modern National Guard / Reserve / Territorial forces where the military role is secondary to the civilian role except in emergencies or war. A possible advantage of this option is that in a time of emergency, military command and control covers the entire crew.

2.3.4 Some of the Crew Citizen-Soldiers

This is getting to the lower end of military capability in that only some of the crew will have some military training as a secondary role. It is also reflective of some modern societies (like the US) that have a small percentage of people in the National Guard / Reserves as opposed to 2.3.3 where all crewmembers are reservists. It is the "lowest cost" option which still provides some explicit military capability.

2.3.5 A Completely Civilian Crew

A completely civilian crew, with no military training may maximize the scientific and colonization benefit, but by "zeroing the denominator" maximizes risk. This option assumes a completely benign universe and/or willingly concedes any conflict.

2.4 Secondary Question—Which Authority Is Paramount Military Or Civilian?

This is perhaps actually a paramount question, but is secondary within the structure of this paper simply because of the very involved arguments. Assuming some mix of military and civilian crew, whether full or part-time, how do you decide which authority is paramount and when? It ranges from complete civilian control to military dominant all the time to situation dependent (e.g. existential threat drives a military command). Military and civilian leadership could be vested in the same person or people in any of the scenarios, which simplifies things, or it could be separate. There are good and bad examples of both cases throughout history.

2.5 Artificial Intelligence (AI)/Robots

A ship "crewed" only by AI could still need all the dual-use technology discussed above and be equipped with military only tech if desired. A ship with a mixed crew, human and AI, would devolve back to all the above options for a human crew except when the crew is non-conscious. Assuming unlimited memory or learning capacity, an AI might be able to fill the roles of scientist and military tactician equally well all the time as needed, and thus moot the questions above about percentages and training levels. Assuming also that any AI sent out would have the same level of trust that a human crew would have to use deadly force, there might be no restrictions on weapon use. This is a highly complex and changing area of research. Robots on the battlefield today are non-intelligent and limited, but the situation is changing rapidly. Non-intelligent systems can already make "trigger" decisions today, with varying levels of human supervision. Barring a binding international agreement, we will cross a threshold of autonomous, lethal AI robots at some point in the foreseeable future.

3. Conclusion

Based on the narrow but deep data set of human experience, venturing forth into interstellar space defenseless is not prudent. Some degree of military weaponry and crew training/programming should be planned into interstellar missions. Options range from minimal added cost in terms of time and resources to making the military mission primary. Even "low end" options must be planned for, will take resources and require much thought.

Why worry about the specific military aspect of missions at this point? Beyond the obvious hardware decisions, some other aspects of design could be significantly affected; redundancy, habitat layout, medical facilities, even the AI and supporting software might be substantially different if military considerations are in place. Early in design decisions to maximize military value will be much more cost effective than later decisions. Also to be considered is how much information on our civilization to carry on board and our communications method back to Earth. Even the course trajectory needs to be considered, though currently it is difficult to conceive of how these things could in practical terms be "hidden". This paper only skims the surface of the topic and outlines a min-max-min framework for looking at the problem. The hope is to stimulate further discussion and work on military aspects of interstellar travel.

References

1. Truth in advertising; I hold the Rank of Major General in the United States Air Force (Reserve) and Texas Air National Guard, and have participated in most of the U.S. conflicts of the last few decades, none the less I am not a war "hawk". Also, nothing in this paper should be construed as representing the official position of the United States or Texas Government, the Department of Defense, or the US Air Force. The opinions and any errors are mine and mine alone.

2. I point readers to the recent 100 Year Starship conference http://symposium.100yss.org/technical-tracks and the Icarus Interstellar Starship Congress conferences http://www.icarusinterstellar.org/congress-livestream/ as starting points.

3. "Treaty on Principles Governing the Activities of States in the Exploration and Use of Outer Space, including the Moon and Other Celestial Bodies", United Nations Office for Disarmament Affairs, available at http://disarmament.un.org/treaties/t/outer_space/text, entered into force: 10 October 1967.

4. Gat, Azar, "War in Human Civilization", Oxford University Press, 822p., 2006, is meticulous in detailing the endemic war in human history. Steven A. LeBlanc, "Constant Battles: Why We Fight", St Martin's Griffin, 2003, also excellent and covers the subject in a much shorter book – both refute the myth of the "peaceful savage" Rousseau-ish thinking.

5. Pinker, Steven, "The Better Angels of our Nature: Why Violence has Declined", Penguin Books, 2012.

   "Steven Pinker: The Surprising Decline in Violence", TED Talk at http://www.ted.com/talks/steven_pinker_on_the_myth_of_violence.html , 2007.

   Morris, Ian, "War! What is it good for?", Farrar, Strauss and Giroux, New York, 2014.

6. Clausewitz, Carl von, "On War", M. Howard & P. Paret eds. & transl., Alfred A. Knopf, New York, 1993.

7. The potential for civilization population leveling or decline (without outside drivers) pertains to the demographic transition - the idea that as conditions improve (prosperity, technology, education) societies/countries move through four stages. Initially for pre-industrial countries birth and death rates are high, pretty much cancel each other out, and total population is stable. In stage two, as development increases, mortality (death rate) falls, while birth rate remains high (a generation or two typically). In stage three the death rate is low and birth rate falls. Taken together stage two and three result in a significant population increase (explosion). These cascading population booms (first in the industialized world followed by other regions) resulted in the rapid rise in world population in the 19th through 21st centuries. The fourth stage is low birth rate and low death rate (post transition) resulting in a new equilibrium population. More recently a fifth stage is postulated in which the birth rate falls and remains below replacement rate resulting in slow population decline. Assuming the entire world eventually reaches "stage 5", world population will reach a peak and then start a long slow decline. There is not enough data by far to make a guess at whether there is a "floor" to declining population. This overall scheme is reflected in UN projections of a maximum world population of around 10 Billion in 2100 (note these numbers are very sensitive to poorly defined variables). (Reference needed here, not just a statement. This may be best inserted into the main body of the document.)

8. Heather, Peter J., "Holding the Line: Frontier Defense and the Later Roman Empire", pp. 227-246, in Victor Davis Hanson, Ed. "Makers of Ancient Strategy: From the Persian Wars to the Fall of Rome", Princeton University Press, 2010.

9. Hawking, Stephen, "Into the Universe with Stephen Hawking", *Discovery Channel*, 2010.

10. Diamond, Jared, "To Whom It May Concern", *New York Times Magazine*, December 5, 1999, p. 68-69.

11. Brin, David, "Shouting at the Cosmos… Or How SETI has Taken a Worrisome Turn Into Dangerous Territory", Lifeboat Foundation (http://lifeboat.com/ex/shouting.at.the.cosmos), 2006-2008.

12. Niven, Larry, "The Warriors", in "Man Kzin Wars", Baen Books, 1988.

13. While it is possible to use communication systems as jammers or to improvise decoys, the effectiveness is likely to be so low as to be negligible. Militarily, today all effective jamming and decoy systems are purpose built. (Reference needed here, not just a statement.)

14. General Dynamics Corp., "Nuclear Pulse Vehicle Study Condensed Summary Report", NASA, available at; http://ntrs.nasa.gov/archive/nasa/casi.ntrs.nasa.gov/19760065935_1976065935.pdf , 1964.

15. Obousy, R. K., "A Review of Interstellar Starship Designs", Presented at the One Hundred Year Starship Conference in Orlando, available at http://www.icarusinterstellar.org/wp-content/uploads/2012/05/A-Review-of-Interstellar-Starship-Designs9-27-11.pdf , October 2011.

Minimal Requirements for a Fully Autonomous Health Care in a Closed System

Natalie Bordag, Ph.D. and Kurt Zatloukal, M.D.

Abstract

Long-term human space flights pose several yet unsolved challenges such as adequate health care is of utmost importance for mission success. At the same time such flights provide a unique experimental environment for quantum leap innovations with the potential to revolutionize general health care. Space flights prohibit access to classical health care and therefore must have a fully autonomous health care system suitable for all known and yet unknown (e.g. resistant pathogens) disease conditions since health checks cannot exclude occurrence of diseases or health problems in the long-term. Since a starship cannot host all competencies of a classical health care system, new concepts are needed. An autonomous health care system starts with highly sensitive and specific diagnostics which flows into integrated data analysis yielding medical care instructions and outcome predictions coupled to drug assembly and personalized dosage up to robotic surgery. Although for most parts building blocks already exist or are on the immediate horizon, diagnostics remain the fundamental bottleneck and utilization of metabolomics is proposed. The metabolome is in itself the integration of the cascade of interactions that take place between environmental factors, individual (epi-)genetic makeup and his transcriptome, proteome and microbiome. Therefore metabolomics can detect essentially in real time the adaptive, multiparametric response of the organisms to pathophysiological stimuli even before clinical symptoms manifest and would be a perfect base for stand alone, autonomous health care systems.

Keywords

autonomous, health care, metabolomics, computational modelling, personalized, prevention

1. Introduction

For long-term space flights, health care cannot be provided on a remote basis. Therefore a fully autonomous health care system needs to be established. Such a health care system has to be suited to address all known and yet unknown disease conditions since health checks cannot exclude eventual occurrence of diseases or health problems in a long term perspective. Since it is not feasible to establish all competencies and infrastructure of a classical health care system, new concepts are needed. Furthermore, the shelf life of drugs is limited makes it necessary to produce drugs from stable precursors and requires capacities to produce new drugs for new diseases conditions (e.g. antibiotics or vaccines for newly emerging or drug resistant pathogens).

Building blocks for an autonomous health care system do already partially exist or are on the immediate horizon. Technologies such as specialized non- or minimally invasive diagnostic, computational modelling of disease conditions utilizing massive omics data and biosensor information, computer-designed drugs composed of individualized mixtures of functional components (drug precursors), stem cell-based regenerative medicine or robotic surgery will form the foundation for an autonomous health care system. Diagnostics are the starting point for any health care and are currently also the one fundamental bottleneck. Many diseases, ranging from metabolic syndrome over various cancer types, can be well treated in cases of sufficiently early detection. However, to cover the spectrum of current diagnostics a huge range of technologies and demanding infrastructure (e.g., for medical imaging) are needed and would for several major diseases still not achieve the necessary specificity and sensitivity. Therefore medical diagnostics in a closed system must integrate different relevant, easier obtainable and more sensitive parameters to inform on the health or disease status of individuals thereby giving guidance for personalized prevention or therapy. The approach should improve diagnostics up to a level where disease onset is safely detected before manifestation of clinical symptoms and irreversible organ damage occurs. Disease prevention and stabilizing management of unpreventable diseases is in a long-term enclosed environment such as space flight of critical importance for mission success.

2. Metabolomics a Key Pillar for Diagnostics

Metabolomics is a technology that provides the information on the cascade of interactions that take place between environmental factors, the individual genetic makeup and his microbiome which results in a highly informative metabolic state as defined by the sum and relative ratios of metabolites (small molecules <1.500 Da which are the products of metabolic reactions and define the functional state of an organ or whole body) (Figure 1). In this context, metabolomics has a preferential role with respect to other omic technologies because of its ability to detect essentially in real time the adaptive, multiparametric response of the organisms to pathophysiological stimuli or (epi-)genetic modifications.

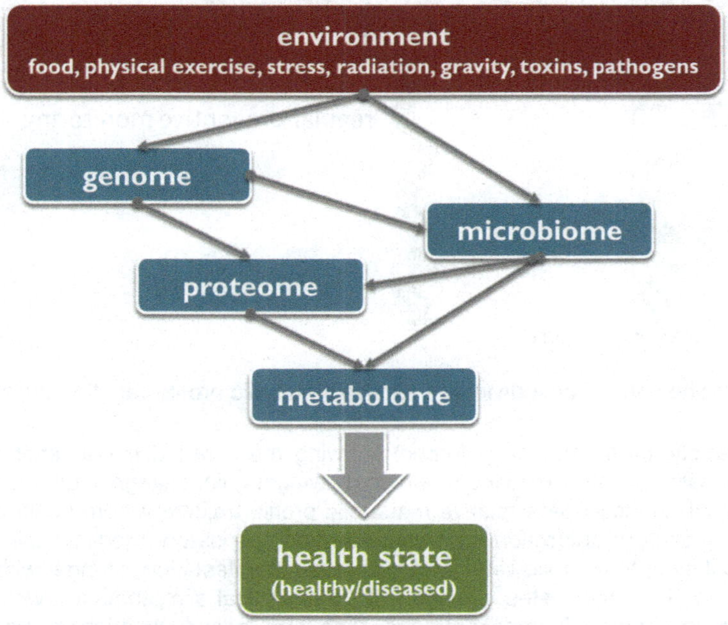

Figure 1: Schematic of the interaction of environment with genome, proteome, microbiome and metabolome which is the closest imprint of the individual's health state.

A specific advantage of metabolomics is that analyses can be performed on biological samples which can be collected by non invasive procedures (e.g. urine, saliva, exhale condensate) and further processed in automated work flows. Currently the stand alone and wide spread use of metabolomics for routine diagnostics is hindered by the demand for very high quality sampling often only possible in hospitals. Therefore the development of a sampling devise allowing for non invasive (e.g. urine, saliva, exhale condensate) self sampling by the patient or space traveler constitutes a major breakthrough towards personalized, stable, specific and most of all preventive health care. Another advantage of metabolomics is that the metabolic fingerprint is highly specific for a person and can only be interpreted reasonably in the context of the personal metabolic trajectory. This avoids several concerns typically discussed for genetic information and essentially eliminates the risk of misuse of data, e.g., by insurance companies or employers.

2.1 Metabolomics for Disease Prevention

The earliest available signs of any disease onset are alterations of the individual's metabolome, often resulting from compensatory mechanisms which start long before clinical symptoms and the disease manifest themselves. A specific and reliable detection of these early signs should allow for effective prevention even with mild counter measures like lifestyle changes or low dose treatments (Figure 2).

For metabolomics, two complementary analytical techniques exist, namely high-field proton nuclear magnetic resonance (1H-NMR) and liquid or gas chromatography coupled mass spectrometry (LC /GC MS). 1H-NMR spectra provide the overall metabolic profile of essentially all abundant (c>1 μM) metabolites in one sample, yielding a "metabolic fingerprint" which is highly specific for a person. LC-/GC-MS dissects and quantifies with high accuracy many, but not all, of the metabolites in a sample and provides the sum of those quantifications as metabolic profile in the form of a table with values per metabolite per sample. We propose exploiting the advantages and the added value from combining both technologies to establish the best possible protocol for monitoring the individual's metabolic trajectory changes.

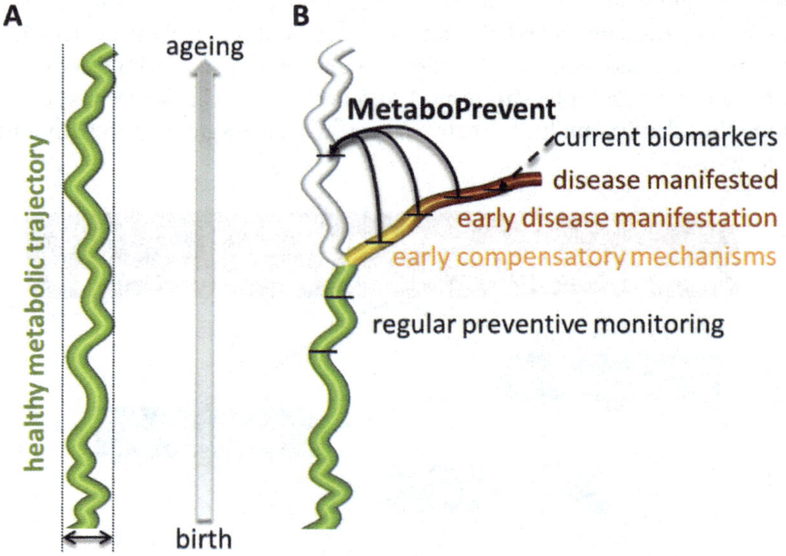

Figure 2: Schematic of an individual's relative metabolic profile variation during life time.

(A) Healthy metabolic profile trajectory (green) showing the intra-individual variability caused by factors, such as ageing, smaller health events (common cold), weight gain, change in dietary habits and others.
(B) Development of an individual's relative metabolic profile trajectory from healthy (green) to manifested disease (brown). Trajectory aberration start with early compensatory mechanisms (yellow) prior to manifestation of clinical symptoms, followed by early disease manifestation (orange) with first clinical symptom manifestation ending in a manifested disease with clear clinical symptoms (brown). Trajectory aberrations change the individual's metabolic profile stronger than intra-individual variability. Early detection allows for mild interventions and resumption of the healthy trajectory (black arrows). Different diseases may result in different metabolic profile trajectory aberrations (not shown).

2.2 Individualized Health Monitoring with Metabolite Patterns or Whole Spectra

Statistical robust detection of subtle metabolic changes from early compensatory mechanisms before disease onset based on single biomarkers is virtually impossible. However, the combination of several metabolites into complex biomarker patterns results in good statistically performance enabling early detection in the first place. The technical feasibility and relevance of biomarker patterns has been demonstrated with MetaMap® Tox, a rat plasma data base for early and systematic drug safety testing with currently more than 100 toxicological patterns deduced from more than 500 reference substances [2, 1]. The statistical power of metabolic patterns to detect early stages has been demonstrated for drug-induced toxicities and injuries[2]. A metabolic pattern contains 2-35 metabolites and improves statistical outcome in terms of sensitivity and specificity tremendously when compared to single biomarkers. This approach has been extended to use the whole information of metabolome NMR spectra to characterize the health status of individuals. NMR spectra have been shown to be highly specific for individuals and any deviation from this individual spectrum was indicative for a major lifestyle change or onset of disease[5, 4, 3].

2.3 Higher Specificity and Sensitivity with Personalized References

Currently the variability between individuals prohibits development of metabolomic biomarkers of sufficient sensitivity and specificity to detect early compensatory mechanisms (Figure 3).

The rationale for personalized approaches is the observation that intra-individual variability is at least twofold lower than the variability between individuals[6]. Personalized assays would allow much higher sensitivity at equally high specificity. Studies have shown that individual metabolic fingerprints exist in the urine of healthy volunteers[7] and tend to remain sufficiently stable over time, and stability is not compromised by interfering factors like aging, lifestyle changes, smaller health events (common cold), weight gain, change in dietary habits and others[8, 9]. Early prediction of disease states based on referencing changes in metabolome to previous data from the same person (when he or she was still healthy) has also been shown for diabetes[3] or coeliac disease[5, 4].

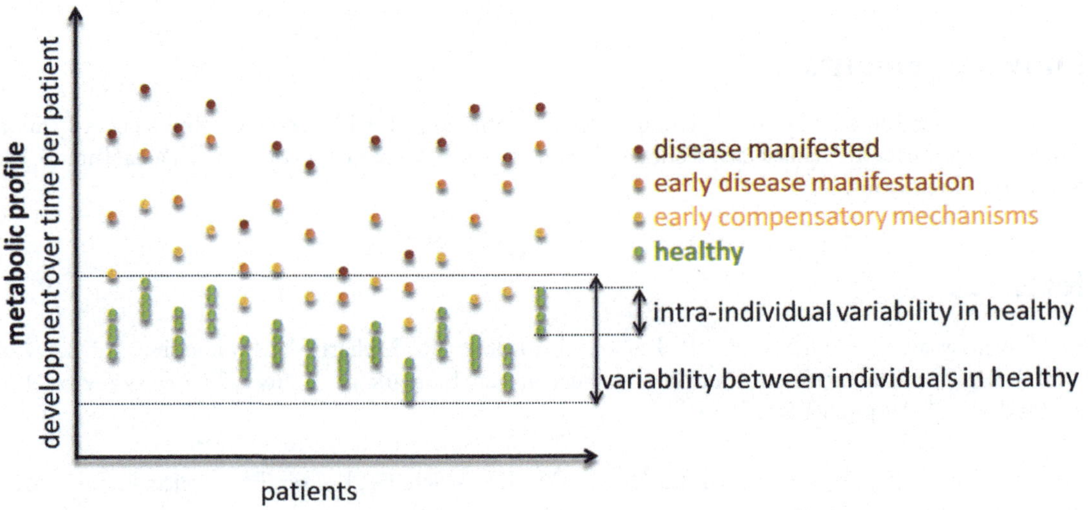

Figure 3: Schematic of development of metabolic profiles over time from healthy to disease manifested for various patients. The range of variability is larger for a healthy population than the intra-individual variability. Specific and sensitive detection of early disease manifestation is only possible with personalized cut offs requiring multiple measurements during lifetime. Personalized normalization and analysis allows specific and sensitive detection even of early compensatory mechanisms before disease manifestation.

2.4 The Virtual Human

The second step in an autonomous health care system after the establishment of metabolic diagnostics (e.g. via non invasive self sampling) is the automated measurement and analysis. The virtual human is the integrated computational model of all parameters impacting an individual's health status directly predicting disease/health status, and thereby giving individualized medical guidelines with treatment and therapy management instructions.

Currently, automated push-button metabolic measurement solutions and data analysis already exist, although not in the health care sector. They are able to deliver reliable targeted and untargeted screening results for complex biological samples like wine[10] or juice including automatic quantification of most decisive parameters, comparison to reference values and verification as well as prediction. A plethora of modeling techniques applied to metabolic data exist (e.g., elastic net, penalized logistic regression, random forest, support vector machines, knowledge discovery by accuracy maximization[11]) and move automated data modeling for the virtual human into reach. The metabolome is an ideal starting point for the virtual human as it comprises less complex data than other omics technology but is simultaneously in itself already an integration of the multiparametric response to pathophysiological changes. Addition of further data sources, such as proteomics, trascriptomics, metagenomics or environmental exposure data is envisioned by creating in a stepwise manner an increasingly complex and more reliable computational model of the human. Principal feasibility has already been demonstrated in integrated modelling of transcriptomics and metabolomics data in the context of metabolic liver disease[12].

3. Conclusions

Long-term human space flights pose yet unsolved health care challenges, and at the same time provide a unique experimental environment for quantum leap innovation of health care in general. An interdisciplinary think tank could work on the specification of key components of a fully autonomous health care system which can be established in a starship environment. Such a process could lead to a research agenda which has the potential to generate highly innovative concepts for addressing global health challenges related to ageing populations, remote locations, emerging and re-emerging infectious diseases and financial constrains in funding health care. Since innovation in medicine and health care is often hampered by complex regulatory processes, lack of flexibility in routine health care, traditional training and career models, the experimental environment of a starship could avoid several of these innovation roadblocks. Many building blocks exist or are on the immediate horizon but overcoming singular bottlenecks, optimizations of existing buildings blocks and most importantly integration of these buildings blocks into one whole autonomous system are needed.

Acknowledgements

This work was funded in part by the Christian Doppler Laboratory for Biospecimen Research and Biobanking Technologies. We sincerely thank Hans Lehrach, Berlin as well as Claudio Luchinat and Paola Turano, Florence for stimulating discussions.

References

1. Van Ravenzwaay, B., Cunha, G. C.-P., Leibold, E., Looser, R., Mellert, W., Prokoudine, A., … Wiemer, J. (2007) The use of metabolomics for the discovery of new biomarkers of effect. *Toxicology Letters,* 172(1-2), 21–28 doi:10.1016/j.toxlet.2007.05.021

2. Mattes, W. B., Kamp, H. G., Fabian, E., Herold, M., Krennrich, G., Looser, R., … Piccoli, S. P. (2013) Prediction of clinically relevant safety signals of nephrotoxicity through plasma metabolite profiling. *BioMed Research International,* 2013, 202497 doi:10.1155/2013/202497

3. Wang, T. J., Larson, M. G., Vasan, R. S., Cheng, S., Rhee, E. P., McCabe, E., … Gerszten, R. E. (2011) Metabolite profiles and the risk of developing diabetes. *Nature Medicine,* 17(4), 448–53 doi:10.1038/nm.2307

4. Bertini, I., Calabrò, A., De Carli, V., Luchinat, C., Nepi, S., Porfirio, B., … Tenori, L. (2009) The metabonomic signature of celiac disease. *Journal of Proteome Research,* 8(1), 170–7 doi:10.1021/pr800548z

5. Bernini, P., Bertini, I., Calabrò, A., la Marca, G., Lami, G., Luchinat, C., … Tenori, L. (2011) Are patients with potential celiac disease really potential? The answer of metabonomics. *Journal of Proteome Research,* 10(2), 714–21 doi:10.1021/pr100896s

6. Gruden, K., Hren, M., Herman, A., Blejec, A., Albrecht, T., Selbig, J., ... Jeras, M. (2012) A "crossomics" study analysing variability of different components in peripheral blood of healthy caucasoid individuals. *PloS One,* 7(1), e28761 doi:10.1371/journal.pone.0028761

7. Assfalg, M., Bertini, I., Colangiuli, D., Luchinat, C., Schäfer, H., Schütz, B., & Spraul, M. (2008) Evidence of different metabolic phenotypes in humans. *Proceedings of the National Academy of Sciences of the United States of America,* 105(5), 1420–4 doi:10.1073/pnas.0705685105

8. Bernini, P., Bertini, I., Luchinat, C., Nepi, S., Saccenti, E., Schäfer, H., ... Tenori, L. (2009) Individual human phenotypes in metabolic space and time. *Journal of Proteome Research,* 8(9), 4264–71 doi:10.1021/pr900344m

9. Heinzmann, S. S., Merrifield, C. A., Rezzi, S., Kochhar, S., Lindon, J. C., Holmes, E., & Nicholson, J. K. (2012) Stability and robustness of human metabolic phenotypes in response to sequential food challenges. *Journal of Proteome Research,* 11(2), 643–55 doi:10.1021/pr2005764

10. Godelmann, R., Fang, F., Humpfer, E., Schütz, B., Bansbach, M., Schäfer, H., & Spraul, M. (2013) Targeted and nontargeted wine analysis by (1)h NMR spectroscopy combined with multivariate statistical analysis. Differentiation of important parameters: grape variety, geographical origin, year of vintage. *Journal of Agricultural and Food Chemistry,* 61(23), 5610–9 doi:10.1021/jf400800d

11. Cacciatore, S., Luchinat, C., & Tenori, L. (2014) Knowledge discovery by accuracy maximization. *Proceedings of the National Academy of Sciences of the United States of America,* 111(14), 5117–22 doi:10.1073/pnas.1220873111

12. Pandey, V., Sultan, M., Kashofer, K., Ralser, M., Amstislavskiy, V., Starmann, J., ... Wierling, C. (2014) Comparative Analysis and Modeling of the Severity of Steatohepatitis in DDC-Treated Mouse Strains. *PloS One,* 9(10), e111006 doi:10.1371/journal.pone.0111006

2015

finding earth 2.0

100YSS PUBLIC SYMPOSIUM 2015

100 YEAR STARSHIP™
canopus 310 LY

Finding Earth 2.0 explores the technologies and processes to definitively find a planet outside our solar system capable of supporting Earth-evolved or terran life, as well as the impact of the indisputable discovery of an Earth analogue or Earth 2.0, so to speak.

Presenters, speakers, and attendees at the Symposium asked questions and sought answers to what specific capabilities and systems—scientific, technical, and societal—will be needed over the next five to twenty-five years, not to merely suggest or catalog Earth analogue candidate exoplanets, but to identify at least one definitive Earth 2.0.

Finding Earth 2.0 has profound implications for technology, knowledge, and systems across the spectrum of human experience, skills, capacities, perspectives, and ambitions with each new milestone reached. All are a true part of a bold, audacious journey.

"FINDING EARTH 2.0 HAS PROFOUND IMPLICATIONS ACROSS THE SPECTRUM OF HUMAN EXPERIENCE . . ."

- Dr. Mae Jemison

A Framework for the Design of Ethical Robotic Probes

David Burke

Abstract

One of the challenges inherent in sending a robotic probe to an exoplanet is that of interacting effectively with any potential life forms it encounters. In such a situation, how important is it that the probe, as an ambassador from Earth, acts morally and ethically? How do we even begin to define what those terms mean in this context?

We've just started to wrestle with similar challenges here on Earth. In recent years, there has been increasing interest in the problem of how to program robots to behave in a moral and ethical fashion when interacting with humanity (Asimov's "Three Laws of Robotics" being an early and well-known example). Various approaches have been proposed, ranging from reducing moral behavior to a set of irreproachable rules for the robot to enforce, to providing robots with a "moral calculus" module that computes the greatest good for the greatest number, to instilling robots with virtuous qualities (courage, truthfulness, kindness, etc.)

Whereas all of these approaches have merit, robot morality is still a complex, unsolved problem. And yet, building ethical robots on Earth is a simpler problem to tackle than the scenario above, where we can't easily make assumptions about the exoplanet's life forms: we can't presume they have human-like emotions or rational faculties; we will know very little about their evolutionary history, and we have no idea how they will conceptualize or perceive our probe.

In this paper, we discuss how people might constructively begin to think about this challenge, how to merge the various ethical perspectives mentioned above, and to suggest some next research steps towards an meta-ethical framework that would enable effective moral interactions between the probe and any denizens of an exoplanet it encounters.

Keywords

robotics, ethics, robot probe, thermodynamics, prime directive

1. Introduction

Given the challenges of sending humans on interstellar journeys, first contact with some life form on an exoplanet is overwhelmingly likely to be through a robotic probe. And without some as-yet undiscovered method of faster-than-light communication, the probe is going to be on its own, functioning autonomously without meaningful human guidance. Hence, the probe is a de facto ambassador for humanity, and the stakes for its behavior are very high. In such a situation, can we ensure that the probe will act morally and ethically?

It is easy to construct fanciful situations such as the probe arriving on an exoplanet that contains intelligent life, and the probe is under pressure to take sides on in some sort of political dispute. Such scenarios are very common in science fiction: as an example, consider the television show 'Star Trek' and the so-called 'Prime Directive', which states that humanity must not interfere with the natural development of evolution and society on other worlds. Fictional starship captain Jean-Luc Picard once stated the reasoning behind such thinking: "Whenever mankind interferes with a less developed civilization, no matter how well-intentioned that interference may be, the results are invariably disastrous." [1] (on the other hand, that interference makes for compelling drama, which is why the Prime Directive was violated so frequently.)

In recent years, there has been increasing interest in the problem of how to program robots to behave in a moral and ethical fashion when interacting with humanity, to say nothing of interactions with extraterrestrial life. Isaac Asimov's "Three Laws of Robotics" is an early and well-known example of exploring the consequences of robot ethics programming (and like 'Star Trek's Prime Directive, it was ostensible violations of these three laws that generated the plots for Asimov's stories). Various approaches have been proposed, ranging from reducing moral behavior to a set of irreproachable rules for the robot to enforce, to providing robots with a "moral calculus" module that computes the greatest good for the greatest number, to instilling robots with virtuous qualities (courage, truthfulness, kindness, etc.)

Whereas all of these approaches have merit, robot morality is still a complex, unsolved problem. And yet, building ethical robots on Earth is a simpler problem to tackle than the scenario above, where we can't easily make assumptions about the exoplanet's life forms: we can't presume they have human-like emotions or rational faculties; we will know very little about their evolutionary history, and we have no idea how they will conceptualize or perceive our probe. In this paper, we discuss how people might constructively begin to think about this challenge, how to merge the various ethical perspectives mentioned above, and to suggest some next research steps towards an meta-ethical framework that would enable effective moral interactions between the probe and any denizens of an exoplanet it encounters.

2. The Challenge in Historical and Contemporary Context

Before specifically tackling the challenges of designing and specifying ethical behavior for robots, it is worth leveraging the rich tradition of philosophical thought about ethics as a guide to our computational instantiations.

Very roughly, we can sort ethical thinkers into three different schools of thought about how morality fundamentally works:

1. Deontological: These are those philosophical and religious thinkers who gravitate towards the idea that there is (or should be) a set of rules or natural laws that describe our ethical obligations towards one another. Kant is an exemplar of this tradition, in which people's actions are judged by their intention to conform with natural law.
2. Consequentialism: In this school of thought, morality is essentially a matter of cost & benefit, the "greatest good for the greatest number". Moral acts are judged by their consequences on society; Jeremy Bentham was an early exponent of this philosophical worldview.
3. Virtue Ethics: In this school of thought, there are cardinal virtues to which all humans should aspire to: humility, courage, prudence, and so forth. If we can mold people's characters so that they adopt these virtues, then moral behavior will naturally result. This was Aristotle's view, for instance.

The very fact that there continue to be (at the very least...) three schools of thought on the subject is evidence that we don't really know the answer. It is very easy to design scenarios which illustrate the unintended consequences of a slavish adherence to any of these worldviews. For instance, for any rule that might be put forth as a categorical statement, say, "Don't kill", it isn't hard to come up with an exception scenario.

The potential weaknesses that lurk in the consequentialist approach were explored by the philosopher Philippa Foot in her "Trolley Problems" – a famous set of ethical thought experiments. [2] In these scenarios, you are asked to take actions that sacrifice innocent lives to save others, and these demonstrate that the "greatest good for the greatest number" is more complicated than you might initially think. All of the various schools of though suffer from these kinds of difficulties, and the way forward is even less clear once you start imagining that you need to generate the kind of formal behavioral specification that would be required if you were to direct a machine's actions. As a result, the field of 'machine ethics' is an active and contentious one. [3,4,5]

This issue isn't just an abstract, philosophical one, because the future is coming at us very quickly. DARPA (Defense Advanced Research Projects Administration) funded the roboticist Ronald Arkin to look at how to design architectures for battlefield robots. In his final report, Arkin concluded that there is reason to believe that robots can be more ethical in combat than humans, since they don't suffer from fatigue, anxiety and other emotions that could lead to irrational actions in battle.

The popular press has also shown great interest in the subject, with headlines like this appearing in the last couple of years:

- "What if your self-driving car decides that one death is better than two – and that one is you?"
- "Killer Robots with Automatic Rifles Could Be on the Battlefield in 5 Years"

The smart money is on the bet that semi-autonomous and autonomous robots will be increasingly ubiquitous in the near future, and that we will be relying on them to make sensible decisions in domains that have a serious impact on human welfare. And yet, there is no consensus as to the underlying framework to support robot morality. How do we avoid the potential negative consequences of this future? And if we suggest a way forward, is it applicable to the autonomous exoplanet ambassador scenario?

In the rest of this paper, we will unveil a framework to support thinking about morality as a general principle, abstracted away from any particular human concerns. To do this, we will begin by looking at ethics as an abstraction of energy flows. Looking more carefully at the phenomenon of energy flows, we will find that universe is full of persistent, far-from-equilibrium gradient reduction mechanisms, life in particular. And this insight will lead us to a preliminary formulation for a proposed Prime Directive for robotic probes. This new formulation can serve as a meta-ethical framework and a foundation for further investigation.

3. Ethics as a Resource Design Problem

In an effort to bring some clarity to the topic of human morality, we're first going to step back and take a naturalistic approach: imagine that we are extraterrestrial beings trying to scientifically understand the concept of human morality: what problem, exactly, is ethics trying to solve in the first place? In this way we divorce the issue from the usual human-centered emotional associations with Right and Wrong, Good and Bad, Heaven and Hell, and so forth.

In our view, human morality is the question of how human beings live in groups and manage resources so as to mediate the natural tension between the individual and the group. That is, which resources does the individual keep, and which resources are shared with the group (and the details of that sharing). Anthropologists and biologists describe human beings as prosocial animals – it is unnatural for us to live alone for long periods of times, and we have evolved to live in groups. But that doesn't mean it is easy, and morality is the challenge of maintaining that balance through the allocation of resources.

Early in human evolution, the resources in question were what we call 'direct' resources – material resources such as food, shelter, clothing, and access to mates. Other resources, such as land, and social status, can be thought of as indirect: if you control land, you control the ability to dispense material resources, and in the case of social status, you can use your prestige or role in a group to influence other people to give you resources (for example, commoners paying tribute to a king simply because of his role).

It is fascinating to consider how resources have become increasingly indirect and abstract as societies have evolved. We now live in a staggeringly complex social world, filled with abstractions such as financial derivatives, or elaborate rituals such as professional sports. In a modern culture, these resources matter a great deal – witness how some of the most wealthy and prestigious members of our society have roles such as hedge fund managers or professional athletes. Also, it can be argued that in a society in which basic material resources such as food and shelter have been satisfied, abstract resources such as social status, fame, and prestige become more important.

These supposedly abstract resources can generate very material consequences, however: for instance, think of the amount of time and energy spent fighting for the future direction of the Hugo Awards during 2015 as a result of the "Sad/Rabid Puppies" campaigns. Clearly, the battles weren't about the value of the material trophy itself – they were about the argument as to who is deserving of the prestige that comes with winning an award that is considered to confer high status (and the associated financial consequences of incremental book sales, invitations to speak at events, and so forth). When one side of such a controversy is accusing the other side of unethical behavior, it is because what is at stake is the prospect of future resources. Again, morality is fundamentally about the question of resource allocation, where those resources matter to somebody or some group.

4. Resources, Flows, and Thermodynamics

It has been often said that "Energy is the Ultimate Resource". So if we're going to talk about morality as being essentially a matter of resource allocation among organisms (humans in particular), then it is worth thinking about the energy, and its potential applicability to the problem at hand.

The science of thermodynamics can be thought of as the science of energy flows. Classical thermodynamics is often considered to be a well-mined, even dull subject that centers on heat engines, steam tables, and the like. It is also sometimes thought of as a depressing subject, due to the 2nd law of thermodynamics, which supposedly predicts an inevitability to our universe dying a heat death after it has run out of usable energy. However, the science of thermodynamics is now undergoing a renaissance, manifested in the sub-field of "nonequilibrium" thermodynamics or "open system" thermodynamics, which includes the study of self-organized systems that evolve and gain in complexity, seemingly (but not really) in violation of the 2nd law.

One of the key lessons from this new research is that flows, whether these flows comprise energy, matter, or information, are the key to understanding how systems evolve. Over time, systems (both natural and man-made) tend to acquire better configurations to provide more access for the currents that flow through them. This insight has been captured with the following characterization, which is known as the Constructal Law:

"For a finite-size system to persist in time (to live), it must evolve in such a way that it provides easier access to the imposed currents that flow through it." [9]

This design principle illuminates why we see similarities in nature between river deltas (flows of water) and lungs (flows of oxygen), or for the similarities in shape between human-designed systems such as airport terminals (flows of people) and heat sinks (flows of heat).

A complementary way of looking at the nature of flows is to say:

"Nature abhors a gradient" [10]

This aphorism explains why hurricanes and tornadoes form - to reduce heat gradients in the atmosphere. Their complex structure (their spiral swirls, in this case) allow them to dissipate energy faster than if they were simple, amorphous structures.

An explosion can be thought of as a violent gradient reduction (for example, when a match triggers the dissipation of the potential energy inherent in a canister of gasoline), and so that raises the question – if gradient reduction is the name of the game, why don't we see more explosions in nature? The answer has to do with the dimension of time – gradient reduction is not just a phenomenon that needs to take place at one moment in time, and the results of an explosion can render a system incapable of doing any future gradient reduction. For example, if you induce your automobile to explode, you certainly reduce the fuel gradient that existed in the gas tank. But now your automobile is no longer capable of reducing future fuel gradients.

This line of thinking is very suggestive when it comes to understanding the phenomenon of life – all living creatures are continually reducing gradients (turning the energy from food into a combination of muscle action and the waste products), and it has been proposed that life arose in the first place because living organisms are more effective at gradient reduction over time than any non-living entities. Through reproduction, we ensure that even when our bodies finally succumb to entropy and die, we have created replacement gradient reducers by bearing children. Life as energetic flow that persists through time has the name autopoesis. [11] If you're looking to define meaning of life, you could do much worse than to claim that the meaning of life is simply gradient reduction: Life on Earth persists because it tends to effectively reduce the gradient between the hot sun and its cold surroundings.

5. Gradients and Far-from Equilibrium Conditions

As mentioned previously, the most exciting development in the science of thermodynamics is the research being done on "open" thermodynamic systems. In these systems, sufficient energy continually comes into the system so that the system can support complex, self-emergent behavior. These open systems are also known as "far-from-equilibrium" systems. One of the insights from this work is the realization that gradient reduction and far-from-equilibrium conditions are duals of each other – neither phenomenon is causal with respect to the other. Both natural and designed energy flows can induce gradients that persist at far-from equilibrium conditions, and these conditions can induce further gradients for reduction.

The concept of ecological succession is a good example of this. In ecological succession, we have a cyclical process with a duration on the order of a hundred years. Consider a meadow that has just experienced a fire that destroyed virtually all the existing plant life. If we observe this meadow over the next 100 years, we will observe various phases. In the first phase, taking place over the first five years or so, the meadow is settled by 'pioneer species' – hardy grasses, mostly. After these have been established, we begin to see the appearance of intermediate species – a collection of grasses and shrubs with the addition of some species of fast-growing trees such as pines. As the decades pass, a 'climax community' arises as the meadow is populated by mature trees such as oaks. This climax community persists until another major fire signals the beginning of a new cycle.

We can characterize the process of ecological succession by observing the complexity of the system and the energy flows it contains. As ecological succession occurs, we observe:

- The system moves farther from thermodynamic equilibrium
- The system contains greater amounts of free energy (i.e., potential energy)
- System components have longer energy residence times
- The system contains a hierarchy of energy flows, creating a more complex overall structure.

Similar phenomena can be observed if you take a look at a modern technological society. Structures such as dams and flying aircraft are operating at far-from-equilibrium (for instance, equilibrium for the dam/water system is for the dam to fall apart, and water on each side to reach the same level; for the airplane, it is violating the ever-present gravity gradient). Energy flows and gradient reduction can even contribute insights to the study of potential life on other planets. As an illustration of this, there is the fascinating story of how James Lovelock was consulting to NASA in the 1960's during a time when robotic missions to Mars were being planned in order to look for evidence of life. Lovelock insisted that life on Mars was exceedingly unlikely: the fact that the Martian atmosphere existed in chemical equilibrium, unlike the Earth's atmosphere, which is in a constant far-from-equilibrium state, was a strong piece of evidence against the hypothesis that life exists on Mars.

The key takeaway here is that complex energy flows and the associated gradient reductions are inextricably linked to far-from-equilibrium conditions. This is as true of human-made artifacts as it is to natural ones such as the ecological state of a meadow, or the physiological state of an organism. Energy flows, persistent gradient reduction and complex structures all go together such that where you have one, the other two are always present.

6. Conclusion: A Prime Directive for Robotic Probes

Now we'd like to put the pieces together, and make explicit the implied argument that gradient reduction and its associated concepts can provide guidance in the design of ethical robotic probes. Retracing the steps, our argument thus far goes like this:

1. Ethics is about the allocation of resources
2. Resources are, fundamentally, energy flows – ones that matter to an organism
3. Both natural and human-made energy glows can induce gradients that persist at far-from-equilibrium conditions
4. Persistent gradient-reducing energy flows give rise to complex structures that tend to induce those flows.

Therefore, to act ethically in presence of persistent energy flows is to respect the gradients that matter.

Morality is simply respecting existing gradients. A sophisticated intelligence (of the kind we hope a robotic probe will be armed with) in this context means the ability to detect diverse kinds of gradients: radar, pressure, ultrasound, temperature, radiation, capacitance, humidity, infrared, and so forth. The more gradients the probe can sense, the more aware it will be of which ones are likely to matter: those that are persistent, and whose interruption lead to a violation of the integrity of other flows.

In terms of morality, a simple corollary to the principle we have just distilled becomes our "first draft" of a proposed Prime Directive:

Prime Directive for Robotic Probes: Do not interfere with the maintenance or growth of persistent energy flows on any world you visit.

As we become more skilled at the awareness of and the management of the energy flows on our own world, we will no doubt refine this draft Prime Directive into something more nuanced, and equip our robotic ambassadors to the stars with the wisdom necessary to make mutually beneficial first contacts with the other inhabitants of this universe.

References

1. "Prime Directive" retrieved from Wikipedia.org

2. Edmonds, David, *Would You Kill the Fat Man? The Trolley Problem and What Your Answer Tells Us about Right and Wrong*, Princeton University Press, 2014

3. Anderson, Michael & Anderson, Susan Leigh, editors, *Machine Ethics*, Cambridge University Press, 2011

4. Lin, Patrick & Abney, Keith, editors, *Robot Ethics: The Ethical and Social Implications of Robotics*, MIT Press, 2011

5. Wallach, Wendell, *Moral Machines: Teaching Robots Right from Wrong*, Oxford University Press, 2010

6. Arkin, Ronald, *Governing Lethal Behavior in Autonomous Robots*, Chapman and Hall/CRC Press, 2009

7. "What if your self-driving car decides one death is better than two – and that one is you" appeared in the online *Washington Post*, October 28th, 2015

8. "Killer Robots With Automatic Rifles Could Be on the Battlefield in 5 Years" appeared in *WIRED* online, October 18th, 2013.

9. Bejan, Adrian & Lorente, Slyvie, "Design with Constructal Theory, Wiley, 2008

10. Schneider, Eric & Sagan, Dorion, *Into the Cool: Energy, Flow, Thermodynamics, and Life*, University of Chicago Press, 2006

11. Maturana, Humberto & Varela, Francisco, "Autopoiesis and Cognition: The Realization of the Living", D. Reidel Publishing Company, 1980

12. Lovelock, James. *Gaia: A New Look at Life on Earth*, Oxford University Press, 2000

What's Right for Life:
Can Life Exist Without Water?

Morgan L. Cable, Ph.D., Jonathan Lunine,
Jack Beauchamp, and Christophe Sotin

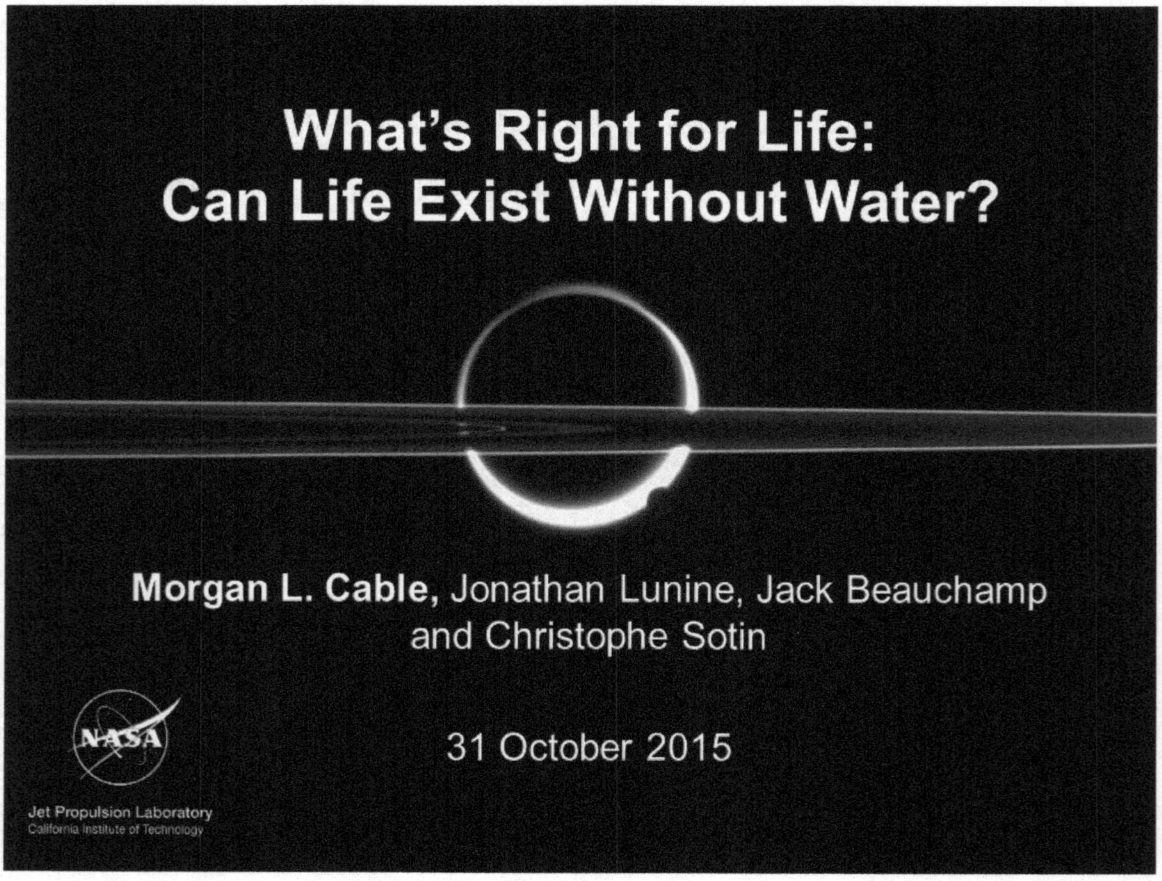

Keck Institute for Space Studies
KECK INSTITUTE FOR SPACE STUDIES
Earth
Planetary
Keck
Astrophysics
Engineering
Generously supported by the W. M. Keck Foundation
Home
Visit
Programs
Events
Workshops
Lectures
Short Courses
Symposia
Special Events
People
Archive
About
Follow Us
Calendar
Search
WORKSHOP #1
LECTURE
MAP
WIKI
Contact Us:
(626) 395-6630
(626) 395-3570 - fax
Email Us
Don't Follow (Just) the Water: Does Life Occur in Non-Aqueous Media? - Part II
September 14-17, 2015
California Institute of Technology - Pasadena, CA 91125
Team Leads
Jack Beauchamp
California Institute of Technology
Christophe Sotin
Jet Propulsion Laboratory
Jonathan Lunine
Cornell University
Description
Workshop Overview:
The opening workshop in July 2015 addressed non-aqueous environments that meet the overall requirements for habitability (energy, organics, liquid). After some discussion the workshop settled on three such environments:
1. terrestrial petroleum deposits isolated from liquid water;
2. the methane-ethane seas of Titan, and
3. CO2 above and below the critical point.
With respect to (2), the team decided after much deliberation that the solubilities and reactivities in the Titan seas were too low for significant chemistry to occur, and so a broader range of environments - informed by Cassini data - were selected for consideration in the next workshop. Environment (1), the petroleum, is of interest provided samples not in intimate contact with water (or at least geologically pasteurized after exposure, followed by isolation). One such potential samples exists in the lab of a workshop participant (Prof. Orphan), but a careful protocol would need to be developed to open and test the sample. Further, there may be existing data from laboratories on petroleum samples that may shed some light on the question of contamination by aqueous organisms. Environment (3) was developed only after discussion about the extent to which planetary modelers needed to be specific about the T-P conditions and geologic settings.
http://kiss.caltech.edu/new_website/workshops/life2.html

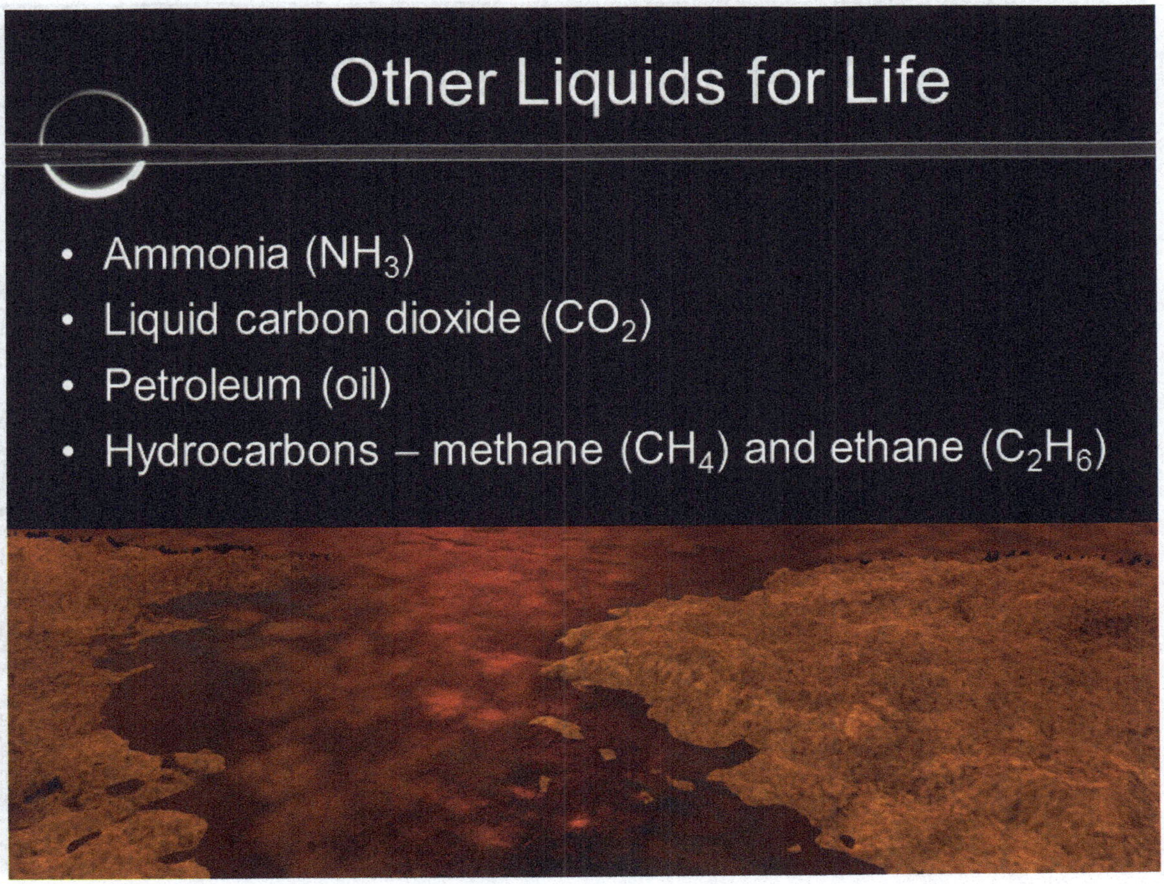
Other Liquids for Life
• Ammonia (NH_3)
• Liquid carbon dioxide (CO_2)
• Petroleum (oil)
• Hydrocarbons – methane (CH_4) and ethane (C_2H_6)

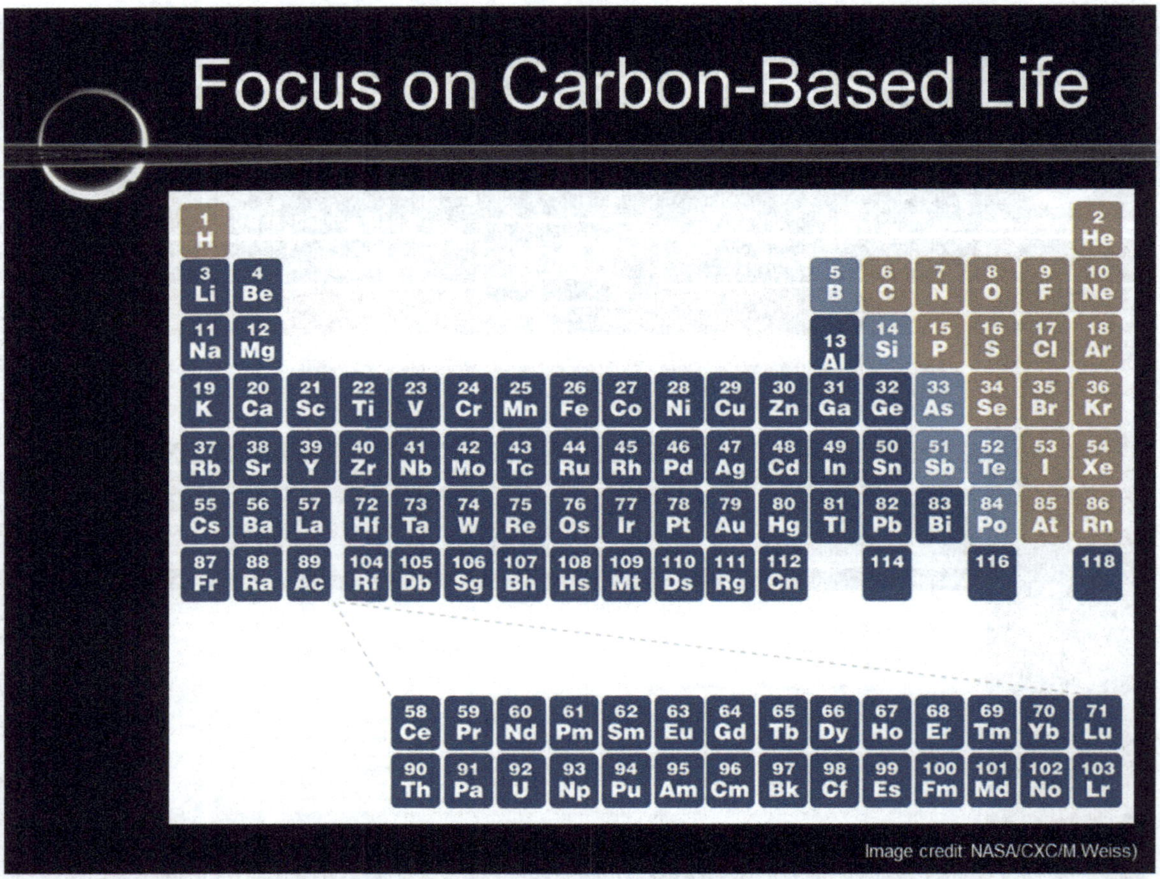
Focus on Carbon-Based Life
1 H
2 He
3 Li
4 Be
5 B
6 C
7 N
8 O
9 F
10 Ne
11 Na
12 Mg
13 Al
14 Si
15 P
16 S
17 Cl
18 Ar
19 K
20 Ca
21 Sc
22 Ti
23 V
24 Cr
25 Mn
26 Fe
27 Co
28 Ni
29 Cu
30 Zn
31 Ga
32 Ge
33 As
34 Se
35 Br
36 Kr
37 Rb
38 Sr
39 Y
40 Zr
41 Nb
42 Mo
43 Tc
44 Ru
45 Rh
46 Pd
47 Ag
48 Cd
49 In
50 Sn
51 Sb
52 Te
53 I
54 Xe
55 Cs
56 Ba
57 La
72 Hf
73 Ta
74 W
75 Re
76 Os
77 Ir
78 Pt
79 Au
80 Hg
81 Tl
82 Pb
83 Bi
84 Po
85 At
86 Rn
87 Fr
88 Ra
89 Ac
104 Rf
105 Db
106 Sg
107 Bh
108 Hs
109 Mt
110 Ds
111 Rg
112 Cn
114
116
118
58 Ce
59 Pr
60 Nd
61 Pm
62 Sm
63 Eu
64 Gd
65 Tb
66 Dy
67 Ho
68 Er
69 Tm
70 Yb
71 Lu
90 Th
91 Pa
92 U
93 Np
94 Pu
95 Am
96 Cm
97 Bk
98 Cf
99 Es
100 Fm
101 Md
102 No
103 Lr
Image credit: NASA/CXC/M.Weiss)

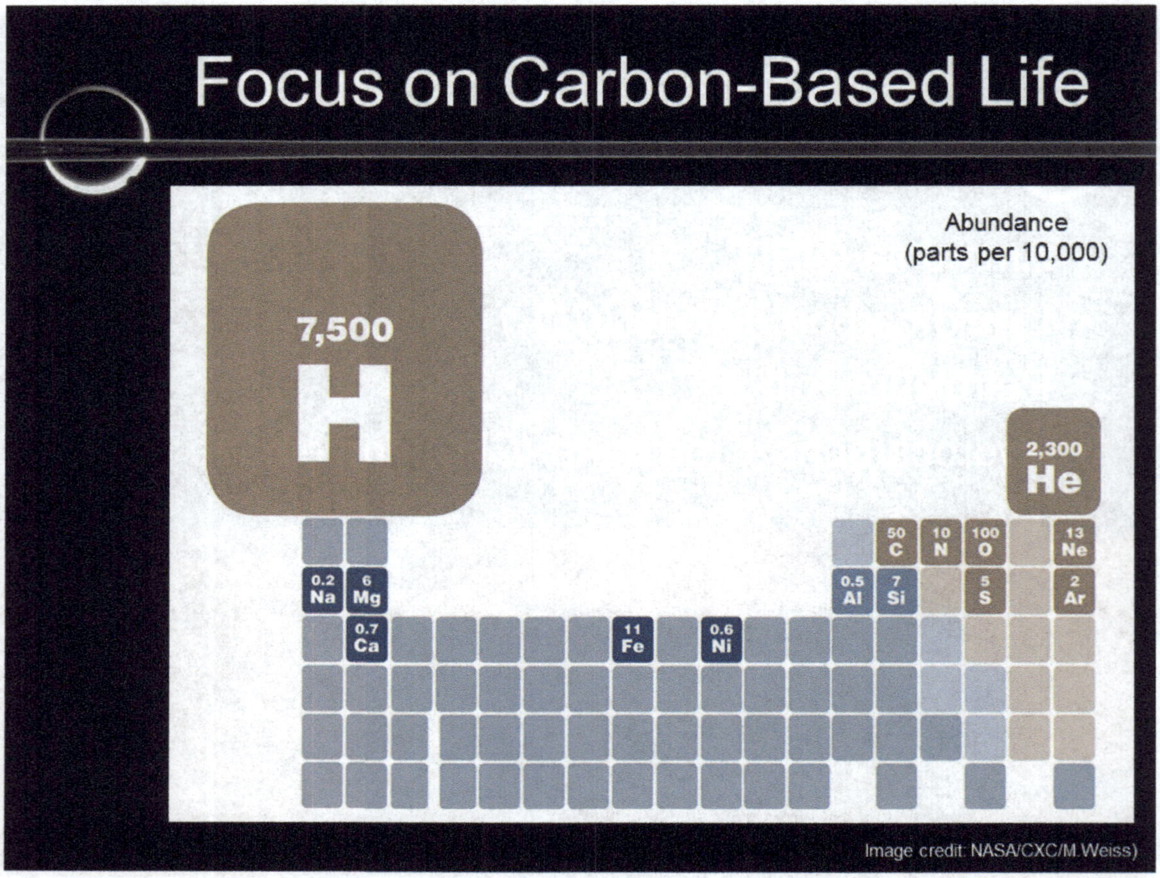
Focus on Carbon-Based Life
Abundance
(parts per 10,000)
7,500
H
2,300
He
50 C
10 N
100 O
13 Ne
0.2 Na
6 Mg
0.5 Al
7 Si
5 S
2 Ar
0.7 Ca
11 Fe
0.6 Ni
Image credit: NASA/CXC/M.Weiss)

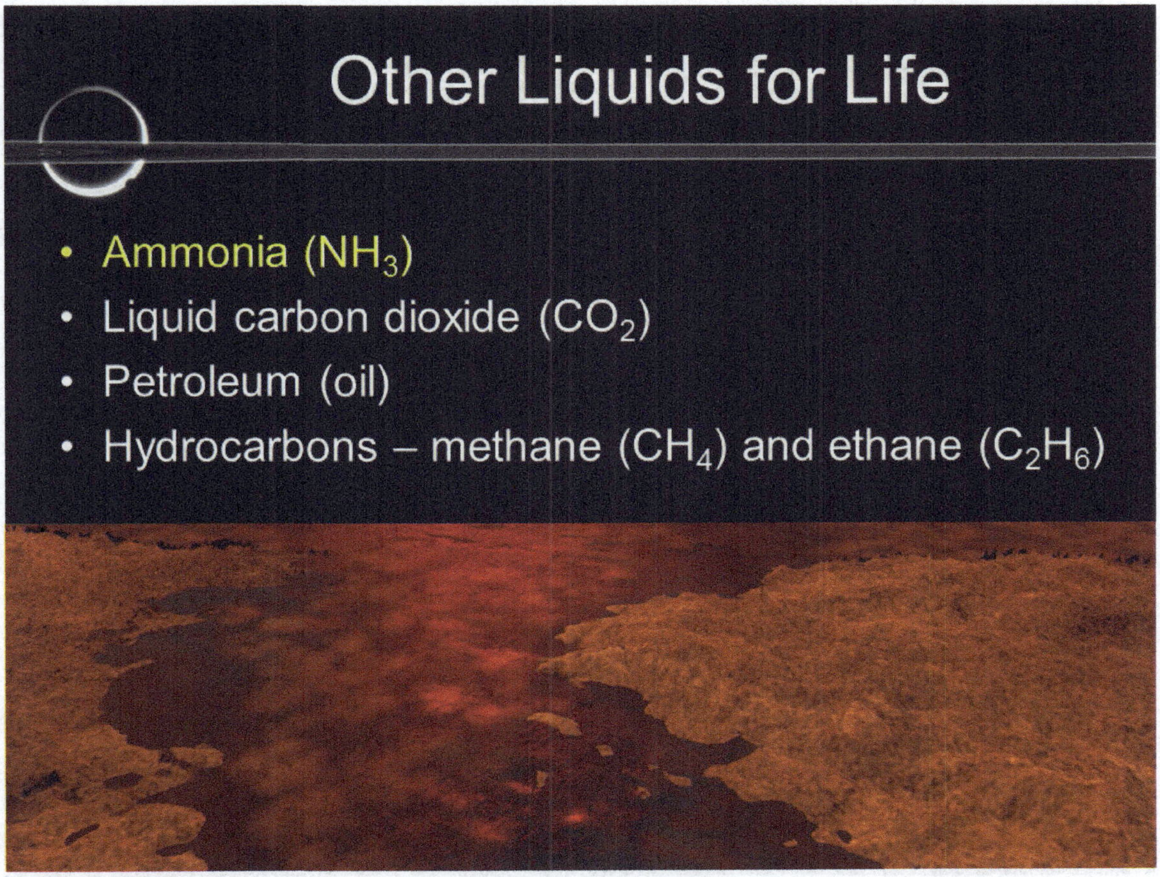
Other Liquids for Life
• Ammonia (NH_3)
• Liquid carbon dioxide (CO_2)
• Petroleum (oil)
• Hydrocarbons – methane (CH_4) and ethane (C_2H_6)

Ammonia
NH_3
Image credit: NASA-JPL/Caltech

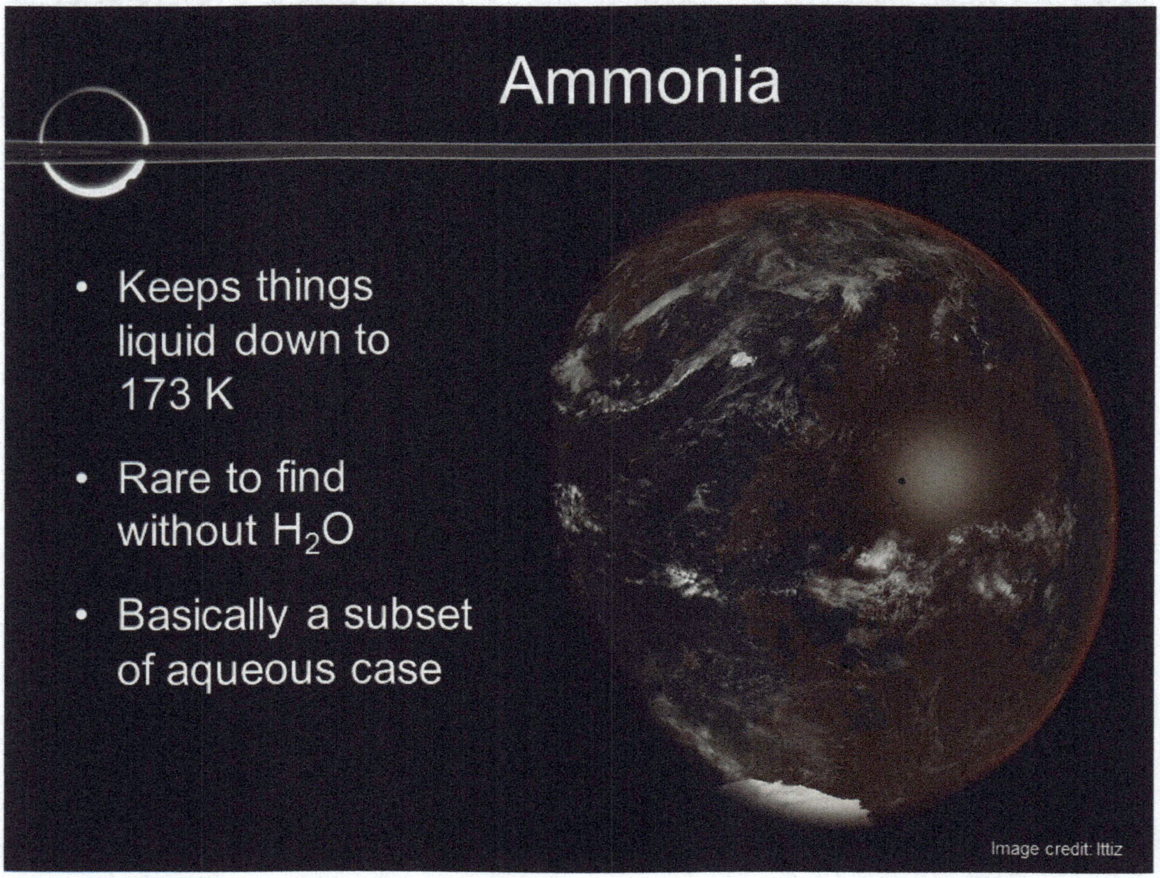
Ammonia
Keeps things liquid down to 173 K
Rare to find without H_2O
Basically a subset of aqueous case
Image credit: Ittiz

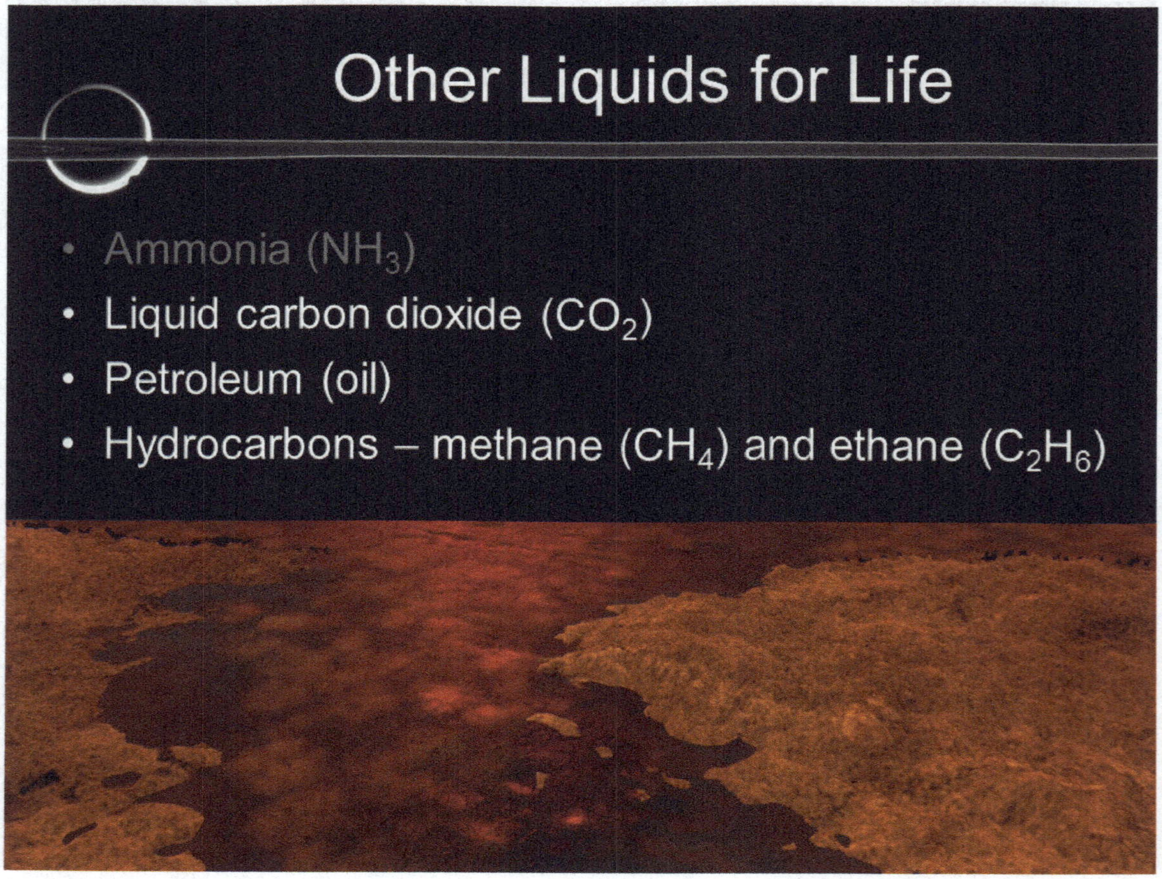
Other Liquids for Life
Ammonia (NH_3)
Liquid carbon dioxide (CO_2)
Petroleum (oil)
Hydrocarbons – methane (CH_4) and ethane (C_2H_6)

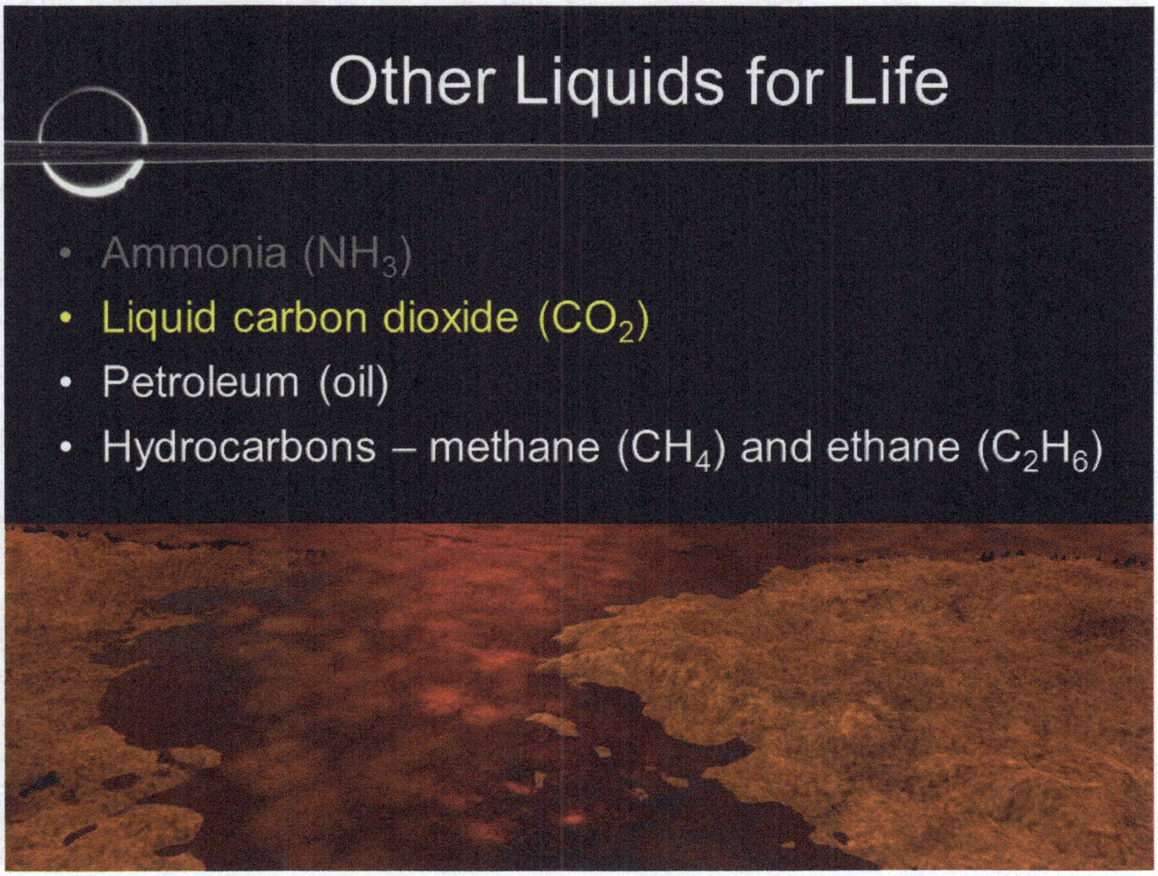
Other Liquids for Life
• Ammonia (NH_3)
• Liquid carbon dioxide (CO_2)
• Petroleum (oil)
• Hydrocarbons – methane (CH_4) and ethane (C_2H_6)

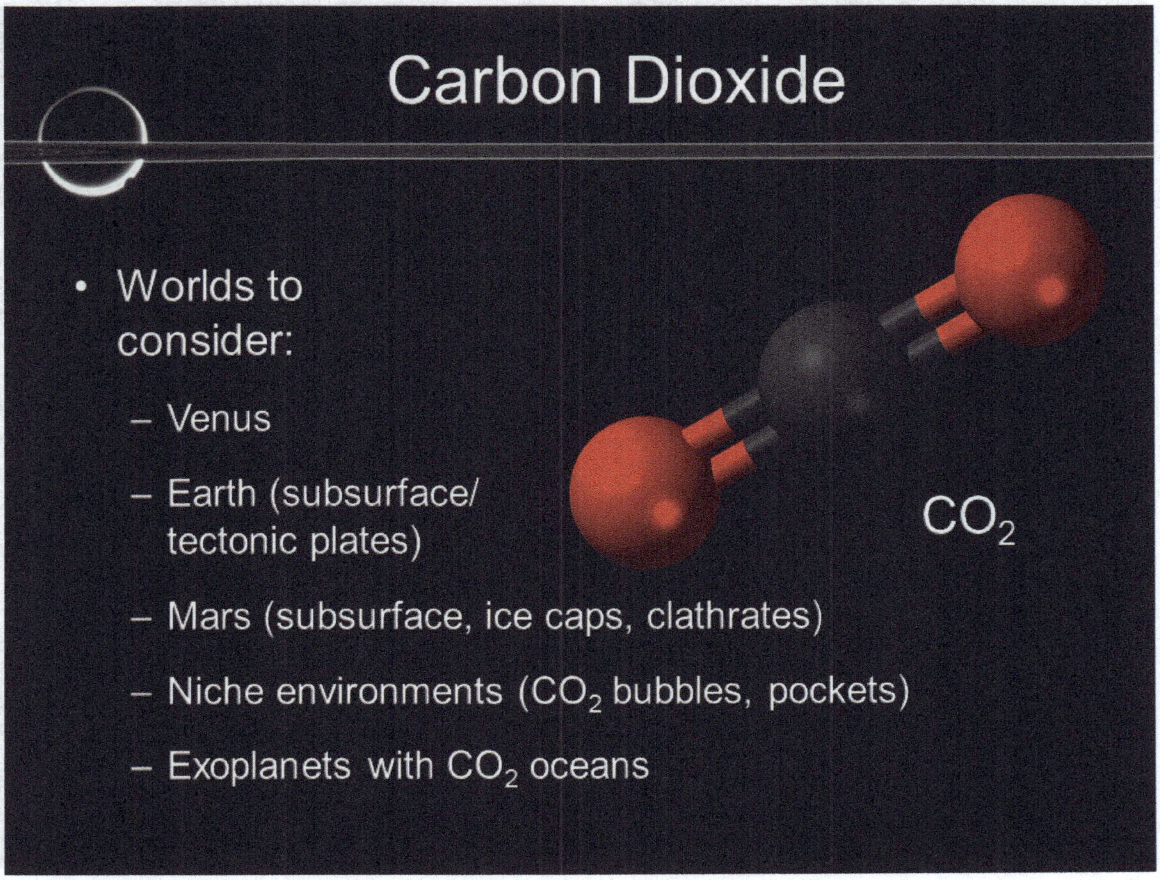
Carbon Dioxide
• Worlds to consider:
– Venus
– Earth (subsurface/ tectonic plates)
– Mars (subsurface, ice caps, clathrates)
– Niche environments (CO_2 bubbles, pockets)
– Exoplanets with CO_2 oceans
CO_2

Carbon Dioxide
• Common in planetary atmospheres
• Comes in two liquid flavors:
– Liquid CO_2
– Supercritical CO_2
Image credit: NASA, ESA, & G. Bacon/STScI, STScI-PRC14-06

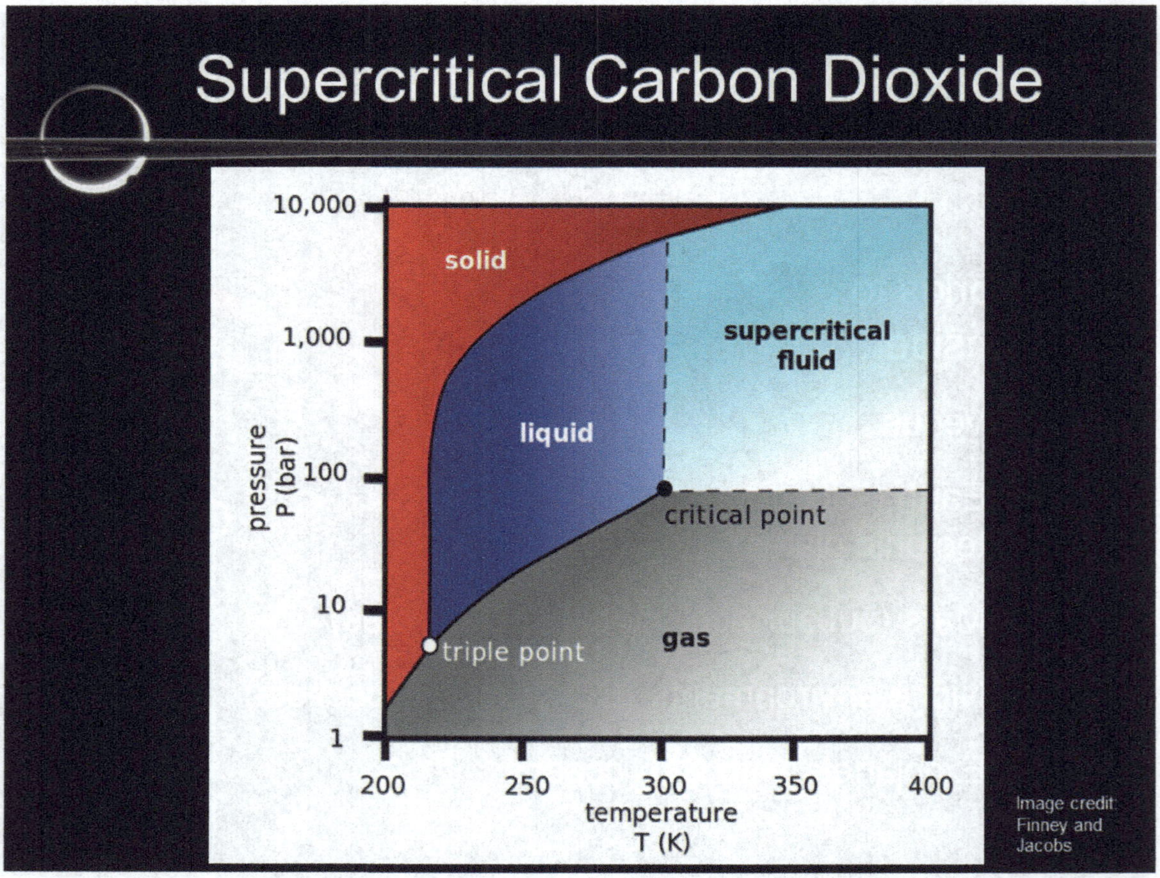
Supercritical Carbon Dioxide
10,000
1,000
100
10
1
pressure P (bar)
solid
liquid
supercritical fluid
critical point
triple point
gas
200
250
300
350
400
temperature T (K)
Image credit: Finney and Jacobs

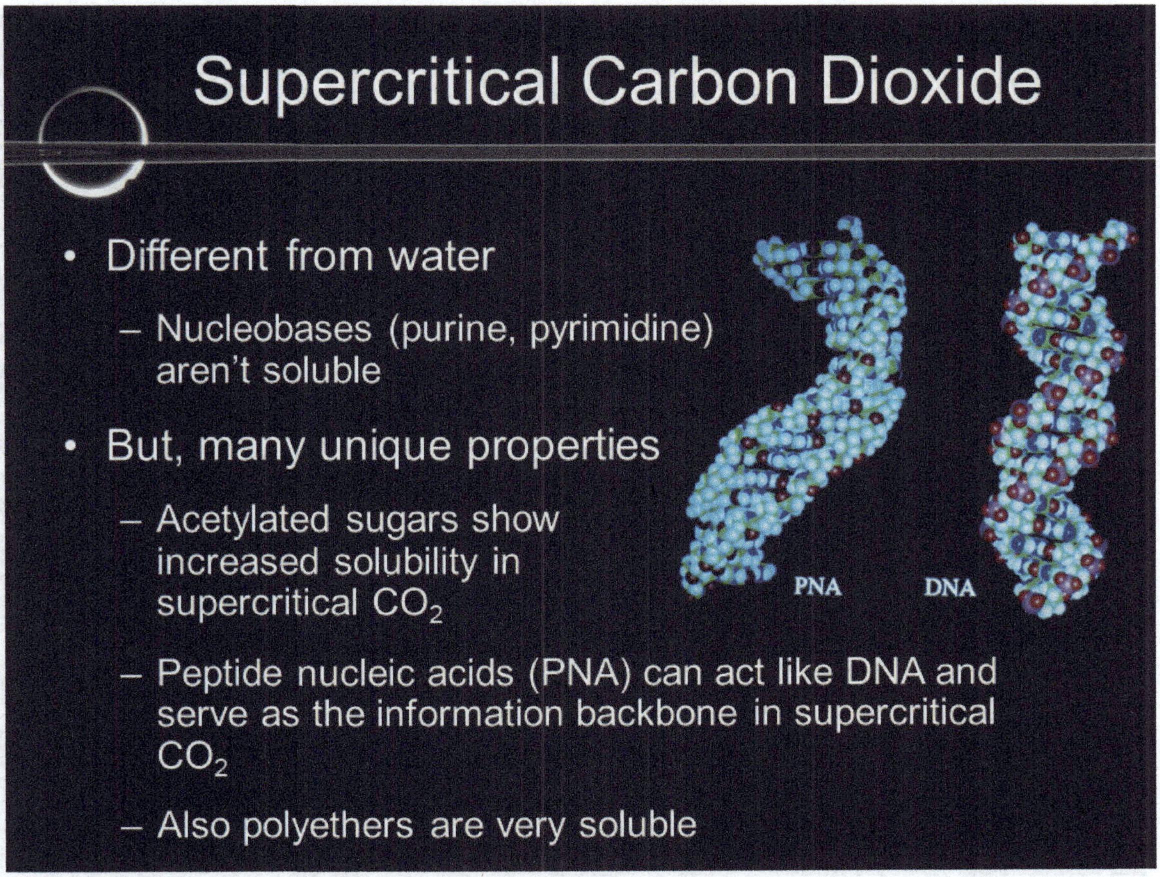
Supercritical Carbon Dioxide
• Different from water
– Nucleobases (purine, pyrimidine) aren't soluble
• But, many unique properties
– Acetylated sugars show increased solubility in supercritical CO_2
– Peptide nucleic acids (PNA) can act like DNA and serve as the information backbone in supercritical CO_2
– Also polyethers are very soluble
PNA
DNA

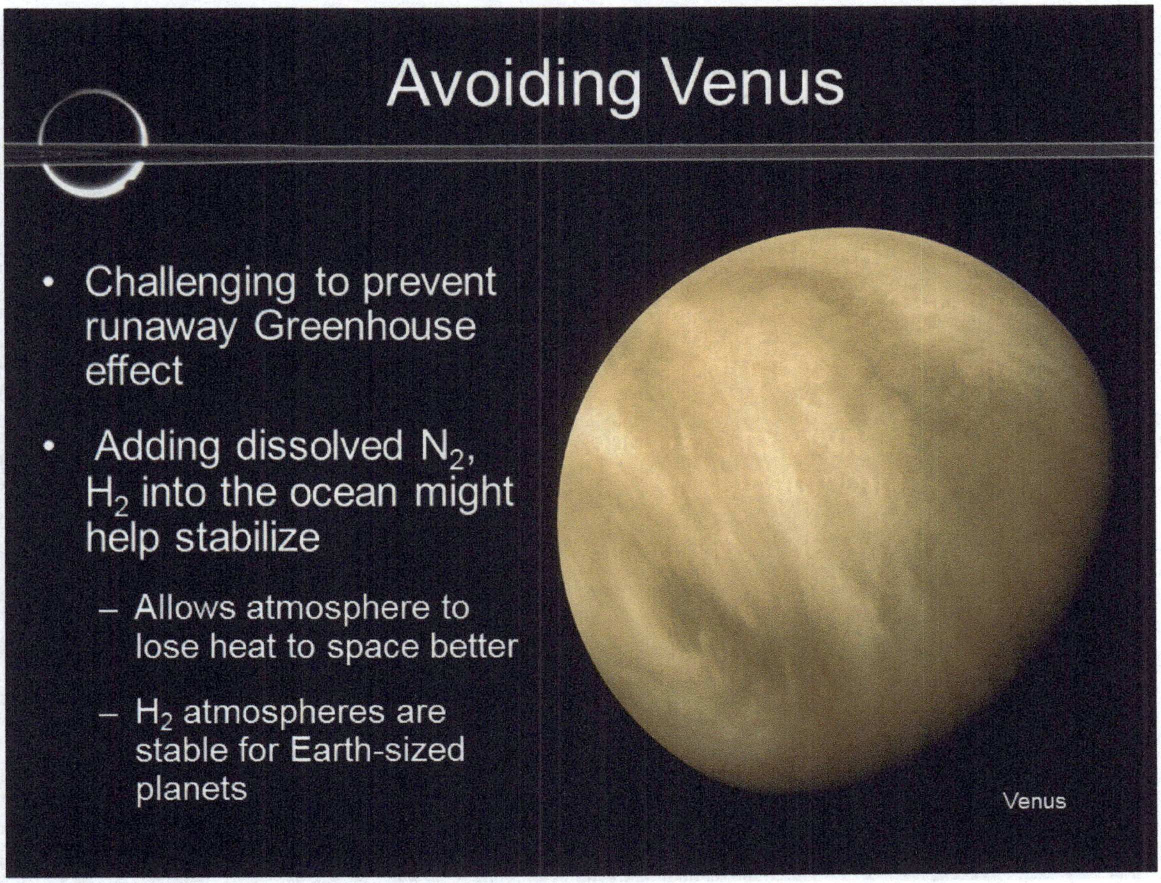
Avoiding Venus
• Challenging to prevent runaway Greenhouse effect
• Adding dissolved N_2, H_2 into the ocean might help stabilize
– Allows atmosphere to lose heat to space better
– H_2 atmospheres are stable for Earth-sized planets
Venus

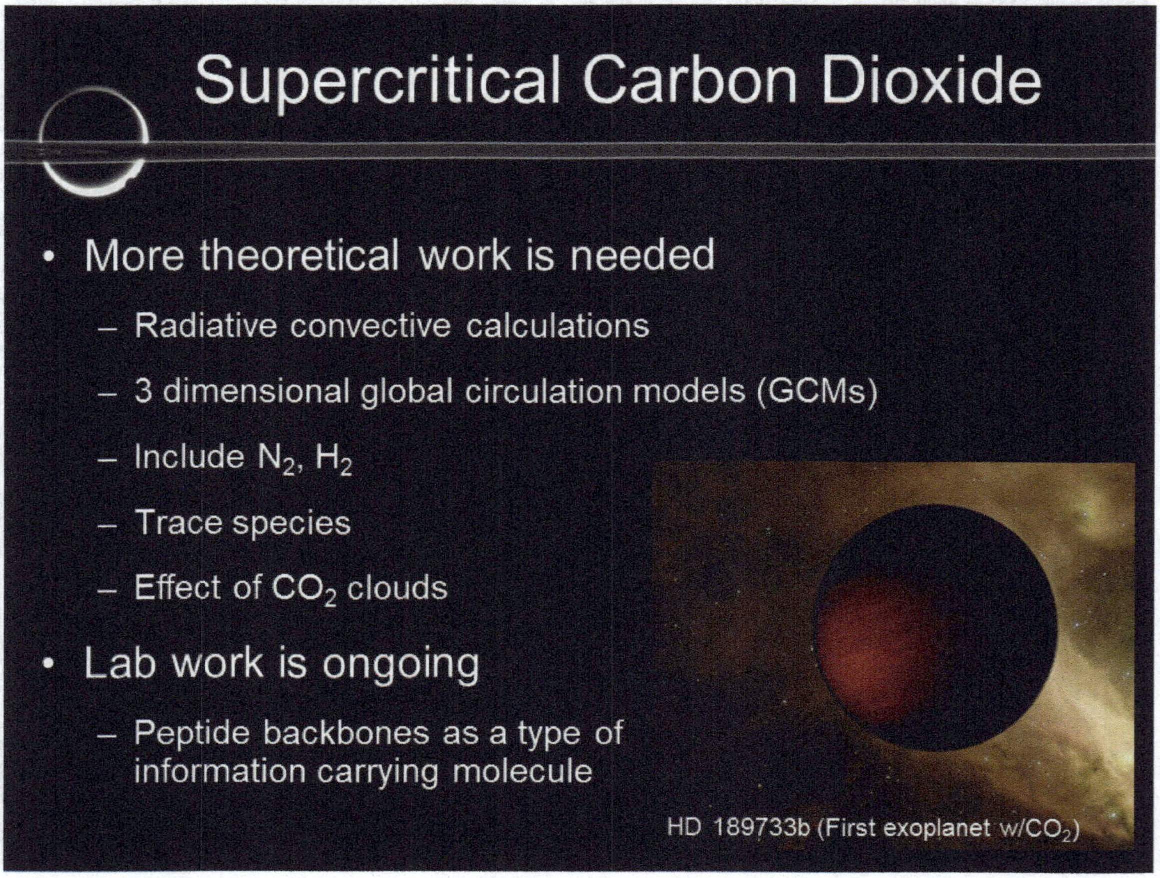
Supercritical Carbon Dioxide
• More theoretical work is needed
– Radiative convective calculations
– 3 dimensional global circulation models (GCMs)
– Include N_2, H_2
– Trace species
– Effect of CO_2 clouds
• Lab work is ongoing
– Peptide backbones as a type of information carrying molecule
HD 189733b (First exoplanet w/CO_2)

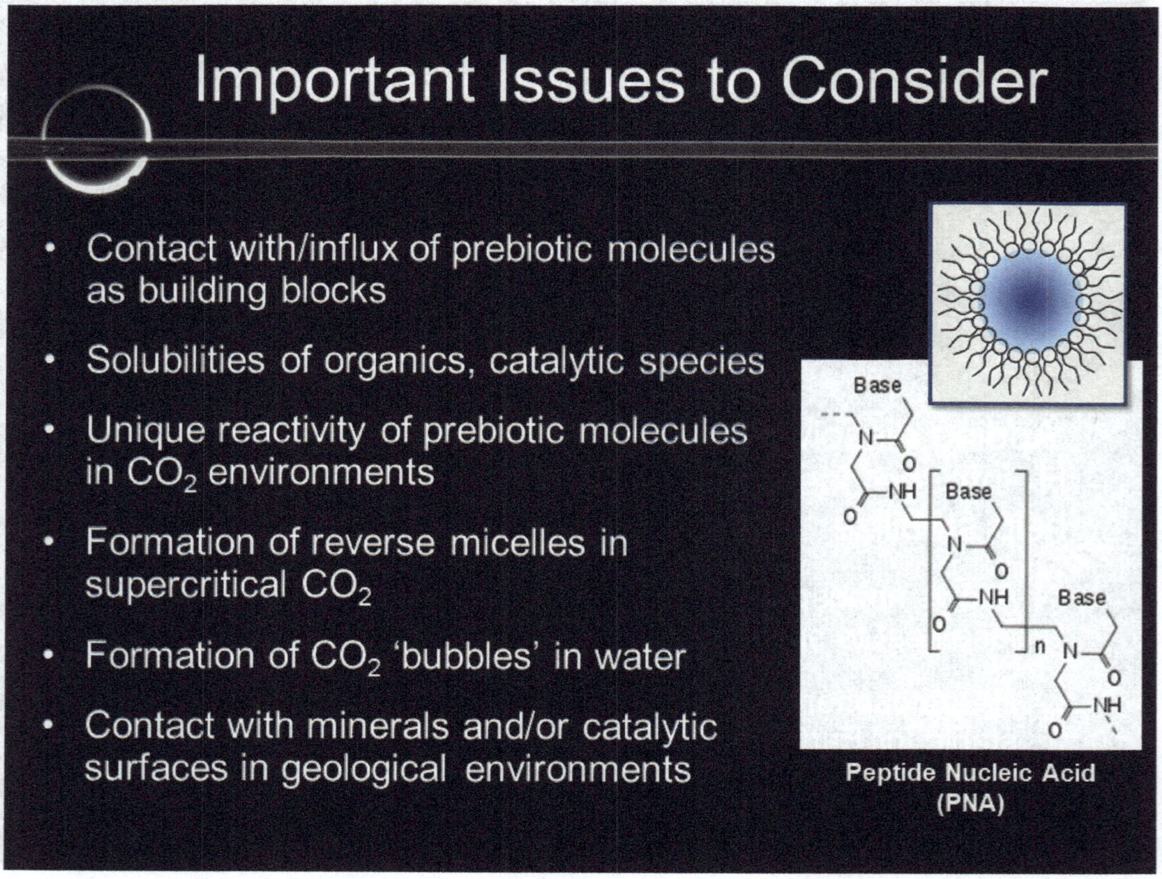
Important Issues to Consider
• Contact with/influx of prebiotic molecules as building blocks
• Solubilities of organics, catalytic species
• Unique reactivity of prebiotic molecules in CO_2 environments
• Formation of reverse micelles in supercritical CO_2
• Formation of CO_2 'bubbles' in water
• Contact with minerals and/or catalytic surfaces in geological environments
Base
N
O
NH
O
Base
N
O
NH
O
n
Base
N
O
NH
O
Peptide Nucleic Acid (PNA)

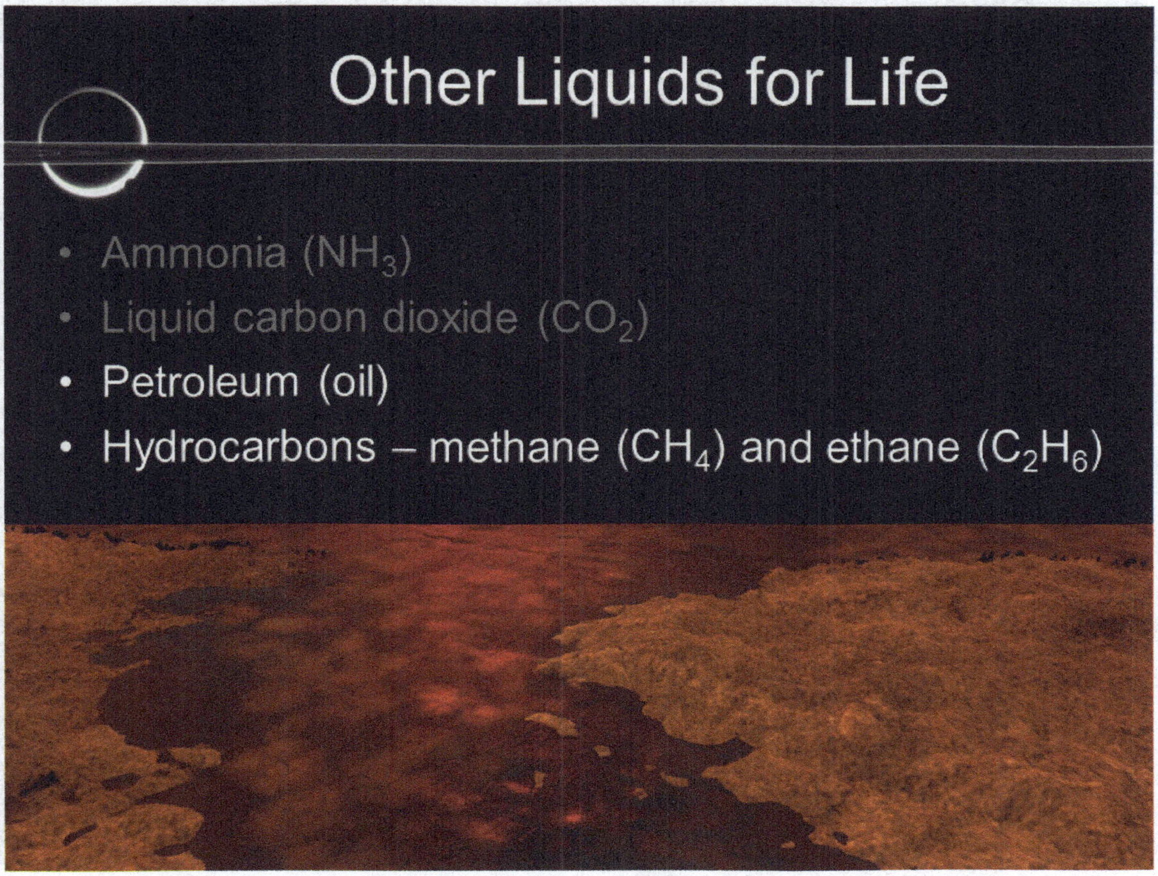
Other Liquids for Life
• Ammonia (NH_3)
• Liquid carbon dioxide (CO_2)
• Petroleum (oil)
• Hydrocarbons – methane (CH_4) and ethane (C_2H_6)

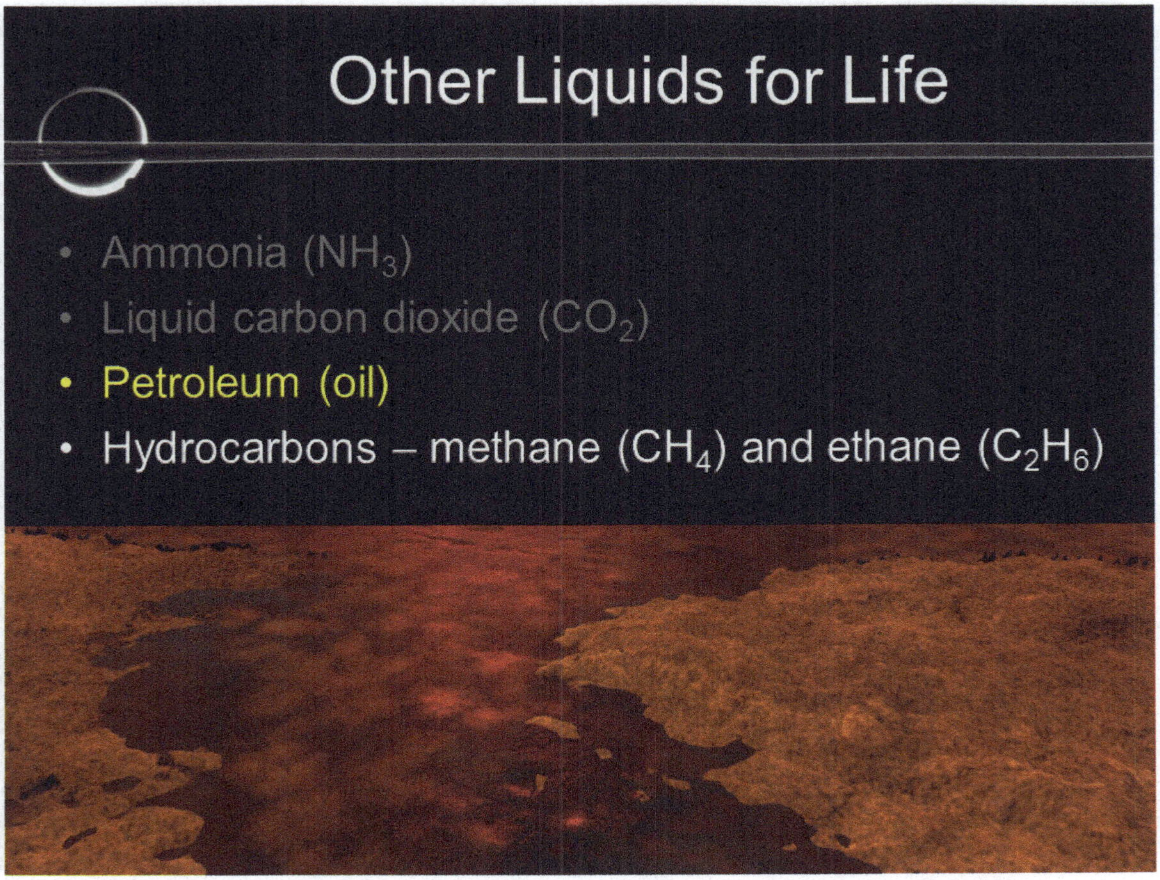
Other Liquids for Life
• Ammonia (NH_3)
• Liquid carbon dioxide (CO_2)
• Petroleum (oil)
• Hydrocarbons – methane (CH_4) and ethane (C_2H_6)

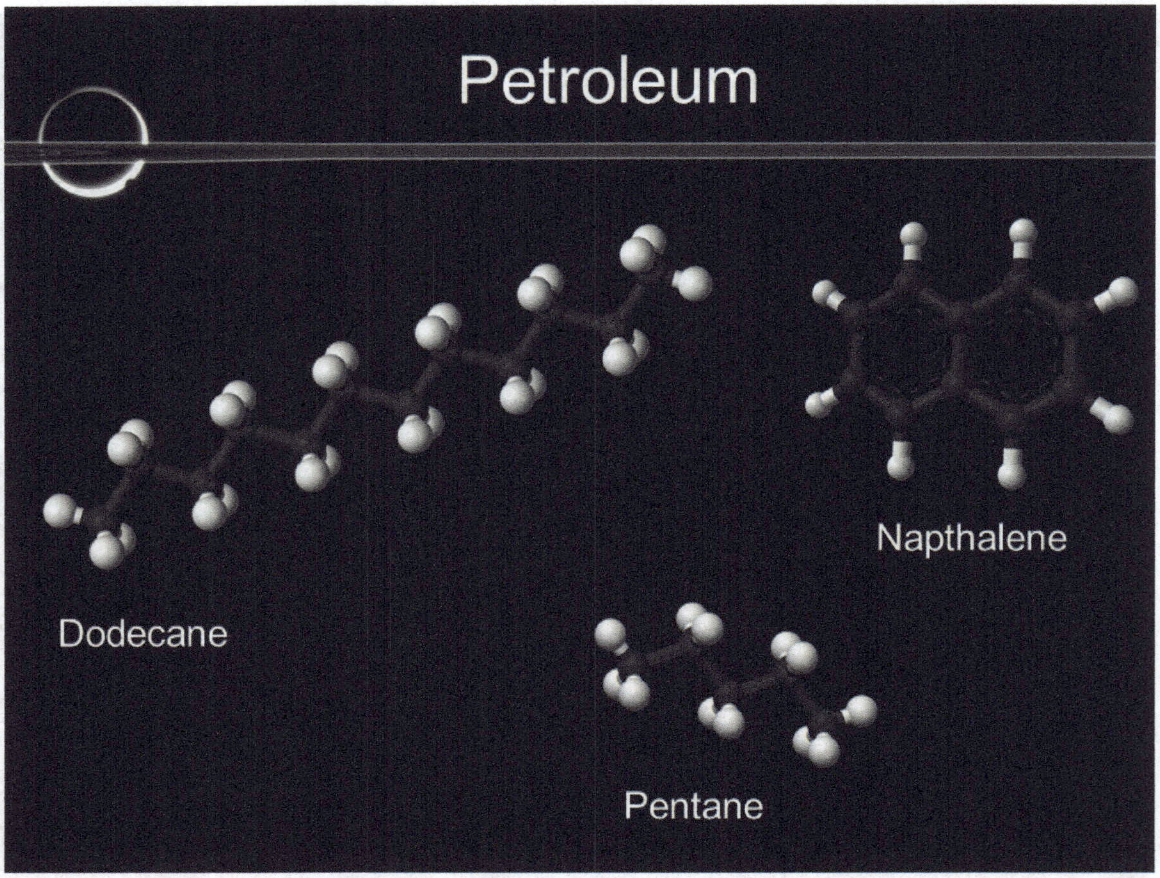
Petroleum
Dodecane
Napthalene
Pentane

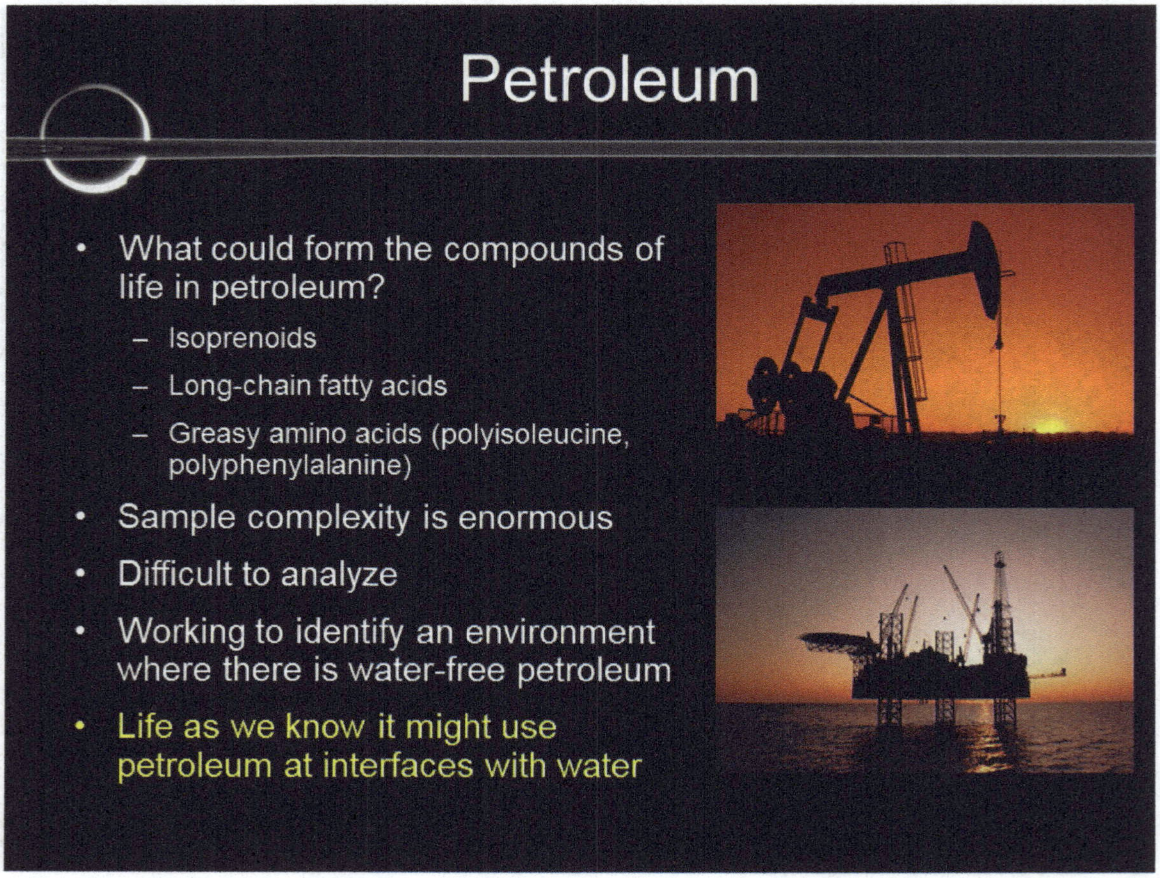
Petroleum
• What could form the compounds of life in petroleum?
– Isoprenoids
– Long-chain fatty acids
– Greasy amino acids (polyisoleucine, polyphenylalanine)
• Sample complexity is enormous
• Difficult to analyze
• Working to identify an environment where there is water-free petroleum
• Life as we know it might use petroleum at interfaces with water

Petroleum

- What could form the compounds of life in petroleum?
  - Isoprenoids
  - Long-chain fatty acids
  - Greasy amino acids (polyisoleucine, polyphenylalanine)
- Sample complexity is enormous
- Difficult to analyze
- Working to identify an environment where there is water-free petroleum
- Life as we know it might use petroleum at interfaces with water

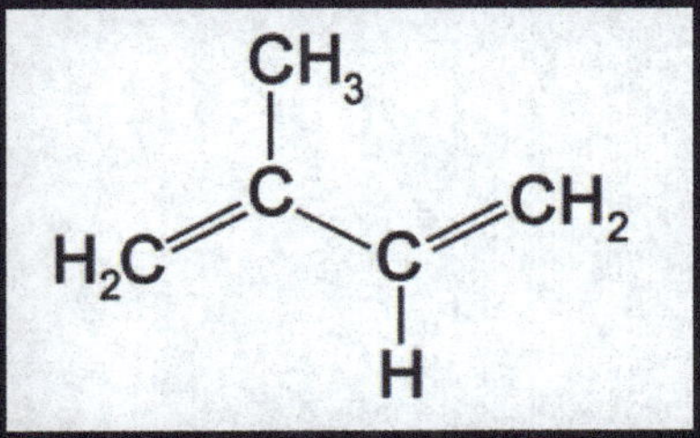

Isoprene

Oleic Acid (C18)

Petroleum

- There could be life in petroleum, but finding it will be challenging
- Look for petroleum deposits on Earth where contact with water has been broken for some amount of time
- Redox and metabolic experiments to identify:
  - Petroleum-associated water-based life as we know it
  - Hydrocarbon-based life as we don't know it
- Baseline abiotic processes in lab, then look at a biological process and see what the deviation is

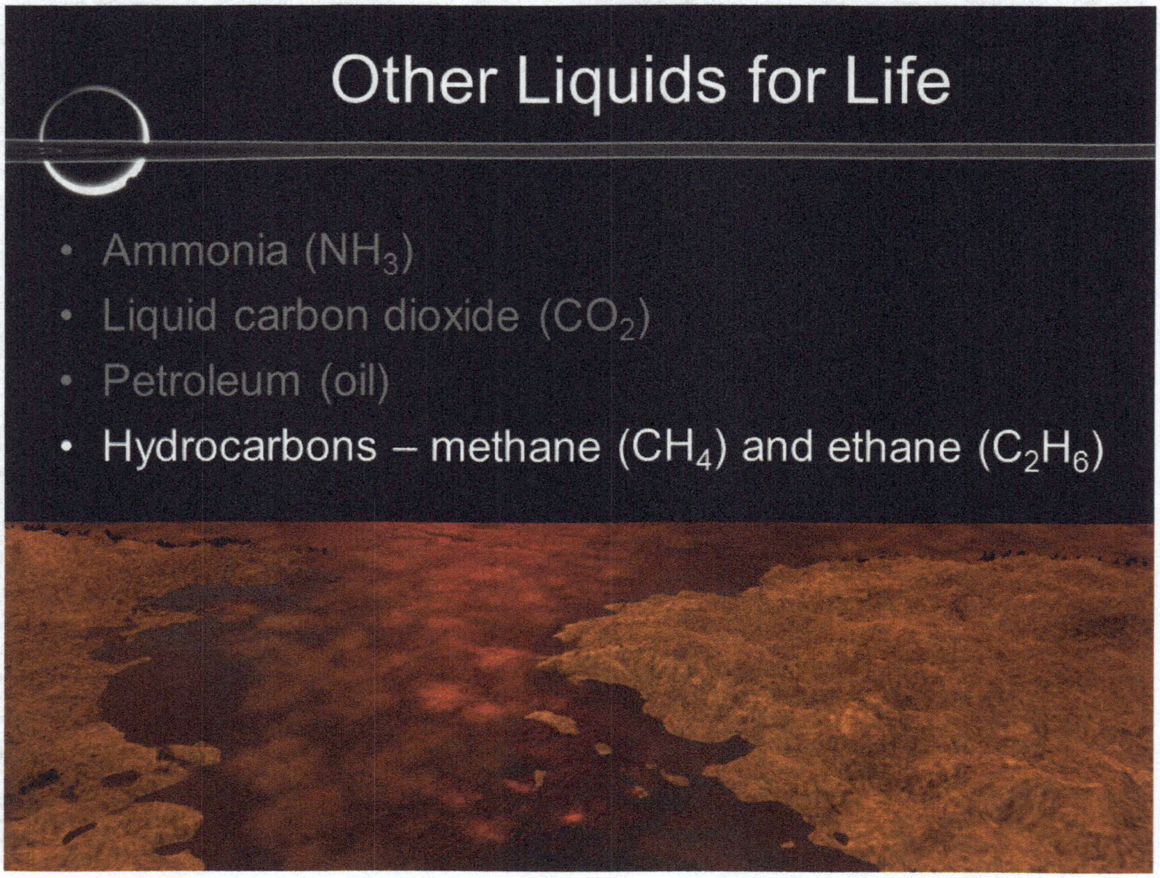
Other Liquids for Life
• Ammonia (NH_3)
• Liquid carbon dioxide (CO_2)
• Petroleum (oil)
• Hydrocarbons – methane (CH_4) and ethane (C_2H_6)

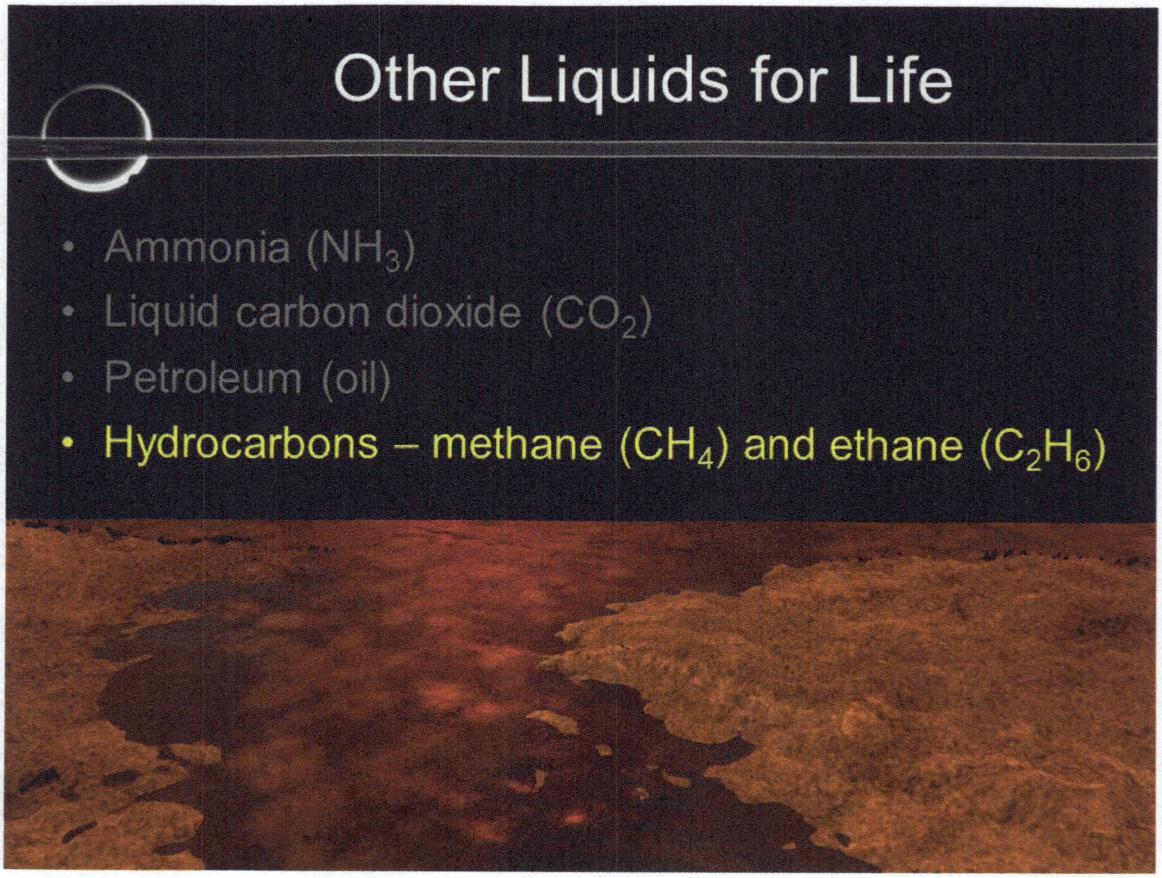
Other Liquids for Life
• Ammonia (NH_3)
• Liquid carbon dioxide (CO_2)
• Petroleum (oil)
• Hydrocarbons – methane (CH_4) and ethane (C_2H_6)

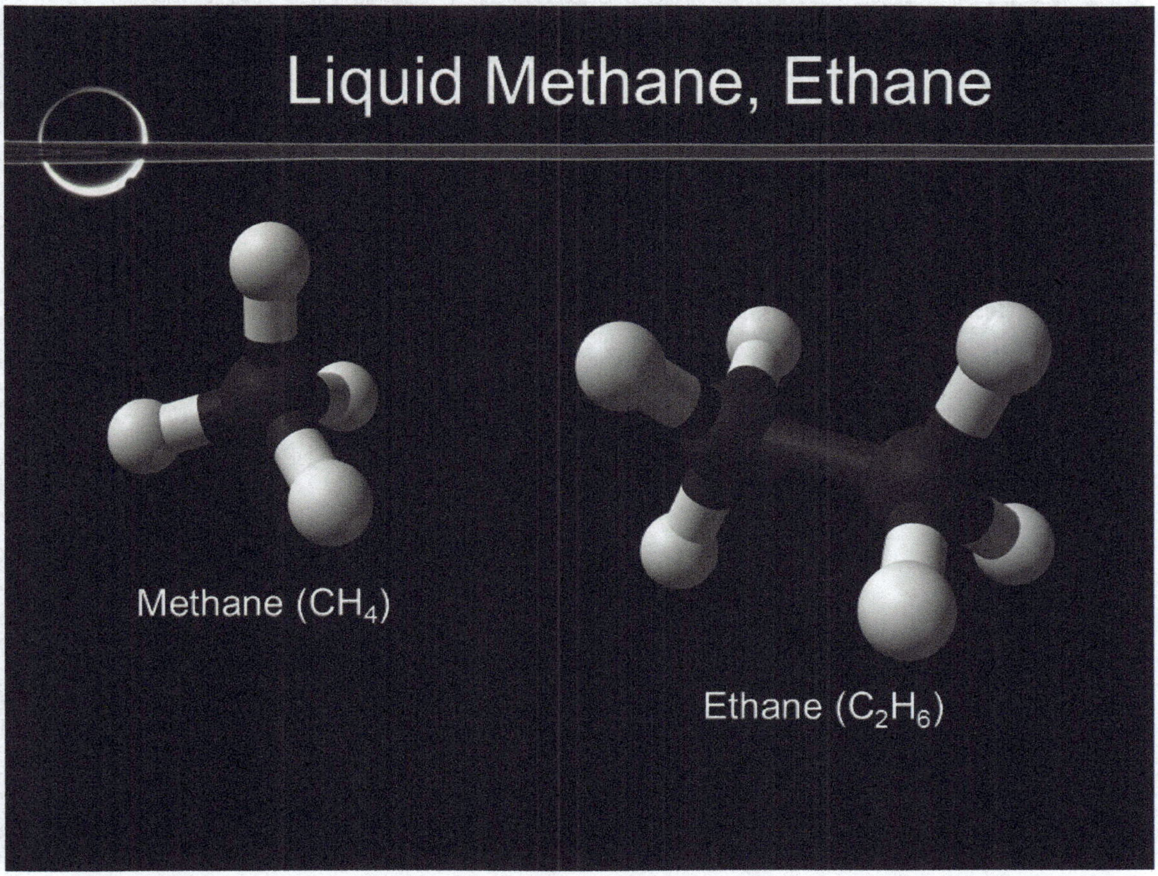
Liquid Methane, Ethane
Methane (CH_4)
Ethane (C_2H_6)

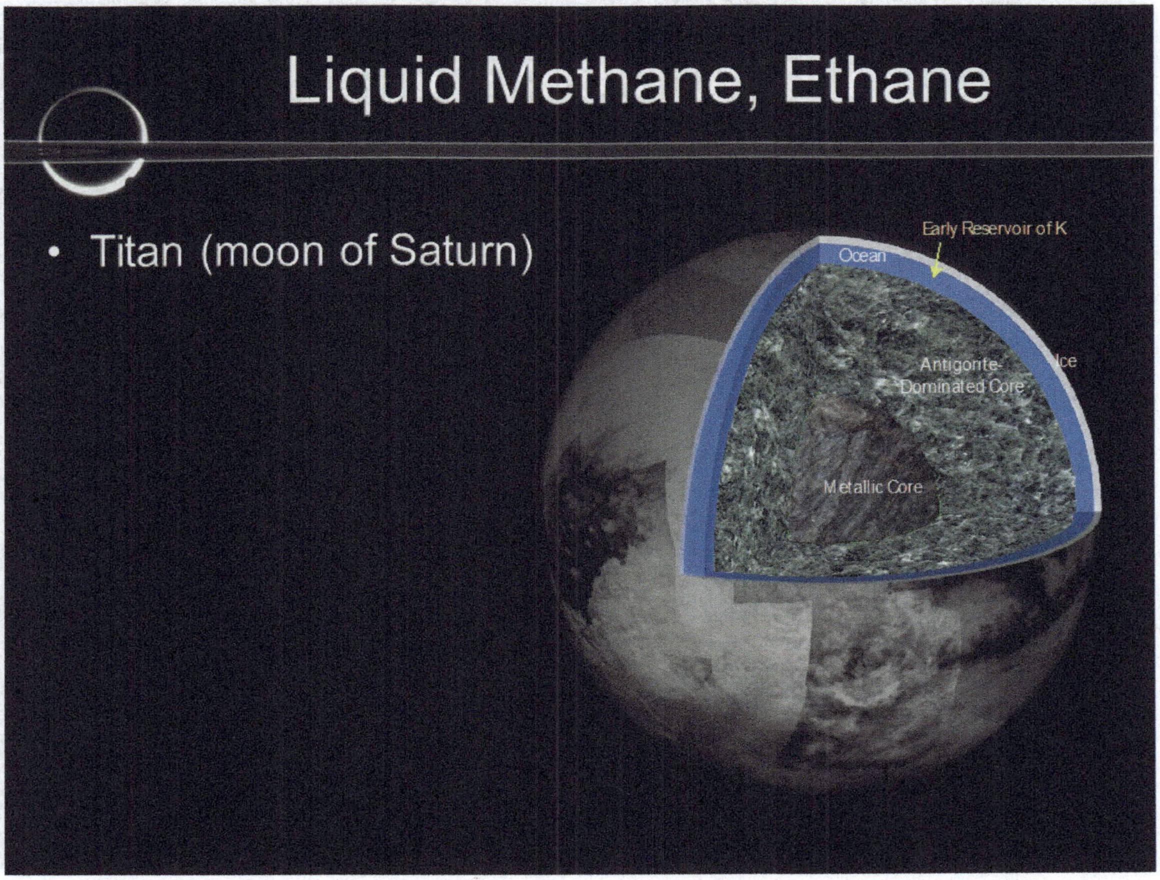
Liquid Methane, Ethane
• Titan (moon of Saturn)
Early Reservoir of K
Ocean
Ice
Antigorite-
Dominated Core
Metallic Core

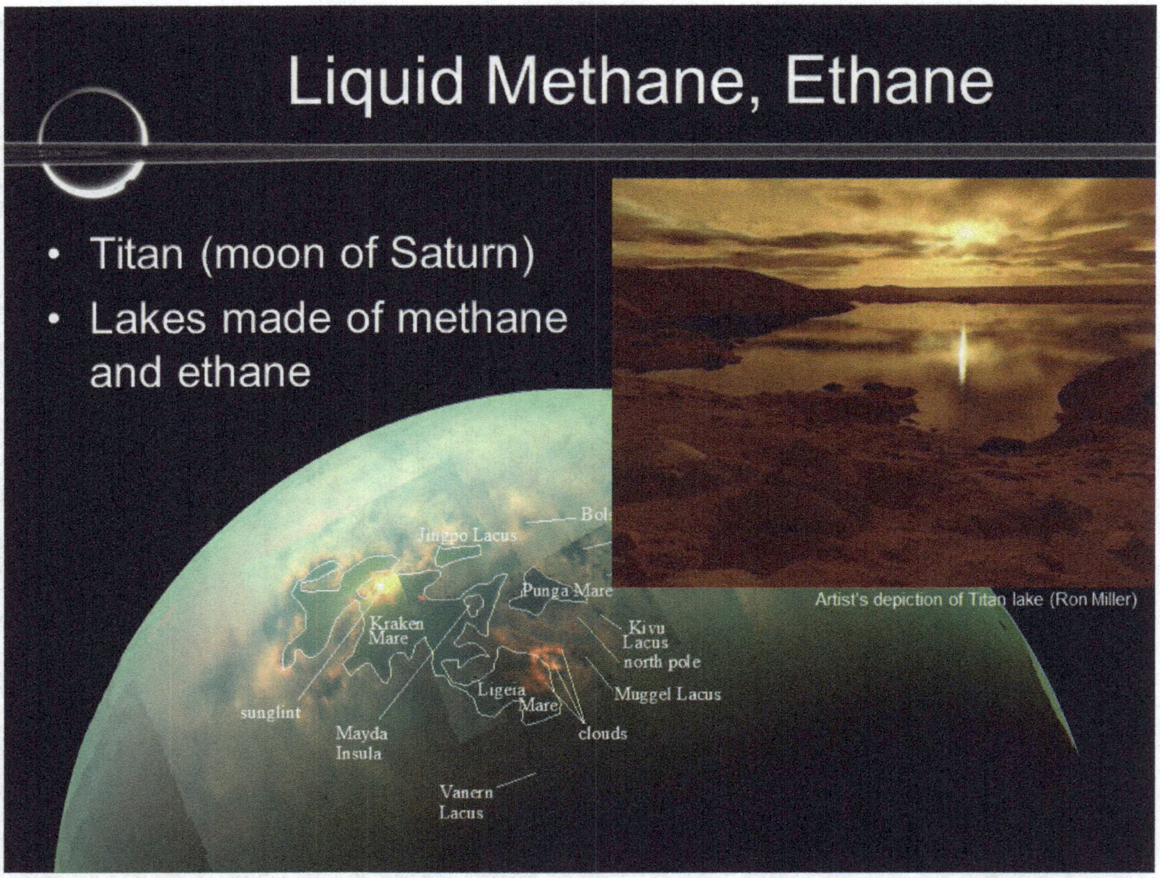
Liquid Methane, Ethane
• Titan (moon of Saturn)
• Lakes made of methane and ethane
Jingpo Lacus
Punga Mare
Kraken Mare
Kivu Lacus
north pole
Muggel Lacus
Ligeia Mare
clouds
sunglint
Mayda Insula
Vanern Lacus
Artist's depiction of Titan lake (Ron Miller)

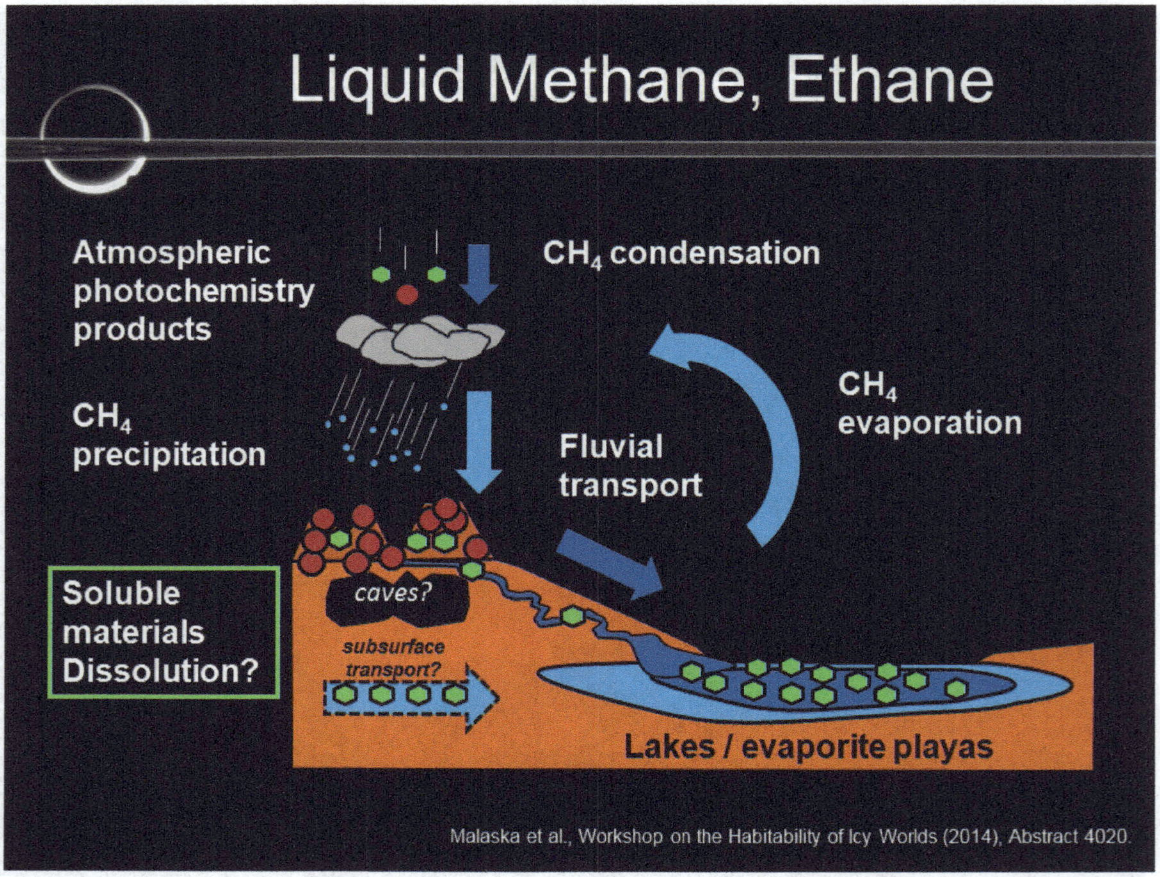
Liquid Methane, Ethane
Atmospheric photochemistry products
CH4 condensation
CH4 precipitation
CH4 evaporation
Fluvial transport
Soluble materials Dissolution?
caves?
subsurface transport?
Lakes / evaporite playas
Malaska et al., Workshop on the Habitability of Icy Worlds (2014), Abstract 4020.

Liquid Methane, Ethane

- Solubility of most molecules in liquid methane and/or ethane is poor

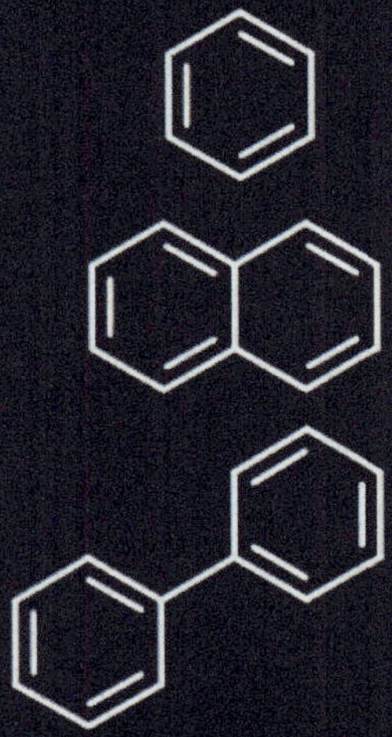

Compound	Solubility in Liquid Ethane at 94 K (mg/L)
Benzene	18.5 ± 1.9
Napthalene	0.159 ± 0.003
Biphenyl	0.039 ± 0.006

Sucrose in water (25 °C): 3,750,000 mg/L
Stearic acid (~olive oil): 3 mg/L

Malaska et al. Icarus 2014.

Looking to the Shorelines

Solid-Phase Chemistry

- Carbamation

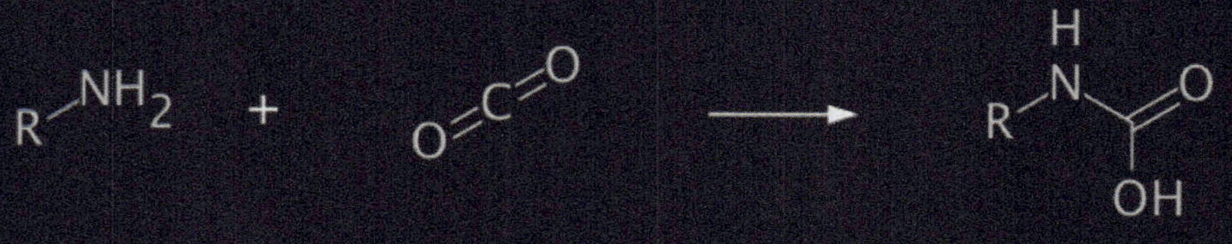

Carbamic acid

- Imine, ether formation
  - Polyimine could form a catalytic site
- Azide and alkyne polymerization
  - HCN polymerization

Looking for Organic/Ice Interfaces

RADAR – effective dielectric constant is low ε = 1.6 – 2.0 [1]
Consistent with organics with voids

H_2O ice	ε = 3.1
Hydrocarbon material	ε = 2.0 – 2.4
Avg. Titan surface	ε = 1.6 – 2.0

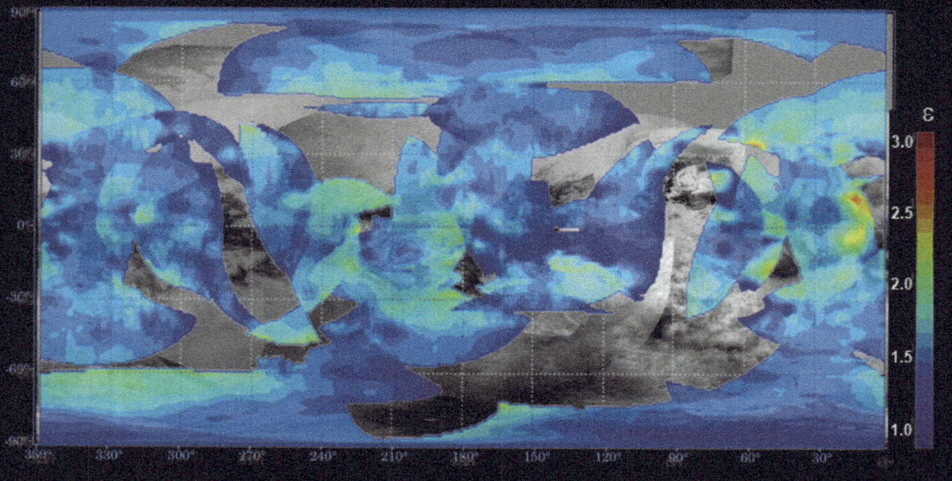

[1] Janssen et al., Icarus 200 (2009) 222-239

Low Temperature Considerations

- Weaker forces dominate
  - Noncovalent interactions
- We need to rethink our requirements for life

Earth	Titan	Interaction/Bond Strength
Ionic Bond	Covalent Bond	Permanent or semi-permanent
Covalent Bond	Hydrogen bonds, π-bonds	Can be made/broken on timescales relevant for life (ATP, proteins, etc.)
Hydrogen bonds, π-bonds	van der Waals forces	Loose associations that help hold together secondary structures

Prebiotic versus Protobiotic

- Instead of looking for life itself, target the mechanisms that lead to prebiotic chemistry (aka, Protobiotic chemistry)
  - Geologic patterning
  - Chemical gradients
  - On Earth, this led to more advanced chemistry, which eventually led to life
- How would we see this?
  - Look for differences in orientation and association of chemical subunits
  - Domination of one type of material on Titan (might change with seasons)

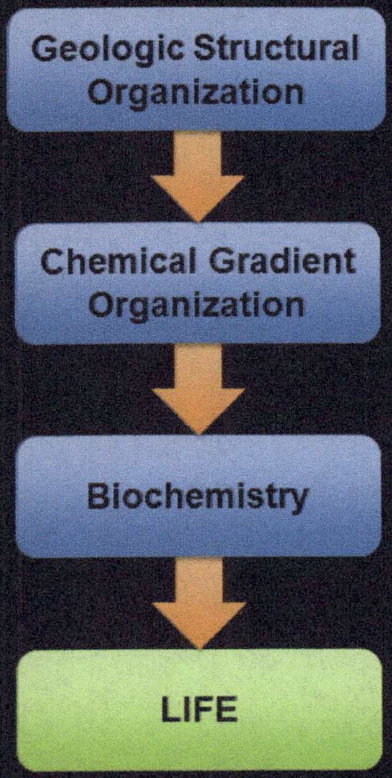

Overall Study Findings

- Mars' charter is explicitly, "Follow the Water"
- Our conclusions have found that you may find interesting things apart from water
- Seek out the interfaces!
- Weak interactions may be the key

NASA shouldn't just be looking for areas with liquid water

Happy Halloween from Cassini!

Image credit: NASA-JPL/Caltech

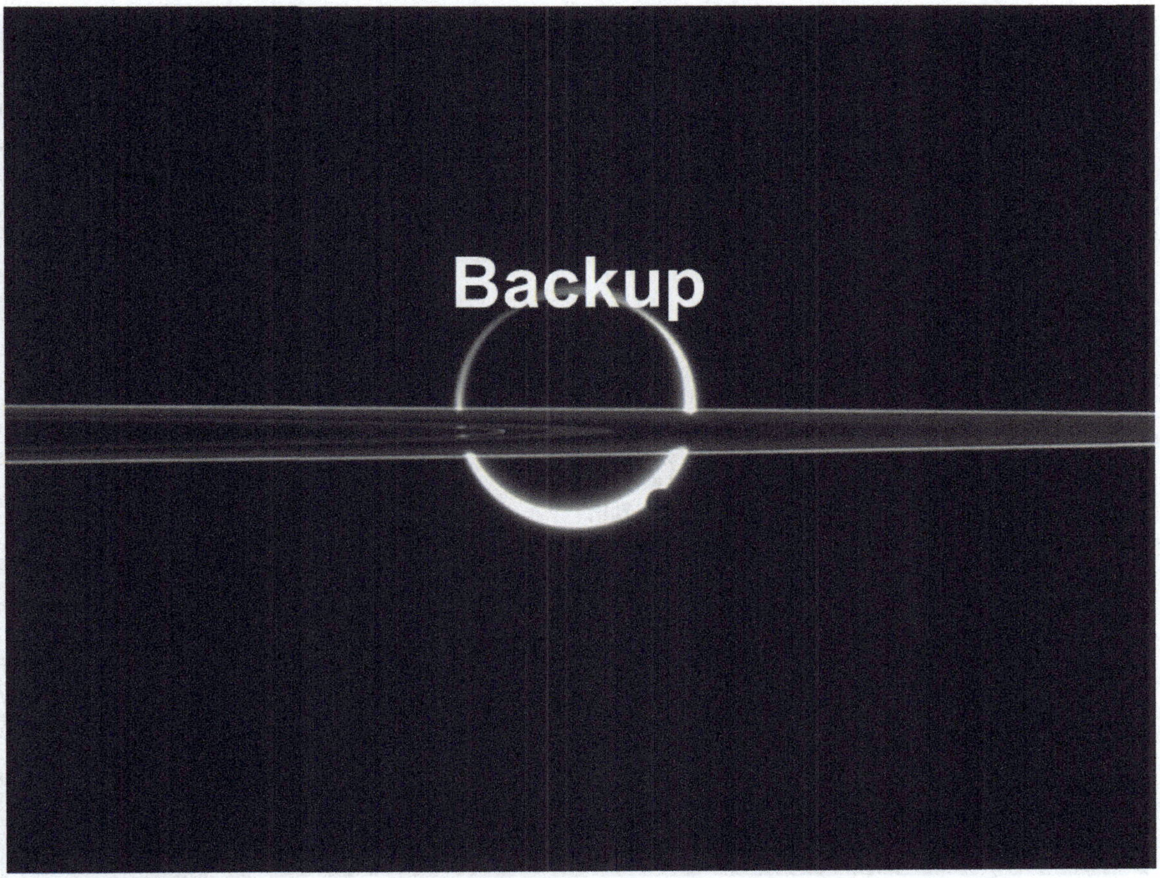
Backup

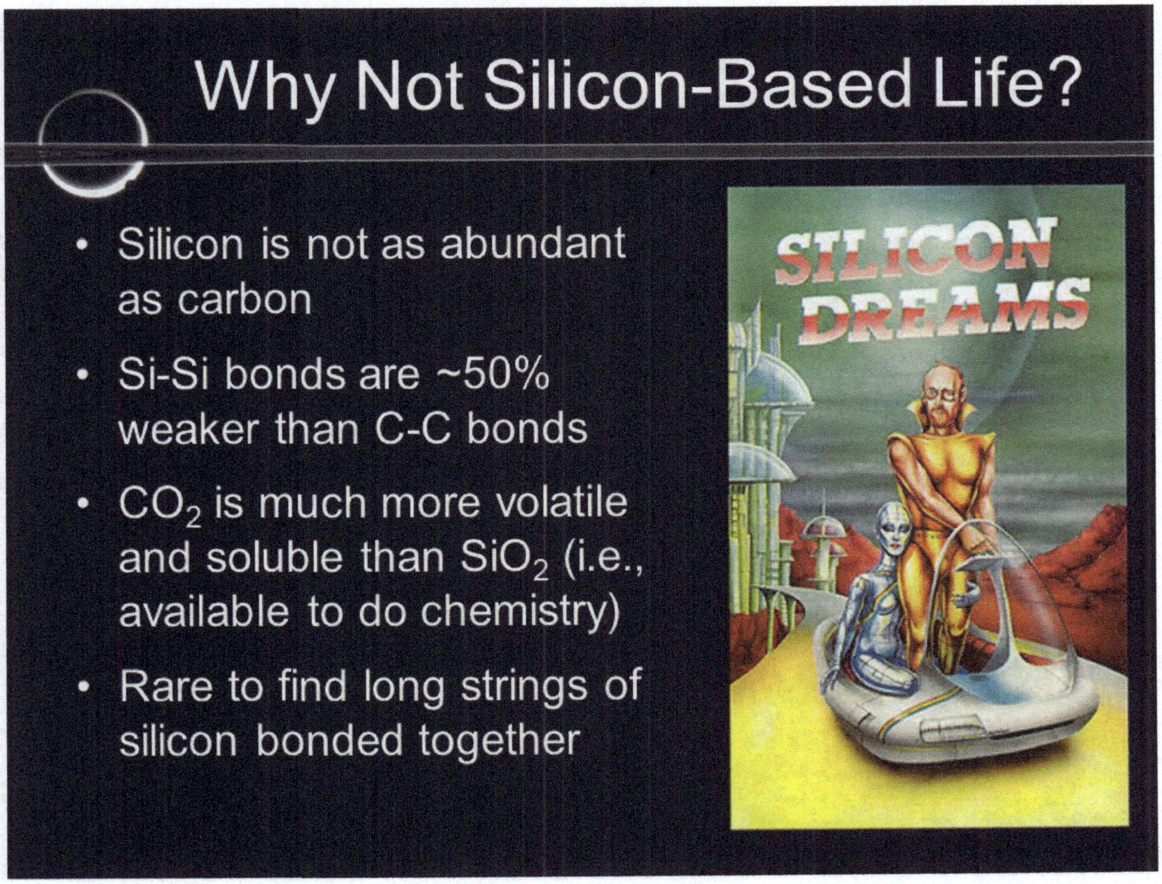
Why Not Silicon-Based Life?
• Silicon is not as abundant as carbon
• Si-Si bonds are ~50% weaker than C-C bonds
• CO_2 is much more volatile and soluble than SiO_2 (i.e., available to do chemistry)
• Rare to find long strings of silicon bonded together
SILICON DREAMS

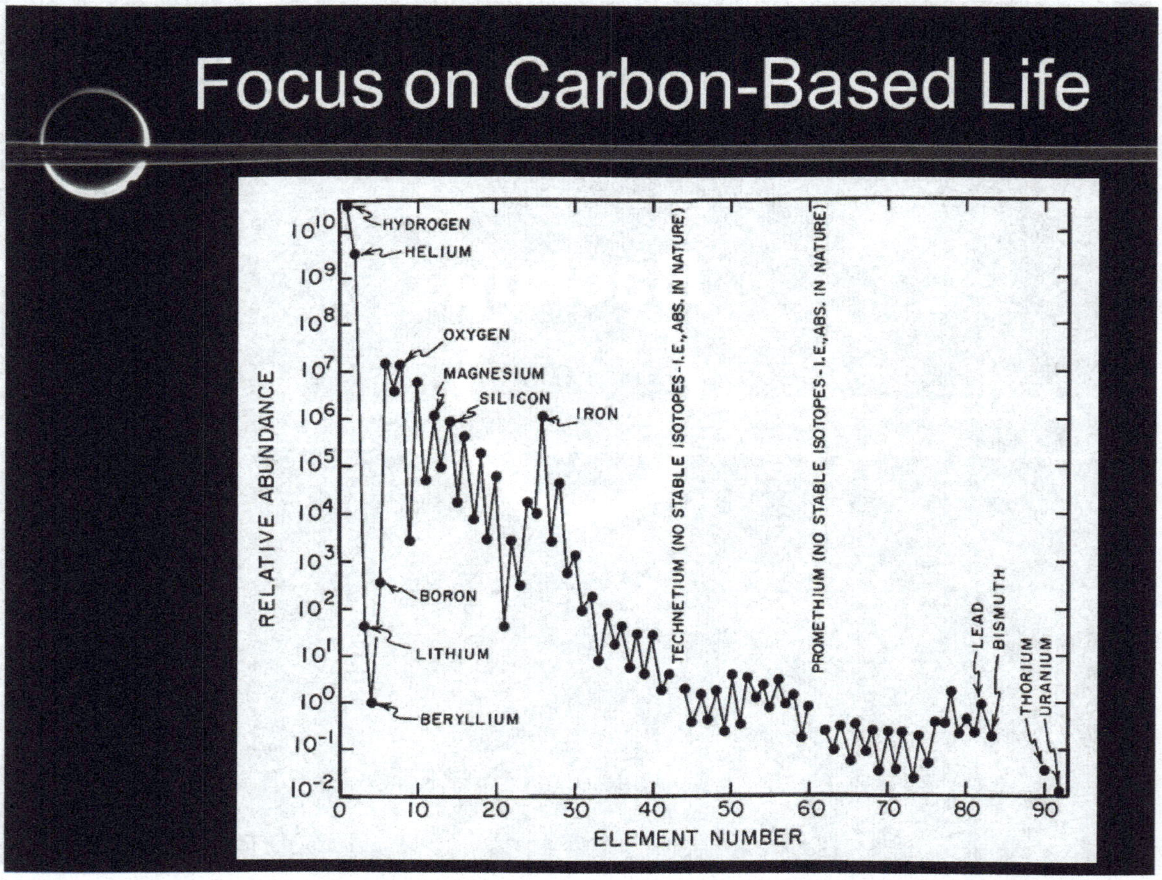
Focus on Carbon-Based Life
RELATIVE ABUNDANCE
ELEMENT NUMBER
HYDROGEN
HELIUM
OXYGEN
MAGNESIUM
SILICON
IRON
BORON
LITHIUM
BERYLLIUM
TECHNETIUM (NO STABLE ISOTOPES-I.E.,ABS. IN NATURE)
PROMETHIUM (NO STABLE ISOTOPES-I.E.,ABS. IN NATURE)
LEAD
BISMUTH
THORIUM
URANIUM

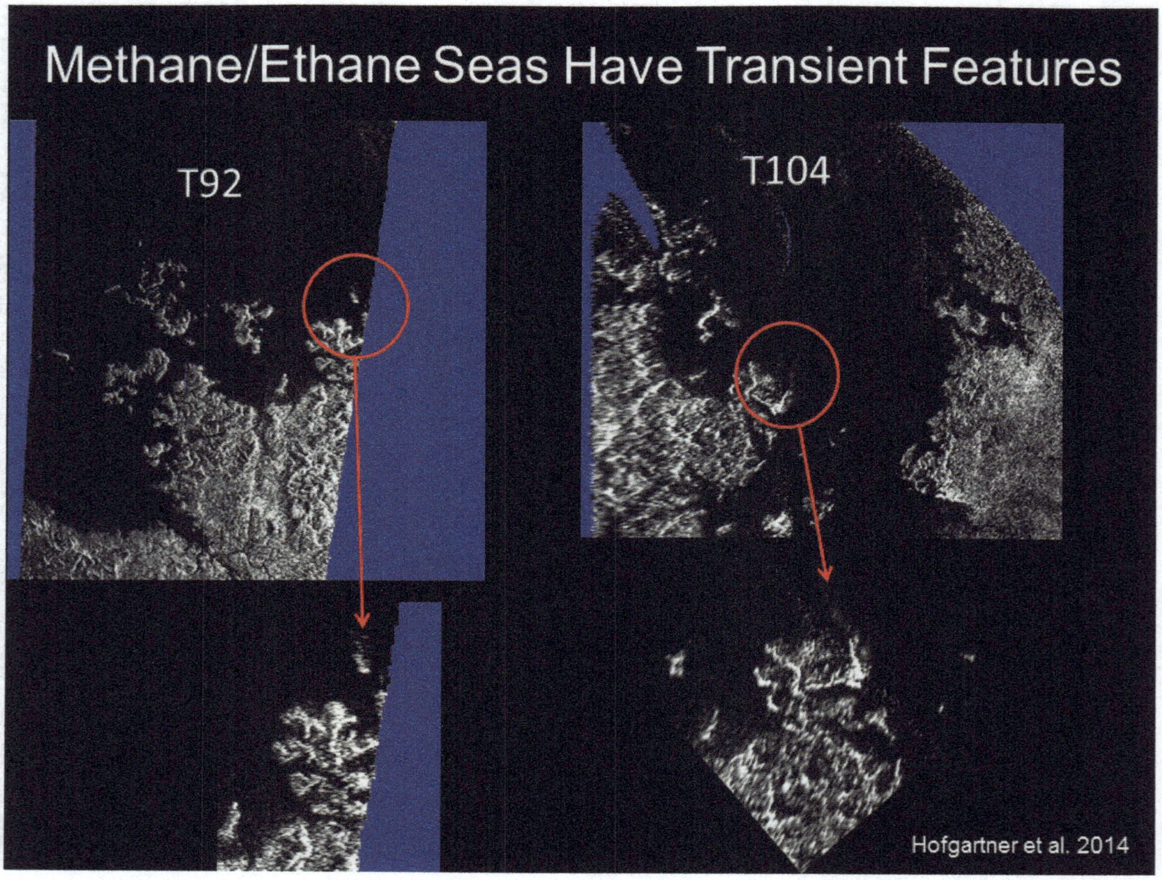
Methane/Ethane Seas Have Transient Features
T92
T104
Hofgartner et al. 2014

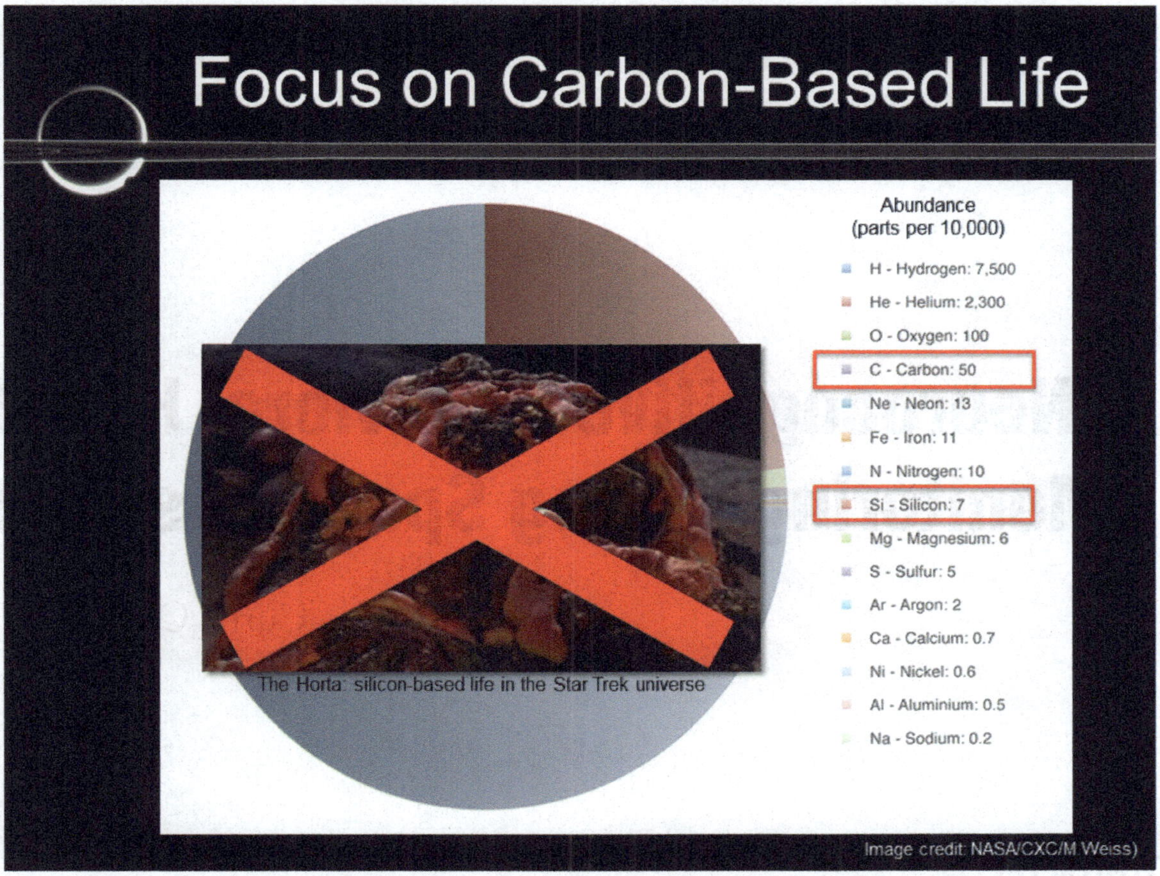
Focus on Carbon-Based Life
Abundance
(parts per 10,000)
H - Hydrogen: 7,500
He - Helium: 2,300
O - Oxygen: 100
C - Carbon: 50
Ne - Neon: 13
Fe - Iron: 11
N - Nitrogen: 10
Si - Silicon: 7
Mg - Magnesium: 6
S - Sulfur: 5
Ar - Argon: 2
Ca - Calcium: 0.7
Ni - Nickel: 0.6
Al - Aluminium: 0.5
Na - Sodium: 0.2
The Horta: silicon-based life in the Star Trek universe
Image credit: NASA/CXC/M.Weiss)

Neurocognitive Behavioral Health Monitoring During Space Missions

Curtis Cripe, Ph.D.

Abstract

Previous mission reports suggest that long duration space crews are at risk for psychosocial/behavioral issues that are affecting crewmember cognitive performance, moods, work performance and overall mission success. Behavioral health evidence indicates crewmembers are experiencing a myriad of stressors related to reduced neurocognitive performance that include lapses of attention, mood changes, cognitive performance decline, sleeping problems, and waning motivation. When viewing these reports from a behavioral medicine lens, there is a need to monitor, predict, and remediate reduced brain based neurocognitive changes in real time, especially during long duration space missions. With this in mind, a need exists for a space qualified Neurocognitive Predictive Performance Surveillance - Screening and Remediation tool (NPPS-SR) to monitor onboard behavioral health changes. The NPPS-SR tool can be used to predict neurobehavioral performance changes aiding mission control teams in structuring onboard mission activities based upon crewmember's cognitive resiliency. Further, the NPPS-SR tool can be used to strengthen neurocognitive resiliency changes based upon real-time neurocognitive performance markers. To date, brain-based neurometric literature contains over 6,100 publications related to cognitive performance (listed on Pubmed), all derived from earth based analog clinical populations. More critically, it is vital to note such publications have common NASA roots stemming from original early astronaut selection research. More importantly, clinically focused similar NPPS-SR tools exist and are presently used to successfully monitor, predict and aid in remediation of neurocognitive performance in earth based analog populations that include concussion, TBI, depressions, substance abuse, and learning issues.

Keywords

Behavioral Health, Radiation effect, Microgravity effect, Cognitive Rehabilitation

1. Introduction

An astronaut's resiliency while performing during the complex and dynamic conditions in the extreme environment of space is critical to the success of any manned space mission. Equally crucial, though less intuitive to harnessing the full benefits "stemming" from space exploration spin-offs, is the continued engagement of the credible leadership qualities of each crew member upon their return to earth and their subsequent reintegration into society.

This is achievable only by way of continued evidence of their exemplary performance in post flight leadership roles back on earth. Preserving opportunities to successfully demonstrate such leadership qualities by proactively protecting and preserving neurocognitive performance, will further reflect the value of manned space travel in the eyes of the world. Such due diligence would include accounting for potential neurocognitive declines due to exposure to long term space environmental effects.

From a systems engineering perspective, mission success requires many mission subsystems to function together flawlessly as an integrated unit. Many of these subsystems are mechanical in nature (i.e. spacecraft infrastructure) that support the mission. However, the most mission critical and vulnerable subsystems are the astronauts themselves; their requirements and needs must be addressed with care.

When viewing an astronaut from a systems' perspective, two major categories musts be addressed by the systems engineering and mission planning/operations teams: 1) physical health; and 2) behavioral health. With this framework in mind, this article will focus on an astronaut's onboard behavioral health from a behavioral medicine perspective. Behavioral medicine is an interdisciplinary field that integrates knowledge about human behavioral expressions by incorporating the knowledge from the fields of neuroscience, psychology, social sciences, occupational sciences, and medicine that are relevant to healthy behavioral expressions, which in our case are the astronaut's ability to perform both mission tasks and their interrelationship abilities with each other in space.

For long duration missions the impact of negative behavioral health responses is a concern, especially those due to space related brain changes affecting neurocognitive performance. These concerns include impacts on mission performance during critical tasks, fragmentation of crew cohesiveness, and ground team/crew communication disconnects. Additionally, these concerns include possible reductions in an astronaut's cognitive thinking abilities (reasoning, decision and judgmental thinking, perception, memory, attention and general information processing), and/or possible brain based emotional regulation disturbances affecting an astronaut's sense of self-agency (sense of self control) and utility, leading to potential disruptions in attention and performance. For example, disruptions resulting from radiation effects could impact mission safety and success by inducing distractions to learning and one's affinity to access memory, (including speed of memory recall and/or speed of cognitive processing), especially during time-critical reactions and tasks such as a docking sequence, EVA, deorbit burn, or landing sequences.

Deep Space Readiness Consortium, (DSR 30), is a focus group of subject matter experts dedicated to assisting in characterizing the scope and strategic positioning of potential countermeasures to assess, track, and reinforce a healthy and safe operational environment conducive to sustaining productive and thriving behavioral performance. The anticipated 2030 Mission to Mars (DSR30), with its focus of discussion centering, in part, around brain adaptation during long duration missions have contemplated questions around the effects of exposing crews to altered gravity environments. During the discussions several questions have arisen as they relate to on board behavioral health during long duration space missions:

- Is there a real possibility that neurophysiological changes might occur?
- Would negative neurophysiology changes place the mission and crew in life threatening conditions?
- If an Astronaut's visual perceptual abilities were distorted during docking or landing what might happen?
- If an Astronaut's memory retrieval abilities were distorted, slow or delayed what might happen?
- If an Astronaut's social perception abilities were distorted or dysfunctional in a multicultural, confined space during high pressure situations what might happen?
- How can the risk of a Crew-to-Ground communication disconnect be minimized and managed?
- What would happen if the ground Mission Planning Teams and/or the Mission Crew were unaware of neuro-cognitive changes and are there potentially behavioral markers that might link or map back to neurocognitive effects that could serve as an" early warning system"?
- Can we anticipate these possibilities and protect against them?

2. Psychosocial Difficulties from Mission Reports

From articles summarizing previous mission reports, it is clear that many of the above questions are possible and need to be addressed for long term space missions. During the early years of manned space exploration, missions were comparatively short and crews were mostly comprised of males, generally from similar cultures and backgrounds [1]. Accordingly, the impact on astronaut behavior and performance due to psychological and interpersonal factors was considered minimal and generally not addressed by either mission planning or operations teams. However, recently with the introduction of longer term missions on orbiting space stations, missions have become longer with crews drawn from multinational sources, often consisting of mixed genders and cultural backgrounds.

American and Russian anecdotal mission reports are indicating the presence of psychosocial issues occurring in space crews. When viewing these issues from a psychosocial model, it appears that these issues are impacting space missions by inducing detrimental effects on cognitive performance, crewmember moods and work performance [1–14]. However, behavioral health evidence from near earth short and long term mission reports also suggests many crewmembers are experiencing a myriad of stressors that are related to reduced neuro functional performance, especially those involving longer-durations of space environment of four to six months that might be intensifying observed psychosocial difficulties in its place.

These difficulties include lapses of attention, mood changes, cognitive performance decline, work performance decline, emotional lability, psychosomatic symptoms, sleeping problems, irritability toward crewmates and/or mission control staff, and a decline in robustness and motivation [4, 10-14]. When viewing many of these difficulties from a neurobehavioral lens, the evidence suggests that they might be derived from exposure to microgravity and space related radiation or other environmental factors, which in turn is effecting a crew member's neuro-cognitive performance abilities [1-14], therefore modifying crew member's emotional behavioral responses.

The astronaut health literature contains several articles addressing various aspects of crew related psychosocial difficulties. A recent meta-analysis by Kanas and Manzey [3] summarizes four potential sources of possible reduction in neurocognitive performance that might induce related neurobehavioral problems. These sources include: (1) physical factors, derived from acceleration, microgravity, radiation, and changes in light/dark cycles; (2) habitability factors, that are derived from vibrations, noise, temperature, light, and air quality; (3) psychological factors, derived from isolation, danger, monotony and workload; and finally (4) social or interpersonal factors, that may include gender issues, cultural effects, crew size, leadership issues and personality conflicts. Nevertheless, little attention on how these sources of possible reduction in neurocognitive performance exists and how it might affect mission critical events or how to monitor these effects in flight.

3. Why Is Neurocognitive Performance Important--

Self Agency, Ego Strength, and Cognitive Resiliency

Neurocognitive performance is intimately related and correlated to self agency, ego strength and cognitive resilience. Self agency refers to the subjective awareness that one is controlling their actions, volitionally. It is the "I" inside of ourselves that is executing our movements and thinking our thoughts. It is the source of how we creatively access our internal reserves to uniquely make things happen in the world. Ego strength, as a concept, depicts how an individual preforms in life by managing internal and external stresses and maintaining emotional stability, without loosing self agency. Yet, ego strength, as is self agency, also is dependent upon cognitive resilience [15]. Clinically, ego strength is frequently used to described a patient's ability, in the face of distress and conflict, to perform and maintain their internal sense of self integrity (i.e. be themselves). High or strong ego-strength individuals approach challenges with an attitude that they are able to overcome problems and grow stronger as a result. In our case, an astronaut's strong internal sense of ego-strength will allow him or her access to natural internal resources required to cope with problems, including finding new ways of dealing with struggles "in the moment" - all without losing a sense of self agency (i.e. won't panic under pressure). This is an extremely important quality needed in space. In contrast, weak ego-strength individuals view challenges as something to avoid and in many cases, reality can seem too overwhelming to deal with "in the moment". Weak ego-strength failures during moments of high reconciliation demands, such as during mission critical task or those affecting crew group cohesiveness could compromise crew performance.

Baumeister [15], in his Ego-Resource Depletion model, described how one's ego strength fluctuates depending upon the amount of cognitive effort exerted during the day. The model offers an explanation on the interrelationship between ego strength and ego-resources, and neurocognitive constructs that include self-regulation, executive function, and control interact. Baumeister proposes that self-regulation, like many other cognitive domains, will fatigue with extended effort. In his model, ego-resource capacities are not fixed, but fluctuate throughout the day and rely on internal resources that by their nature are limited and finite. Cognitive neuroscience studies support this view by demonstrating that many neurocognitive constructs, such as self-control and executive function, are mediated by fatigue [16-21]. Subjectively, cognitive fatigue is often observed not only as a decrease in ego strength, but with a corresponding increase in negative neurobehavioral expressions such as impulsivity, grumpiness, and/or narrowed myopic executive decision-making processes (reduced executive function capacity). Furthermore, evidence suggests that ego depletion detrimentally affects not only self-control and executive function, but also our general cognitive resilience as well, thus affecting performance on many tasks. From resiliency and coping studies in earth based analog populations that include learning challenged children, adults with Traumatic Brain Injury (TBI), and those in substance abuse recovery, the author has found this concept to be not only true, but also a key ingredient in the treatment remediation plan for individuals with brain-based challenges [22,23,24].

Cognitive resilience is a psychological construct that depicts the capacity to overcome the negative effects of setbacks involving associated stresses on cognitive functions that affect performance. There is a direct effect on an individual's ego strength and self agency depending on their cognitive resilience [18]. Intrinsically, cognitive resilience can be understood as a continuum of cognitive functionality. Cognitive processes that become overwhelmed and ineffective are on one end of the continuum. In contrast, the other end of the continuum represents an area where there are few or no negative effects due to stress on cognitive performance. Matching where a person lies along the cognitive resilience continuum with the task at hand is significant.

For long duration missions, gaining a measure of an astronaut's cognitive resilience is vital for mission success: 1) to predict performance ability (or risk) during higher levels of mission demands; 2) to predict when cognitive resources might need to be replenished (rest, or performing a battery of cognitive enhancement exercises); 3) to provide a quantifiable measure of the astronauts' general Ego Strength. Presently, in clinical practice the author has successfully applied measures of cognitive resilience, based upon specific brain measurements as real time neuro responses, to both child and adult earth based analog populations. These methods have consistently and quantifiably aided these earth analog populations in identifying and in recovery from learning and brain development issues, TBI, and substance abuse [22,23,24]. If similar objective brain response measures were obtained during a mission, this added data will give mission control teams a window into how to assist structuring onboard crew mission activities as well as assigning similar brain based exercises as used in earth based analog populations to help ensure mission success.

4. What are Possible Neurocognitive Performance Markers

From a classic neuropsychological model perspective, neurocognitive functions are considered fundamental building blocks that make up neurobehavioral expressions or behaviors. Behaviors range from task execution to interpersonal or mood expressions, to general or specific thinking abilities. Classically, neurocognitive functions are classified into categories such as executive function, perceptual function, memory function and self regulatory function, each with corresponding interconnected neurological structures that define a specific neuro network. It is important to note that the brain operates as an integral unit with each network either receiving or transmitting information to another network. In other words, most neurocognitive functions interact and influence each other to receive information, perceive, store, learn, recall, feel and create a thought and then to execute a task. This action occurs in a precise sequence with precise timing.

When using the right instrument, this action can be observed with quantifiable variables through measurements of intensities (strengths) that include speed, power, connectivity, timing and phase relationships within neural network elements via the EEG. Thus from a behavioral medicine or neuroengineering perspective, all neurobehavioral expressions or behaviors can be considered performance expressions derived from the interaction between individual brain subsystems. Yet, operating as an integral unit to produce the desired behavior. Moreover, each subsystem and subsequent expression is measurable. For this reason, to predict astronaut performance and onboard behavioral expressions, it is important to monitor neurocognitive performance of many brain structures and predict their performance ability effect on task performance. In other words, if the neurocognitive functions that support an astronaut's memory system should become slow to respond or fail to respond, it is important to know how it will impact other aspects of their mission performance.

To date, changes in neurocognitive and neurobehavioral responses, due to microgravity effects, sleep circadian cycle disruption, and radiation exposure, are not well understood, but have been observed and reported in the space literature. For example, recent ISS experimental reports indicate that an astronaut's visual perceptual processing abilities are affected during medium duration missions [25]. These effects include: orientation illusions, errors in sensory localization, postural imbalance, changes in vestibular, reflexes, space motion sickness, face recognition errors, negative impacts on mental rotation abilities, and errors in judgments of direction and special orientation. Report summaries indicate that effects on visual perceptual processing abilities are due to the brain's reduced ability to adjust to a microgravity field.

Specifically, as can be seen in figure 1, event-related recordings of EEG signals in the frontal and occipital brain areas showed that phase-locking (a measure of neuro network connectivity) of specific oscillations were suppressed in weightlessness for 3D tunnel images but not 2D images. Meaning that in a microgravity field, the brain processes 3-D images differently, but not 2D images. More importantly, the report demonstrated the brain on Earth processed 3D information using a directional top-down control from the frontal to the occipital areas. However, in weightlessness, the fronto-occipital top-down informational control was altered into a diverging flux originating from the central brain areas towards the frontal and occipital areas almost equally. Further when observing astronaut 3D drawing from mission experiments, they resemble drawing from children with learning issues know to originate due to cerebellar issues, either structural or with lack of primitive reflex suppression [25]. In our case, a more likely conclusion would be as a result of reflex disturbances due to microgravity effects.

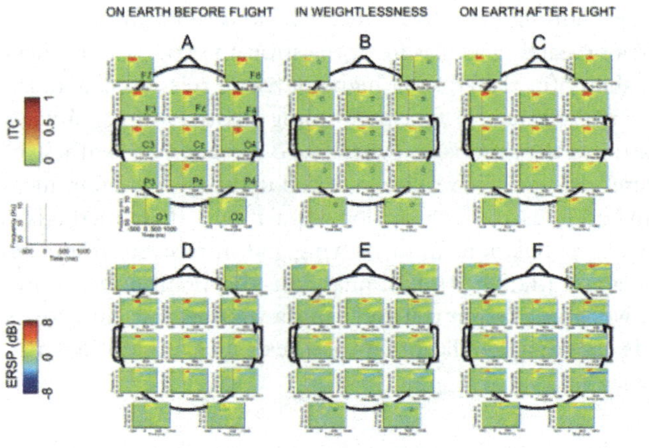

Figure 3. Effect of microgravity on the topographical representation of ITC and ERSP. ITC are represented in the upper part (*A–C*) and the ERSP in lower part (*D–F*). Grand average (n = 5) triggered by the 3D-tunnel presentation on Earth before flight (*A, D*) in weightlessness (*B, E*) and on Earth after flight (*C, F*). Each map corresponds to a single recording channel (from F7–F8 to O1–O2) disposed on the scalp. Statistical significance (Friedman ANOVA) $p<0.05$ is indicated by an asterisk.
doi:10.1371/journal.pone.0082371.g003

Figure 1.

Clinically, similar observations have been reported in adults with blood flow reduction due to heart issues, and children with learning issues. Of concern within these populations is that similar neurocognitive performance changes have been shown to impact both memory recall (precursor to dementia) and the brain's natural self-regulatory abilities as they relate to information processing abilities. These compromised neuro abilities effect objective thoughts and subjective emotional interpretations, moods, and cognitive resilience [25-35]. For long term missions, where there is a chance of the brain to naturally adapt to the microgravity environment, these adaptations need to be considered for crew onboard performance and reintegration upon Earth's return.

Equally important, if not more important are possible changes to the brain structure and its performance due to radiation effects. Radiation is known to effect hippocampal performance by reducing its neurogenesis ability and thus affecting cognitive resiliency [41-46]. Hippocampal neurogenesis aids in the ability to adapt to environment changes by providing the necessary building blocks for neuroplasticity. Neuroplasticity is used to expedite and take shortcuts in accelerating the use of amount of information acquired in every second by learning from what has transpired from past actions. Neuroplasticity through neurogenesis is required to make new neuro connections and adapt to changes that are deeply rooted in existing neuronal connections. With this in mind, the hippocampus is one of the key brain structures responsible for learning and memory and is involved in reaction times (or delays) required to react to an event [36-46]. Both positive and negative experiences exhibit notable affects on the hippocampus, altering memory, tendency towards anxiety, and an individual's perception of their environment and how they handle stress [36-40].

It is important to note that hippocampal performance is considered a key structure in mediating stress and supporting cognitive resilience [36-40]. Compromised hippocampal performance has been implicated in affecting the ability to cope with changes in the environment, as well as creating anxiety, depression, and other performance-inhibiting disorders [41,42]. Of concerns for the 2030 Mission to Mars are routine behaviors that occur during the 3 years in space, which to some may be considered negative and stressful. More importantly, not often accounted for, are indirect hippocampal performance effects that either directly or indirectly affect task performance that ultimately has an effect on an astronaut's cognitive reserve, cognitive resilience, emotional resiliency and general ego strength.

5. Is Neurocognitive Performance Measureable in Space?

The short answer is yes and the instrument to do so is presently onboard the ISS. What is required is the development or a modification to existing clinical tools to an on board software package that monitors targeted neurocognitive systems and subsystems. The preferred methodology that the author is proposing is based upon a neurometric methodology which had its original roots in the selection of the astronauts in the original NASA space program.

During the late 1950s, a UCLA team led by Dr. Ross Adey, pioneered a numerical method of using raw EEG measurements as part of a NASA study directed at astronaut selection [46-48]. Similar to our needs today, the thrust of the study was to determine the strength of the neuro-performance of astronaut candidates. Study results were used in initial astronaut selection, as well as in future NASA pilot training studies, which is where the author was first introduced to the methodology in graduate school.

It is important to note, Dr. Adey's team's initial study developed many of the essential quantitative foundations required to digitize and use the EEG as an analytical tool, still used today in clinical settings. The initial UCLA database included several hundred subjects based upon UCLA faculty and students and those who were candidates for the early NASA space exploration. In the 1970s, two Swedish neurologists, Drs. Milos Matousek and Ingemar Petersen, published the first clinically oriented normative database which allowed clinicians to compare a child's brain wave activity to a normative standard [49,50]. This study was replicated and verified by E. Roy John's lab at New York University in 1975 [51-56]. Beginning in the 1990s, co-registration of qEEG measurements with other instruments that include: Positron Emission Tomography (PET), Magnetic Resonance Imaging (MRI), CT Tomography, Single Photon Tomography (SPECT) occurred providing equivalent numerical markers that allows cross registration between instruments.

From these efforts, EEG neurometric measures have grown to include several non-invasive neuro-electric neuroimaging methodologies such as EEG Tomography (tEEG), that provides recordings with a time resolution of less than 1 millisecond with a 3-dimensional spatial resolution of about 1 cubic centimeter. To date, this methodology has been applied to many clinical conditions such as ADD, ADHD, schizophrenia, compulsive disorders, depression, epilepsy, TBI and a wide number of clinical groupings of patients, with well over 6,100 studies currently listed on Pubmed. The tEEG methodology, which includes insights in to the neuro network Functional Effective Connectivity is proposed as the primary analysis methodology.

Within the last ten years, tEEG measurements have been applied to many practical applications outside of the clinical realm that include many Department of Defense (DOD) projects. These projects are focused on teamwork, workload monitoring, emotional content processing, and others, all applicable to onboard neurobehavioral monitoring.

For example, Stevens and colleagues [21] showed that team work neurophysiological synchronies measuring individual member workload and engagement measures are interlinked with each team member. More importantly, these measures can be used as a general measure of team cognition and have been demonstrated as EEG based neurophysiologic indicators of team communication metrics that are complementary to conventional subjective paper measures of team cognition. Equally important, researchers from around the world are demonstrating that there are several effective EEG based neurophysiologic indicators of emotional and task performance engagement that are consistent across populations and cultures.

6. How Can We Monitor Neurocognitive Performance?

As noted earlier, cognition functions from various cognitive domains that act as an integrated system. These domains include, but are not limited to, arousal, perception, attention, memory, learning, thinking, mental organization, affect (feeling) and expression, plus executive functions. Recent cognitive neuroscience has begun to establish how many neurocognitive functions operate to produce neurobehavioral expressions in terms of cognitive endophenotypes describable in neural network terms [29-33]. Core cognitive endophenotypes of brain-behavior models are broadening our understanding of healthy brain function and many mental disorders by explaining the importance of unconscious schemas, motivational processes or learning and reinforcement principles, and how they relate to both proper brain function and psychopathology [36-42].

At present research is primarily focused on identifying dysfunctional cognitive endophenotypes found to underlie common symptoms in many mental disorders. Impulsivity is an example of a meaningful cognitive endophenotype that has been recognized in several categorical disorders such as ADHD, anti social behaviors, substance use and more [57]. In this manner, cognitive endophenotypes offer clinicians the ability to target specific cognitive dysfunctions that contribute to the mental disorder rather than treating symptoms, by using targeted brain exercises to strengthen the areas of the brain (neuro network) that define the endophenotype [30–33,57-63]. Another example of a neurocognitive endophenotype is working memory, which is correlated with hippocampal performance. Fig 2 is a slide presented at the 2015 100 YSS Symposium depicting working memory operations and the network level (Brodmann areas are specific locations in the brain that define its neuro network) and its possible radiation affects.

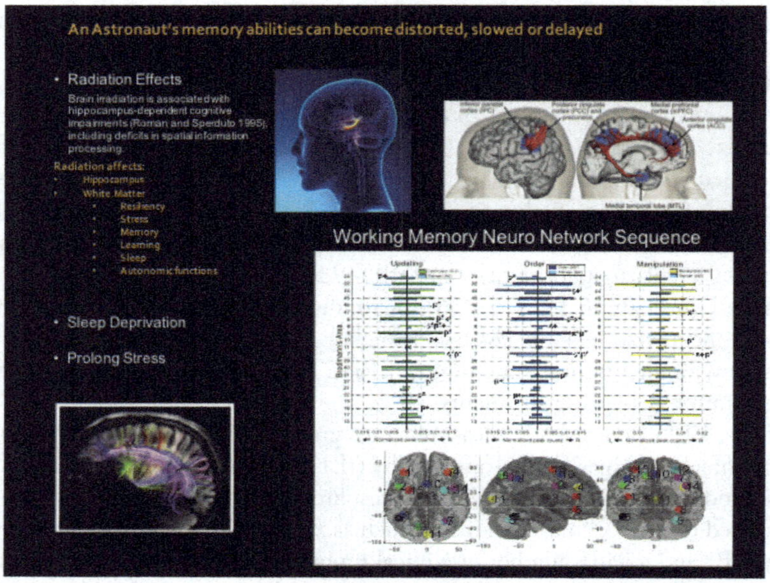

Figure 2. Depicts working memory neuro network and areas of the brain that may be affected by radiation.

Based upon functional connectivity neurometrics Figure 3 depicts an image that highlights the processing strength of the functional connectivity of the hippocampus with 44 Brodmann areas. This image was derived from a patient who was presenting with working memory issues and mood dysregulation that was affecting not only his cognitive resiliency, but his willingness to remain employed as a managing executive at his work place. Note the the light blue colors in the figure indicating weak or low Functional Connectivity between the Brodmann area and the hippocampus from each hemisphere and the brighter orange colors indicating over active or hyper connectivity. Green areas represent age appropriate levels of functional connectivity.

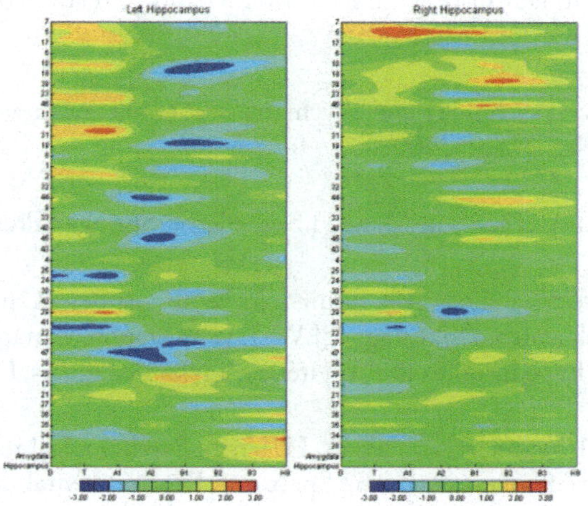

Figure 3. Functional Connectivity Neurometrics depicting hippocampus functional connectivity between 44 Brodmann brain areas. Green areas represent normal response connectivity; Blue areas represent hypo connectivity; Orange areas represent hyper connectivity.

7. In Conclusion--

Neurocognitive Predictive Performance Surveillance - Screening and Remediation tool (NPPS-SR)
Onboard Crew Monitoring and Cognitive Remediation Tool

In conclusion, the success of any manned space mission depends not only upon an astronaut's ability to perform and cope in space, but also equally to continue to demonstrate their natural leadership qualities upon return to earth and reintegration into society. Research suggests that long duration space crews are at risk for psychosocial/behavioral issues within the crew environment that may affect cognitive performance, crewmember moods and work performance. Behavioral health evidence conveys that crewmembers are experiencing a myriad of stressors that are related to reduced neurocognitive performance. These difficulties include lapses of attention, mood changes, cognitive performance decline, work performance decline, emotional lability, psychosomatic symptoms, sleeping problems, irritability toward crewmates and/or mission control staff, and a decline in robustness and motivation.

When viewing many of these issues from a neurobehavioral lens, there is a need to monitor, predict, and remediate reduced neurocognitive changes in real time, especially during long duration space missions. To support the viability of such a tool, to date, EEG based neurometric related literature contains over 6,100 publications listed on Pubmed, all derived from earth based analog populations. More critically, it is vital to note all such publications have common NASA roots stemming from original early astronaut selection research. Additionally, clinically focused neurocognitive performance tools have now evolved to successfully aid in the remediation of earth based analog populations including concussion, TBI, depressions, substance abuse and learning issues.

With this in mind, the author is suggesting there is a need for the creation of a space qualified Neurocognitive Predictive Performance Surveillance - Screening and Remediation tool (NPPS-SR) to monitor onboard behavioral health. The NPPS-SR tool can be used to predict neurobehavioral performance in order to aid mission control teams in how to assist structuring onboard crew mission activities, which will aid crew members in strengthening possible neurocognitive resiliency based upon real time neurocognitive performance markers.

References

1. R. North, Human requirements for long-duration missions: Antarctic and Arctic stations, planetary surface operations, and space transportation vehicles, Revista Portuguesa de Medicina Militar 40 (1991).

2. J.R. Ball, Safe Passage: Astronaut Care for Exploration Missions, Institute of Medicine, National Academy of Sciences, Washington, DC, 2001.

3. N. Kanas, D. Manzey, Space Psychology and Psychiatry, second ed., Microcosm Press, El Segundo, CA, and Springer, Dordrecht, The Netherlands, 2008.

4. D.A. Shayler, Disasters and Accidents in Manned Spaceflight, Springer/Praxis, Chichester, UK, 2000.

5. P. Suedfeld, Canadian space psychology: the future may be almost here, Canadian Psychology 44 (2) (2003) 85–92. J. Rivolier, R. Goldsmith, D.J. Lugg, A.J.W. Taylor, Man in Antarctica. The Scientific Work of the International Biomedical Expedition to the Antarctica, Taylor & Francis, London, 1988.

6. G.M. Sandal, T. Bergan, M. Warnche, R. Værnes, H. Ursin, Psychological reactions during polar expeditions and isolation in hyperbaric chambers, Aviation, Space and Environmental Medicine 67 (1996) 227–234.

7. P. Suedfeld, Invulnerability, coping, salutogenesis, integration: four phases of space psychology, Aviation, Space and Environmental Medicine 76 (2005) B61–B66.

8. Sweet, T, Panda,N., Hein, A, Das,S, Hurley,S, Olschowka,J. , Williams,J., & O'Banion, (2014) M, Central Nervous System Effects of Whole-Body Proton Irradiation, Radiation Research 182, 18–34

9. Denisova, N., Shukitt-Hale, Rabin & Joseph (2002) Brain Signaling and Behavioral Responses Induced by Exposure to 56Fe-Particle Radiation Radiation Research158 725-734

10. Heuer, H.; Manzey, D.; Lorenz, B.; Sangals, J. Impairments of manual tracking performance during spaceflight are associated with specific effects of microgravity on visuomotor transformations. Ergonomics 2003, 46, 920–934.

11. Johannes, B.; Salnitski, V.P.; Polyakov, V.V.; Kirsch, K.A. Changes in the autonomic reactivity pattern to psychological load under long-term microgravity—twelve men during 6-month spaceflights. Aviakosmicheskaia i Ekologicheskaia Meditsina 2003, 37, 6–16.

12. Manzey,D.;Lorenz,B. Poljakov,V. Mental performance in extreme environments: Results from a performance monitoring study during a 438-day spaceflight. Ergonomics 1998, 41, 537–559.

13. Manzey, D.; Lorenz, B.; Heuers, H.; Sangals, J. Impairments of manual tracking performance during spaceflight: More converging evidence from a 20-day space mission. Ergonomics 2000, 43, 589–609.

14. Whitmore, M.; McQuilkin, M.L.; Woolford, B.J. Habitability and performance issues for long duration space flights. Hum. Perferm. Extrem. Environ. 1998, 3, 64–74.

15. Baumeister R.F., Heatherton T.F. (1996). Self-Regulation Failure: An Overview. Psychol Inq. 1996; 7:1–15.

16. Baumeister, RF., et al. (1994). Losing control: how and why people fail at self- regulation. Academic Press.

17. Muraven M, et al. Self-control and alcohol restraint: an initial application of the self- control strength model. Psychol Addict Behav. 2002; 16:113–120.

18. Lavie, N., (2010). Attention, Distraction, and Cognitive Control Under Load. Current Directions in Psychological Science 2010 19: 143.

19. Thush, C., et al. (2008). Interactions between implicit and explicit cognition and working memory capacity in the prediction of alcohol use in at-risk adolescents. Drug and Alcohol Dependence, 94, 116-124.

20. Oberauer, K., Su, H., Schulze, R., (2000). Working memory capacity—Facets of a cognitive ability construct. Personality and Individual Differences, 29, 1017-1045.

21. Stevens, R., Berka, C. Behneman, A, Wolgeuth, T, Lanb, J, Buckles, R. (2011) Linking models of Team Neurophysicologic Synchronies for Engagement and Workload with Measures of Team Communication. 20th Conference on Behavior Representation in Modeling and Simulation (BRIMS), Sundance, UT

22. Cripe, C. (2014) "Psychology's New Design Science: Theory and Research, (Eds.) S. Imholz and J. Sachter, Univ of Illinois/Common Ground Publisher - Curtis Cripe, Chapter 6 "Neuroengineering – Brain Recovery Methods – Applied to Substance Abuse Recovery"

23. Cripe, C. (2006) 'Effective Use of LENS Unit as an Adjunct to Cognitive Neuro-Developmental Training', Journal of Neurotherapy, 10: 2, 79 — 87

24. Gunkelman, J., Cripe C. (2008) Clincial Outcomes in Addiction: A Neurofeedback Case Series. Biofeedback 36.4, 152-156

25. Cheron G, Leroy A, Palmero-Soler E, De Saedeleer C, Bengoetxea A, et al. (2014) Gravity Influences Top-Down Signals in Visual Processing. PLoS ONE 9(1): e82371. doi:10.1371/journal.pone.0082371

26. Donovan, M. H.; Yazdani, U; Norris, R. D.; Games, D; German, D. C.; Eisch, A. J. (2006). "Decreased adult hippocampal neurogenesis in the PDAPP mouse model of Alzheimer's disease". The Journal of Comparative Neurology 495 (1): 70–83.

27. Mu, Y; Gage, F. H. (2011). "Adult hippocampal neurogenesis and its role in Alzheimer's disease".Molecular Neurodegeneration 6: 85

28. Jacobs, B. L., H. van Praag, F. H. Gage (2000). "Depression and the Birth and Death of Brain Cells".American Scientist 88.

29. Emmelkamp, et. al. (2014). Advancing psychotherapy and evidence-based psychological interventions. Int. Jrl. Methods Psychiatr. Res. 23(Suppl. 1): 58–91.

30. Lambert M.J. (2007). What we have learned from a decade of research aimed at improving psychotherapy outcome in routine care. Psychotherapy Research 17(1), 1–14.

31. Insel T., et al. (2010). Research Domain Criteria (RDoC): toward a new classification framework for research on mental disorders. American Journal of Psychiatry.

32. Hofmann S.G. (2008). Cognitive processes during fear acquisition and extinction in animals and humans: implications for exposure therapy of anxiety disorders. Clinical Psychology Review 28, 199–210.

33. Hofmann S.G., Sawyer A.T., Fang A., Asnaani A. (2012). Emotion dysregulation model of mood and anxiety disorders. Depression and Anxiety 29, 409–416.

34. Prigatano, G.P. & Schacter, D.L. (Eds.) (1991). Awareness of deficit after brain injury: Clinical and theoretical issues. New York: Oxford University Press.

35. Heatherton, T. and Wagner, D. (2011). Cognitive Neuroscience of Self-Regulation Failure. Trends Cogn Sci. 15(3): 132–139.

36. Neves, G; Cooke, S and Bliss, T (2008). "Synaptic plasticity, memory and the hippocampus: A neural network approach to causality". Nature Reviews Neuroscience 9 (1): 65–75 doi:10.1038/ntn2303.

37. Becker S (2005). "A computational principle for hippocampal learning and neurogenesis". Hippocampus 15(6): 722–38. doi:10.1002/hipo.20095

38. Wiskott L, Rasch MJ, Kempermann G (2006). "A functional hypothesis for adult hippocampal neurogenesis: avoidance of catastrophic interference in the dentate gyrus". Hippocampus 16 (3): 329–43.

39. Aimone JB, Wiles J, Gage FH (June 2006). "Potential role for adult neurogenesis in the encoding of time in new memories". Nat Neurosci. 9 (6): 723–7.

40. Shors TJ, Townsend DA, Zhao M, Kozorovitskiy Y, Gould E (2002). Neurogenesis may relate to some but not all types of hippocampal dependent learning. Hippocampus 12 (5): 578–84.

41. Opendak, M; Gould, E (2015) Adult neurogenesis: a substrate for experience-dependent change Trends in cognitive sciences DOI: http://dx.doi.org/10.1016/j.tics.2015.01.001

42. Opendak, M; Gould, E (2011) New neurons maintain efficient stress recovery Cell stem cell -

43. Sweet TB, Panda N, Hein AM, Das SL, Hurley SD, Olschowka JA, Williams JP, O'Banion MK (2014). Central nervous system effects of whole-body proton irradiation. Radiat Res. 182, 18-34.

44. Cherry JD, Williams JP, O'Banion MK, Olschowka JA (2013). Thermal injury lowers the threshold for radiation-induced neuroinflammation and cognitive dysfunction. Radiat Res. 180, 398-40

45. Cherry JD, Liu B, Frost JL, Lemere CA, Williams JP, et al. (2012) Galactic Cosmic Radiation Leads to Cognitive Impairment and Increased Ab Plaque Accumulation in a Mouse Model of Alzheimer's Disease. PLoS ONE 7(12): e53275. doi:10.1371/journal.pone.0053275

46. Adey, W.R., Walter, D.O. and Hendrix, C.E. (1961). Computer techniques in correlation and spectral analyses of cerebral slow waves during discriminative behavior.Exp Neurol., 3:501-524

47. Adey, W.R. (1964a). Data acquisition and analysis techniques in a Brain Research Institute. Ann N Y Acad Sci., 31;115:844-866.

48. Adey, W.R. (1964b). Biological instrumentation, electrophysiological recording and analytic techniques. Physiologist., 72:65-68.

49. Matousek, M. & Petersen, I. (1973a). Automatic evaluation of background activity by means of age-dependent EEG quotients. EEG & Clin. Neurophysiol., 35: 603 -612.

50. Matousek, M. & Petersen, I. (1973b). Frequency analysis of the EEG background activity by means of age dependent EEG quotients. In P. Kellaway & I. Petersen (Eds.), Automation of clinical electroencephalography (pp. 75-102). New York: Raven Press.

51. John, E.R. Functional Neuroscience, Vol. II: Neurometrics: Quantitative Electrophysiological Analyses. E.R. John and R.W. Thatcher, Editors. L. Erlbaum Assoc., N.J., 1977.

52. John, E.R. Karmel, B., Corning, W. Easton, P., Brown, D., Ahn, H., John, M., Harmony, T., Prichep, L., Toro, A., Gerson, I., Bartlett, F., Thatcher, R., Kaye, H., Valdes, P., Schwartz, E. (1977). Neurometrics: Numerical taxonomy identifies different profiles of brain functions within groups of behaviorally similar people. Science, 196, :1393-1410.

53. John, E. R., Prichep, L. S. & Easton, P. (1987). Normative data banks and neurometrics: Basic concepts, methods and results of norm construction. In A. Remond (Ed.), Handbook of electroencephalography and

clinical neurophysiology: Vol. III. Computer analysis of the EEG and other neurophysiological signals (pp. 449-495). Amsterdam: Elsevier.

54. John, E.R., Ahn, H., Prichep, L.S., Trepetin, M., Brown, D. and Kaye, H. (1980) Developmental equations for the electroencephalogram. Science, 210: 1255– 1258.

55. John, E. R., Prichep, L. S., Fridman, J. & Easton, P. (1988). Neurometrics: Computer assisted differential diagnosis of brain dysfunctions. Science, 293, 162-169.

56. John, E.R. (1990). Machinery of the Mind: Data, theory, and speculations about higher brain function. Birkhauser, Boston.

57. Robbins T.W., Gillan C.M., Smith D.G., de Wit S., Ersche K.D. (2011). Neurocognitive endophenotypes of impulsivity and compulsivity: towards dimensional psychiatry. Trends in Cognitive Science 16, 81–91.

58. Miller E, Cohen J. (2001). An integrative theory of prefrontal cortex function. Ann Rev Neurosci; 24: 167–202.

59. FitzGerald, M., Carton, S., O'Keeffe, F., Coen, R. & Dockree, P. (2012). Impaired self- awareness following acquired brain injury: current theory, models and anatomical understanding, The Irish Journal of Psychology, 33:2-3, 78-85.

60. Carroll, E. & Coetzer, R. (2011). Identity, grief and self- awareness after traumatic brain injury, Neuropsychological Rehabilitation: An International Journal, 21:3, 289-305.

61. Barco, P.P., Crosson, B., Bolesta, M.M., Werts, D., & Stout, R. (1991). Training awareness and compensation in postacute head injury rehabilitation. In J. Kreutzer & P. Wehman (Eds.), Cognitive rehabilitation for persons with traumatic brain injury. Baltimore: Paul H. Brookes.

62. Bandura, A., (1991). Social Cognitive Theory of Self-Regulation. Organizational Behavior and Human Decision Processes 50, 248-287.

63. Heatherton, T. and Wagner, D. (2011). Cognitive Neuroscience of Self-Regulation Failure. Trends Cogn Sci. 15(3): 132–139.

Interstellar Embryos through Interstellar Teens

Negotiating Human Development through Education in an Off-World Context

Janet de Vigne, MEd

Abstract

Bringing up Terra-based children is fraught with difficulty; guiding children through development in an extra-terrestrial context may prove even more so. Neuroscience is increasingly feeding into education in the 21st century, so what are the neurobiological game changers today, and where might education be headed in the next thirty years? This paper presents recent findings that impact on child development and classroom practice, examining issues such as managing the teenage brain and its attraction to risk; potential neurological losses should we choose not to include, for example, linguistic diversity or music; and the development of computer or child-driven education systems. Along the way we will discuss issues of culture and society and their merits and demerits: what do we choose to keep? What do we choose to lose, in terms of Terran culture and mores (Shakespeare, Milton, the Bible)? How do we encourage the best of being human in an extra-terrestrial context? How do we maintain human rights and guarantee liberty? What definition of liberty will we deploy, particularly in the case of children?

Keywords

education in space, off-world education, children's rights, technology and education, neuroscience and education, child development, brain development

1. Introduction

The idea of sending humans off world is gathering pace as Mars One recruits and trains volunteers to carry the race into space with no chance of return. Where there are humans, there must therefore be reproduction, be it natural or mechanically aided, or there would be no point. The survival of the species must depend on its capacity to reproduce and educate the new generation to carry us forward. This paper will argue that ensuring a healthy, successful community of humans will depend on the ways in which children are educated, focusing specifically on the period of adolescence, popularly conceived as problematic. If we wish to avoid already imagined scenarios of tyranny, such as the chaos of The Lord of the Flies (1954) [1] or Total Recall (1990) [2], considering the possible roots of the exercise of control might be wise. I suggest that an equitable society supported by the principles of social justice and democracy would be the one to aim for, as shared responsibility for the success (meaning non-implosion and destruction) of a small, off-world community must mean the practice of quasi-utopian ideals of conflict resolution, listening to each other, valuing difference, and practising tolerance. It would be wise to capitalise on the strengths of the community, as the risks arising from disaffection would endanger it.

2. The Context

Twenty-first century Western society is really not very nice to our teenagers [3]. Managing this crucial period of human development is a trial for all concerned. Middle-class parents might choose to ignore it by deploying the "keep them busy" tactic, hoping that extracurricular activities will ease their child into adulthood by distracting them from any possibly inconvenient behaviours. Other parents (from any and all backgrounds) might view aspects of the teenager with increasing bewilderment–what is happening? A "normal," polite, respectful, and chatty child turns overnight into a monosyllabic, raging monster, full of resentment and irrational disaffection. This period is universally recognised and caricatured in numerous comedic performances, in the UK (and now on YouTube™) successfully by British comedian Harry Enfield in his portrayal of "Kevin the Teenager"–a spotty, strangely dressed creature, constantly screaming at his gentle, long suffering, and increasingly perplexed parents, desperate to 'fight' where there is no conflict, and convinced of the injustice being meted out to him with no real evidence for the same [4]. Off-world irrationality—behaviour and thinking—is greatly to be feared. It doesn't bode well on earth either, but there appears to be space for teens to work it out without compromising the existence of the entire human race. (Adult irrationality is of course another matter.)

Attempting to control irrational behaviour usually does not work; the consequences of any attempted control might well be worse than the initial problem. Managing it may be a better option, by which I mean providing ways in which issues can be played out with little or no risk to the social and physical environment. Rites of passage—practices perpetuated to this day by older societies than ours—have a role here, as does heightened play: the value of Shakespeare in emotional growth and Theory of Mind (effective team building) development (El Sistema)? More of this anon.

Why, then, is the period of adolescence so dangerous? What is really happening in terms of biological development, and how might it be possible to manage this in a positive way? Only considering management, and thereby conceptualising the teen years as a problem, may not be the best way forward. If instead we viewed this time as a positive phase, where teenagers might perform a valuable role in the building of a new society, we might find a more helpful and acceptable way of negotiating the turmoil.

2.1 The Neurobiology of Adolescence

Many people use the surge in neurobiological activity of the teenage brain to explain—and write off—the more unacceptable facets of teen behaviour. The real story is of course more complex, and it should be possible to see every supposedly risky aspect as having a positive side. Teenagers love risk; how could this be built upon to benefit the community? Similarly, the desire to negotiate a social world, to compete, the vulnerability to peer pressure, and the unpredictability of endocrinological impulses including sex hormones—perhaps the most notorious area of development at this stage—all these might in fact be regarded as positives. Some consider that this period is an evolutionary necessity, occurring in order for the young adults to be encouraged to leave the community and go off to forge their own destiny without the confines of parental influence. Other mammalian behaviour seems to substantiate this, because of course ours is not the only species to go through adolescence. Understanding our neurobiology holds the key.

Advances in medical technology mean that it is now possible to observe the brain without physically interfering with it. As a consequence, it has been possible for a while to measure the development of the brain, and blame teenage behaviour on this in a rather determinist manner. We know now that the brain develops from back to front, and that the mechanisms of the frontal lobes act as the "brakes": the consideration of consequences of any given action. With an undeveloped "consideration" function, a teenager is unlikely to be able to make an accurate assessment of the consequences of his or her actions. Extrapolating from this, it is easy to fear the unpredictability of an irrational act that would imperil an entire off-world community, and in our time, this is seen as the most dangerous aspect of teenage development. Is it really as simple as this, though? Strengthening the cognitive-control mechanism occurs throughout adolescence, and there is a maturational gap between this and the remodeling of the brain's socio-emotional reward system. What is now better understood is the relationship between peer related stimuli sensitising the brain's reward system to the value of risk—this is (in part) a social process [5]. The picture becomes more complex as we realise the dynamic developmental relationship between behavioural practice and neurobiological development. (Susan Greenfield noted something similar in the development of the hippocampus in London taxi drivers. "Experiences, then, are reflected in the strength and extension of brain connections." [6]) Recent research by Casey and Caudle (2013) shows that teenagers perform better in terms of making decisions than adults and children, unless they are being subjected to peer pressure. In other words, when in the company, exclusively, of their peers, teenage risk taking becomes dangerous [7]. Albert et al's research demonstrates the attraction of risk among peer groups: "It's no fun doing it on your own." [8] There is a hugely increased propensity for risk taking when in the company of other adolescents. It seems that they are not able to negotiate "socio-emotional and choice related incentive cues" [9] and although high-risk behaviour shows high levels of testosterone (boys) and estradiol (in girls), levels of estradiol are lower than those of testosterone [10].

The potential here for good and ill is clear; I suggest that it would be possible to use this function to the benefit of the community by mixing groups—say of pilots—but allowing teens the responsibility of making decisions. The problem that then arises is the old issue of when and where a human should be considered mature—I suppose that my argument here is that such old definitions need to be suspended if and where the function of an adolescent might outstrip that of an older adult. There are obvious philosophical implications. It might be, however, that the dynamic relationship between cause and effect could have a very positive effect on development.

2.1.1 Where the hormones, there moan I . . .

It seems, therefore, that "specific hormones impact specific neural systems to influence specific behavioural tendencies" [11]. No surprises here, but what is interesting is that the level of the hormone may not be related to the biological signal that it gives. In adolescence there is a "neuro-behavioural nudge" towards social bonding. The oxytocin and vasopressin system, shown to be central to this, and development of social cognitive processes such as the development of trust, changes at puberty. Even if these changes appear to be small initially, over time they may have a greater impact. To explore this in greater depth, even though there is not much human research in this area, there appears to be a reflexive relationship between the social environment and hormonal levels. Research by Peper and Dahl (2013) shows a relationship between the mediating effect of the environment on behaviour and on levels and production of testosterone [12]. These effects can begin positive or negative trajectories that will manifest in later life—effective management of these to the benefit of the individual concerned then seems crucial:

Interactions between socio-affective processing and cognitive control 'heat up' considerably during puberty in the ventral striatum. If these interactions are positive, good things will ensure; if they are negative, and even if they begin in a small way, they may result in depression and substance abuse. These effects will not be 'one offs' they will 'cascade' across adolescence. [13]

Of further interest is the evidence that appears to show a relationship between socio-emotional hormonal activity and processing of guilt and shame—these seem to take place in the same area within the anterior temporal lobe [14]. As behavioural drivers, the toxic potential of guilt and shame is well documented, particularly in the growing field of psychosomatic trauma management [15].

2.2.2 Teenagers and mental health

The importance of the mental health of teenagers cannot be underestimated. It is acknowledged by health care professionals in the UK, that we do not care for our adolescents in a good way. Bearing in mind the positive and negative trajectories indicated by the reflexive relationships between environment and hormone production, it should be noted that one in five teens are likely to have mental health problems, and of these, fifty percent will continue into adulthood with the same issues. To a degree and with readily accessible help available to adults who can afford it, earth-based problems stand a chance of effective management, hopefully leading to resolution. Is this a statistic that could be so readily borne in space? I suggest not—and the research would indicate that it is possible to minimise the risks if we put our minds to it. Effective negotiation of this crucial developmental phase might include: watching for signs of depression (these may manifest themselves as irritability), managing peer interaction (balancing time and needs to mitigate the possible negative effects of negative peer pressure), being open about the neuro-behavioural nudge to the wider social world, and making informed and relaxed discussion spaces for the topic of sex. Safe spaces for effective anger management will be a big part of this, as emotions will be running high. We will need highly skilled emotional intelligence-related practices to negotiate a successful way through feelings, irrationality, and risky behaviour.

Decisions will need to be made, then, about what to include in a possible learning environment for teens. Topics and curricula that promote social cohesion while building trust and team work (Why not mention love and friendship—essential in a small community?) and encouraging the practice of conflict resolution and forgiveness, as well as personal strategies for releasing negative emotions are perfectly possible. The risk at the moment is that in the rush to ensure competence in the wide variety of scientific functions that will be necessary for survival, we will forget the human skills inherent in emotional intelligence. Knowledge has been advancing in these areas also, particularly in organisational development and conflict resolution. It is my contention that the period of adolescence will throw these into sharp relief, and it will be the responsibility of everyone in the community to help the teens in through and out the other side safely.

3. Strategies and a Curriculum for Emotional Intelligence

One of the most effective media for helping children work through emotion successfully is heightened play. This has been demonstrated in the work of Cicely Berry, director and voice coach with the Royal Shakespeare Company, resident in the UK. In the DVD Where Words Prevail, Berry works with a group of street children, mostly teenagers, in Rio de Janeiro. Using the language of Shakespeare, heightened language that makes it safe to articulate very strong feelings, students clearly and powerfully express the reality of their lived experience, achieving catharsis [16] or, if you prefer, an emotional detox. This type of practice could be very valuable in a small community—a safe space in which to articulate issues that could cause disruption, or a place to simply let off steam. What then shall we choose to take with us in terms of culture and cultural practice? Which great global authors would help to manage heightened emotion in this context? (Decisions about language will also need to be made, as at the moment we seem to be assuming that everything will be in English).

Music, too, is a powerful force for team work and social cohesion, shown perhaps most notably in the work of Jose Antonio Abreu and the "El Sistema" movement:

In my view, the root cause of all our social problems is exclusion. We must fight to ensure that a larger number of people – everyone, if that's possible - have access to this wonderful world – the world of music, the world of the orchestra, the world of song, the world of art. [17]

Abreu is very clear that the El Sistema method of teaching children from all backgrounds to play popularly named classical music (in the Western tradition) is the way forward for social cohesion and the practice of equality. Many other benefits are associated with music—again, it acts as a mechanism of emotional catharsis while teaching team work, listening, discipline, attentiveness, and fluency in the language of music, arguably as good for the brain as fluency in another language. But might this be a misplaced attempt to include an imperialist imposition in a small, off-world community? It would appear not. A practice that brings the community together to communicate in a different way from the exchange of information, including their emotions, might be an excellent way to encourage positive brain development in the teen. Might there be room on a spacecraft for the instruments? Why not? Many claims are made for El Sistema, [18] among them reduction in crime levels, social cohesion and better learning. Accessibility is open; there are no pre-existing entrance requirements. Although some criticise the system for its discipline, the movement is pretty much universally accepted as a very good thing, and has now been instituted in Scotland where it has revolutionised several socio-economically deprived areas. Given that the root cause of social unrest is the resentment coming from exclusion, Abreu seems to have achieved his aim in focussing communities on the positive advantages of music. Over 310,000 children have so far benefitted from El Sistema.

4. Conclusion

Paying attention to human development to a degree not currently practised on earth will be essential in an off-world context if the species is to survive. Psychology and neuro-biology has much to offer in terms of understanding the most difficult period in the trajectory of human development, but as yet the relationship of emotions and hormones has not been made explicit, and it is clear that the socio-affective influences the socio- cognitive, and vice versa, to the detriment or benefit of all concerned. To help teenagers achieve their best potential will benefit not only them, as individuals, but also, clearly, work to the benefit of the entire group—it is not only possible to enable this, but it should be considered as part of the role of every person present. Would this were the case on earth today! Teenagers and all interstellar humans will need every skill we have and are developing—it is a concern that with a focus on scientific and engineering competence, these skills will not be considered as teachable. It is possible to learn skills and capitalise on talent, but the thorny issues of a successfully negotiated adolescence and an emotionally intelligent emergent adult should not be left to chance. To do so would be to place the species at risk, if we really are relying on interstellar embryos to carry humanity into the stars.

References

1. Golding, W. "The Lord of the Flies," London: Faber and Faber Limited, 1954.

2. Verhoeven, P., Director. Total Recall. [Film]. USA: Carolco Pictures, 1990.

3. St. Louis, Connie. "Life as a Teenager," London: BBC Radio 4 http://www.bbc.co.uk/radio4/science/lifeasa-teenager.shtml, 2015.

4. Enfield, Harry. "Kevin the Teenager," Tiger Aspect for BBC Comedy, London, 1997.

5. Albert, D., Chein, J., and Steinberg, L. "The teenage brain: Peer influences on adolescent decision making," Current Directions in Psychological Sciences, vol. 22, no. 2, pp. 114-120, 2013.

6. Greenfield, S. "Mind, brain and consciousness," The British Journal of Psychiatry, vol. 181, no. 2, pp. 91-93, 2002.

7. Casey, B.J., Caudle, K. "The teenage brain: Self control," Current Directions in Psychological Science, vol. 22, no. 2, pp. 82-87, 2013.

8. Albert, D., Chein, J., and Steinberg, L. "The teenage brain: Peer influences on adolescent decision making," Current Directions in Psychological Sciences, vol. 22, no. 2, pp. 114-120, 2013.

9. Albert, D., Chein, J., and Steinberg, L. "The teenage brain: Peer influences on adolescent decision making," Current Directions in Psychological Sciences, vol. 22, no. 2, pp. 114-120, 2013.

10. Peper, J.S., Dahl, R.E. "The teenage brain: Surging hormones—brain-behaviour interactions during puberty," Current Directions in Psychological Science, vol. 22, no. 2, pp. 134-139, 2013.

11. Peper, J.S., Dahl, R.E. "The teenage brain: Surging hormones—brain-behaviour interactions during puberty," Current Directions in Psychological Science, vol. 22, no. 2, pp. 134-139, 2013.

12. Peper, J.S., Dahl, R.E. "The teenage brain: Surging hormones—brain-behaviour interactions during puberty," Current Directions in Psychological Science, vol. 22, no. 2, pp. 134-139, 2013.

13. Crone, E.A., Dahl, R.E. "Understanding adolescence as a period of social-affective engagement and goal flexibility," Nature Reviews Neuroscience, vol. 13, pp. 636-650, 2012.

14. Zahn, R., Moll, J., Krueger, F., Huey, E.D., Garrido, G., and Grafman, J. "Social concepts are represented in the superior anterior temporal cortex," Proceedings of the National Academy of Sciences, vol. 104, p. 6430–6435, 2007.

15. van der Kolk, B. "The Body Keeps the Score," London: Allen Lane (Penguin), 2014.

16. Berry, C. Where Words Prevail [DVD], London: Sojourner Media, 2005.

17. El Sistema - trailer for the upcoming film. [Film]. Venezuela: UmbrellaClassical https://www.youtube.com/watch?v=276oR_tEmbs, 2008.

18. Majno, M. "From the model of El Sistema in Venezuela to current applications," Annals Of The New York Academy Of Sciences, vol. 1252, "The Neurosciences and Music IV: Learning and Memory," pp. 56-64, 2012.

Loading, Please Wait:
Long-Distance Crisis Communications

Jaym Gates

Abstract

Communication in space, barring intense further advancements, will require a great deal of forward-thinking and planning, much like a military campaign. For the first few missions, the inhabitants of any expedition will be advance operating groups, nearly cut off from any support or rescue. They will need two-part communications plans: the ability to handle the immediate issues on board with only the resources at hand, and a longer-term communication with the home base.

This paper addresses some of the communications challenges presented by interstellar travel, and possible ways to address them, based on current military and space travel procedures.

Keywords

communication, systems, interstellar, space, community

1. Introduction

"In the beginning was the word..."

Communication forms the foundation of life. From the simple "reproduce" command, to complex interactional communications, it is what keeps the world moving. Humans, in particular, have built their supremacy on complex communication systems. The biggest leaps in human advancement have been on the back of communication breakthroughs: the written word, the printing press, the telephone, the internet. At the same time, those were brought about by the pressures of a growing, globalizing civilization. Communication and human evolution go hand-in-hand, an infinite loop.

Now, we find ourselves at the edge of a major cultural evolution, one which will, by necessity, drive a new evolution in communication, as well. And no, that isn't a question, but a certainty. Not only do we not have the technology or systems in place, we are only now beginning to put together the understanding of what that system will need to do.

Nor is it a question of whether or not we will, as a species, move onward from this planet, only a question of when, how, and what we will be leaving behind.

> "If we are to send people, it must be for a very good reason - and with a realistic understanding that almost certainly we will lose lives. Astronauts and Cosmonauts have always understood this. Nevertheless, there has been and will be no shortage of volunteers."
> —Carl Sagan, *Pale Blue Dot: A Vision of the Human Future in Space*

To create the tools to become an interstellar civilization, we have to develop a culture not only accepting of such a future, but one that demands it. We must be obsessed with the final frontier again, in a way that puts to shame any previous such interest. That obsession must infect our media, our creativity, and our very language.

Not if, but when.

And as we develop that culture, we must also be learning, ourselves, how humans communicate. It isn't enough to understand the modern methods of communication,; we must look at earlier examples of communication and community as well.

The successful management—or even survival— of any crisis is dependent on clear communication, and fast response. Calm, organized, prebuilt communications systems are a must if we are to survive the infancy of off-world civilization.

Even simple in-system, off-planet travel will be fraught with never-before-experienced challenges, much less the demands of actual interstellar travel. The conditions our early off-world efforts will lead to higher stress, more pressure, and a more explosive situation during a crisis, and the larger the team, the more stress points and vectors for crisis. In space, the results of those explosive situations could be the death knell of anyone trapped in them.

Crisis, tragedy, isolation, and stress will create a volatile environment, one that does not exist in the understanding of most first-world humans. To adequately prepare explorers, we will need to understand the psychology, communication, and community systems and development of war-torn countries, early pioneers, refugees, prisoners, and high-risk military operational groups. We will need to study and develop the communities that surround our risk-takers, early adopters, and creators.

Our history is filled with grand initiatives and hopeful reaches into the unknown, but nothing like what we're now contemplating. To launch our species into the stars, we will first need to win their hearts and minds, and then develop those hearts and minds to create the systems that will support an interstellar civilization.

And all of that necessitates a robust, agile, and integrated communications system, one that reaches well into both the past and the future. One that will require and has a substantial understanding andof the evolution of technology, psychology, creativity, and science, hand-in-hand.

2. An Overview of Crisis Communications

> "The report covers 15 years of failures and gets highly technical, but it's clear that poor communication, information overload, and bewildering PowerPoint slides take much of the blame for the loss of 13 lives."
> —"Two Misleading Words Triggered GM's Catastrophic Communication Breakdown," Carmine Gallo, Forbes

A crisis is a constantly-evolving geography composed of three primary elements and their interplay: humans, technology, and environment. Successful management of a crisis is achieved through measured, flexible, informed response to that changing landscape. It's all about the information, and the clear flow of that information.

The first 72 hours of any crisis are particularly crucial: Gather relevant information. Create a plan. Spread accurate information. Respond to crisis. Resolve the crisis. Gather information to prevent a repeat of that crisis.

It's all about the information, and the clear flow of that information. The successful management—or even survival—of any crisis is dependent on clear communication, and fast response.

However, fast and agile response is hampered if the proper training and stress-testing has not been doneoccurred in advance. In many cases, there is little to no time to prepare for specific crises, leaving responders scrambling to build or adapt their systems on the fly.

Although we cannot hope to predict even the majority of the crises we will face in the next hundred years, there is still a great deal of preparation and planning that can be done. Planning ahead will reduce stress, fallout, and lag, even for unexpected crises.

There are certain steps that are universal and essential to any crisis scenario:

- Anticipate potential crises
- Develop flexible response plans and resources
- Identify your response team
- Identify and train your figureheads
- Establish and stress-test notification and response systems
- Create fallbacks and resource caches, plan for outages and interruptions
- Create analog backups in case of technological failure
- Consider long-tail management

While these things are easy enough to develop for on-Earth crises, interstellar travel poses additional challenges that we, as a species, will not have faced before. Much of our initial process will be trial-and-error. Early Americans, from the Pilgrims to settlers traveling the Oregon Trail, died by the hundreds, despite available resources. In space, if a resource runs out, it is out. Such conditions will lead to higher stress, more pressure, and a more explosive situation during a crisis. However; however, in space, the results of those explosive situations could be the death knell of anyone trapped in them.

Calm, organized, prebuilt communications systems are a must if we are to survive the infancy of off-world civilization, but equally necessary is the mental adaptability to respond quickly and creatively, perhaps without very basic resources.

3. The Many Faces of a Communications System

It is tempting to break communication down into its component parts, to separate it out. Crisis communications, in particular, are often subject to this isolation. Unfortunately, in doing so, we remove a great deal of the development and support provided by the other areas. To create a strong communications environment, there must be consideration for, and communication between, the following areas:

Cultural: As mentioned in the introduction, we must have a culture that is willing to make the necessary sacrifices. We have, for years, focused on the minimization of human risk—through advanced weapons, automation, safety standards, and policy-making—but that will not be an option in the early days of our interstellar exploration. Things will go wrong. We will lose people and machines. We may lose many people, even if we never find another race, even if we do everything 'right'.

Culturally, we must be prepared for this certainty, and we must have the societal and organizational resilience in place to overcome it without losing our hope for the stars.

Community: The strongest support systems are community-based. Those communities may come in many forms—families, businesses, shared interests or backgrounds—and each will respond with micro differences and macro similarities.

As we start to look at generation ships and other large-scale exploratory efforts, communities will need to be educated and empowered to handle themselves in crisis.

Institutional: Institutions will have multidirectional communications patterns to be aware of—internal, internal to external, external to internal. The more central they are to a crisis, the more robust and diversified their plans and systems will need to be.

Personal: The most potential for improper disaster response comes on the personal level. The perception is that only the response or PR teams need to understand the situation or be briefed on how to deal with the press, families, or social media. However, the peripheral characters, due to their lack of coaching, are often the direct targets of those looking for unrehearsed answers.

4. Systems

> The Federal Emergency Management Agency (FEMA) wrote in its 2013 National Preparedness report that during and immediately following Hurricane Sandy, "users sent more than 20 million Sandy-related Twitter posts, or 'tweets,' despite the loss of cell phone service during the peak of the storm." [2]

As we begin to explore the stars, crises will happen, and communication will make the difference between a disaster being the spur to better technology and higher safety standards, or destroying humanity's desire to risk that final frontier. Yet, with so much necessity for it to be robust, we are still hard-hit by very mundane disasters.

We can see this in our recent natural disaster history. During Katrina, the communications infrastructure failed badly, as did local and governmental systems, leaving residents stranded without a line to the outside, or even basic information. The failure only became more critical as agencies began responding. According to the FCC, "more than 2,000 police, fire, and emergency medical service personnel were forced to communicate in single-channel mode, radio-to-radio, utilizing only three mutual aid frequencies." [3] These overloads hamstrung the response and recovery. The report also concluded that many of the problems were caused by a lack of training on the equipment and technology that could have aided the efforts, and called it "perhaps the most widespread critical infrastructure collapse that any advanced country has experienced since World War II".

Hurricane Sandy showed little improvement in the official systems, but a new player had entered the field, something in its absolute infancy during Katrina.

Social media showed its potential for crisis communications management as Sandy began to descend. People around the world watched through pictures and videos. The stress and fear was palpable, the impending sense of doom as the skies darkened and people rushed for final supplies. Reports of flooding, structure damage, and power outages flooded the feeds.

And then, through that long night, we watched through social media as the city went dark and the feeds became quiet. It was intense and terrifying from the outside, too.

But within the city, social media was providing a lifeline. As the communications and emergency response systems crumbled, people used social media to communicate with loved ones . . . and service providers used it to identify the worst damage and stress points. "At one point during the storm, we sent so many tweets to alert customers, we exceeded the [number] of tweets allowed per day," said PSE&G's Jorge Cardenas, vice president of asset management and centralized services.

However, a lack of technological training still hampers both civilians and responders, and the repeated collapse of communications systems during disaster tells us that something needs to change even to cope with our current needs, much less the growing threat of massive natural disasters.

With our current systems, we cannot even dream of leaving our planet in any numbers, much less our solar system. To reach any interstellar capacity, we will need to create massively better communications. To survive interstellar demands, we will need new technology, new systems, and, for the early years, near-autonomy of communications hardware. We will also need to provide widespread training of both staff and civilians in the use of these systems and tools, as well as redundancies and fail-safes.

The temptation, when looking at any sort of evolution, is to make it more complex, more agile, and more adaptable. The problem with complexity, however, is the support system that must be built around it. More humans with more specialized knowledge, more resources, more money, more chance for things to go wrong.

However, with the minimal resources and the extensive strain of distance and time, we will need to consider simplicity and redundancy, too. The more self-sustaining it is, the better its chances of functioning under the extreme demands that will be placed on it.

Sometimes, the strongest communications system is a bullhorn in the hands of a trusted community leader.

5. Community Preparedness

In July, 2014, a car backfired in the dry grass alongside a hill-country highway in Northern California. Within a few hours, most of the region was under mandatory emergency evacuation. The speed and unpredictability of the Sand Fire in the middle of ranch country meant that many people were forced to abandon livestock, pets, and personal belongings unsecured, as well as to leave without adequate resources.

The community had not been faced with such an abrupt crisis in the past, allowing for a close look at the process of a community mobilizing and responding to disaster.

Within three days, the community's page on Facebook had been closed to all outside posts, and was being used entirely to manage housing, supplies, information flow, and animal care, both for the evacuees, and the response teams.

Community members stepped in to provide vital support for the official personnel, using their skills to fill in the gaps, allowing fire teams to focus on defending structures and suppressing a volatile and aggressive fire. Leaders and managers were identified, taking on roles in directing the resources and needs.

About two months later, an arsonist triggered another fire on the other side of town. This one grew to monstrous proportions overnight, eventually burning nearly 100,000 acres within two weeks, and commanding over 8,000 personnel from as far away as Alaska and Hawaii, a larger resource pool than any other wildfire in United States history. The smoke plume was visible from the other side of the state… and from space.

This time, within three hours, the community had rallied the systems tested during the Sand Fire. Posting on the community page was restricted to moderators, who carefully managed daily evacuation notices, evacuee resources, fire movement, air quality, news, and animal care. They rotated shifts, monitoring the official channels, interfacing with city, county, and crisis officials.

These systems supported over three thousand evacuees and their stock and pets, as well as over eight thousand firefighters and their support personnel, over the course of three weeks.

They were so successful that the news organizations were turning to the page moderators for information. As soon as the containment began, the same resources slid smoothly into place to assist the recovery.

None of these people have had any training in crisis response or management.

5.1 Community Command Structures

Any community under pressure will, over time, develop some form of command structure. This may go against established command strategies, or develop as a sub-structure of them. This structure can be a community's best asset or worst liability. It is not always the best-informed or most-competent person who takes the lead, and if someone without the correct skills and knowledge is in charge when a crisis arises, catastrophe is likely.

In such cases, the community will find leaders who are capable of responding decisively and accurately. It is, therefore, wise to ensure that, in a situation as potentially deadly as off-world travel, these assets and liabilities are discovered and developed early on.

Additionally, the question arises: what is the breakdown of civilian versus military assets? If the community is primarily civilian, there will be a longer response time as assets are identified and leaderships are developed, but there is a potential for faster individual response and stronger community involvement in response and recovery.

If it is primarily military, a great deal of that training will be in place already (making an excellent argument for prioritizing military background in the selection process for early missions). Stress will be expected and familiar, even if the circumstances are not.

Better yet is a community with a strong veteran presence. They are already integrated, and bring a wide variety of backgrounds to the table, but have the training and mindset to respond to high-stress situations.

Additionally, this would minimize stress and maximize communication between civilian and military components to any mission. In a crisis, the military component on a mission or ship is the best-suited to response, but that will require clear communication and cooperation with and from the civilian element.

But no matter the strength of the command system, emergency response will fail if it is not coupled with equal and cooperative support from within the community. Training and empowerment must be provided to all levels of the structure.

6. Social Media, News, and Crisis Culture

With the rise and spread of social media, individuals may now have more voice and reach than ever before. While this has many positives, care must be taken to plan for the negatives, as well.

Every individual will have family, friends and loved ones, but they may now also have a strong following on social media. Look, for example, at "Mohawk Guy" from the Mars Curiosity mission. An instant and unexpected celebrity, sought out for interviews, guest appearances, and more.

Similar things will happen in a crisis. We can't predict the spokespeople of a crisis, but we can identify hubs of communication and information. In any group, there are people who connect important resources to create a clear and concise stream of information. These people must be given the resources and training necessary to act as hubs and support structures.

One of the biggest issues we run into in the intersection of media and crisis management is the virulent spread of misinformation. While we've always had, as a race, a tendency to favor sensationalism over fact, we have, in the past, at least had some control over the spread of information, and a relatively small pool of outlets to be monitored.

With the advent of the internet, village gossip became a business, and the number of outlets to monitor grew exponentially. With the advent of social media, those outlets exploded. Everyone's an expert now, and it is harder than ever to control and maintainmaintains an accurate flow of information. While that can be useful—as demonstrated by the Arab Spring, in particular—it is potentially lethal in a crisis, and the misinformation can be difficult or impossible to contain and correct.

During Hurricane Sandy, misinformation and scams were cropping up everywhere, and being spread by well-meaning but uninformed people. FEMA responded by setting up a page to combat and correct these rumors, combing social media to catch as many as possible, but it is never possible to catch them all. Even one uncorrected rumor will add stress to the response teams. [4]

Additionally, in the case of an interstellar disaster, the world will have its eyes firmly fixed on every detail. Speculation, accusation, and misinformation will run rampant, and the resources necessary to counter all of that would simply be impossible to maintain. Overly-direct, negative response can, all too often, appear to be defensiveness, damaging the communication and trust between the parties.

We may be fortunate, and disaster may not happen for decades or longer, but it will happen eventually. In the event of a massive disaster, it is necessary to create an immediate response plan, but to gear the public relations efforts on a much larger scale. A long-term plan to regain trust and support is necessary before a single human leaves orbit again.

The strongest public relations plan will not only prepare for the crisis itself, but understand that the recovery effort is on a much slower timeline than the initial, negative response. Repair may take weeks, months, even years, but handled correctly, it could become a net gain.

7. Conclusion

In conclusion, it is clear that it is not only our technology that holds us back, but our very perception of the importance of communication. Until we are able to take the time to overhaul our communications systems, to really explore and understand them before creating better, stronger resources, we do not dare reach much further than we already have.

For now, we must ask the hard questions about where our priorities lie, and if we are willing to make the cultural and political changes necessary to ensure that our first reach to the stars isn't the last we'll see in our lifetime.

References

1. http://www.scientificamerican.com/article/how-social-media-is-changing-disaster-response/

2. http://www.carlisle.army.mil/DIME/documents/Hurricane%20Katrina%20Communications%20&%20Infrastructure%20Impacts.pdf

3. http://www.carlisle.army.mil/DIME/documents/HurricaneKatrina/Communications.pdf

4. http://www.fema.gov/hurricane-sandy-rumor-control

Dawn of the African Space Era

Mmboneni Muofhe

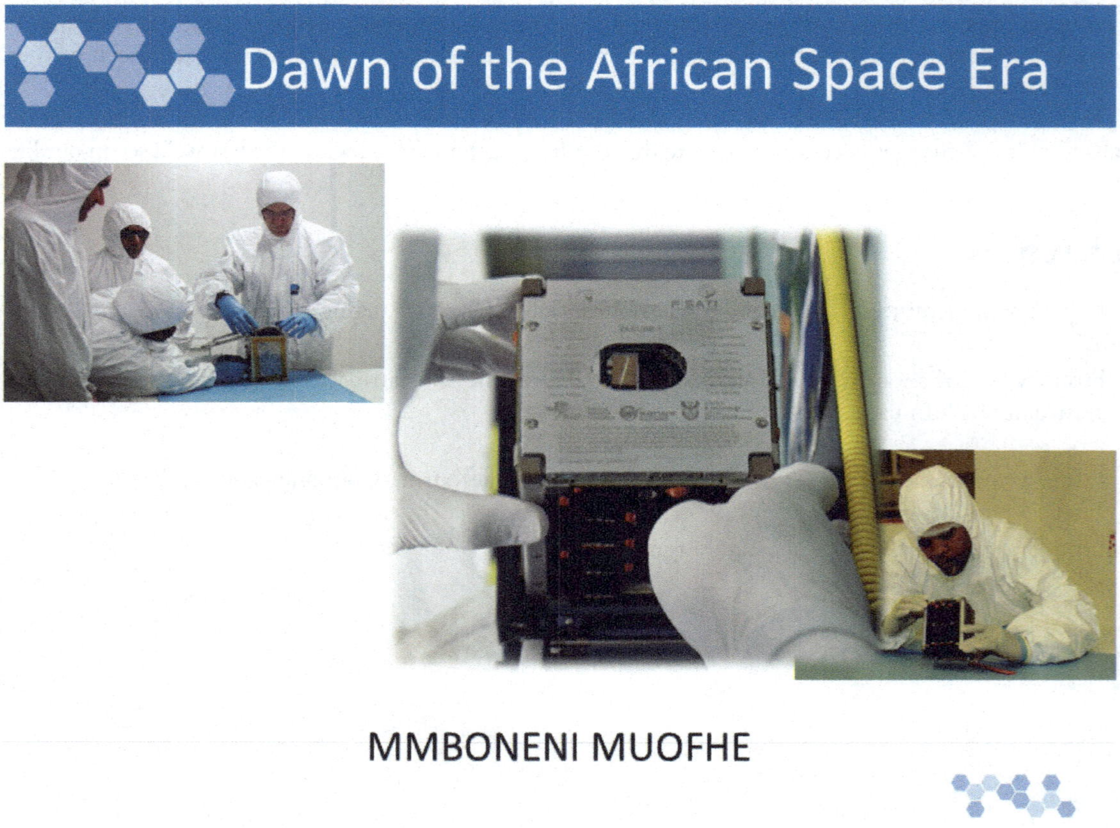

Countries with Space Programmmes

- Nigeria
- South Africa
- Egypt
- Morocco
- Algeria

Towards a continental Agency

- African Space Policy
- African Space Strategy
- Establishing a PAN African Space University
  - South Africa to host
  - Finalising the model
  - Identified possible host institution
  - Negotiating host agreements
  - Planning a needs assessment

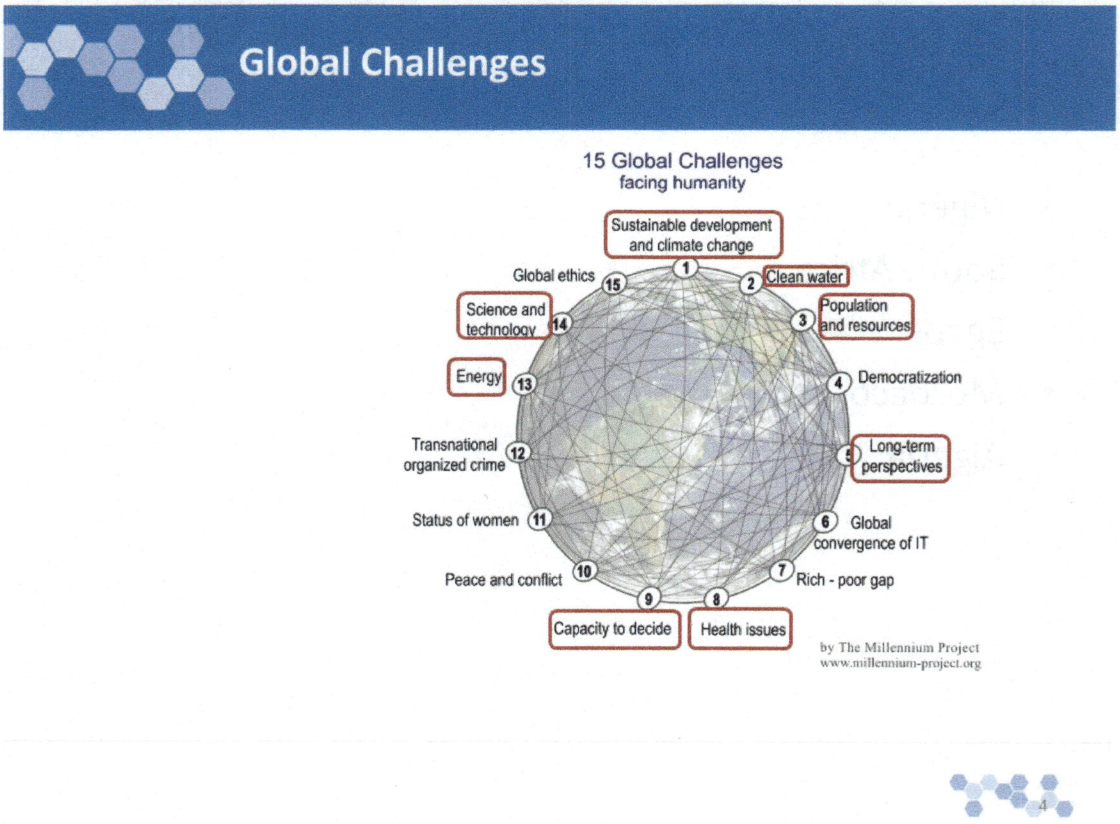
Global Challenges
15 Global Challenges
facing humanity
Sustainable development and climate change
Clean water
Population and resources
Democratization
Long-term perspectives
Global convergence of IT
Rich - poor gap
Health issues
Capacity to decide
Peace and conflict
Status of women
Transnational organized crime
Energy
Science and technology
Global ethics
by The Millennium Project
www.millennium-project.org
4

Building a knowledge economy
Technology Development and Innovation
Farmer to Pharma
Space Science
Energy
Climate Change
Human & Social Science
Human Capital
Centers of Excellence, SA Research chairs initiative, professional development program, etc.
Knowledge Infrastructure
Universities, Science councils, state-owned enterprises, global projects, etc.
DST 10-year Innovation Plan

Tshepiso – What has been achieved

50 Master's graduates

15 Engineers-in-training

> 15 000 learners reached over past 5 years

17 peer reviewed journal papers

60 conference papers

Innovation – Africa Space Innovation Centre (ASIC)

SARChI Research Chair

Hi-tech job creation (8 positions in ASIC)

Equipment and world class infrastructure > R10 million

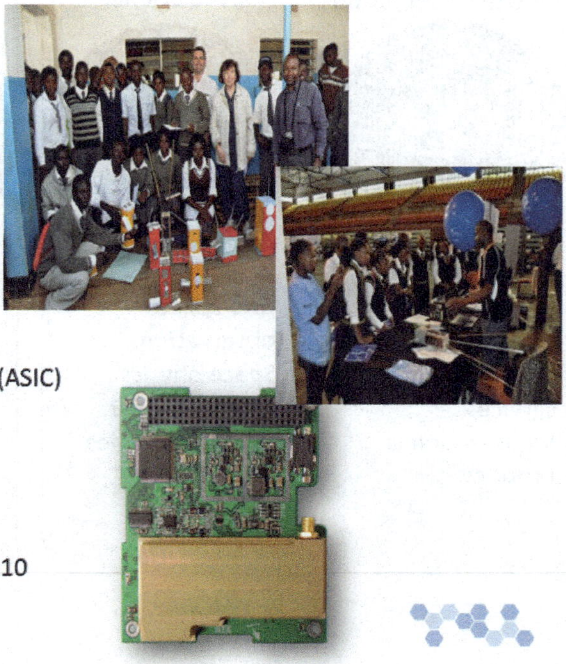

Multi-National Institute

science & technology
Department: Science and Technology
REPUBLIC OF SOUTH AFRICA

sansa
SOUTH AFRICAN NATIONAL SPACE AGENCY

UNIVERSITEIT STELLENBOSCH UNIVERSITY

UF UNIVERSITY of FLORIDA

CAL POLY

UNIVERSITY OF THE WITWATERSRAND JOHANNESBURG

groupe cesi

TECHNOLOGY AND DEVELOPMENT

gt GHANA TELECOM

EADS astrium

CLYDE SPACE

9

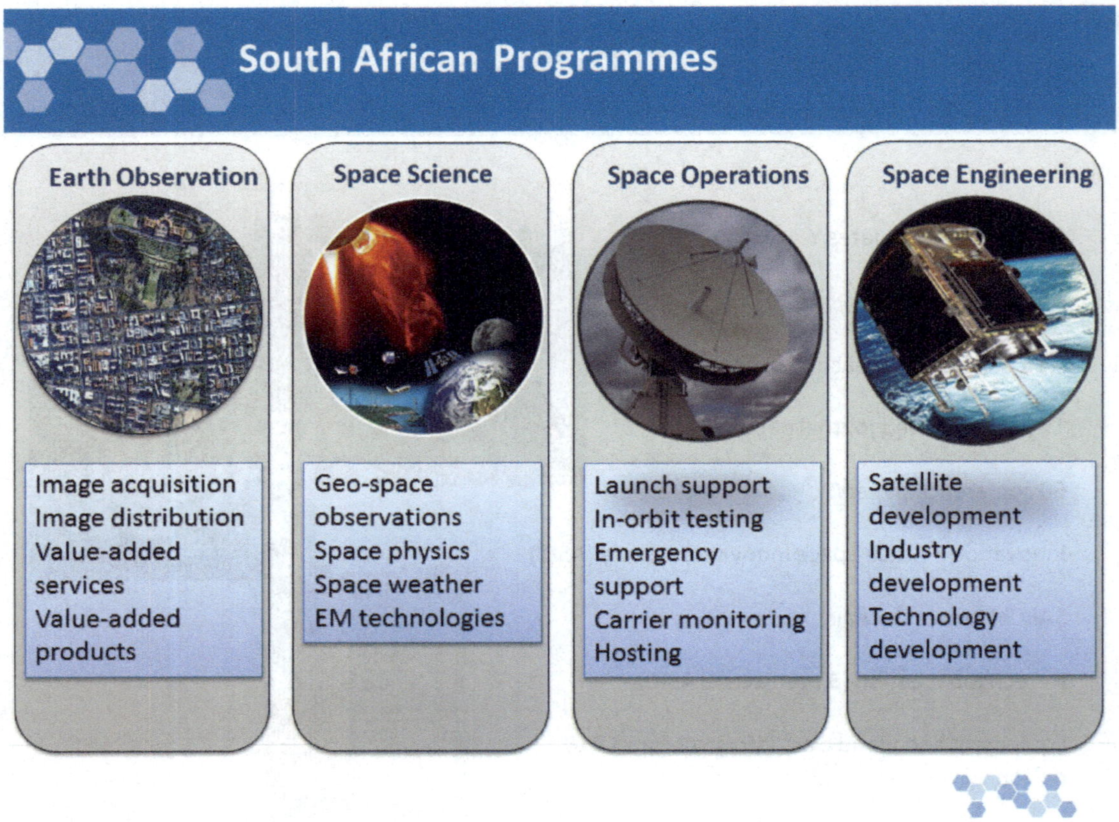

- BUIDING ON A SUCCESFUL CUBESAT PROGRAMME

Student demographics

Current missions: ZACUBE-2

- S-Band Transmitter
- ADCS System
- VHF/UHF Communication System
- Vessel tracking payload
- **Technology demonstrator for Operation Phakisa**
- 5 MP Camera Payload
- Possible IR camera
- Engineering model 2015

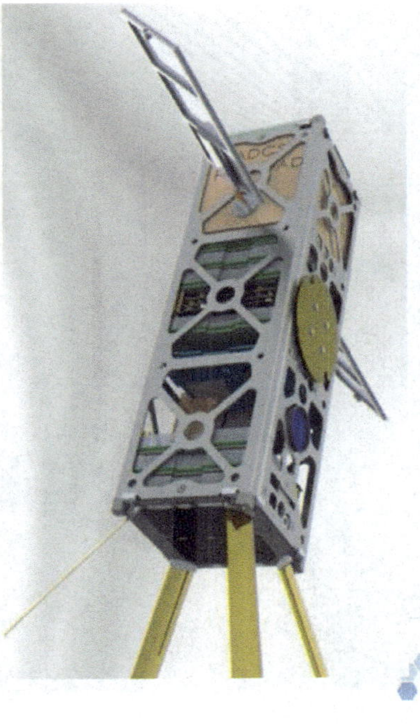

Making STEM cool

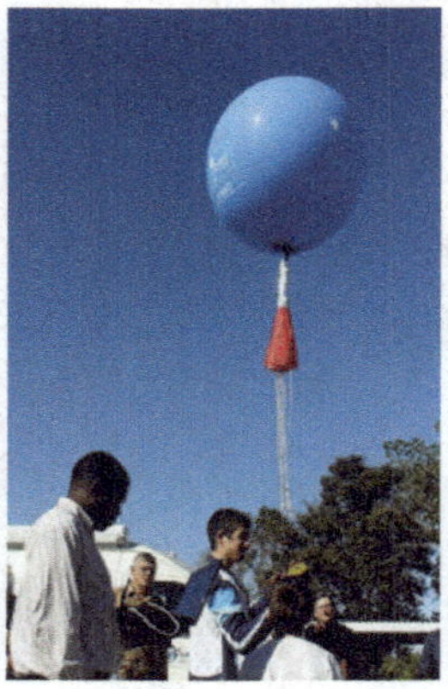

South Africa, 2013 Zambia, 2012

Satellite Development Facilities
FSATI, CPUT

Soldering Area
PCB polulation

Satellite Production Lab
FM Model Area + Clean Room

Clean Room
Class 100 000

Microwave and RF Labs
Additional computer and lecture rooms

Satellite Development Lab
12 Workspaces + Store room

Post Graduate Facilities
48 research spaces for Masters and Doctoral students

Post Graduate Facilities
48 research spaces for Masters and Doctoral students

The earth – our future

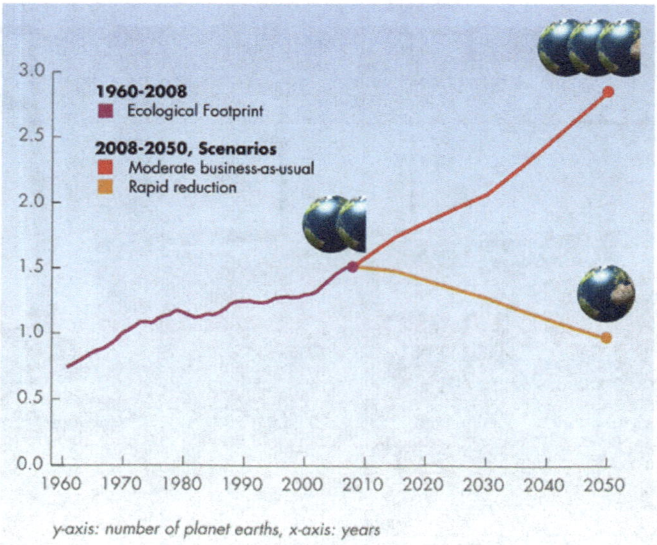

21

Fire Warning Services

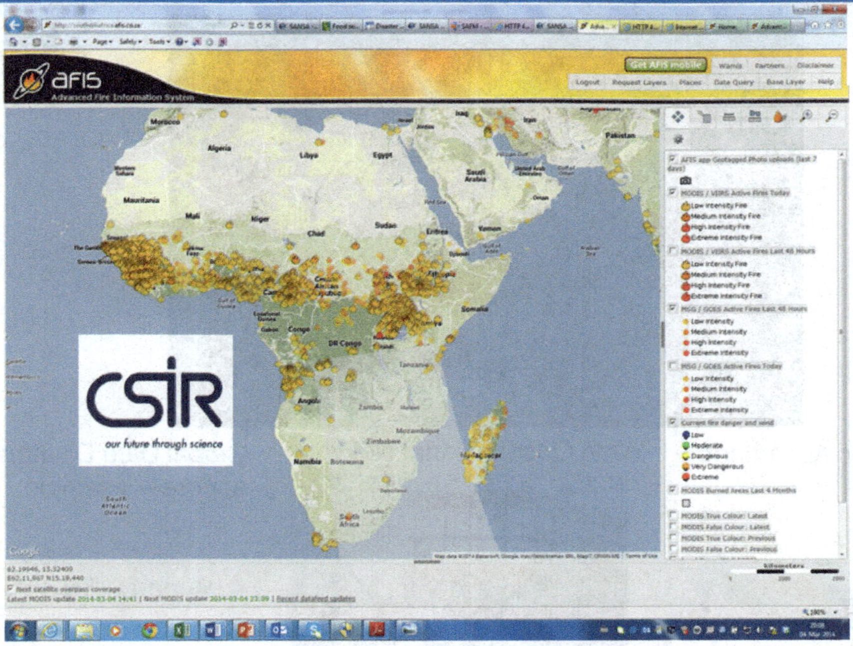

22

Example: Urbanisation Monitoring & Planning

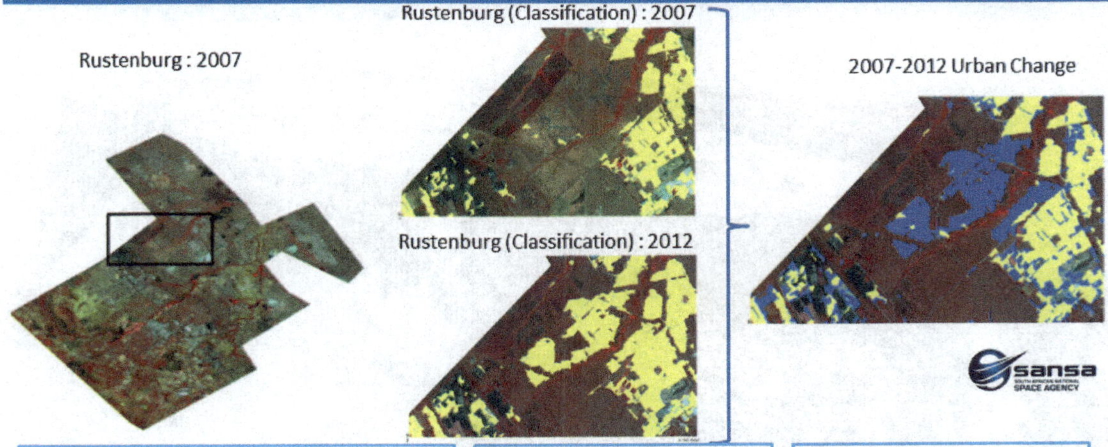

Problem

- Global urban population of 47% (2000), projected to increase to 67% (2050)
- 90% of growth is in developing countries
- Africa has the fasted urbanisation rate @ 3.5%
- Current African urban population of about 33% forecasted to increase to 50% by 2030

Consequences

- Inadequate amenities
- Strain on natural resources
- Social strife and strikes
- Poverty and crime
- Pollution

Space Solution

- Intelligence for urban planning
- Development of SMART cities
- Identification of problem areas

23

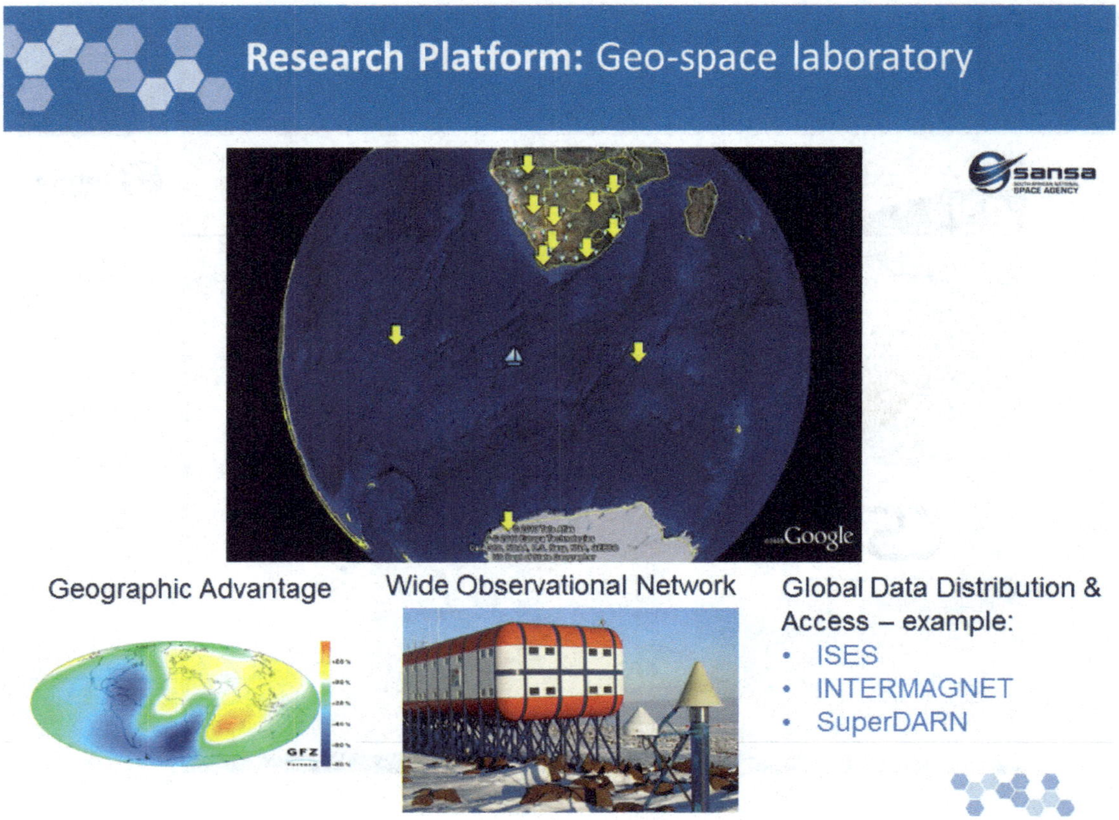
Research Platform: Geo-space laboratory
sansa
SOUTH AFRICAN NATIONAL SPACE AGENCY
Google
Geographic Advantage
Wide Observational Network
Global Data Distribution & Access – example:
• ISES
• INTERMAGNET
• SuperDARN
GFZ

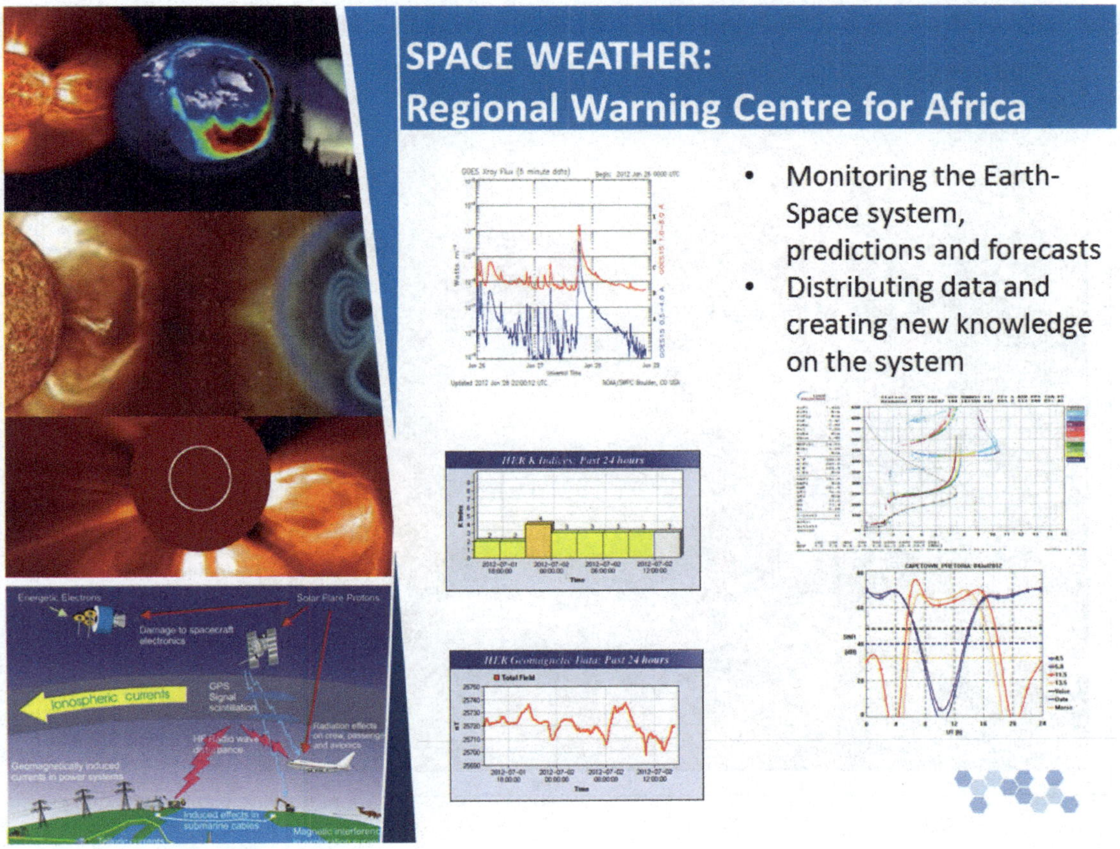

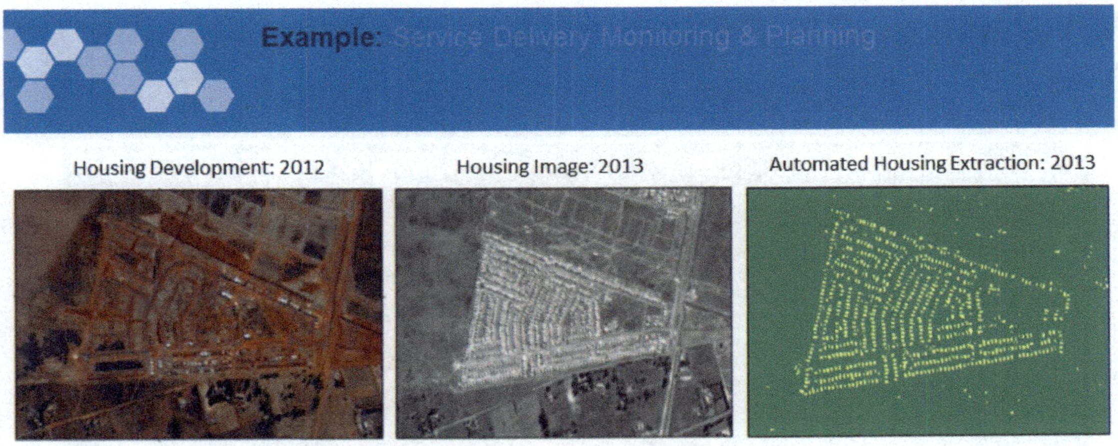

Potential Usage

- Independent report on progress made on infrastructure development
- Information to support population estimates
- Information to support infrastructure planning
- Information to support policies on environmental, natural resources and disaster management.

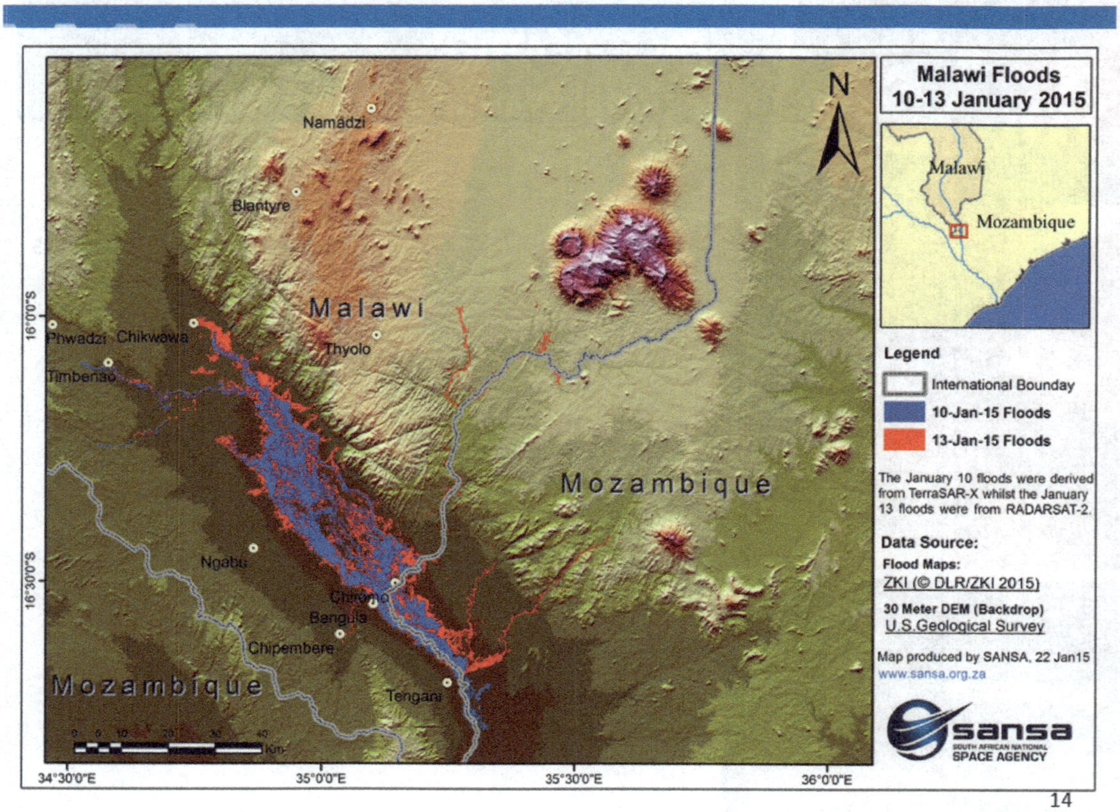
Malawi Floods
10-13 January 2015
Malawi
Mozambique
Namadzi
Blantyre
Chikwawa
Thyolo
Phwadzi
Timbenao
Ngabu
Chiromo
Bangula
Chipembere
Tengani
N
Legend
International Bounday
10-Jan-15 Floods
13-Jan-15 Floods
The January 10 floods were derived from TerraSAR-X whilst the January 13 floods were from RADARSAT-2.
Data Source:
Flood Maps:
ZKI (© DLR/ZKI 2015)
30 Meter DEM (Backdrop)
U.S.Geological Survey
Map produced by SANSA, 22 Jan15
www.sansa.org.za
sansa
SOUTH AFRICAN NATIONAL SPACE AGENCY
16°0'0"S
16°30'0"S
34°30'0"E
35°0'0"E
35°30'0"E
36°0'0"E
Km
14

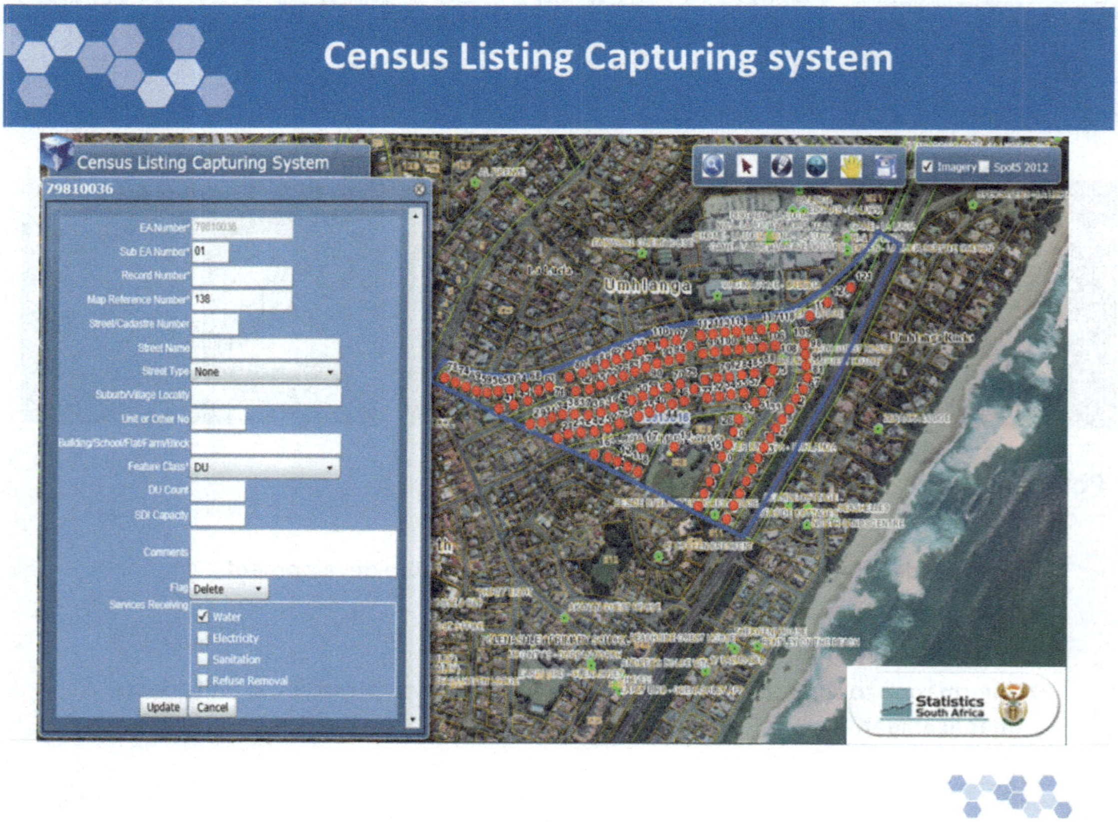
Census Listing Capturing system
Census Listing Capturing System
79810036
EA Number*
Sub EA Number*
Record Number*
Map Reference Number*
Street/Cadastre Number
Street Name
Street Type
Suburb/Village Locality
Unit or Other No
Building/School/Flat/Farm/Block
Feature Class*
DU Count
SDI Capacity
Comments
Flag
Services Receiving
Water
Electricity
Sanitation
Refuse Removal
Update
Cancel
Imagery
Spot5 2012
Umhlanga
Statistics South Africa

Space and election management

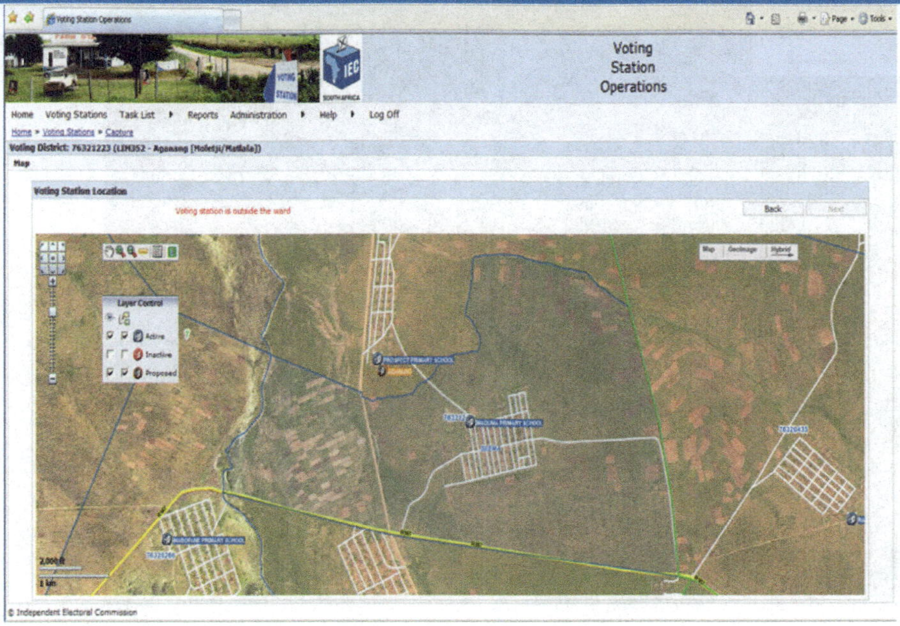

11

Residential Encroachment

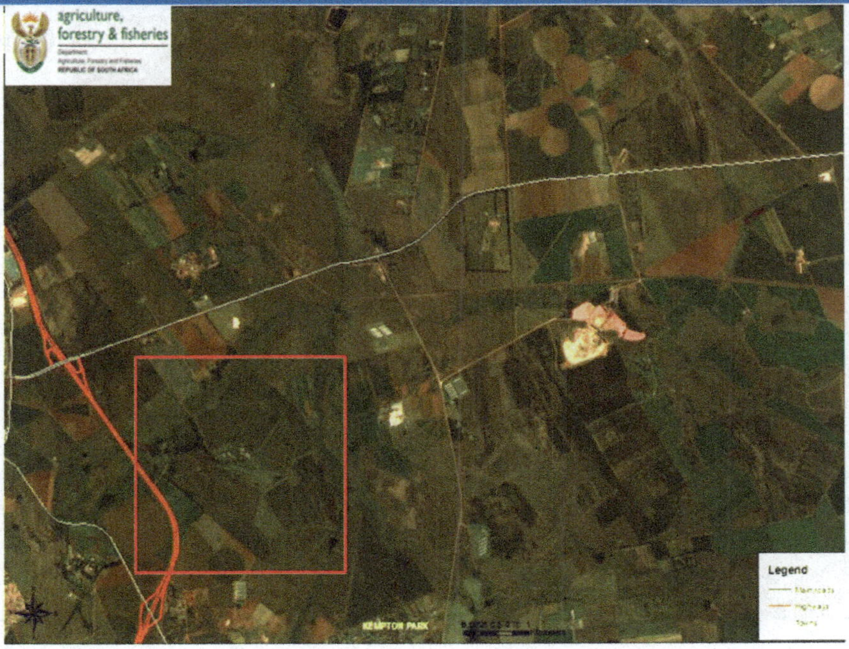

31

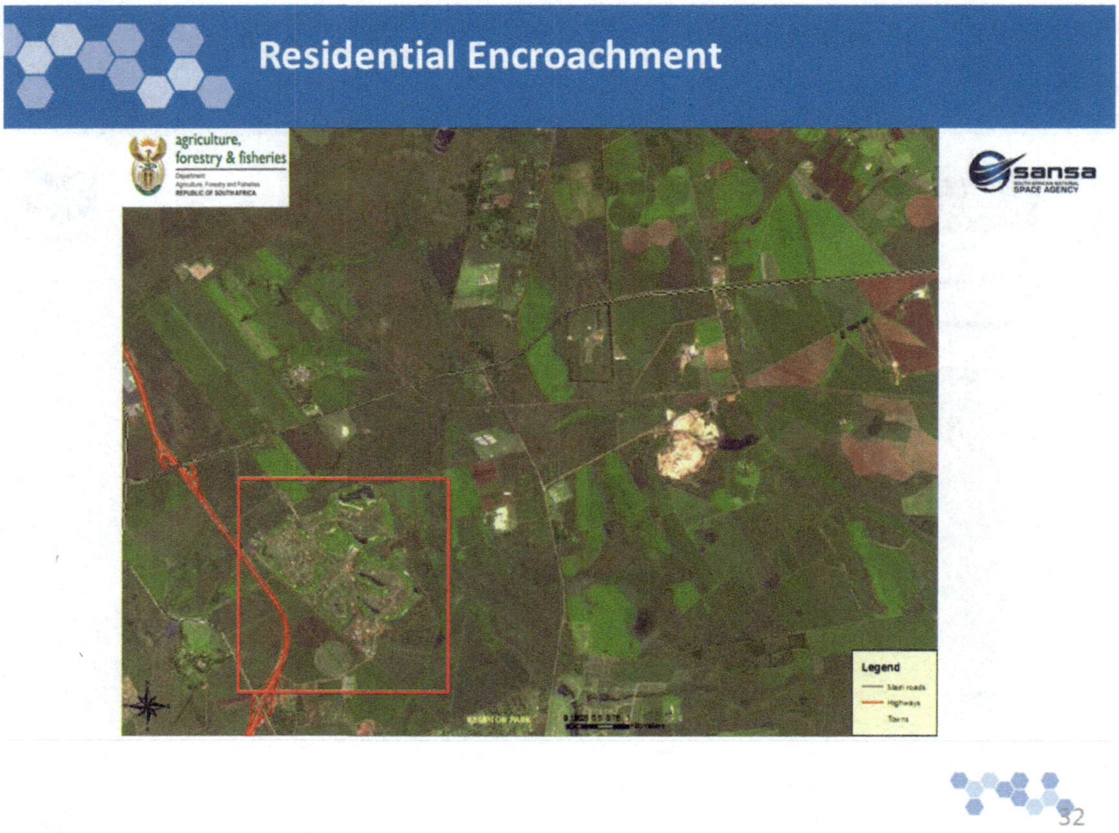
Residential Encroachment
agriculture,
forestry & fisheries
sansa
SOUTH AFRICAN NATIONAL SPACE AGENCY
Legend
32

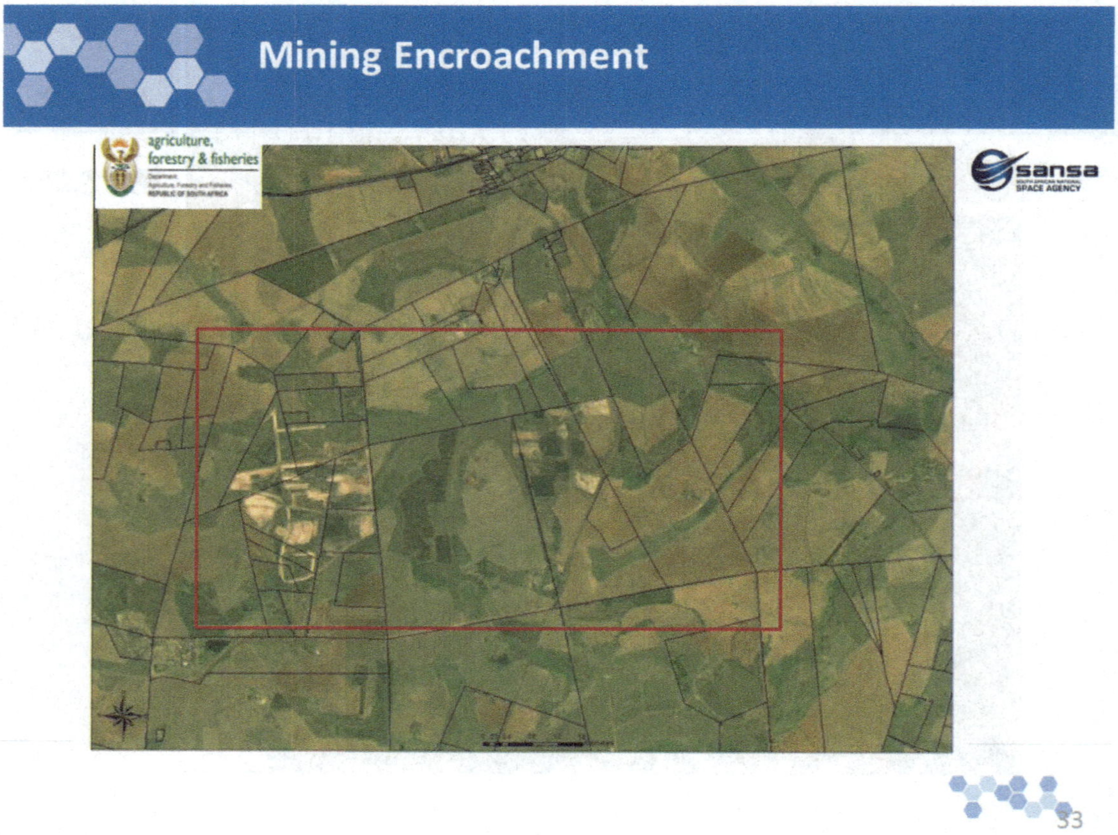
Mining Encroachment
agriculture,
forestry & fisheries
sansa
SOUTH AFRICAN NATIONAL SPACE AGENCY
33

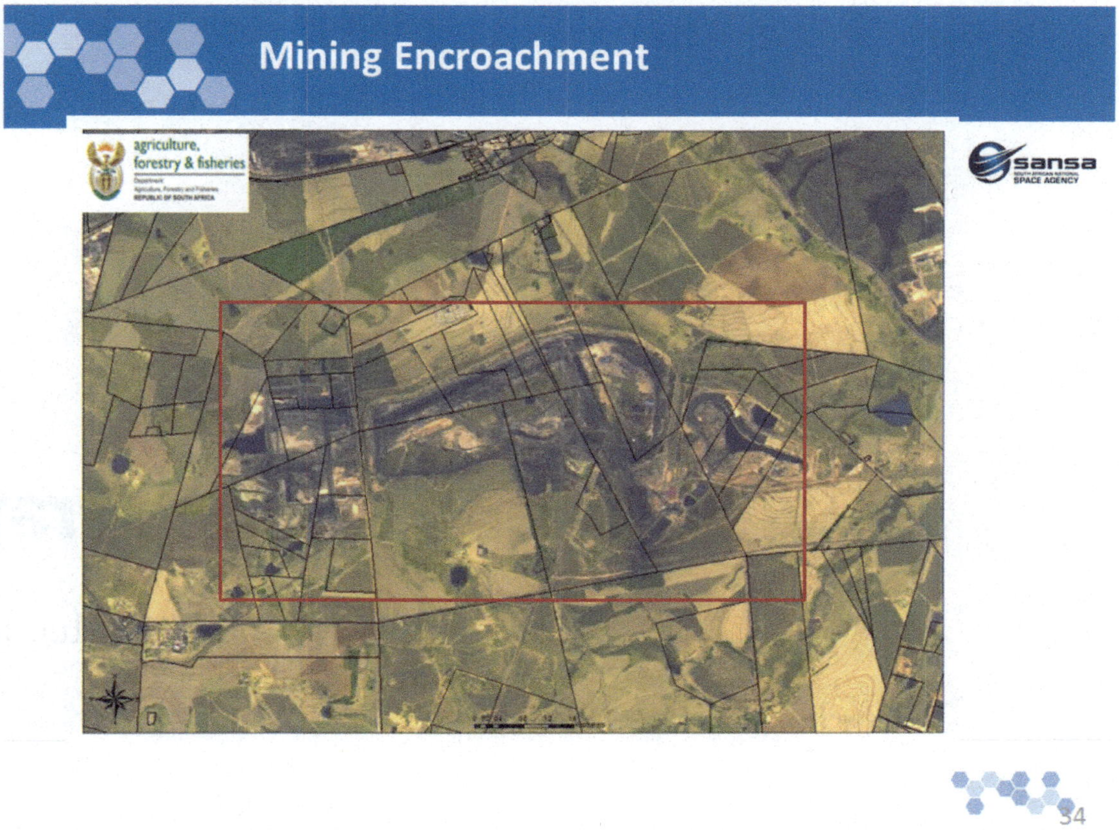
Mining Encroachment
agriculture,
forestry & fisheries
REPUBLIC OF SOUTH AFRICA
sansa
SOUTH AFRICAN NATIONAL
SPACE AGENCY
34

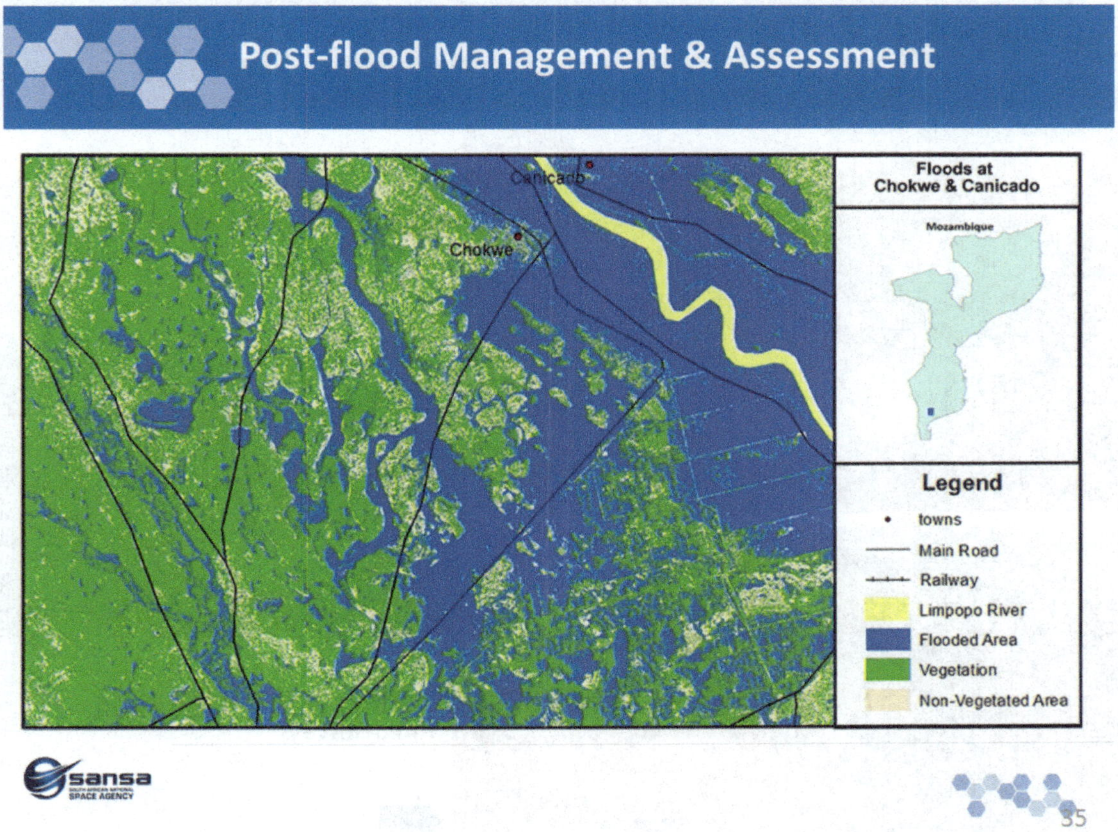
Post-flood Management & Assessment
Canicado
Chokwe
Floods at
Chokwe & Canicado
Mozambique
Legend
towns
Main Road
Railway
Limpopo River
Flooded Area
Vegetation
Non-Vegetated Area
sansa
SOUTH AFRICAN NATIONAL
SPACE AGENCY
35

The Square Kilometer Array

Mmboneni Muofhe

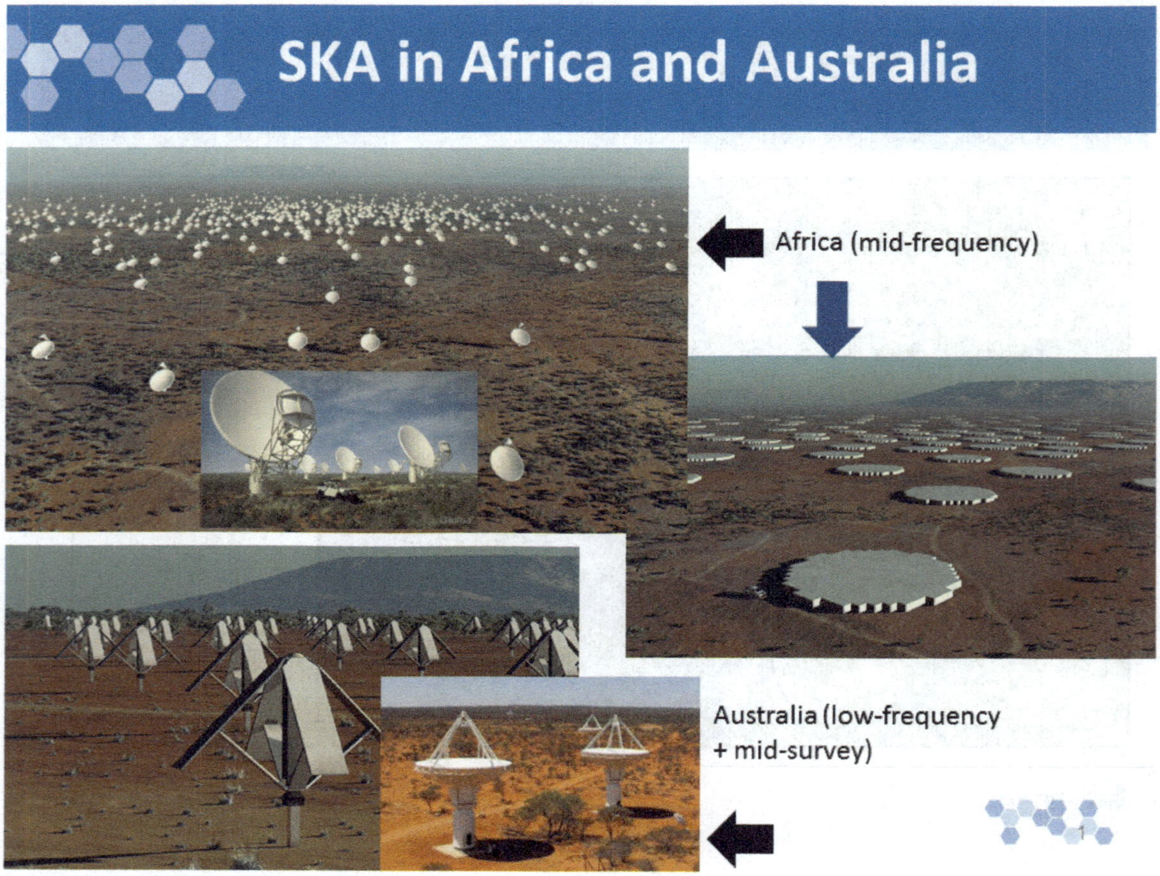

Big Data Africa Programme

- The SKA will be a Big Data machine (exascale) – the ultimate prize will be the discoveries from the data
- Big Data phenomenon –collection of data sets so large and complex that it is difficult to process using traditional methods & technologies.
- Complexity = Function (volume, velocity, variety and veracity)
- Data has become a natural resource (data mining) – value addition is critical and therefore the new set of skills to do this

MeerKAT Data Processing

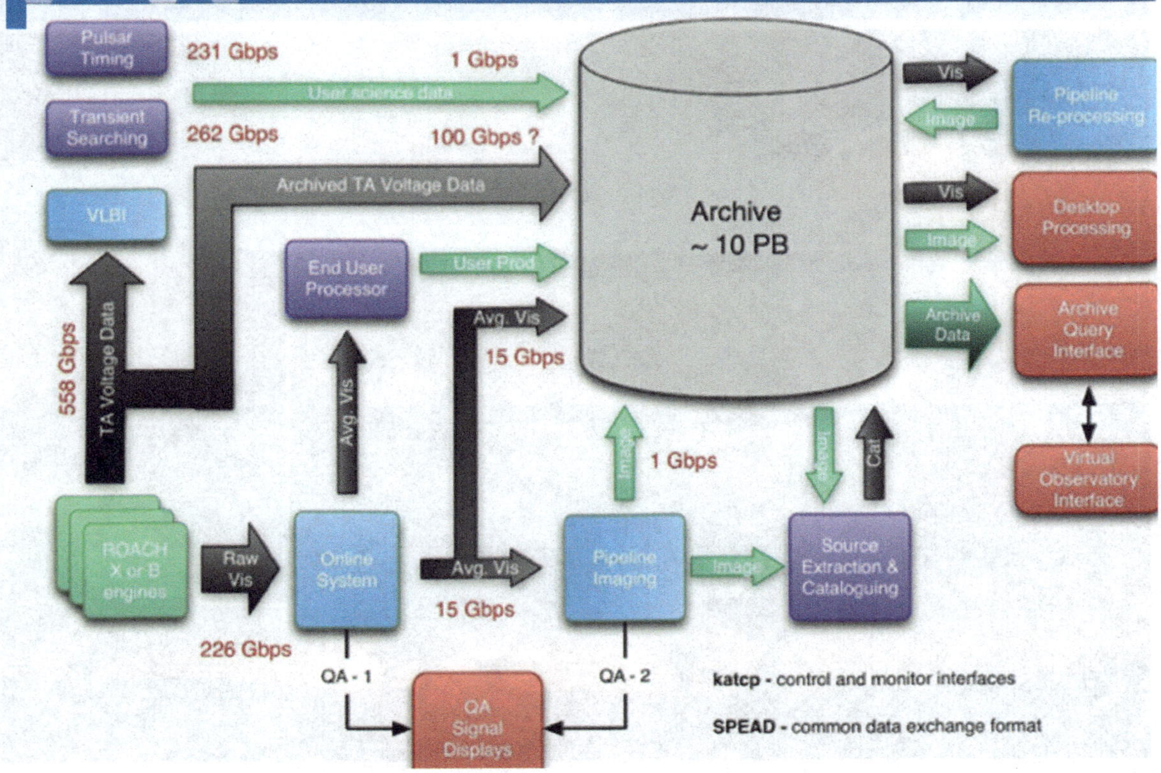

Big Data Initiatives

- Research Chair in Big Data at UCT
- Recently launched a R50m Inter-varsity Institute for Data-Instensive Astronomy (IDIA) on 3 September 2015 – Universities involved include UCT, UCT and NWU
- IBM established a R 700m Research Lab with a big focus on Big Data at Wits University.
- CISCO – set up a R 50m optic fibre research centre at Nelson Mandela Metropolitan University – to find cost-effective solutions for data transport

Starship Alpha

The Case for an Earth-based Proto-starship

Manuel Richey

Abstract

This paper proposes the creation of a self sufficient earth-based colony to simulate the effects of long-term space travel as would be encountered on a generational starship or other space based colony. The purposes for creating such a colony are identified. Issues are identified that can be addressed though the creation of such a colony (as well as those that can't). It is shown that creation of an earth-based colony can address most of the unknowns of long term space flight, leaving only a small list of purely technical challenges.

This paper describes one approach to conducting such an endeavor, what can be learned from it, and how to finance it. It is not assumed that such an endeavor can be planned flawlessly up front. Rather the described approach assumes an initial plan with refinement over time during its execution. Its final outcome (after several years) will ultimately be a self-sustaining colony. Financing for the endeavor is discussed to demonstrate project viability. It will come via NASA sponsorship, corporate donations, volunteer colonists supporting themselves through their own labor, crowd-sourced fundraising, TV contracts and enthusiasm such as that created by the Mars One project. From this 1st generation colony, we'll learn what can be produced and what must be transported; we'll learn solutions to physical and social issues pertaining to colony wellbeing and self sustainment; we'll then build the 2nd generation colony off-planet.

Keywords

proto-starship, colony, prototype, generational, simulation

1. Introduction

When Earth 2.0 is finally identified, a journey for exploration or colonization to this planet will traverse considerable distance and require a significant amount of time. With future technology, this trip may still require several generations to complete. The issues that must be surmounted for the creation of a viable space colony or generational starship are numerous. Examples are: propulsion, shielding, construction, political organization, crew psychological wellbeing, agriculture, local manufacturing, information technology, communication, environmental control, recycling, financing, equipment maintenance, etc. Most of these issues should be modeled in a closed system on earth prior to the construction of an actual starship or large space colony. Attempts have been made to do this in the past, such as Biosphere-2 [1] or HI-SEAS [2]. However, no attempt has yet been made to simulate a large (100+ person) self-sustaining colony over an extended (multi-year) period of time.

The time is right to make an attempt at creating a self-sustaining space colony here on earth; to learn what can be learned before an actual colony of such size is attempted in space. When a self-sustaining colony can be maintained under favorable conditions on earth, whose only imported raw material is electricity, then many of the non-propulsion issues involved in the creation of a generational starship or space station will have been addressed.

Can such a colony be constructed today and maintained using existing technology with limited finances? Would such an endeavor be worth attempting? This paper says YES to these questions, describing who, what, when, where, why and how. The biggest of these questions are why and how, but let's start with what.

2. What is an Earth-based Proto-starship?

Before a generational starship or large colony is developed in space, it should be prototyped on earth where adjustments can be made and the consequences of failure and catastrophe are reduced. This prototype is an earth-based proto-starship (Starship Alpha). It is stationary facility that emulates the living area of a generational starship or space colony. Indeed, in this paper we use the term proto-starship and earth-based colony interchangeably. A truly successful self-sustaining proto-starship or colony is a pressure vessel that contains and supports a sufficiently large group of humans with only one imported raw material, and that is electricity. Everything else is recycled. An information link to the outside world would also be provided, but we won't treat information as a raw material in this paper.

3. Why Create an Earth-based Colony

The concepts of a self-sustaining space colony for off planet expansion, or of a generational starship to reach a distant home, are well known and popularly written about in fiction. But is either of these concepts really even feasible? We don't yet know. We won't know for sure until we have successfully built one. We cannot predict how many failures will be required before a successful approach is concluded. Even if we think we have solutions to all of the problems we can anticipate, our efforts will undoubtedly be hampered by problems we cannot anticipate.

The purpose for creating a colony on earth to simulate either a generational starship or a space colony would be simply to test out theories and hypothetical solutions to the kinds of problems such a colony would encounter in the hostile environment of space. It makes sense to first encounter these problems in a friendly environment, where they can be overcome with less risk. The advantages of construction cost, the risk to life and limb, and the reduced consequences of failure all decisively indicate that the very first space colony should be built and populated on earth.

There are two approaches used for mission planning in space. The typical approach is top down, where all possible problems are identified upfront, and possible solutions are brainstormed. The best of these solutions are then selected for implementation. A less favored approach is bottom up or empirical. Under this approach, a problem is presented to a heavily vested group, which solves the problem via an initial approach followed by iterations to correct deficiencies. Due to the nature and risk associated with space flight, the top down approach to mission planning is generally used. Though both approaches have strengths and weaknesses for addressing problems, the bottom up approach of solving problems empirically is seldom available for endeavors in space. An earth-based space colony presents the opportunity for a bottom up approach to solving many of the problems associated with long duration space flight.

The technology exists to create this colony today. We need not wait. For many issues, this won't be an attempt to test already arrived at solutions, but rather a test bed for creating solutions to problems both foreseen and unforeseen. Not all problems that the colony could encounter need be solved at the outset. Many of these problems will be solved as they are encountered by the colony itself. Both the problems and the solutions can then be documented and reused. As Werner Von Braun said, "One good test is worth a thousand expert opinions."

As an example, the technical aspects of atmospheric management are not addressed by this paper, but would be addressed during this colony experience. The colony provides the laboratory and the rats (colonists) for those who are working on the technical aspects of atmospheric management. This is done outside of the starship by the support team. This starts with normal temperature and humidity control as is found in any building. However, it gradually transitions to a completely closed atmosphere through means which are not described here. When the starship atmosphere finally becomes self contained and remains so for a long duration, then a solution for atmospheric management has been achieved. Much of this activity can be performed by external support personnel without actually disrupting the colonists or risking their health, and can be completed long after the colony is initiated.

So what aspects of long duration space flight are addressed by creating a prototype colony on earth? That question is better answered by asking what aspects cannot be addressed with a prototype colony. Here is that surprisingly small list:

- Starship propulsion. Creating a stationary earth-based colony will not help further our knowledge on propulsion in space.
- Power Generation. This problem may benefit from simulation with an earth-based colony, just to evaluate the maintenance capability of the colony and to provide measured data on electricity usage. However, the actual problem of power generation can efficiently be solved without the creation of such a colony. Commercial power would probably be used just to reduce the cost of supporting the earth-based colony.
- Construction in Space. Though much can be learned about the physical facilities required to support a long term colony in space by first building an earth-based colony, many aspects of space construction cannot be addressed by such an endeavor. After all, one advantage of creating a prototype colony on earth is the low construction cost as opposed to the extremely high construction cost of building a structure capable of withstanding the harsh environment of space.
- Shielding. An earth-based colony is already adequately shielded against radiation, so it is not a natural candidate as a platform for radiation testing. Shielding from micrometeoroids can likewise not be tested with an earth-based colony.
- Gravitational effects. An earth-based colony would naturally use simulated artificial gravity (i.e. real gravity) with the assumption that there are no significant differences between artificial gravity and the real thing. The validity of that assumption must be verified by other means.
- Atmospheric leakage. Though an earth-based colony could be constructed within a vacuum in an attempt to address this issue, doing so would raise the cost of the facility and would not be an economical approach to solving this technical issue.

Most other issues pertaining to long term human space flight can be studied and addressed with the creation of an enclosed long term earth-based colony.

3.1 What Will Be Learned From An Earth-based Colony?

A list of issues involved in creating a space colony on earth are listed in Table 1with an indication of how an earth-based colony can help address the issue. Obviously, entire books (or even libraries) could be written on each of these issues, and we won't have room to describe any of them in detail. The table does however list the issues, identify them as technological, logistical or human related, and the last column is color coded to identify which issues can be addressed with an earth-based colony. In the third column of Table 1, red shading indicates that an issue cannot be helped by creation of an earth based-colony. Green indicates that the issue is almost identical for both a space colony and an earth-based colony. Yellow indicates that the issue is similar, but not almost identical between a space colony and an earth-based colony. We will start this table with the five previously identified issues that cannot readily be addressed by an earth-based colony.

Table 1: Issues with Space Colonization

Issue with Long-term Space Flight	Type(1)	Can an Earth-based Colony Help?
Starship propulsion	T	Cannot help (earth colony is stationary)
Power Generation	T	Some help, (maintenance & data on power usage)
Construction in Space	T/L	Shouldn't help (not cost effective)
Radiation Shielding	T	Shouldn't help (wouldn't test with real humans)

Long-term Effects of Artificial Gravity	T	Cannot help (cannot defeat real gravity on earth)
Atmospheric Leakage	T	Shouldn't help (not cost effective)
Colony Organization and Government	H	Almost identical to issue faced by space colony
Law Enforcement and Justice System	H	Almost identical to issue faced by space colony
Medical issues	L/H	Almost identical to issue faced by space colony
Crew Psychological Wellbeing	H	Almost identical to issue faced by space colony
Colony Economy	H	Almost identical to issue faced by space colony
Organization of Labor	H	Almost identical to issue faced by space colony
Agriculture	T/H	Almost identical to issue faced by space colony
Population Management	H	Almost identical to issue faced by space colony
Potential Life Threatening Catastrophes	T/H	Much lower risk than issue faced by space colony
Recycling	T/H/L	Almost identical to issue faced by space colony
Social Issues	H	Almost identical to issue faced by space colony
Colony Communication	T/H	Almost identical to issue faced by space colony
Physical Facility Maintenance	L/H	Similar to issue with space colony
Manufacturing	T/L/H	Almost identical to issue faced by space colony
End of Life Issues	H	Almost identical to issue faced by space colony
Religious Issues	H	Almost identical to issue faced by space colony
Documentation of Lessons Learned & Procedures	H	Would directly benefit a space colony
IT Infrastructure	T/L	Almost identical to issue faced by space colony
Atmospheric Management	T	Almost identical to issue faced by space colony
Family Life	H	Almost identical to issue faced by space colony
Ranching	H	Almost identical to issue faced by space colony
Colony Termination	H	Not similar to issue faced by space colony
Moral Issues	H	Similar to issues faced by space colony
Water Treatment	T	Almost identical to issue faced by space colony
Clothing and apparel	T	Almost identical to issue faced by space colony
Colonist selection approach	H	Almost identical to issue faced by space colony
How do you handle people who want to quit the colony?	H	Not applicable to a space colony
*Spillover to non-space industries	T	May help other types of colonies (ocean, arctic, etc). Lessons Learned from extreme recycling may benefit all of mankind.
Risk to life and health	H	Greatly reduced from that of a space colony
Criteria for measuring colony success	H	Different criteria for success from a space colony

Note 1: Each issue is categorized as to type as follows: T=Technical, L=Logistical, H=Human.

One may ask, "Why are so few of these issues color coded red?" In other words, "Why are so many of the issues of long-term spaceflight also applicable to an earth-based colony?" The answer is because the vast majority of the issues have to do with keeping the crew alive and prosperous while residing inside a pressure vessel over a long period of time. If you are a crew member living inside a pressure vessel, where that vessel actually resides has little impact on your daily life since you live inside the vessel. Thus, if you are on a starship in route to a distant solar system, or if you are in a simulator here on earth, for 99% of the voyage, your daily life would be the same.

Obviously, since the deltas for pressure, temperature and radiation are much less for a simulator on earth, it will cost a lot less to construct an earth-based simulator than it would a starship.

Table 1 presents a large number of closed system colony issues at a very high level, but does not solve them. As was mentioned previously, the proposed colony would serve as a laboratory for testing solutions to the problem of long duration space travel. The approach presented here is that the colonists and mission planners start with a solution to each of these problems and that the colony then be constructed and inhabited, and that colony life begin based on the initial solutions. As deficiencies are observed in colony life, adjustments will be made and implemented by the colonists with the help of an external support crew as needed. An earth-based colony probably won't start as a fully self sustaining colony, but after years of adjustment, it will hopefully achieve that status.

For an earth-based colony, failure is accounted for and recovery is built into the system. Furthermore, the consequences of failure are non-catastrophic. If one aspect of colony life does indeed fail, the simulation can continue and allow other aspects to succeed. To illustrate this (again using the atmospheric example), should a viable approach to atmospheric recycling prove too difficult to achieve during the lifespan of the colony, the colony can carry on using regular atmosphere and continue testing and refining other important aspects of space colony life.

To summarize, an earth-based self sustaining colony need not start with solutions to the problems of long term space travel. Rather, it develops and tests solutions, and ultimately ends with them.

4. When Should We Create An Earth-Based Colony?

Now that we understand why such a colony should be developed, let's move on to the question of when it should be developed. It should be developed at least two generations before a generational starship is actually launched. The first generation of a colony would be hand-picked from a large group of volunteers. The second generation of colonists will not be hand-picked, but will inherit their status. It is reasonable to assume that the second generation will not be of the same caliber as the first. Therefore, it is important to observe the second generation closely to determine colony viability when both motivation and capability have decreased.

5. How Should an Earth-based Colony be Developed?

Now let's focus on how an earth-based colony could be created, funded and managed. Bear in mind that there are myriad ways to create an earth-based colony. This paper describes one approach to demonstrate that creating such a colony is viable. The path we'll take to accomplish this is as follows:

- Estimate an appropriate colony size
- Select a starship approach
- Outline a top level design of a proto-starship and facility
- Price the facility and effort

We'll then present a table describing how numerous issues with long term space colonization can be addressed by creating an earth-based colony, and conclude with a discussion of financing.

5.1 Colony Size

The first question that must be addressed for an earth-based colony is its size. How many people are we talking about for creation of a space colony? That of course depends on the duration of the colony, the cost of construction, and the probable list of willing colonists. It also drives most other decisions made about creating such a colony.

Obviously, a colony on the order of 5,000 people or more would be best to assure genetic diversity and trade specialization [3]. However, the cost of moving that many people (and the materiel necessary to support them) into space would be astronomical. To date, economic expenditure has been the limiting factor on space exploration, and it is reasonable to assume that this will continue to be the case. So for the sake of economics, let's assume that a minimum sized colony is desired for a generational starship. Just to drive a nail in the coffin of the big colony approach though, let's point out that given the choice of sending one large colony on an interstellar voyage or numerous smaller colonies, the numerous smaller colonies approach would win to mitigate the catastrophic effect of failure en route.

From a purely biological point of view, anthropologist John H. Moore estimated that a population of 150–180 would permit a stable society to exist for 60 to 80 generations — equivalent to 2000 years [4]. We can safely assume that this colony size could be reduced through the use of artificial insemination and the transport of a genetically diverse supply of sperm from earth to a distant destination. But let's keep the 150 number in mind as we look at other approaches to arriving at a minimal sized generational starship.

A minimal colony must maintain sufficient skill set to both survive and thrive while isolated from the rest of the human population. Though much information will be available to the colony via the onboard intranet, thorough expertise is still probably required in several fields. The critical fields for maintaining high levels of skill are medical, engineering, manufacturing and environmental/agricultural. A minimum quantity of skilled professionals for these essential fields is three, to allow mentoring of replacement personnel and to accommodate the unexpected loss of these personnel. As an example, three people capable of performing surgery should be planned to be available to the colony at all times. Should one (or even two) of these people die, replacements can be trained by study and through mentoring from the remaining surgeon. These high-skilled personnel will not be deep experts in sub-fields of their discipline, but rather broadly proficient across their discipline. Specialized knowledge is available when needed via the colony intranet and via communication with the outside world.

Only a certain percentage of the colony can be considered productive workers at any given time since many will be attending school and some will be retired and beyond their working years. Let's take a stab in the table below at a minimum sized colony based on critical professions and see what number that leads us to.

Table 2: Critical Professional Roles

Occupation	Quantity
Surgeon/Doctor	3
Health Professional: Doctor/Dentist/Optometrist	3
Electrical/Computer Engineer	3
Software Engineer/IT Specialist	3
Manufacturing/Mechanical Engineer	3
Agricultural/Environmental Expert	3

Data from the 2010 U.S. census [5] indicates that 35% of the population is of educational age (younger than 24) and 11% is of retirement age (older than 70). The remainder of the population is of working age. If we make the assumption that only 25% of that population will have the capacity (motivational, mental, emotional, etc.) to achieve competency in these fields and then pad by 10%, a minimum sustainable colony size would be:

$$CP = \text{minimum colony population} = 1.1 * 18 / ((100 - 35 - 11) * .01 * .25) = 147 \text{ individuals.}$$

Having reached approximately the same number twice, let's assume the use of a transported bank of genetic material to ensure genetic diversity, then round this last number up to 150 individuals and set that as our minimum target colony population.

5.2 Facility

We now need to discuss a colony facility. In fact, we will do a top level design and calculate the cost of such a structure, because that cost is important to determining the plan's viability.

Many approaches to space habitat design are possible, but we need to select one approach to demonstrate feasibility. So here is an example approach that we select for the facility for this colony. We call it the long hallway approach to starship design. We'll arrange four long hallways with rooms on either side of the hallway, three stories high. The four hallways are linked together with cables or structural ties in a cylindrical manner as shown in Figure 1. Artificial gravity is provided by a rotating cylindrical arrangement in space, and is simulated by real gravity in a parallel arrangement on earth. During acceleration or deceleration, such as during the departure or arrival phases of a generational starship, the hallways can be repositioned to provide artificial gravity by aligning the hallways with the axis of acceleration. The same figure shows how the hallways would be set up for an earth-based colony. Admittedly, this approach is taken because it maps well to an earth-based colony.

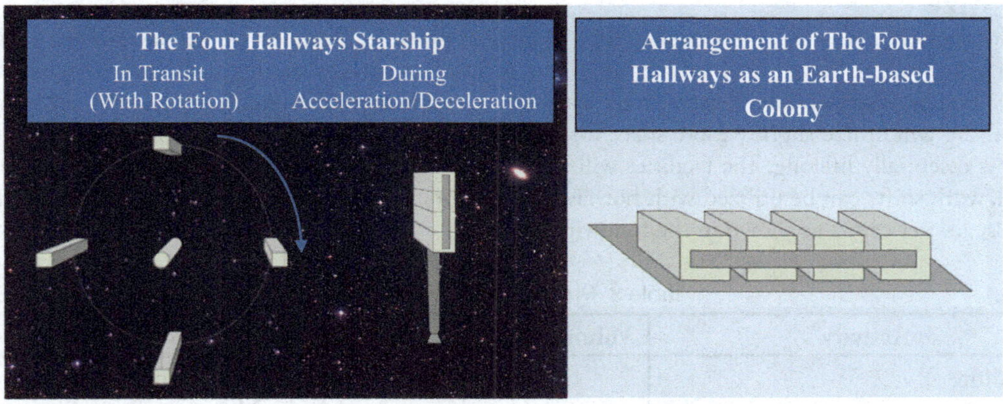

Figure 1: Space Station Approach, Credit: Background Image from European Southern Observatory http://www.eso.org/public/images/eso0116a/

The hallways are redundant to accommodate disaster recovery. As an example, if a fire started in one hallway, the occupants could be evacuated to other hallways until the atmosphere is reconditioned and the damaged hallway reclaimed. Should a meteorite puncture the hull of a hallway and kill all of the occupants, then people from the other hallways may be able to repair the hull breach, reclaim the damaged hallway and increase the birthrate to eventually repopulate the empty hallway.

Another advantage of this approach is that in space the hallways could possibly be constructed from re-purposed launch vehicles. We typically build rockets in stages and then drop the stages as their fuel becomes exhausted. The fuel constitutes most of the weight of a rocket stage, so why not reuse the hull? It could then be repurposed in orbit to become a single hallway of a multi-hallway space station or starship.

Let's describe this approach just a little further. Each hallway would contain a generic facility and a specialty facility. Each hallway's generic facility would have the following services:

- Domestic living area
- Hygiene Areas
- Agricultural Area
- Gym/Exercise Area
- Recreational Area with inside 2 story high park facility.
- Cafeteria/Meeting Area
- Life Support Systems
- Two standard docking ports (one on each end) each with an attached escape vehicle capable of holding 1/6th of the colony (i.e. a little more than ½ the population of each hallway).
- A trolley docking port for transportation to other hallways.
- A small but well thought out trolley system will serve to transport personnel from one hallway to another. Each Hallway would also have one specialty facility for a total of four specialty facilities. Although actual equipment and supplies would be dispersed throughout the hallways to allow for disaster recovery, each hallway would focus its remaining area on creating a single specialty facility. These would be as follows:
- School/Social: Classrooms with a small theatre facing the meeting area.
- Hospital/Geriatric: A room for surgery, a warehouse for medical storage, offices for doctor/dentist/optometrist visits & a pharmacy.
- Factory/Recycling: Additive and subtractive manufacturing, recycling center, and a warehouse area
- Administration/Control : Administrative offices, IT and engineering labs, and a control room

To simulate the conditions that would be found in an actual space colony, the starship habitat should be as compact as possible, while still providing sufficient space for a contented and self sustaining crew. So what is that volume? How much room should we allocate for the inhabited area of our colony?

5.3 Size

Again, for economic reasons, this area should be as small as possible. However, we will not assume a level of compression as is typically found on a submarine, where techniques such as sharing beds (hot-bunking) are utilized to reduce domestic space, and cramped facilities are designed with knowledge that the crew's mission is short in duration. We will utilize slightly more space for psychological wellbeing, with the knowledge that the colonist's mission is essentially lifelong. The facilities will still be Spartan by terrestrial standards though. Since all light is artificial, work shifts can be utilized with hot-swapping of office and production stations. With that thought, we allocate space per capita as shown in the following table.

Table 3: Physical Volume per Colonist

Activity	Volume in m3	Comment
Agriculture	50	This number was taken from an analysis by So, et al. on evaluating the Mars One Project [6]
Living Quarters	8	Assumes a private dormitory for each colonist of 2x2x2 meters.
Public Space	8	Assumes a commons area for meals and meetings with 2x2x1meters/person and a tall park-like open natural area of 1x1x4 meters/person.
Manufacturing/Lab	2	Assumes manufacturing and lab facilities of 1x1x2 per person. These stations may be shared on work shifts.
Mobility	4	This is for a central hallway area of 0.5x2x4 per person
Kitchen and Gym	1	Assumes 1x1x2 per 8 people for a kitchen area and an equal amount for an exercise area.
Escape Vehicle Area	1	Should accommodate one person seated for a total area of 0.5x1x2 per person
Storage	2	Assumes 1x1x2 per person for storage of manufactured products and supplies being transported and emergency supplies
Office	3	Assumes a 2x2x3 meter office area for 1/4th of the colonists. This also includes a hospital area
Hygiene Space	1	Lavatories and Laundry Facilities. 1x2x2 toilet/8 people and 1x2x2 shower/8 people.
Maintenance	8	This includes electrical cabinets, waste water treatment, plumbing, electrical wiring, duct work, recycling, atmospheric management and soil storage and management facilities. Assumed 2x2x2 in area/person.
Total	*88*	*In cubic meters*

If you multiply this space by the total planned population, and add a 15% margin of error then the total internal volume for our target population is:

CV = internal volume for colony = 88 x 150 x 1.15 = 15180 cubic meters

With the assumption of three stories, each with an internal hallway that is 1 meter wide and two high, and rooms on each side of 2.5 meters deep and 2 meters high, you have a module width of 6 meters with a height of 6 meters. The total length for four such hallway structures can be calculated:

15,180m3 / (6m (width) x 6m (height) x 4 (hallways)) = 105.4 meters/hallway

5.4 Layout

The most appropriate external facility to house an earth-based colony would be a modern warehouse or manufacturing facility. These are typically divided into a front end office area and a back end warehouse or factory, consisting of a single large open area and loading ports for trucks in the back. The front lobby and office area would provide a public face to the colony project and house the small support staff. The large warehouse area would house the colony itself. This colony's living facility would be constructed within the warehouse area, which is typically open and usually about 20 to 30 feet high. The colony living facility could be constructed to resemble the layout of the inhabited area of a similar structure in space. The warehouse would provide a controlled environment that will ultimately bring down the cost of the colony structure housed within it. It would also serve to shield the colony from external influences.

The four hallways give a warehouse area of about 110m x 40m. If you add 20 meters on the front for administrative space and 10 meters on the back for truck berths and storage, your warehouse would occupy 7700 square meters and could be laid out as shown in the following figure.

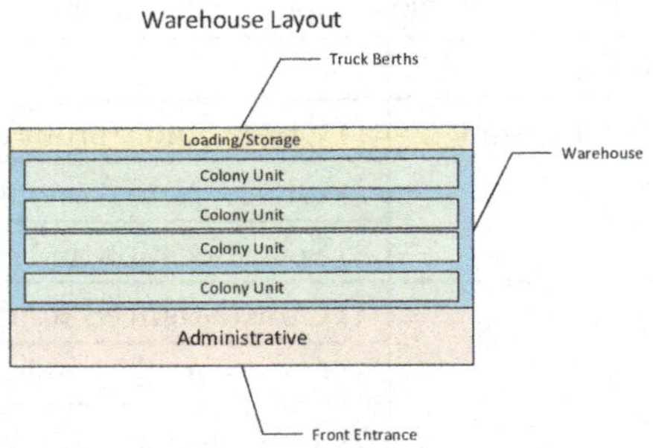

Figure 2: Layout of External Structure Housing Colony

The front lobby and administrative area of the facility houses the external management of the colony (i.e. Ground Control if you will). It would contain a male and female dormitory, a cafeteria, a recreation room, a front lobby and numerous workstation cubicles. The dormitories and cafeteria are used to house the college interns who will constitute the majority of the ground crew. This is elaborated on in the finances discussion later in the paper.

5.5 Facility Cost

We now need to provide a rough estimate for the cost of this facility for determining feasibility. Let us assume that warehouse space is leased and that an adequate existing location and facility can be identified. Warehouse leasing costs vary dramatically by location, but numerous facilities are available ranging from $3 to $10 per square foot per year. Numerous locations offering $5/sqft/yr are currently available in the Midwest [7]. We'll use that as our target.

Though the warehouse can be leased, the colony will need to be constructed. Units could be prefabricated and transported to the site where they can be assembled. If we use construction cost data from the Midwest [8], we would select a cost of from 70-$150/ft2. We'll stay on the cheap side of this, knowing that the inside facility need not stand up to the weather, that its foundation is already in place and that each hallway can be prefabricated. For this estimate, we'll assume that the indoor colony facility cost's $80/ ft2. This puts the total lease and upfront cost of constructing the colony facility at:

Annual Lease Cost = 7700m2 x 10.76 ft2/m2 x $5/ft2 yr = $413,490/yr

Upfront Internal Construction = 9m x 105m x 2 floors x 4 units x 10.76 ft2/m2 x $80/ft2 = $6.533 million dollars.

This is of course a rough order of magnitude calculation, but it is very useful for determining the feasibility of an earth-based colony. We'll use this dollar amount later when discussing the finances of the operation. This does give us an idea though that the plan is not at all unreasonable. When compared with the $6 billion [9] that is anticipated for the Mars One initial crew, this is quite economical and should have a higher probability of success.

The facility estimate doesn't include specialized equipment for recycling air, water and other materials. It also excludes needed medical and production equipment. Corporate donors will be sought for obtaining this equipment.

Now if we had made different choices for our space colony, then the corresponding earth-based colony would appear different. A spacecraft designed for constant acceleration (i.e. accelerating to the journey's halfway point and decelerating past that point) might have a large number of small stories as in a small high-rise building. A colony with a large population (>1000 people) might become a town constructed in a small valley that curves upwards at both ends and is domed overhead.

6. Approach to Starting an Earth-based Colony

So how should an earth-based colony be created? The answer is "in phases", with each phase requiring significantly more effort and resources than the previous. A next phase would not start until the previous phase had been successfully completed. This approach, outlined in Table 4, reduces risk and allows for early termination of the project should it become unviable.

Table 4: Colony Startup and Financing

Phase	Purpose	Activities
1	Organize and Sponsorship	1) Create a non-profit organization
		2) Obtain NASA sponsorship
		3) Select location and obtain state & local govt sponsorship
2	Raise Funds	4) Crowd-sourced fundraising
		5) Corporate donations
		6) TV contract (A Mars One idea)
		7) Colonist application fee (small) and sponsorship (larger)
3	Construct and Recruit	8) Select initial volunteers
		9) Plan and construct one hallway
		10) Operate initial hallway for a year with pioneer crew
4	Complete	11) Continue fundraising
		12) Construct three additional hallways
		13) Recruit additional colonists
5	Ongoing Operation	14) Continuous fundraising
		15) Facility maintenance
		16) Colony refinement and Analysis

6.1 Financing an Earth-based Colony

Of course, funding for basic research is currently difficult and government spending is particularly constrained. In fact, the advantages of this whole concept may seem pretty obvious, but no one is yet doing it (except in Hollywood). The reason it hasn't yet been attempted is primarily financial. So how can such an endeavor as that described by this paper be funded in today's economic environment? Here is a seven step process to achieve funding today:

1. NASA Sponsorship
2. State Sponsorship
3. Crowd-sourced Fundraising Campaign
4. Corporate Sponsorship
5. TV contract
6. Recruit Volunteers with Sponsorship
7. Donations

Let's now look at each of these steps in more detail.

6.1.1 NASA Sponsorship (This is step 1)

This is step 1 because this step provides legitimacy to the project. It costs the government little, but can be leveraged by the colony to obtain significant additional funding. Of course, what NASA gains by sponsorship is a significant voice in how the colony is constructed and operated without having to finance the endeavor itself. The fuel for this whole financial approach is enthusiasm (or hype), and that starts with NASA sponsorship.

6.1.2 State Sponsorship (Step 2)

Once NASA sponsorship is achieved, a suitable location for the colony facilities must be obtained.

- Place all 50 states in competition for this NASA project
- Possible donation of Land (maybe part of a space port)
- Selection process of U.S. state to host colony (e.g. no property taxes, inexpensive real estate, willing to exempt local laws such as standards for room occupancy, etc)

6.1.3 Crowdfunding Campaign (Step 3)

Armed with a noble cause, and exploiting the angle that the government would accomplish this if NASA were adequately funded, a well planned Kickstarter campaign should be able to raise a significant amount of money. Assuming that a successful Kickstarter campaign can be assembled and became a top performer, millions of dollars of funding could be brought in by this method. An average of the top 10 Kickstarter projects earned $8,678,228 [10]. None of those campaigns supported a cause as noble as the one described by this paper. None of them were endorsed by NASA. None of them would further knowledge and serve mankind the way this campaign would. Let's assume that Kickstarter and other crowdfunding campaigns can be mounted that will achieve the average of the top 10 to date.

6.1.4 Corporate Sponsorship (Step 4)

While the crowdfunding program is proceeding, corporate sponsors should be sought. For corporations, supporting this project could be a big publicity windfall. This may involve bragging rights on TV commercials, as well as in other advertising endeavors. Donations could be monetary in nature, or could be made in the form of essential equipment. Corporate donors would be recognized as sponsors and have the right to advertise their sponsorship.

- Facility Donations (e.g. donations of mock B330 modules or mock Orion Cockpit for station use.)
- Hardware, IT and component donations from various companies
- Donations of medical and manufacturing equipment
- Corporate Partners could provide equipment for experimentation (e.g. air and water treatment companies). They would receive recognition as sponsors.
- Will create a monthly publication (A magazine for all sponsors) to keep them informed and provide free sponsor advertising. Will also maintain a website and Facebook page to keep everyone (inside and outside) informed of what is happening.

It should also be noted that corporations may wish to sponsor the project just to obtain some influence on its direction and test out their equipment. After all, this is a lab. A target amount for corporate sponsorship should be set and solicited. For the purposes of this effort, that target amount for corporate sponsorship is set for $1 million dollars. The Mars One project has raised close to this amount from corporate sponsors [11], and Mars One has a much higher probability of failure.

6.1.5 TV Contract (Step 5)

Speaking of Mars One, here is an idea borrowed from them. According to the US Bureau of Labor Statistics, we spend much more on the entertainment industry than we do on space exploration or research, about $2,500 per family per year [12]. A TV contract would provide a way to funnel some of that money back into productive activity. This activity would create celebrity, then use celebrity to generate funding.

The colony would be something new, highly unusual, extremely challenging, but with a noble purpose. These characteristics provide the ingredients for an excellent TV show. If well produced, a program based on this colony should easily appeal to a television audience who may wish to participate, but which can only do so by proxy. As a TV show, the project's celebrity impact will generate revenue from TV proceeds and from other sources related to its celebrity status. Some would even argue that the worse off the colony becomes, the better TV it makes, but we won't go there.

TV contracts for reality series vary from a small stipend to $100K per episode with the higher number for highly successful series [13]. For the purposes of this analysis, let us assume a TV contract would generate $50K/episode and run for 30 weeks/year for eight years. This would result in $1.5 million annual income from TV revenue. This is not an unreasonable assumption with good production and NASA sponsorship. After all, it was Alfred Hitchcock that said "drama is life with the dull bits cut out". There would surely be drama within the confines of a colony and ground crew, even if no one is being killed. This approach may entail providing cameras in the conference rooms and recreational facilities that would not be present in an actual space habitat. But this natural drama could be molded by a skilled editor into a successful TV show and the cameras would be useful for psychological analysis.

The possibility for licensing and merchandising may also exist. Some examples would be T-shirts, cups, toys and sponsor clubs. Once the main TV contract has expired, and due to the celebrity status of the project, publicity income will trickle in, albeit at a much slower pace. Let us assume a small amount of continuing income from articles, interviews, licensing and speaking fees of $100K/yr over the duration of the project. Maybe even a book deal or two would pop up. At the least, a TV contract would serve as a draw for volunteer colonists and ground crew.

6.1.6 Recruit Volunteers (Step 6)

Due to the actual size of the colony and support staff, this project would become extremely expensive unless volunteer labor is used. The proposed staffing approach in this paper relies on using purely volunteer labor for continuing operations. The obvious reason for using volunteers is to drastically control personnel costs. Due to the number of people involved, this project only becomes economically viable if personnel costs are kept low. Here is how this would work:

- No employees allowed. All volunteers are motivated solely by the cause of human advancement. The volunteer colonists are being asked to sacrifice numerous comforts for the sake of learning. Like the colonists, the support staff should also be making sacrifices.
- Volunteer colonists support themselves through their own labor within the station.
- Each candidate colonist would pay a small fee (>$40) for their application to be considered and processed, but each finalist would obtain a sponsor to pay a sponsorship fee ($5,000) to sponsor that colonist. The colonist is already making a significant sacrifice, so the financial sponsor would typically be a corporation or club or relative. That sponsor would keep sponsorship rights for that colonist for the duration of the colony unless they sell those rights. If the colony is successful, then individual colonist sponsorships could even be sold at a profit to other entities.
- The support crew is made up almost entirely of college interns who are offered free room and board at the facility, but no salary. They cook and clean up afterwards. They take out the trash and vacuum. They perform facility maintenance. They do technical analysis and staff the front desk. They even run the project alumni program to solicit funds from previous interns.
- Retired volunteers to provide professional services such as Project Administration, intern supervision, HR, facilities, accounting and legal.
- Personnel on-loan from corporate sponsors and from NASA may be paid, but would not be financed by the project.

The use of college interns for staffing the ground crew may appear to be a liability, but it will actually prove to be the opposite. The interns will rotate every semester and provide each intern with space program experience. In the long run, this will benefit the entire space effort. They will bring a constantly regenerating enthusiasm to the project that could not be obtained in any other way. Their long term support will spill over into financial support as they become project alumni.

Though it may seem important for continuity to have a core paid staff, having one would decrease the dedication of the volunteers. They will give more if everyone is giving, and no one is getting. For this reason a highly paid CEO is out of the question. It is better to settle for a qualified retired CEO who will donate his time for a worthwhile project, than to have the very best that money can buy.

Are there organizations that thrive based on this principle of volunteerism? Surely there are, but they are mostly religious [14]. What the religious organizations have that produce prodigious amounts of volunteer labor is a noble cause; a cause considered worthy of sacrifice by their volunteers. This project would likewise be a noble cause and considered worthy of sacrifice by many people. Volunteers will come pouring out of the woodwork to participate. The challenge will come in selecting appropriate volunteers.

Of course, an established project is required to recruit volunteers, but such projects must somehow become established. Some initial salaried personnel may be required to set up the project. These would be phased out though when the project is well established.

6.1.7 Continuing Donations (Step 7)

The project would be set up as a non-profit organization, and as such can easily solicit tax exempt donations. A website and Facebook page requesting sponsors similar to those supporting the 100 Year Starship program would be created and some money would trickle in. Benefits to sponsorship would be VIP visits to the facility, occasional phone calls with colonists, and other typical benefits that sponsors to non-profit organizations receive (i.e. T-shirts, coffee mugs, monthly magazine, etc.). An alumni association for interns would also be created with similar benefits.

Let us target a modest amount of $250,000/year in donations. This is not entirely unreasonable. After all, donations for the largest 100 non-profits ranged from 118 million per year up to over six billion dollars per organization [15].

6.2 Balance Sheet Overview

We now need to assemble all of this information and demonstrate financial feasibility. Here's is a back of the napkin balance sheet for the project:

Table 5: Colony Balance Sheet

Item	Expense	Income	Balance
	STARTUP		
Expenses			
• Facility	$6,533,472		
• Initial Staff Salary	$1,250,000		
• Furnishings for Front Office	$500,000		
• Colony Equipment and Furnishings (See Note 1)	$1,500,000		
Income			
• Crowdfunding Campaign		$8,678,228	
• Corporate Donations		$1,000,000	
• Colonist Sponsors		$750,000	
Total Startup Balance	*$9,783,473*	*$10,428,228*	*$644,756*
	CONTINUING OPERATION (per year)		
Expenses			
• Salaries	$0		
• Facility Lease	$413,490		
• Utilities (Electricity, Natural Gas, Water, Phone/Data, Waste removal).	$50,000		
• Cafeteria Expense	$70,200		
• Taxes	$0		
• Profit	$0		
• Miscellaneous Expenses	$50,000		
Income			
• TV Contract (See Note 2)		$0	
• Interest on Investments (return-4%/yr)		$265,790	

• Continuing Donations		$250,000	
• Continuing TV and Publicity Income		$100,000	
Total Continuing Operation Balance	*$583,490*	*$615,790*	*$32,100*

Note 1: Much of the furnishings and equipment should not be purchased, but rather manufactured by the colonists; thus ensuring that all essential parts can be replaced and recycled. This expense is only for raw material and specialty items.

Note 2: Half of this income together with the remaining startup income is invested, and an annual return is used to cover annual expenses. This return is shown in the row titled "Interest on Investments". The other half of this income is placed in an endurance fund that is provided to the surviving colonists on termination of the colony. Hopefully this endurance fund could be increased.

This rough balance sheet shows the project starting in the black and continuing in the black year after year. Admittedly, this balance sheet is only a ROM estimate, and the funding approach is unorthodox. But it is also creative and viable. We have demonstrated that finances need not hinder the founding of an earth-based colony. So what's stopping us?

7. Conclusion

In this paper we have described the reasons for creating an earth-based self-sustaining colony. In a nutshell, it would facilitate confronting and solving most of the issues of long term space flight. There are of course numerous approaches to creating a successful and useful earth-based colony. This paper has described one approach to creating and financing such a colony, thus demonstrating that it would be of value, and that it can be accomplished.

The enthusiasm generated by the Mars One project indicates that the time for such an endeavor is upon us. From this 1st generation colony, we'll learn what can be produced and what must be transported. We'll develop new approaches to environmental maintenance and recycling. We'll learn new approaches to production and farming in an isolated and confined space. We'll learn solutions to physical and social issues pertaining to colony wellbeing and self sustainment; we'll then build the 2nd generation colony (Starship Beta) off-planet.

About the Author

Manuel Richey (manuel.richey@honeywell.com) is a principal engineer at Honeywell International in Kansas where he has worked for over 30 years. He is also a computer science instructor at Fort Scott Community College and holds eleven U.S. patents.

Bibliography

1. University of Arizona, "Fast Facts," 2014. [Online]. Available: http://b2science.org/who/fact. [Accessed 15 August 2015].

2. E. A. Moore, "Six humans step into one-year isolation dome," USA Today, 31 August 2015.

3. C. M. Smith, "Estimation of a genetically viable population for multigenerational interstellar voyaging: review and data for project hyperion," Acta Astronautica, p. 16–29, April-May 2014.

4. CNN, "Report: Make Deep Space Travel a Family Affair," 9 March 2002. [Online]. Available: http://www.cnn.com/2002/TECH/space/03/09/family.spacetravel/. [Accessed 14 August 2015].

5. United States Census Bureau, "Patterns of Metropolitan and Micropolitan Population Change: 2000 to 2010," 6 May 2013. [Online]. Available: http://www.census.gov/population/metro/data/pop_pyramid.html. [Accessed 15 August 2015].

6. S. Do, K. Ho, S. S. Schreiner, A. C. Owens and O. L. de Weck, "An Independent Assessment of the Technical Feasibility of the Mars One Mission Plan," September 2014. [Online]. Available: http://dspace.mit.edu/handle/1721.1/90819. [Accessed 14 August 2015].

7. LoopNet, "Search Properties for Lease," 15 August 2015. [Online]. Available: http://www.loopnet.com/xNet/MainSite/Listing/Search/SearchResults.aspx#/KS/Industrial/For-Lease/c!ARYC$BAQ. [Accessed 15 August 2015].

8. Brown Wegher Construction, "How Much Does a Construction Project Cost?," 2015. [Online]. Available: http://www.brownwegher.com/cost-of-construction-chart/. [Accessed 15 August 2015].

9. Mars One, "How much does the mission cost?," 2015. [Online]. Available: http://www.mars-one.com/faq/finance-and-feasibility/how-much-does-the-mission-cost. [Accessed 15 August 2015].

10. N. Zipkin, "The 10 Most Funded Kickstarter Campaigns Ever," 30 March 2015. [Online]. Available: http://www.entrepreneur.com/article/235313. [Accessed 14 August 2015].

11. Mars One, "Support Mars One's Human Mission to Mars," 11 March 2015. [Online]. Available: http://www.mars-one.com/donate. [Accessed 15 August 2015].

12. United States Departement of Labor, "Economic News Release," 9 September 2014. [Online]. Available: http://www.bls.gov/news.release/cesan.nr0.htm. [Accessed 15 August 2015].

13. R. Haidet, "Pay Scale For Reality Hosts & Stars," 24 August 2011. [Online]. Available: http://realitytvmagazine.sheknows.com/2011/08/24/pay-scale-for-reality-hosts-stars/. [Accessed 29 8 2015].

14. J. DiSanto, "Penn Research Shows That Mormons Are Generous and Active in Helping Others," 12 April 2012. [Online]. Available: http://www.upenn.edu/pennnews/news/penn-research-shows-mormons-are-generous-and-active-helping-others. [Accessed 15 August 2015].

15. M. Hrywna, "The NPT 2014 Top 100: An in-depth study of America's largest nonprofits," The NonProfit Times, pp. 1-8, 1 November 2014.

A Roadmap to Interstellar Travel

The Societal Challenges

Abigail Sherriff and Chris Welch

Abstract

An interdisciplinary roadmap is developed to show the preparations needed over a nominal 100-year timeframe to launch a slower-than-light, self-sustaining worldship to carry humans over many generations to other star systems. The project was carried out by Astra Planeta, one of two team projects for the International Space University's 2015 Master of Science in Space Studies. This paper discusses the societal aspects, particularly the challenges of society on and off the worldship and necessary mission planning leading to launch.

The project, as a whole, draws on information from past studies of interstellar missions, and a preferred concept for the worldship is identified. Strategies are outlined for the necessary development and follow a logical progression of enabling missions and projects. An interdisciplinary approach considers the technological and societal challenges leading up to the launch of a worldship, as well as operating it over hundreds to thousands of years.

Through this framework, many major topics were cultivated, including: onboard infrastructure, operational concepts, necessary ship subsystems, onboard societal issues, and mission planning. For civilization in particular to grow and flourish, planning for the mission and the onboard society must be taken into special consideration. The onboard society research examines topics such as education, religion, language, and ethics. Mission planning takes into account the conditions for enabling the construction of a worldship including political considerations, cultural changes, financing, and international partnership.

Since expanding civilization and humanity is arguably the central reason for interstellar travel, this paper also addresses these issues in the context of the proposed mission. The end result is a framework to achieve interstellar travel in the near future with an approach that is financially feasible, technically capable, and culturally desirable.

Keywords

Culture, education, financing, language, religion, society

1. Introduction

This paper addresses the societal challenges as part of a project to create a conceptual design and nominal 100-year roadmap of a slower-than-light, multigenerational worldship to begin establishing humanity across the universe. The project as a whole is the result of the work of the Astra Planeta team, composed of twenty-two students from the International Space University (ISU) 2015 Masters of Science in Space Studies class. Each student comes from a different part of the global community, with the entire team representing four continents and seventeen different disciplines. The project made an effort to leverage these professional and intercultural differences inherent in the team to help aid the creation of an interdisciplinary roadmap. This paper provides a snapshot of the societal aspects drawn out by the work performed by the team. The full report is the culmination of a dedicated effort to critically evaluate existing literature to develop a roadmap for the worldship and should be consulted for a more comprehensive view of the project [1].

Before delving into the specifics of the societal aspects of the research, general background, overview, and outline of the project are provided. In the sections following, a more in depth discussion of the planning and onboard society is provided. References to the 'project' refer to the team project as a whole and following sections detail the societal challenges addressed.

1.1 Definition of a Worldship

An "interstellar worldship," or "worldship" for short, is a spacecraft designed to travel over vast distances at non-relativistic speeds (slower than light: < 0.10c). The term is reserved for spacecraft designed to carry a population of 100,000 people or more to a destination, which may take hundreds to thousands of years to arrive. A worldship is a generational ship in which many generations of people live their whole lives and die onboard. It is not a ship where the occupants are asleep or hibernating and it does not only carry human "seeds" for growth at the destination.

1.2 Aims and Objectives

An attempt has been made to offer an interdisciplinary approach to tackling the challenge of interstellar travel. The concept and roadmap provide a step towards overcoming some of the developmental challenges of interstellar travel. This project aims to contribute to this effort by furthering the state of knowledge of interstellar travel and establishing the foundation for future work and research to be accomplished. This requires that several challenges be discussed in addition to the technical ones. These include societal challenges onboard the worldship as well as throughout the developmental and construction phases. Political, economic, and legal aspects hold barriers for success that must be discussed, in addition to strategies for education and public outreach. The roadmap and suggestions are meant to showcase a preferred mission concept that is technologically and financially feasible, as well as culturally desirable. Given that the future cannot be predicted, the final goal of this project was not to create a detailed design of a worldship, but to identify key technologies and milestones that should be pursued for a preferred mission concept.

1.3 Constraints and Assumptions

In order to provide some context into which the preferred worldship concept could fit, some constraints and assumptions were discussed. Speculative technology such as warp drive, wormholes, and faster-than-light travel were not considered. Furthermore, it was assumed that the energies involved with fast travel (velocities greater than 10% the speed of light: 0.10c) would be prohibitive to achieve with near-future technology. Also, the population onboard the worldship is assumed to be alive and fully awake; thus, dedicated seed ships and cryogenic sleep are not considered. Finally, one of the project aims was to identify a preferred concept for a worldship, not to develop a detailed design.

1.4 Why 100 Years?

To most people a hundred years is a relatable period of time. One can certainly conceive of living that long. Thus, a hundred year time frame was chosen in part because it is approachable to people. It is not so far in the future as to be uninteresting, yet it is far enough that it seems a feasible timeframe. It also serves as a constraint, limiting the use of technology to those that currently exist or will in the near- future.

While the future cannot be predicted, logical and useful statements may be made about it. This allows humans to forecast alternative futures. One of these alternative futures could be the construction and launch of a worldship to explore our stellar neighborhood. Significant changes can happen in a hundred years, as is evidenced by the exponential growth in population and technology over the last one hundred years. To someone living at the beginning of the 20th century, the progress over the following hundred years would have seemed fantastical at the time. While this project makes no effort to predict the future, it considers the realm of what may be possible in the next hundred years. This gives the project scope and forces a realistic consideration of what a worldship may look like without relying on tenuous physics or incredible breakthroughs. The use of the word nominal when describing the timeframe was chosen carefully as it expresses uncertainty. It is a goal, meant to provide scope and boundary to the project as a reasonable starting point.

1.5 Mission Concept

In order to define a preferred concept for a worldship, an interstellar mission must first be conceptualized. That mission is then used to set the boundaries and focus efforts on creating a roadmap to the conceptual worldship. Thus, the proposed interstellar mission is as given:

> *The interstellar mission will consist of a single worldship directly travelling to a single stellar system within 25 ly from the Sun [2].*

Due to the large separation between stars, a voyage to a single stellar system is considered. As this will take hundreds to thousands of years, it is outside the scope of this project to consider what the worldship inhabitants will do when they arrive. A self-sustaining worldship would have the capability to leverage a stellar system's resources to manufacture the necessary components to settle a planet, construct a space station, or continue the voyage to another stellar system. Furthermore, due to the high energy required to accelerate a worldship, distances further than 25 light-years (ly) were considered to be unreachable in hundreds to thousands of years and are not considered. A fleet of several starships was considered as it offered several advantages, such as redundancy. However, the extra complexity and costs of constructing and managing a fleet was seen as prohibitive.

2. Planning for the Worldship

The road to the launch of the worldship is full of hurdles and challenges both technological and societal. The societal challenges range from economical to cultural. There are many steps along the way that must be taken to become the interstellar society necessary to travel to a distant stellar system. Some of the major challenges and possible solutions are outlined.

2.1 Governance

Determining the political governance structure onboard the worldship is a major concern that must be developed during the planning and construction of the worldship to ensure that it is fully primed. Hill discusses three motivations for future space exploration, including a corresponding governance model for each [3]. First is a global emergency, such as an approaching large asteroid. The corresponding governance model would be similar to an economically controlled government where decision-making is highly centralized. The second is driven by free enterprise and follows the market economy model, which will be chaotic with private companies and multinational organizations working towards personal gain or prestige. In this scenario, each company must be independent of governmental support in order to be a true free enterprise, complicating the development of a worldship governance structure. Third, and most likely according to Hill, will be a hybrid approach between the first two scenarios. Governments will handle the first and most costly long-term efforts, while commercial organizations take up the short-term activities through incentivized financing. This is similar to the mixed economic and governance model that has been successful in the creation of the Internet or air transportation. In the beginning, the government and military took on the costly and risky endeavors, and once the technology was proven and costs were reduced, the commercial market took over. Each of these scenarios offers a possible catalyst that will lead us off of Earth and into interstellar space. According to Crawford, a space-faring society must have the preconditions of "survival, geopolitical stability, resource availability, and moral justification," so a world government must be in place before the human race can become truly space-faring.

While the idea of an effective world government may seem utopian to some, the idea of an Interstellar City where world governments, commercial enterprises, and the international community work together to develop the worldship may also prove to be a model worth exploring. An Interstellar City would ensure flexibility, as cities are well known for their ability to grow and change and can significantly contribute to all factors relating to the worldship.

The Interstellar City would be a centralized community that can assist in stimulating open and rational debate in a highly politicized area. It allows the centralization of an economic system and labor and would be central hub for major worldship experimentation, research, manufacturing, and governing. The city would be a place where all participating nations and nationals could live and contribute to the creation of an interstellar culture and provide a platform to facilitate collaboration. This city could be an autonomous, sovereign city-state that would be separate from the direct supervision of any single nation-state. Instead, it would be bound by its own constitution, rules, legal structures, and even economy. The city would be religiously and ethnically mixed, have its own currency, and be a center of technological and scientific expertise. Examples of autonomous cities in the past include Danzig, Tangier, or Trieste. These cities inspired creativity and fostered innovation [4]. The present day example of the Vostochny Spaceport that is being built in the Russian Far East is intended to be a state-of-the art "science city" [5] and could serve as a primary model for developing the Interstellar City. An Interstellar City on Earth will also serve as an analog for an Interstellar City on the moon, and finally an Interstellar City on the worldship. More importantly, it will serve as an important base to test governance models.

While the continuation of the human race, exploration, and expansion may be motivation enough for some countries, others may require bigger motivations. The idea of an exclusive, technologically advanced, and scientifically driven Interstellar City will serve as an additional motivation for the worldship project. All participating countries will be allowed to send a certain number of citizens to the City every year. All of the patents, technology, and spin offs that result from the innovation within the city will be the property of the city and the individuals that reside in it. There will be an internal economy that trades with other nations. Participating nations can expect to have a portion of their financial contributions returned to them in contracts, similar to the European Space Agency's Industrial Policy of geographical return [6]. In summary, national and personal prestige, access to technology, and generous return on investment schemes will all serve as tools to motivate nations and citizens to join the Interstellar City.

2.2 Financing the Worldship

Before humanity can realistically achieve interstellar travel, the society of Earth must adapt. One of these key societal aspects is economy. No single nation can realistically fund a worldship on its own, and no project of this scale has ever been attempted. For the construction of the worldship to go ahead, appropriate funding must be collected through international partnerships. This financing must take into account a global change towards the stabilization of world politics and a lessening of economic and social instabilities. Utilizing existing financing models paves the way to devise realistic approaches of achieving financing an interstellar worldship.

There are three complementary ways in which a project of this magnitude could be financed and developed. The following three-pronged approach aims to tackle both the financing and technology development aspects of the project:

1. International Interstellar Fund (IIF) similar to International Monetary Fund (IMF) Quota system
2. Incentive financing to further technological progress
3. Additional sources of funding:
   A. Crowd Funding Platforms
   B. Leveraged Finance

The International Interstellar Fund (IIF) being the main driving force through which the bulk of financing for the project will be carried out. The IIF will serve as the primary body for coordinating financing efforts for the duration of the project. The IIF will use a proportional equality system, or "criterial equality," which stipulates that though the voting rates and funding requirements may vary for each participating state, the calculation method for such rights will be the same for all, depending on distinct, predetermined criteria. An example of this is the allocation of votes in the Board of Governors of the IMF. This is the organization's highest decision-making body and allocates votes to participating countries' decisions based on their financial contributions to the Fund in terms of IMF quotas [7]. To keep the IIF mechanisms simple, the payment requirements will be based on this quota system.

Considering that there are 188 countries that are members of the IMF, applying the IMF model not only simplifies the process for member states, but will also ensure a smooth transition by taking advantage of existing protocols. Initially, the fund will consist of twenty founding members, which include the leading and emerging space nations. After initial founding, these countries will form a committee to establish a governance structure similar to the IMF. Using this model, the IIF body could require members to pay a contribution based on the weighted average criteria outlined below. Thereafter, an open invitation to all nations will subsequently be made allowing new members to be added upon payment of the required membership quota. Aiming for a $500 billion fund within 20 years of founding could be a realistic proposition if membership numbers reach two-thirds of the IMF. The benefits and incentives for participation will need to be emphasized to ensure successful financing, and this is where lobbying and outreach will play an active role. One of the incentives for participation in the IIF could include the production of energy in space. A portion of the output of this energy would correspond to the level of investment made. Space-based solar power systems could revolutionize energy consumption and help drive the economic case for the participating nations. The case for this is made unequivocally in the 1975 NASA Summer Study [8].

IMF's quota system is used to determine a country's financial and organizational relationship, based on these three aspects:

1. Subscriptions or quota share – the maximum financial resources a nation is required to pay upon joining
2. Voting power or voting share – member's voting power
3. Access to financing – the level of available financing for the nation usually capped at 200 percent of the nation's quota

Whereas IMF countries contribute funds to a pool through the quota system from which countries with payment imbalances can borrow, the IIF will serve as a body that utilizes this pool to directly invest in the infrastructure and technological development needs of the interstellar worldship project. The IMF aims to improve the economies of its members through the provision of loans. The IIF will aim to facilitate and foster economic activity for the member countries, advancing their technologies and delivering societal progress in pursuit of its long-term mission of interstellar travel.

Second to the IFF will be incentive financing, which is a way of enticing the development of technologies through award programs. This financing approach can play a strong supporting role that will help drive forward innovation. It is through incentive financing that much of the cost reductions and technological innovations have historically been achieved. An example of this is the Orteig Prize which was a $25,000 ($340,067 as of 2015) prize awarded to Charles Lindbergh for the first non-stop flight across the Atlantic [9]. There is also the modern equivalent of the Orteig Prize known as the X-Prize. These types of prizes help promote competition and drive technological innovation. Involving the commercial space ventures in this way adds to the compelling case for the mission and ensures there are added economic benefits of which can be taken advantage. Prize money for these awards could come from the IIF, with a focus on ensuring equal involvement from member states.

Additional funding sources, though miniscule in comparison, are also necessary. One type of additional funding that has seen recent proliferation is online Crowd Funding. A Crowd Funding Platform (CFP) is a form of funding that takes advantage of the general public or "the crowd" to raise funding for projects, otherwise known as campaigns. CFPs can be utilized to help support development costs and, in addition, foster public outreach.

A second type of alternative funding is leveraged finance, or borrowed capital. Utilizing leveraged finance to support technology development programs could support the interstellar project through difficult periods of the project. By leveraging debt financing, the project can offset any future costs associated with the project and increase the return on investment. The IMF model, for example, supplements its quota resources through borrowing if it believes that they might fall short of members' needs. A number of member countries and institutions stand by to lend additional resources to the IMF. This type of financing could support additional, unforeseen costs and help support the continuation of technology developments.

In order to fulfill the dream of interstellar travel, many economic issues must be resolved. A full-scale solar system-wide economy should be taking shape, in which the international community plays an integral part in its development and growth. A society capable of building a starship will likely already be a space-faring society, with a solar system-wide economy to draw upon, utilizing all the resources and energy sources available in that system [10]. It is therefore difficult to imagine a situation where interstellar worldships are being developed prior to the existence of a solar system-wide economy.

2.3 Public Outreach and Education

To successfully develop and construct an interstellar, multigenerational worldship, Earth's population must be supportive and educated. Gaining this support will be one of the major challenges on the road to becoming an interstellar society, so effective public outreach is of the utmost importance. Once the public is onboard with the project, they can be educated on the mission objectives and thus influence their governments and officials to support the interstellar project. Overall, public cooperation will be necessary, as in the days of Apollo, where outreach and educational programs reached people of all ages.

There are several challenges associated with an interstellar worldship that are otherwise not present in other space programs. The vast scale of the project gives the need for global participation from many nations. Reaching target audiences with contrasting cultural motivations, languages, and practices will make this difficult. Luckily, the worldwide society is becoming increasingly interconnected through the Internet and social media, but these platforms require new methods and strategies. Finally, retaining the public's interest over the lengthy project timeline will be difficult. The public must remain engaged to maintain financial and technical support. Luckily, there are opportunities to be found in the outreach challenges. For example, younger generations feel the need to be active participants, which can be leveraged by including interactive and individually driven outreach programs. NASA does this through programs designed to crowd source technical solutions from employees and the public alike. This increased focus on individuals and connectivity will also help to solve many international issues.

An effective outreach program for the worldship mission must be multifaceted to ensure that the entirety of the global community has access to participate and access information. Outreach must be pursued consistently and cleverly, and must adapt with technology and societal norms. There should be facets of the program that reach individuals in all fields. There may be design contests that reach members of the science, technology, engineering, and math (STEM) fields and there may be digital design competitions to attract individuals that are not so obviously connected to an interstellar mission.

The Interstellar City will be integral in enacting outreach and education programs. The governmental systems, ways of living, and technological advancements within the city will be fascinating to people on the outside. Visiting and learning about the city will pique the interest of individuals and encourage a different way of thinking. As projects such as the Biospheres become more sophisticated, members of the general public may have the opportunity to visit and live within the structures for short periods to experience life as a future explorer. The Interstellar City will also be the hub for reaching the global community.

In the world of today, social networking has become ubiquitous and can be utilized for a broad reach in a short amount of time. To do this, surveys and design contests may be utilized to ignite interest and even generate innovative new technologies. Additionally, social media can be utilized to monitor data trends that may prove vital in evaluating the effectiveness of outreach programs and interstellar campaigns themselves. Another popular outlet is interactive media, or computer games. Massive multiplayer online games (MMOs) such as World of Warcraft and Minecraft are hugely successful and have enticed people to create entire digital worlds. Games with similar mechanics could be utilized to explore mission concepts, psychological tendencies, and encourage new ideas. Computer game design and competitions will encourage passionate people to explore and gather knowledge on interstellar travel. One such game could be based on designing an interstellar ship, its environment, habitats, and preparing tools to survive. Media outlets such as these that work through the medium of the Internet are some of the most effective ways to reach much of the world's population with minimal expense.

By the launch date of the worldship, the world will be very different from today, and maintaining post-launch outreach activities will be essential. This will ensure strong interaction between Earth and the worldship while communication is possible. Also, this will inspire future generations of explorers to venture deeper into the universe and ensure that this is not a one-time project.

In addition to outreach, strong educational systems on Earth leading up to a launch of the worldship are essential to create a society capable of implementing the vision and to bring value to the purpose of the mission.

In order for a mission such as this to take place, an interstellar-minded society willing to take on the challenge must exist. This will require individuals who believe in the value of the worldship mission and are willing to devote their work to a pursuit they may never see carried out. This means that from a very young age students should be encouraged to think in the very long term. This mindset can be created through an interstellar education integrated into national curricula and extensive interstellar engagement programs.

The first priority will be to ensure that there are enough technically knowledgeable personnel on Earth to physically design and create the worldship. This means that education in STEM fields should continue to have an increasingly strong presence in school curriculum. Nebergall notes that a special focus should be placed on the general problem-solving methods rather than learning to solve specific problems. Creative thinking and innovation will be essential to ensure that the people on Earth are quickly adaptable and able to create what is need with the available resources [11].

An interstellar society requires a change in many ways of thinking in addition to traditional knowledge. In order to prepare students to think in this sustainable way, non-interstellar specific courses would also be vital. Examples of this would include project-based learning with regards to recycling and up-cycling. Interdependencies could be modeled through analog examples such as permaculture and biosphere design. Curriculum would include team-building activities, as well as create an environment in which students will learn to be resilient and self-confident.

A growing trend in the information age is the availability of free online education for those who want to learn for the sake of it. An online course initiative will be an effective way to provide information and create awareness between the enthusiast and interested people in our society. Free online short-term courses will gather and transform the approach of interstellar education. Students, scientists, professionals would have opportunities to participate in such type of courses. Educating individuals would provide a nice avenue to leading them directly to participate in many of the outreach programs in a productive way.

Adequate public outreach and education programs that reach the global community of individuals of all ages and backgrounds is essential. An interstellar, multigenerational worldship will require the majority of humanity's support, understanding, and effort to be a success. Fundamentally, this means that the outreach and educational programs should be a prime focus for this mission.

3. Worldship Society

The societal challenges for the citizens of Earth leading up to the launch of the worldship are daunting. Many considerations must be taken into account for what the society on the worldship will be, from the population to language to religion. Specific decisions about the makeup of the worldship society cannot be made at this early stage, but an idea of the challenges and options facing the society as the launch approaches is investigated.

3.1 Population

The size of the population is a very significant determining factor for the size of the worldship and scope of the roadmap, and there are a number of differing views on how to accommodate the size of the population. The current definition of a worldship is 100,000 individuals, but the specifics of achieving that value are flexible. The launch population size and age distribution are two key factors. Decisions about the population must be made as the launch date approaches.

The worldship must eventually be populated with 100,000 individuals, but the population size at launch may vary. At one end of the spectrum is to launch with the smallest population feasible for genetic diversity, usually defined as 10,000 individuals. This would allow the population to grow at a natural rate and apply constraints to that growth as necessary. However, if there were an event early in the journey that wiped out a large number of the population, the remaining population would struggle with genetic diversity and ship maintenance and upkeep. In addition to genetic diversity, another limiting factor on a ship starting out with a smaller population size is knowledge transfer. With fewer individuals at the start, less knowledge will be available to share with the following generations. At the other end of the spectrum is to launch the worldship with the full population size of 100,000 individuals. There would be little issue with genetic diversity or knowledge transfer, but the population would be faced with maintaining population control from the onset of the journey.

No matter the launch population size, there will eventually be a convergence of population versus space. There are a number of methods to achieve the population control necessary at that point. One method includes a reproduction lottery in which a couple can enter for the chance to have a child when a member of the population passes away. Another method is to limit the number of offspring each couple can produce, which would also help maintain genetic diversity. Each of these limited methods would introduce ethical issues of their own, which must be considered in depth by the population as a whole.

A second key factor is the composition of the launch population. The age distribution of Generation Zero must be considered. First, the population can be made up of individuals with the same age distribution as on

Earth, with a specific number of youth, middle-aged, and elderly individuals. This would be the best option in terms of knowledge transfer and range of knowledge. Second, the population can be composed of individuals from one age range, who then self-distribute as subsequent generations are born. In particular, one option is a Generation Zero comprised of the youngest individuals feasible, which would allow for less emphasis placed on reproduction and knowledge transfer at the start.

In addition to age distribution is genetic and cultural composition. Creating a homogenous Generation Zero would lead to a less confrontational society, but would also limit genetic diversity. To mitigate this, a larger population size is advantageous. A heterogeneous Generation Zero would ease the genetic diversity, no matter the population size, but would introduce a number of possible inter-group confrontations. Ultimately, the population selection criteria must be stringent enough to weed out potentially troublesome individuals and create a Generation Zero that best meets the needs of sending humanity to the stars.

3.2 Culture

Humans are fundamentally social beings, so culture is engrained in every part of life. Culture can be seen as the manifestation of human intellect, reflected by social habits, cuisine, values, art, music, language, and religion. A single person is a part of many cultures, from a local community culture to a professional culture. With the rapid interconnectedness of the 21st century, the lines between different cultures are blurring. Where immigration, travel, and social media have brought cultures into contact, these cultures are adapting and changing. Furthermore, as the world's population grows, so does its cultural diversity and interconnectedness.

However, despite the interconnectedness of the world there is still a large amount of disconnect and contention between the world's nations and religions. This will undoubtedly continue to be inherent onboard the worldship, and strife and contention must be dealt with without causing more friction between groups that may lead to growing gaps over multiple generations. A mission of this scale will require a vast change in the global society's outlook and operations. With the multitude of nations participating in the planning and construction of the worldship, there will be nations working together who have never before done so. This will cause a clashing of cultures. Culture is a very fluid concept and therefore is capable of change at a much quicker pace, potentially causing geo-political complications if not properly mitigated [12].

On Earth, there are many examples of people of different geographical cultures coming together in one place to live peacefully. As the society becomes more homogeneous, aspects from the different cultures "melt together" in what is often called a melting pot. For example, the United States has experienced many occurrences of large immigration, and over time, the various cultures start to change and share characteristics. Generation Zero on the worldship will have a wide range of different cultures, and over time, those cultures will melt together. In addition to geographic cultural differences are generational cultural differences. Older generations are generally more rigid in their methods and beliefs whereas younger generations adapt more quickly. If Generation Zero is composed of a wide age distribution, this will come into play as well. Various cultural factors will affect the continuity of cultures over time through cultural reproduction, which will be significant for the worldship.

The establishment of the Interstellar City prior to launch will start the process of cultural melting to mitigate cultural strife onboard the worldship. Here, individuals from around the world will come together and start the cultural melting process. This will push forward the collective thinking and cohesiveness necessary for the worldship. An example of this is Star City in Russia, where most cosmonauts and astronauts from all over the world train. In addition to the training equipment, this facility has its own high school, shops, post office, and much more. In the Interstellar City, people will not just receive their training; they will also build connections, get used to each other, and start to create their own culture.

Ultimately, the culture that develops onboard the worldship will be affected by everything leading up to launch. Culture is an important discussion point in population selection to determine what traditions, religions, ethics, and language are carried onto the worldship. Choosing which cultural identities are allowed and disallowed in the selection process is a point that must be handled with delicacy, and it must be understood that a combination of cultures will be brought onto the worldship and no one cultural identity is inherently superior to another.

3.3 Ethics

The ethical implications to be considered for the worldship are truly global issues of concern and will affect the mission before launch and continue throughout the mission. Ethics and morals are difficult to separate and are different for each individual. For the purposes herein, ethics are defined as concrete global, social rules for how to act [13]. It should be noted that the ethics onboard the worldship will vary from the earthly ethics, and the divide will grow as the mission continues. At this point in the process, some of the implications should be parsed out and discussion should be started, but no finalized decisions can be made.

First to be considered are the ethical implications of sending off a sizeable group of individuals on a mission that only their fairly distant descendants would complete. Not only would Generation Zero be subjected to the remainder of their life lived entirely on a worldship, the next couple of generations would never even have the opportunity to live life on a planet. Although volunteers will populate the first generation, their descendants will have no choice. However, all the descendants are inherently affected by the actions and decisions of their ancestors, so this ethical consideration is manageable.

Another possible ethical contention is what to do with the deceased onboard the worldship. Different cultures handle their deceased in very different ways. Onboard the worldship, the options could include a funeral, cremation, or jettison into space. For a generational worldship, the loss of nutrients and carbons following generations of deaths would not be maintainable, so the bodies of the deceased must be recycled into the ecosystem of the ship through a controlled method. This will not sit well with many of Generation Zero, but they must board the worldship knowing this unavoidable aspect of worldship life. Once the ship is launched it will be about the worldship community's global survival and certain things will need to be done that may seem inconceivable to society now but will be necessary in the new "Wild West" of the galaxy [14]. Further delving into this particular topic are the issues of whether to jettison the bodies of those that have passed away due to some contagion. Considerations such as this must be taken into account additionally to ensure the continuation of the mission and the health of the population.

These considerations bring into focus one of the most important questions for the mission: will Earth continue to support a mission that must carry out deeds necessary for the survival of the mission that are not considered ethical and humane by Earth's standards? An example may be if the population must euthanize an infected portion of itself to continue a healthy existence and not waste resources on members of the population that may in turn endanger the wellbeing of the population as a whole. Along those lines, a long-term prison would be a major strain on the worldship resources, so a death sentence may be a necessary punishment for a severe enough crime, such as a serial killer.

The essence of humanity cannot, at this point, be conclusively determined, but is at the center of all ethical considerations. The ethics of a population are intrinsically tied to the humanity of that population. Although, determining what makes us "human," we have the ability to reflect on that capacity of being human, and therefore will retain the essence of what makes us human, even if it is guided by a different set of guidelines [15].

3.4 Religion

There are few other thoughts and traditions that have persevered as long as religion, making it a topic of great importance with respect to the success of the worldship. Religion can be either a great asset to the mission or one of the most preventive factors from success.

Onboard the worldship, religion may have a positive or a negative influence. The religious options for Generation Zero must be discussed at a global scale before the initiation of the selection process. The results of this discussion will in turn be implemented into the selection criteria. In general, there are three foreseeable options to be discussed, each with its own advantages and disadvantages:

1. Maintain the demographic of religions as on Earth
2. Allow only one religion onboard
3. Allow no religious beliefs and have an atheistic population at launch

First, the current demographic of religions on Earth as they exist may be maintained. The major advantage is a simpler selection process with one less criterion and will enable more breadth in the makeup of Generation Zero. Additionally, the health and psychological benefits found through religious practices will be retained. The major disadvantage is continued religious strife and contention, as found on Earth. This can lead to division onboard the worldship, particularly between the religions.

Second, one religion may be chosen, likely one of the major world religions: Christianity, Islam, Hinduism, etc. Selecting one religion will decrease interfaith strife, if not eliminate it, and maintain the health and psychological benefits of religion. However, it will add to the selection criteria and narrow the possible pool of applicants.

Third, Generation Zero can consist of all atheists who claim no religious beliefs. This follows the lines of the second option of limited, and likely eliminated, interfaith strife and increased selection difficulty. However, both the second and third options will produce strife during the discussions on Earth and during the selection process.

It can be assumed that each religion will abdicate to be selected. In addition, if either one religion or no religion is selected, there is a high probability that individuals that have a strong wish to be included in the worldship population will simply lie about their religious beliefs, complicating the selection process. These individuals will harbor their beliefs until they can pass them down to their descendants, therefore cause the interfaith strife at some point following launch or in a later generation.

Weighing the pros and cons of each option and having the foresight to examine the possible outcomes of each decision is imperative. Regardless of which option is chosen, it will have a direct impact on the population selection. Therefore, such a decision must be made before the start of that process and perhaps before any population selection. The world must come together on some platform to discuss the implications and choose the most advantageous path.

3.5 Language

Language is another major societal consideration. Each individual will bring their own language, which will in turn give each individual their own outlook on life. These different languages could serve as a detriment, yet should also be cherished. Similar to other psychological and cultural issues of in-group and out-group biases, the population could become divided and have difficulty communicating across multiple different languages, especially those that do not have an etiological link, such as English and Mandarin [16]. Therefore, a common language is necessary to aid in collaboration and cohesion throughout the population.

The chosen common language will depend upon the most influential languages of the time. A globally influential language must not only be widely spoken, but also the centrality and connectedness to other languages. Figure 3 depicts the current global language network with book translations as the major data source. Each node depicts a specific language with the size representing the number of speakers; the connections between the nodes represent the number of translations. If the common language for the worldship were to be selected now, the most obvious choices due to their popularity are English, Chinese, or Hindi. Of those three, English has the strongest centrality to the other global languages. This would lead to the selection of English as the common worldship language, similarly to the selection of English as the common language of the aviation industry.

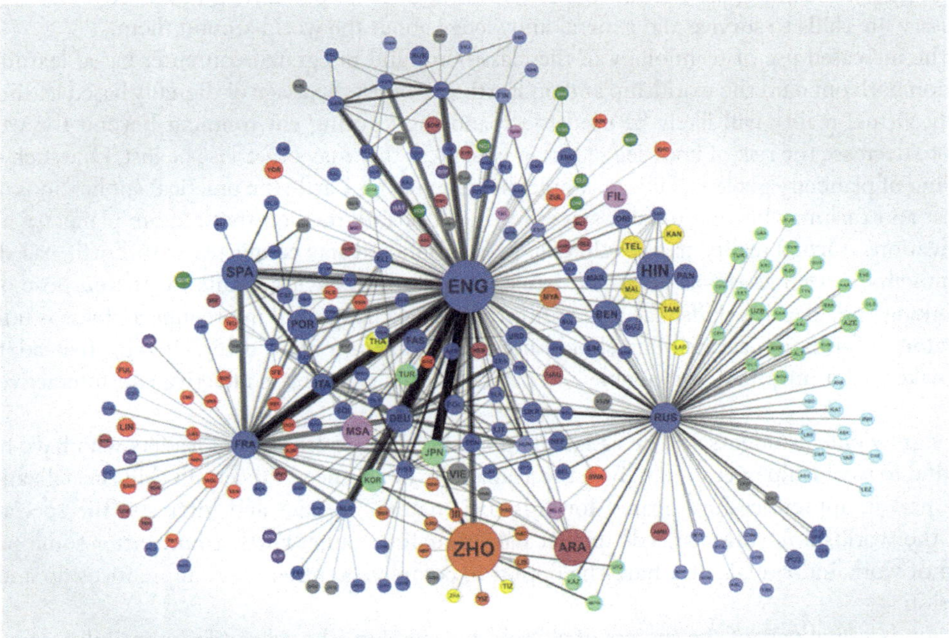

Figure 3: Global language network of book translations [17]

Language drift is another concern. Overall, languages are fairly structured, but the Linguistic Relativity Hypothesis dictates that one thing that will inevitably change will be vocabulary and therefore the population's outlook on the world [18]. For example, an individual from a tropical climate will not have a large vernacular for cold weather. Much of the earthly vernacular people use every day will be irrelevant onboard the worldship. With continued communication with Earth, the drift in language will not be large. However, as the communication delay grows, or if communication stops entirely, the worldship language will likely evolve into something incomprehensible to those speakers not a part of the worldship community. Ultimately, language is not only an important factor for the population selection, but also for the continuation of future communication of an interstellar humanity.

3.6 Education

Education is a key factor in producing competent and effective citizens, so the onboard education must be addressed. As time has progressed, humans have made a continuous record of knowledge in law, agriculture, philosophy, philology, natural history, politics, zoology, medicine, geography, theology, and departments of spiritual, intellectual, and social activity that lie in the range of human knowledge or speculation [19]. A record of human knowledge must be encapsulated in the worldship and accessible to the citizens as, in the words of Cicero, "Those who have no knowledge of what has gone before them must forever remain as children." In addition, an interstellar worldship will have special educational requirements to remain resilient and self-reliant. The educational system onboard must be developed to not only educate citizens about the past, but also prepare them to be effective citizens onboard. The limited population size means that there is a major risk of knowledge loss or skill shortage over time, which is a problem that educational methods must address.

As an interstellar mission would be carried out over multiple generations, many unexpected challenges are likely to arise. It is then argued that the focus of education should not just rely on teaching facts from history, but also on how to think logically to solve problems. Nebergall points out that human nature is accustomed to accepting and using technology without fully understanding the intricacies behind it [20]. As the population's survivability will rely on their capability to repair, recycle, and reuse the resources available, blind reliance on the technological systems of the worldship without a technical understanding could lead to mission failure.

More studies are essential in determining the best systems, configurations and groupings of early childhood educational institutes, primary and secondary schools, research centers, trade schools, laboratories, and universities. Summerford notes that diversity should be encouraged throughout the colony and used the example of a medium sized university on Earth, where, while each dormitory is different, certain halls are all known for something unique [21].

Requirements for technical, cultural, and practical aptitude dictate that a combination of classroom and hands-on settings for learning are needed. Students will need to be put through basic education that will teach them necessary life skills to survive and general knowledge about the world around them.

With the increased use of technology in the classrooms and integrated computer based learning, it is likely that education both onboard the worldship and on Earth will boost the usage of digitally based learning platforms. Additionally, virtual reality will likely be used to expand the learning environment beyond the worldship. This would help to decrease the risk of knowledge loss in vital areas that may otherwise be lost. One such example is an understanding of planetary geology. Unless samples are taken from Earth, the practical applications of this science could be lost and this may be vital to later stages of the worldship mission that require planetary manipulations and identifications. Virtual reality areas could help people onboard stay connected with Earth-based information and environments, as well as expose them to virtual scenarios and environments for the purpose of stimulating creative thinking and adaptive skills. The rate of information exchange and interconnectedness is unlikely to slow down; therefore information accessibility should not be a problem for worldship citizens. Instead, the challenge will be to make useful information available and presented in a way that is effective and interactive for learning purposes.

As they grow older, students will be assigned apprenticeships to learn from leaders who have mastered specific skills vital to worldship operation. Given the availability of and need for certain skill specializations, students would take part in apprenticeships geared towards their natural abilities and interests. The specializations will depend on the worldship's current needs in that particular timeframe. Furthermore, after some predetermined time period of work, individuals may have the option to complete another short apprenticeship and enter into a new profession.

Education is a vital part of the success of a worldship mission. The role and responsibility of education on a worldship will be a community-based effort to relay technical knowledge with craftsmanship, ethical concerns, and teamwork. It will act as a mechanism through which the original Earth culture and intent is preserved. Furthermore, education will play a role of catalyst in conveying knowledge transfer to deal with issues that people may not have dealt with before. Various constraints will still exist but people must be able to adapt to the changes in the new environment.

4. Conclusion

Humanity's destiny and vision has always lain amongst the stars. While early visionaries such as Verne, Goddard, and Tsiolkovsky dreamt and wrote about human travel into interstellar space, current organizations now work to make this a reality. This paper has made an attempt at contributing to this effort by outlining many of the societal challenges to be faced. The societal aspects from the planning and construction stages of the worldship to the actual flight itself have been discussed.

During the road to launch, some aspects include governance, financing, and outreach. Developing a governance system before launch is important. To do this, an Interstellar City can be established. This city will not only be useful to develop a governance system, but also to aid financing efforts, lessen cultural conflict, and begin education and training of future interstellar citizens. To finance the worldship, the primary method will be the establishment of an International Interstellar Fund to gather funds from the earthly nations with additional funding from incentive financing, crowd sourcing, and leveraged finance. Third, a majority of humanity must be invested in this grand project for it to be a success. To do this, an ambitious, multifaceted outreach and education system must be enacted.

There are many layers to society, and special considerations must be taken to each layer of the worldship society. Many of the societal aspects have different options that must be decided closer to launch by the whole global community. First, the starting population size and make-up of Generation Zero must be determined. The starting population may be the maximum of 100,000 or the minimum of 10,000. Additionally, the population may be homogeneous or heterogeneous with respect to age. The individuals that make up this population will come from a diversity of backgrounds, so cultural differences must be taken into account to ensure the smoothest transition possible into the new worldship society. The ethics onboard the worldship will be vastly different from the ethics found on Earth, from death sentences to burials. With respect to this, Earth may have a difficult time continuing to support such a possibly unethical mission. Differences in religion may also have an effect on the global community's support. Before the population can be selected, the religious backgrounds accepted must be considered, whether the population has the religious demographic found on Earth, one religion, or no religion. Each individual of the population, whatever their religion and culture, will not share the same first language. Therefore, a common language must be selected to ensure cohesiveness within the population, but the various languages brought onboard must also be cherished to keep the future generations onboard the worldship connected with their earthly ancestry. Finally, the educational system onboard the worldship should be outlined before launch, and it can be enacted and tested in the Interstellar City. All of these facets and more make up the considerations that must be taken into account for the worldship society. These considerations must be made before the launch of the worldship and are key to the success of the mission.

This paper covers just a small part of the effort to critically evaluate literature in order to create an interdisciplinary roadmap for the development of a slower-than-light, multigenerational worldship for interstellar travel. The worldship concept is one of a self-sustaining ship capable of transporting thousands of humans into interstellar space over many generations. Parameters included non-relativistic speeds and a population of 100,000 individuals living their lives. The full work was created by the Astra Planeta team, composed of twenty-two students from the International Space University's (ISU) 2015 Masters of Science in Space Studies class [22].

Acknowledgements

The author would like to acknowledge the entire Astra Planeta team whose efforts brought this project to fruition:

- Kyle Acierno
- James Bevington
- Shambo Bhattacharjee
- Chaitra
- Guang Chen
- Daphne De Jong
- Lei Geng
- Avishek Ghosh
- Karan Gujarati
- Micah Klettke
- Florin-Cristian Lazar

- Olivier Leblanc
- Yang Liu
- Hameed Monoharan
- Hamza Ragala
- Brian Ramos
- Jean-François Rococo
- Fergus Russell-Conway
- Mansoor Shar
- Abigail Sherriff
- Anderson Wilder
- Peng Zuo

Additionally, the author would like to thank Chris Welch, the advisor for the team and project, and the International Space University for the support throughout the development and completion of the project.

References

1. Astra Planeta, 201Astra Planeta Final Report. [online] Available from: <https://isulibrary.isunet.edu/opac/doc_num.php?explnum_id=731>

2. Astra Planeta, 201Astra Planeta Final Report. [online] Available from: https://isulibrary.isunet.edu/opac/doc_num.php?explnum_id=731

3. Hill, T., 200Space: What Now? The Past, Present, and Possible Futures of Activities in Space. Baltimore: PublishAmerica, LLLP, pp. 4-22

4. Hansen, M.H., 200A comparative study of thirty city-state cultures: An investigation. Kgl. Danske Videnskabernes Selskab.

5. Zak, A., 201The real rocket to nowhere. [online] Available from: <http://www.russianspaceweb.com/vostochny_soyuz.html>

6. European Space Agency, 201ESA, An intergovernmental customer. [http://www.esa.int/] European Space Agency. Available from: <http://www.esa.int/About_Us/Business_with_ESA/Business_Opportunities/ESA_an_intergovernmental_customer>

7. Tsagourias, N., 200Transnational Constitutionalism: International and European Perspectives. Cambridge University Press.

8. Johnson, R.D. and Holbrow, C., 197Space Settlements. A Design Study. NASA, Washington, SP-41[online] Available from: <http://www.unbc.ca/assets/history/courses/201101_nasa_sp_413.space.settlements.a.design.study.pdf>

9. “Raymond Orteig - $25,000 Prize.” Charles Lindbergh - An American Aviator. Spirit of St. Louis 2 Project, 201[online] Available at: <http://www.charleslindbergh.com/plane/orteig.asp>

10. Martin, A., 19World Ships-Concept, Cause, Cost, Construction and Colonisation. Journal of the British Interplanetary Society, 37, p.243.

11. Nebergall, K., 20Becoming an Interstellar Culture. 100 Year Starship 2012 Symposium Conference. Houston, TX: CreateSpace Independent Publishing Platform, pp.291–299.

12. Wooldridge, B., 20Preparing an Interstellar Civilization for the 100 Year Starship: Don't Hold Your Breath! 100 Year Starship 2012 Symposium Conference. CreateSpace Independent Publishing Platform.

13. Arnould, J., February 20Religious & Ethical Questions. Internal Communication.

14. Arnould, J., February 20Religious & Ethical Questions. Internal Communication.

15. Arnould, J., February 20Religious & Ethical Questions. Internal Communication.

16. Kondo, Y., Bruhweiler, F.C., Moore, J. and Sheffield, C. eds., 20Interstellar Travel & Multi-Generational Space Ships: Apogee Books Space Series Burlington, Ontario: Apogee Books.

17. "Global Language Network." Global Language Network. MIT Media Lab Macro Connections Group, 20[online] Available at: http://language.media.mit.edu/visualizations/books

18. Swoyer, C., 20The Linguistic Relativity Hypothesis. [online] Stanford Encyclopedia of Philosophy. Available from: http://plato.stanford.edu/entries/relativism/supplement2.html

19. Kern, A., 20Classical Education In Ancient Greece.

20. Nebergall, K., 20Becoming an Interstellar Culture. 100 Year Starship 2012 Symposium Conference. Houston, TX: CreateSpace Independent Publishing Platform, pp.291–299.

21. Summerford, S., 20Colonized interstellar Vessel: Conceptual Master Planning. 100 Year Starship 2013 Public Symposium Conference. The Dorothy Jemison Foundation for Excellence.

22. Astra Planeta, 20Astra Planeta Final Report. [online] Available from: <https://isulibrary.isunet.edu/opac/doc_num.php?explnum_id=731>

A Roadmap to Interstellar Travel

The Societal Challenges [Poster]

Abigail Sherriff and Chris Welch

Summary

This work presented a thorough and interdisciplinary roadmap to the development and launch of a worldship within the 100 year timeframe. Necessary enabling technological developments were identified and incorporated with existing or emerging technologies.

Editor's Note

Sherriff's and Welch's contribution was an intricate poster laying out a potential roadmap for a generational worldship. Due to the complexity of the design of the poster, it is displayed in two views in the next few pages. On the following page, rotated 90 degrees, the entire poster is presented. Then on the subsequent pages, the poster is provided, magnified, in parts so that the detail of the entry can be viewed and valued. Unfortunately, this does mean that certain words and lines are cut off in the magnified view. However, both views together provide an overall reproduction of their map.

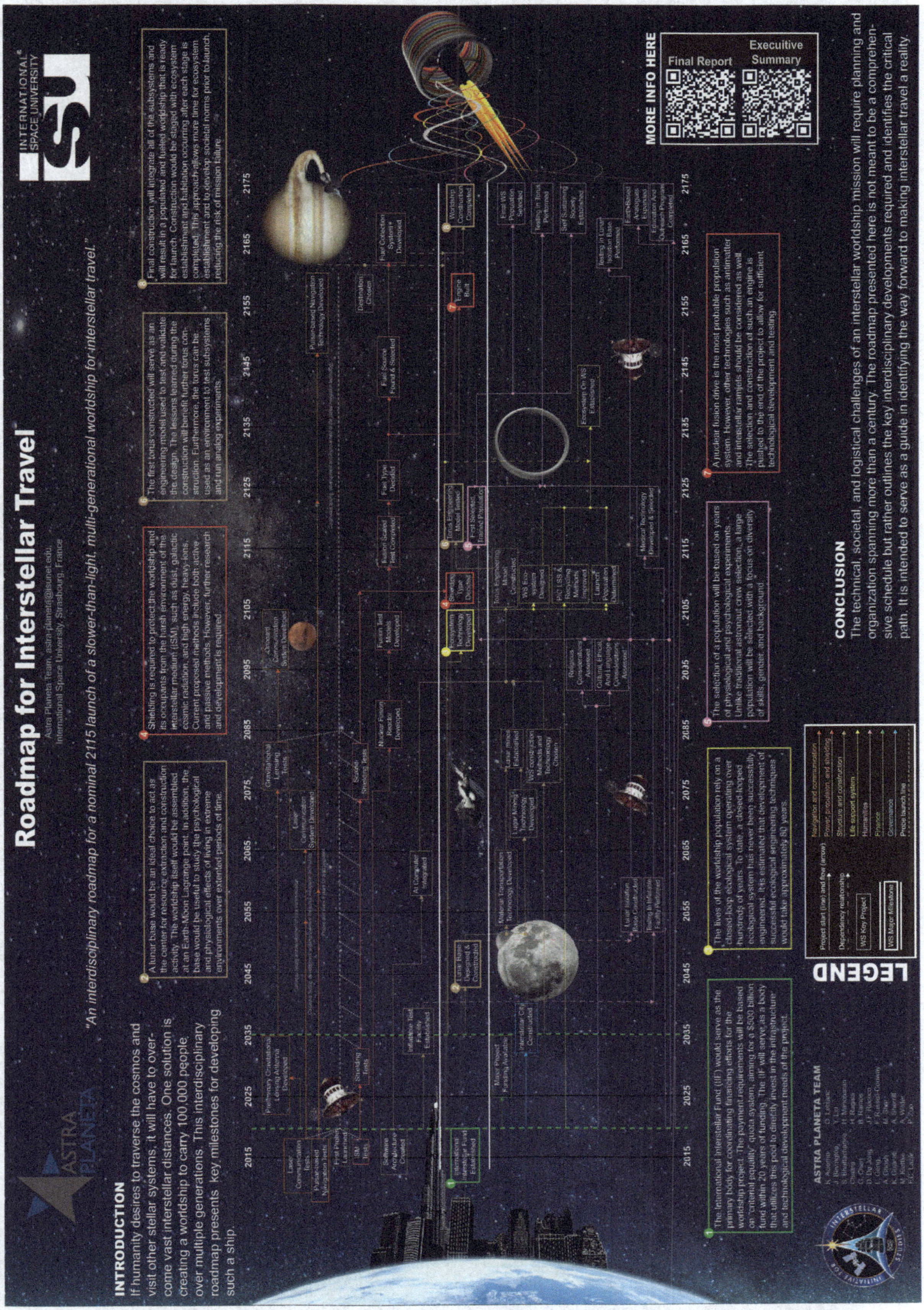
Roadmap for Interstellar Travel
Astra Planeta Team, astra-planeta@isunet.edu
International Space University, Strasbourg, France
ASTRA PLANETA
INTERNATIONAL SPACE UNIVERSITY
ISU
"An interdisciplinary roadmap for a nominal 2115 launch of a slower-than-light, multi-generational worldship for interstellar travel."
INTRODUCTION
If humanity desires to traverse the cosmos and visit other stellar systems, it will have to overcome vast interstellar distances. One solution is creating a worldship to carry 100,000 people over multiple generations. This interdisciplinary roadmap presents key milestones for developing such a ship.
2 A lunar base would be an ideal choice to act as the center for resource extraction and construction activity. The worldship itself would be assembled at an Earth-Moon Lagrange point. In addition, the base would be useful to study the psychological and physiological effects of living in extreme environments over extented periods of time.
4 Shielding is required to protect the worldship and its occupants from the harsh environment of the interstellar medium (ISM), such as dust, galactic cosmic radiation, and high energy, heavy-ions. Current proposed methods include both active and passive methods. However, further research and development is required.
6 The first torus constructed will serve as an engineering model used to test and validate the design. The lessons learned during the construction will benefit further torus construction. Furthermore, the torus can be used as an environment to test subsystems and run analog experiments.
8 Final construction will integrate all of the subsystems and will result in a populated and fueled worldship that is ready for launch. Construction would be staged with ecosystem establishment and habitation occurring after each stage is completed. This approach allows more time for ecosystem establishment and to develop societal norms prior to launch, reducing the risk of mission failure.
2015 2025 2035 2045 2055 2065 2075 2085 2095 2105 2115 2125 2135 2145 2155 2165 2175
1 The International Interstellar Fund (IIF) would serve as the primary body for coordinating financing efforts for the worldship project. The payment requirements will be based on "criterial equality" quota system, aiming for a $500 billion fund within 20 years of funding. The IIF will serve as a body that utilizes this pool to directly invest in the infrastructure and technological development needs of the project.
3 The lives of the worldship population rely on a closed-loop ecological system operating over hundreds of years. To date, a closed-looped ecological system has never been successfully engineered. It is estimated that development of successful ecological engineering techniques would take approximately 80 years.
5 The selection of a population will be based on years of physiological and psychological experiments. Unlike traditional astronaut crew selection, a large population will be selected with a focus on diversity of skills, gender, and background.
7 A nuclear fusion drive is the most probable propulsion system. However, other technologies such as antimatter and interstellar ramjets should be considered as well. The selection and construction of such an engine is pushed to the end of the project to allow for sufficient technological development and testing.
ASTRA PLANETA TEAM
LEGEND
Project start (line) and flow (arrow)
Dependency relationship
WS Key Project
WS Major Milestone
Navigation and communication
Power, propulsion, and shielding
Structure and construction
Life support system
Humanities
Finance
Governance
Probe launch line
MORE INFO HERE
Final Report
Executive Summary
CONCLUSION
The technical, societal, and logistical challenges of an interstellar worldship mission will require planning and organization spanning more than a century. The roadmap presented here is not meant to be a comprehensive schedule but rather outlines the key interdisciplinary developments required and identifies the critical path. It is intended to serve as a guide in identifying the way forward to making interstellar travel a reality.

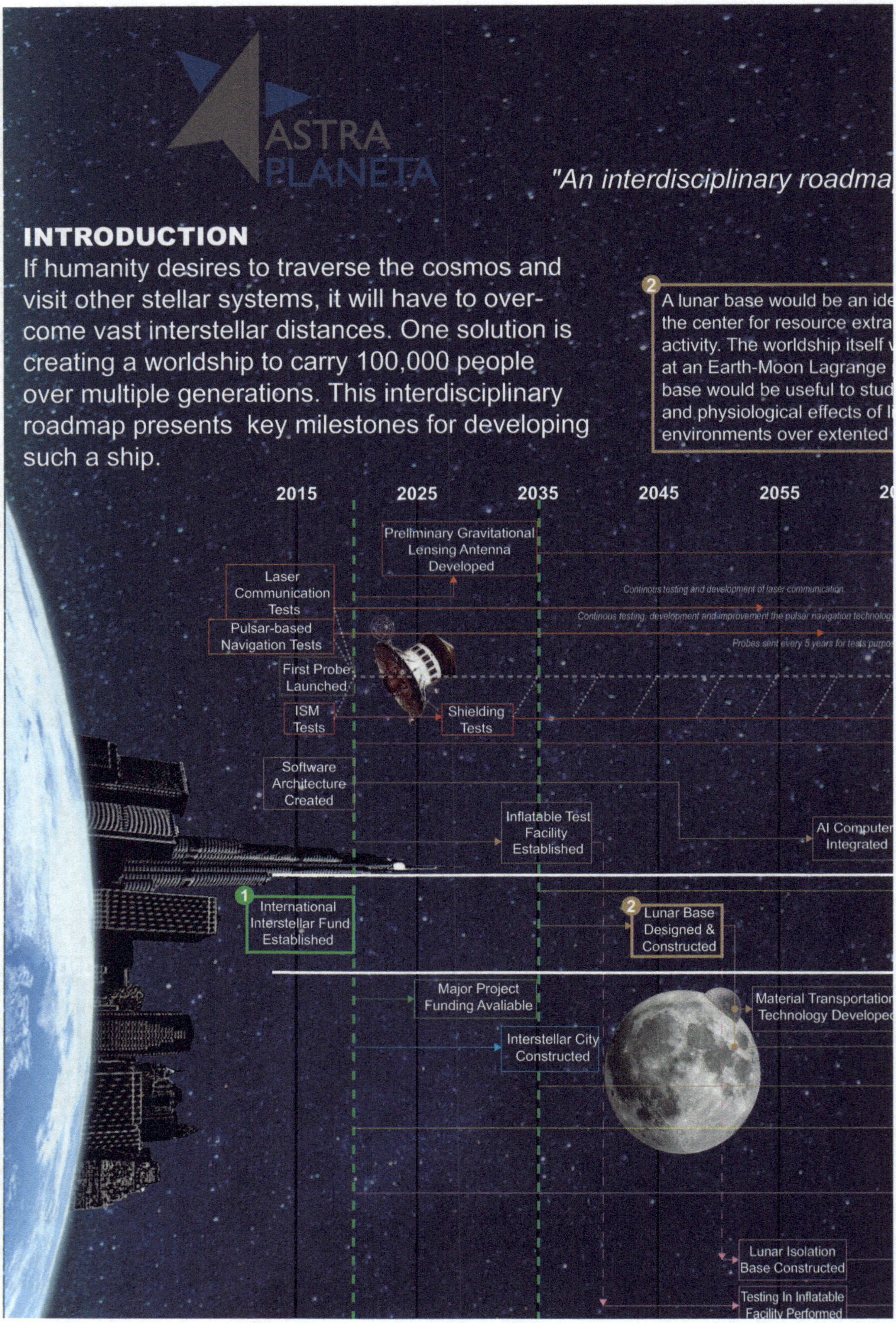
ASTRA PLANETA
"An interdisciplinary roadma
INTRODUCTION
If humanity desires to traverse the cosmos and visit other stellar systems, it will have to overcome vast interstellar distances. One solution is creating a worldship to carry 100,000 people over multiple generations. This interdisciplinary roadmap presents key milestones for developing such a ship.
2
A lunar base would be an ide
the center for resource extra
activity. The worldship itself
at an Earth-Moon Lagrange
base would be useful to stud
and physiological effects of l
environments over extented
2015
2025
2035
2045
2055
Preliminary Gravitational Lensing Antenna Developed
Laser Communication Tests
Continous testing and development of laser communication.
Pulsar-based Navigation Tests
Continous testing, development and improvement the pulsar navigation technolog
Probes sent every 5 years for tests purpos
First Probe Launched
ISM Tests
Shielding Tests
Software Architecture Created
Inflatable Test Facility Established
AI Computer Integrated
1
International Interstellar Fund Established
2
Lunar Base Designed & Constructed
Major Project Funding Avaliable
Material Transportation Technology Develope
Interstellar City Constructed
Lunar Isolation Base Constructed
Testing In Inflatable Facility Performed

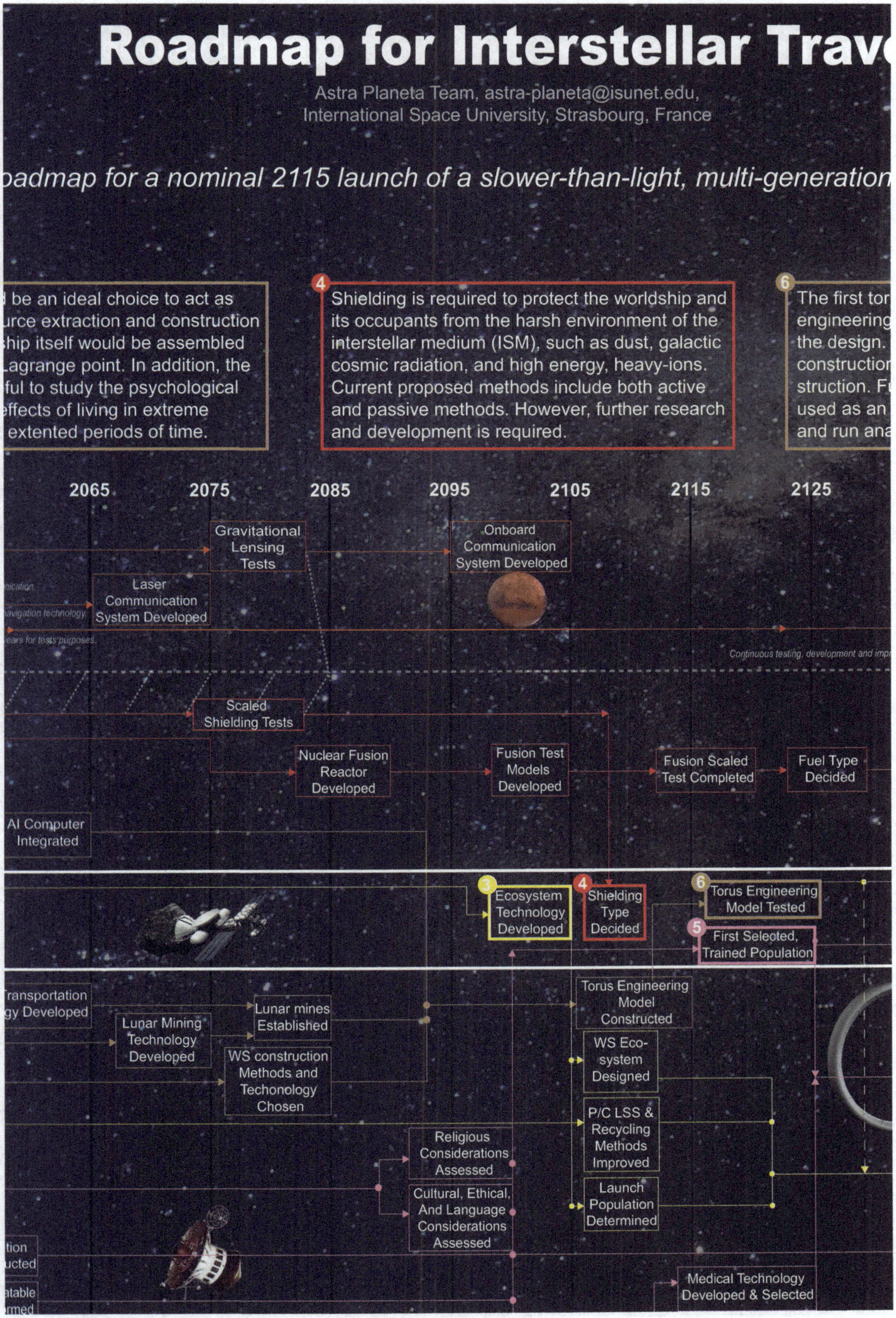
Roadmap for Interstellar Trav
Astra Planeta Team, astra-planeta@isunet.edu,
International Space University, Strasbourg, France
oadmap for a nominal 2115 launch of a slower-than-light, multi-generation
be an ideal choice to act as
urce extraction and construction
hip itself would be assembled
agrange point. In addition, the
ful to study the psychological
effects of living in extreme
extented periods of time.
4
Shielding is required to protect the worldship and its occupants from the harsh environment of the interstellar medium (ISM), such as dust, galactic cosmic radiation, and high energy, heavy-ions. Current proposed methods include both active and passive methods. However, further research and development is required.
6
The first tor
engineering
the design.
constructio
struction. F
used as an
and run ana
2065
2075
2085
2095
2105
2115
2125
Gravitational Lensing Tests
Onboard Communication System Developed
Laser Communication System Developed
Continuous testing, development and imp
Scaled Shielding Tests
Nuclear Fusion Reactor Developed
Fusion Test Models Developed
Fusion Scaled Test Completed
Fuel Type Decided
AI Computer Integrated
3
Ecosystem Technology Developed
4
Shielding Type Decided
6
Torus Engineering Model Tested
5
First Selected, Trained Population
ransportation
gy Developed
Lunar Mining Technology Developed
Lunar mines Established
WS construction Methods and Techonology Chosen
Torus Engineering Model Constructed
WS Eco-system Designed
P/C LSS & Recycling Methods Improved
Religious Considerations Assessed
Cultural, Ethical, And Language Considerations Assessed
Launch Population Determined
Medical Technology Developed & Selected

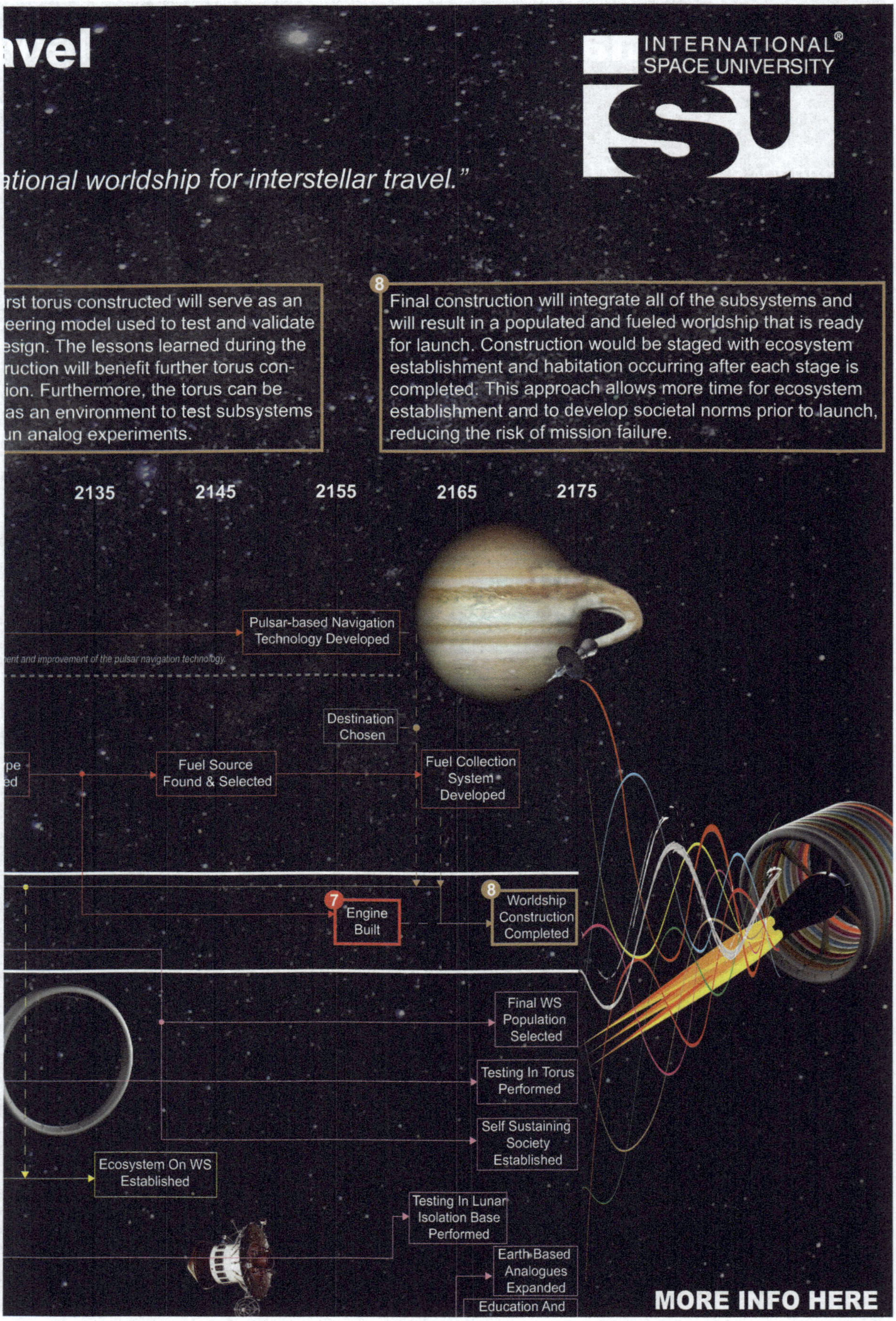
avel
INTERNATIONAL®
SPACE UNIVERSITY
ational worldship for interstellar travel."
rst torus constructed will serve as an
eering model used to test and validate
esign. The lessons learned during the
ruction will benefit further torus con-
ion. Furthermore, the torus can be
as an environment to test subsystems
un analog experiments.
8
Final construction will integrate all of the subsystems and will result in a populated and fueled worldship that is ready for launch. Construction would be staged with ecosystem establishment and habitation occurring after each stage is completed. This approach allows more time for ecosystem establishment and to develop societal norms prior to launch, reducing the risk of mission failure.
2135
2145
2155
2165
2175
Pulsar-based Navigation Technology Developed
Destination Chosen
Fuel Source Found & Selected
Fuel Collection System Developed
7
Engine Built
8
Worldship Construction Completed
Final WS Population Selected
Testing In Torus Performed
Self Sustaining Society Established
Ecosystem On WS Established
Testing In Lunar Isolation Base Performed
Earth Based Analogues Expanded
Education And
MORE INFO HERE

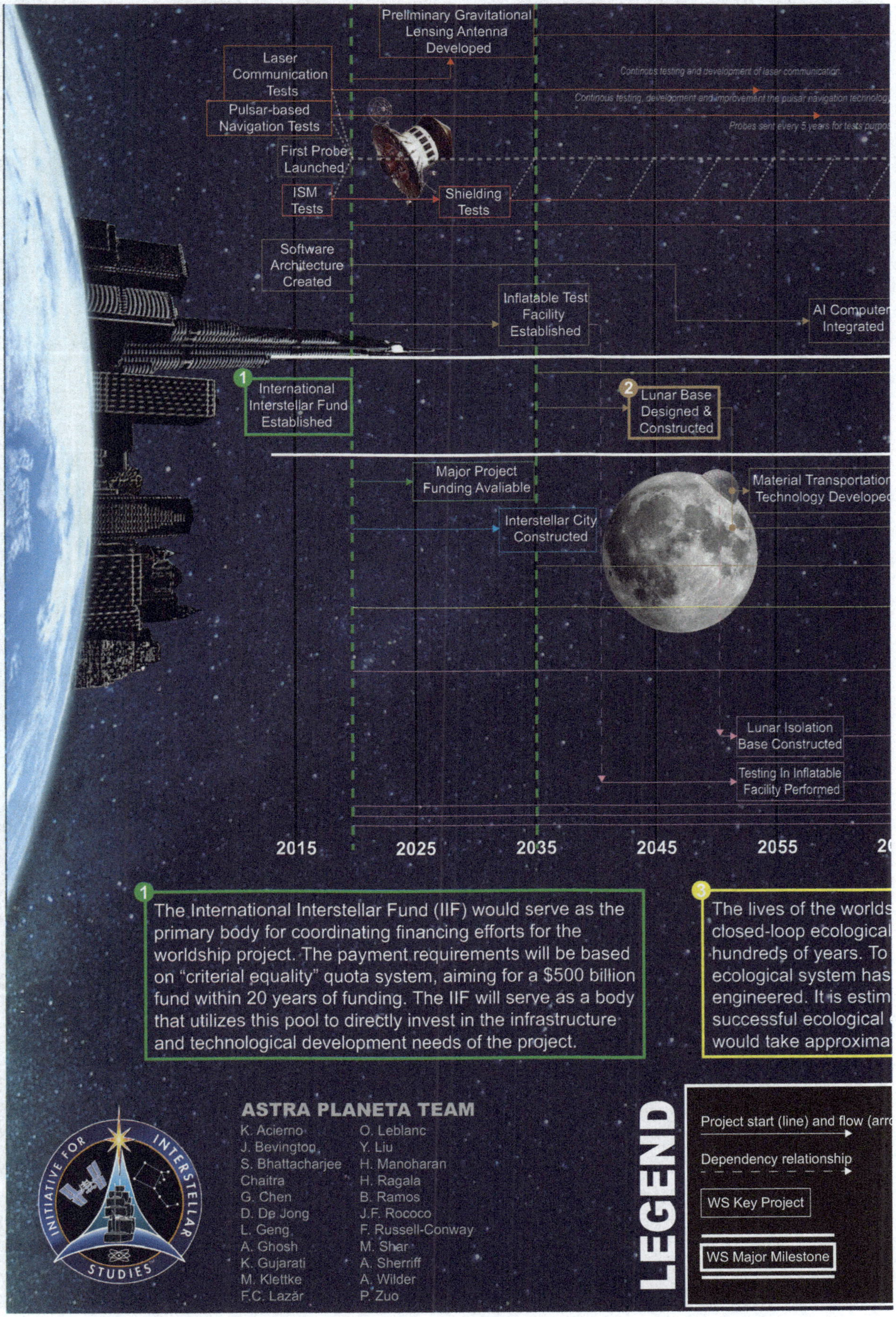

Preliminary Gravitational Lensing Antenna Developed
Laser Communication Tests
Pulsar-based Navigation Tests
Continous testing and development of laser communication.
Continous testing, development and improvement the pulsar navigation technolog
Probes sent every 5 years for tests purpos
First Probe Launched
ISM Tests
Shielding Tests
Software Architecture Created
Inflatable Test Facility Established
AI Computer Integrated
1
International Interstellar Fund Established
2
Lunar Base Designed & Constructed
Major Project Funding Avaliable
Interstellar City Constructed
Material Transportation Technology Develope
Lunar Isolation Base Constructed
Testing In Inflatable Facility Performed
2015
2025
2035
2045
2055
1
The International Interstellar Fund (IIF) would serve as the primary body for coordinating financing efforts for the worldship project. The payment requirements will be based on "criterial equality" quota system, aiming for a $500 billion fund within 20 years of funding. The IIF will serve as a body that utilizes this pool to directly invest in the infrastructure and technological development needs of the project.
3
The lives of the worlds
closed-loop ecological
hundreds of years. To
ecological system has
engineered. It is estim
successful ecological
would take approxima
ASTRA PLANETA TEAM
K. Acierno
J. Bevington
S. Bhattacharjee
Chaitra
G. Chen
D. De Jong
L. Geng
A. Ghosh
K. Gujarati
M. Klettke
F.C. Lazăr
O. Leblanc
Y. Liu
H. Manoharan
H. Ragala
B. Ramos
J.F. Rococo
F. Russell-Conway
M. Shar
A. Sherriff
A. Wilder
P. Zuo
INITIATIVE FOR INTERSTELLAR STUDIES
LEGEND
Project start (line) and flow (arr
Dependency relationship
WS Key Project
WS Major Milestone

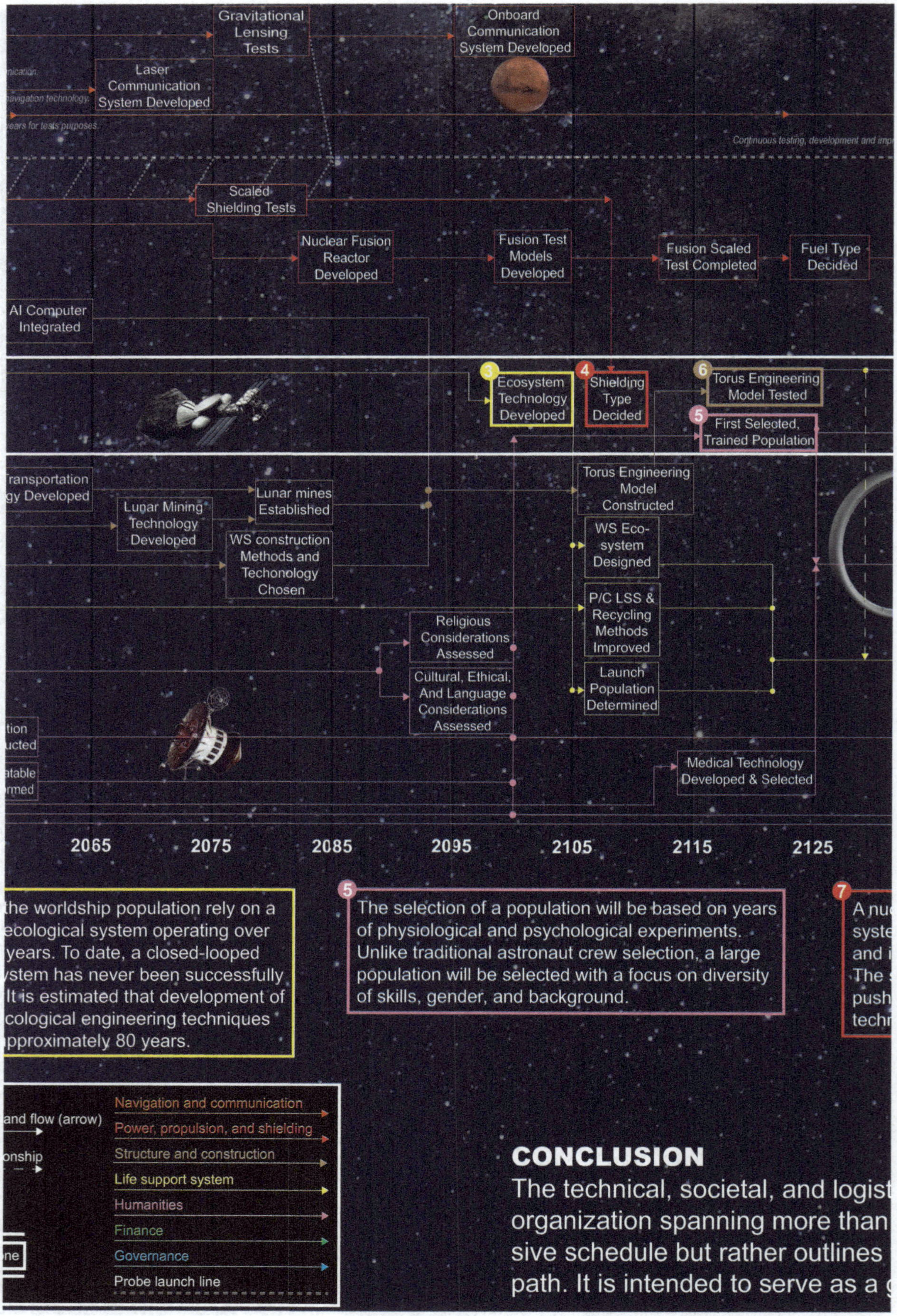

Gravitational Lensing Tests
Onboard Communication System Developed
Laser Communication System Developed
Continuous testing, development and imp
Scaled Shielding Tests
Nuclear Fusion Reactor Developed
Fusion Test Models Developed
Fusion Scaled Test Completed
Fuel Type Decided
AI Computer Integrated
3 Ecosystem Technology Developed
4 Shielding Type Decided
6 Torus Engineering Model Tested
5 First Selected, Trained Population
ransportation gy Developed
Lunar Mining Technology Developed
Lunar mines Established
WS construction Methods and Techonology Chosen
Torus Engineering Model Constructed
WS Eco-system Designed
P/C LSS & Recycling Methods Improved
Launch Population Determined
Religious Considerations Assessed
Cultural, Ethical, And Language Considerations Assessed
tion ucted
atable rmed
Medical Technology Developed & Selected
2065
2075
2085
2095
2105
2115
2125
the worldship population rely on a ecological system operating over years. To date, a closed-looped ystem has never been successfully It is estimated that development of cological engineering techniques pproximately 80 years.
5 The selection of a population will be based on years of physiological and psychological experiments. Unlike traditional astronaut crew selection, a large population will be selected with a focus on diversity of skills, gender, and background.
7 A nu syste and i The s push techn
and flow (arrow)
onship
ne
Navigation and communication
Power, propulsion, and shielding
Structure and construction
Life support system
Humanities
Finance
Governance
Probe launch line
CONCLUSION
The technical, societal, and logist organization spanning more than sive schedule but rather outlines path. It is intended to serve as a g

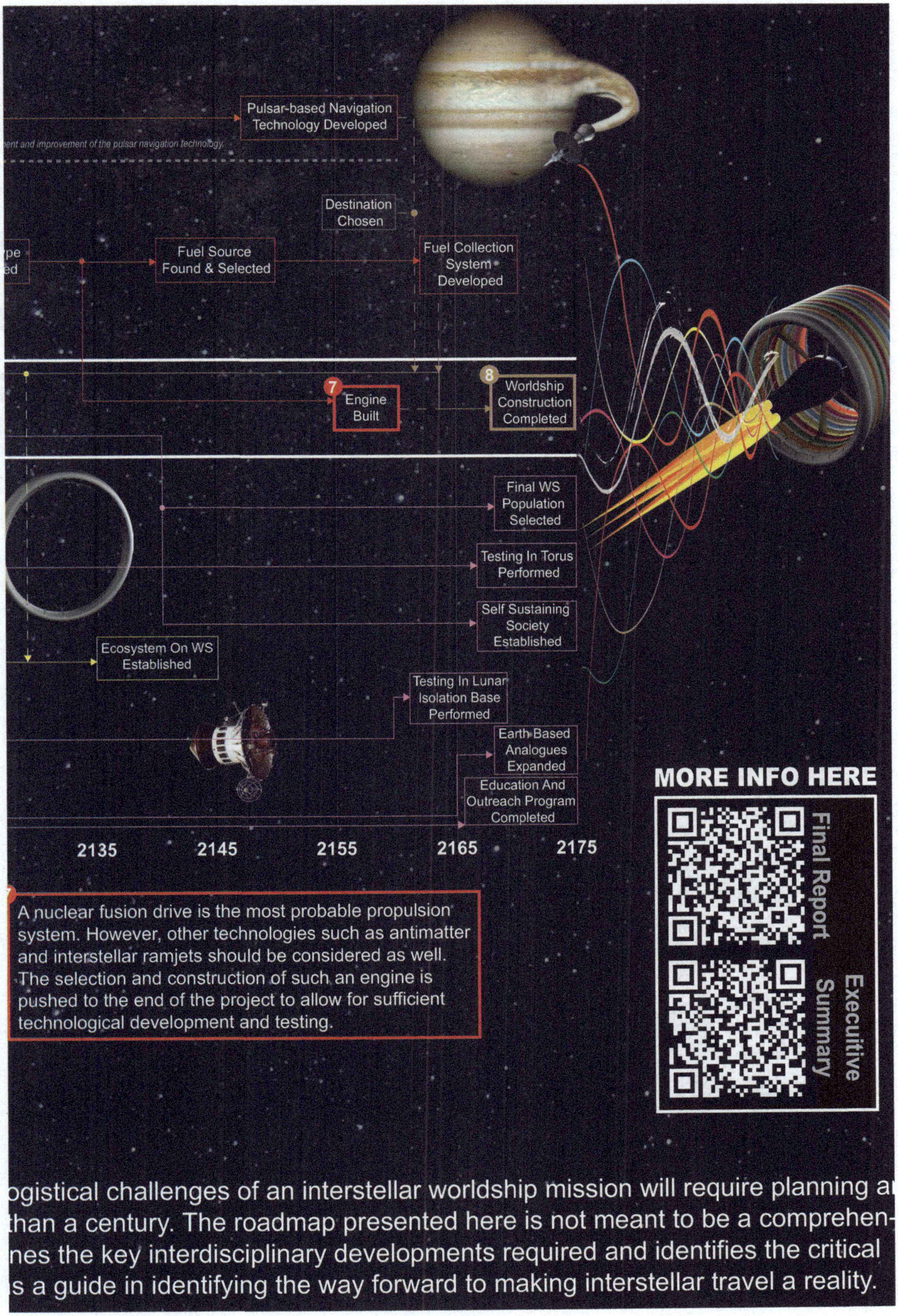
Pulsar-based Navigation Technology Developed
ment and improvement of the pulsar navigation technology.
Destination Chosen
Fuel Source Found & Selected
Fuel Collection System Developed
7 Engine Built
8 Worldship Construction Completed
Final WS Population Selected
Testing In Torus Performed
Self Sustaining Society Established
Ecosystem On WS Established
Testing In Lunar Isolation Base Performed
Earth Based Analogues Expanded
Education And Outreach Program Completed
2135
2145
2155
2165
2175
A nuclear fusion drive is the most probable propulsion system. However, other technologies such as antimatter and interstellar ramjets should be considered as well. The selection and construction of such an engine is pushed to the end of the project to allow for sufficient technological development and testing.
MORE INFO HERE
Final Report
Executive Summary
ogistical challenges of an interstellar worldship mission will require planning a
than a century. The roadmap presented here is not meant to be a comprehen-
nes the key interdisciplinary developments required and identifies the critical
s a guide in identifying the way forward to making interstellar travel a reality.

Technological Approaches That Can Search for Life Close Up in Exotic Destinations

Seth Shostak, Ph.D.

Why bother?

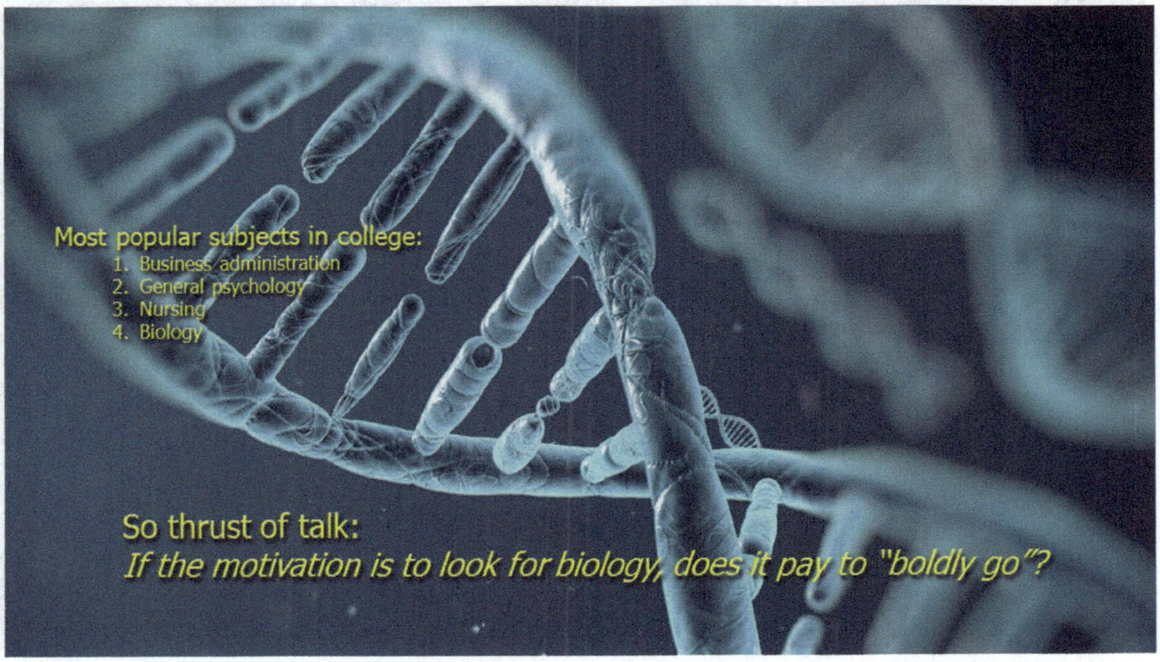
Most popular subjects in college:
1. Business administration
2. General psychology
3. Nursing
4. Biology
So thrust of talk:
If the motivation is to look for biology, does it pay to "boldly go"?

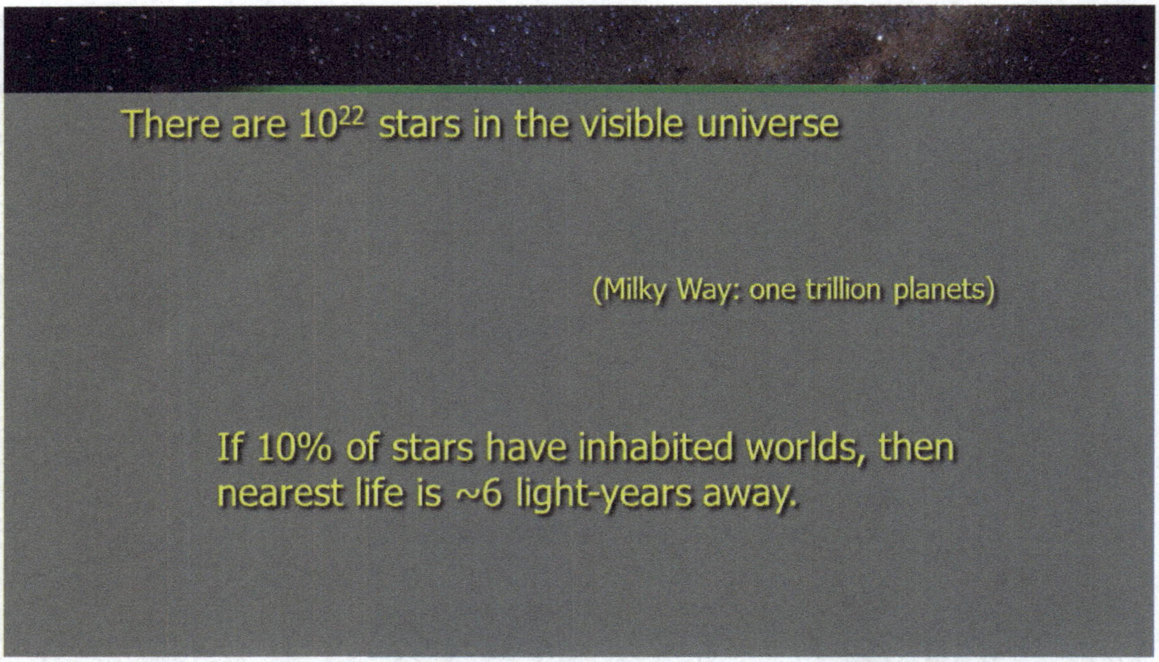
There are 10^{22} stars in the visible universe
(Milky Way: one trillion planets)
If 10% of stars have inhabited worlds, then nearest life is ~6 light-years away.

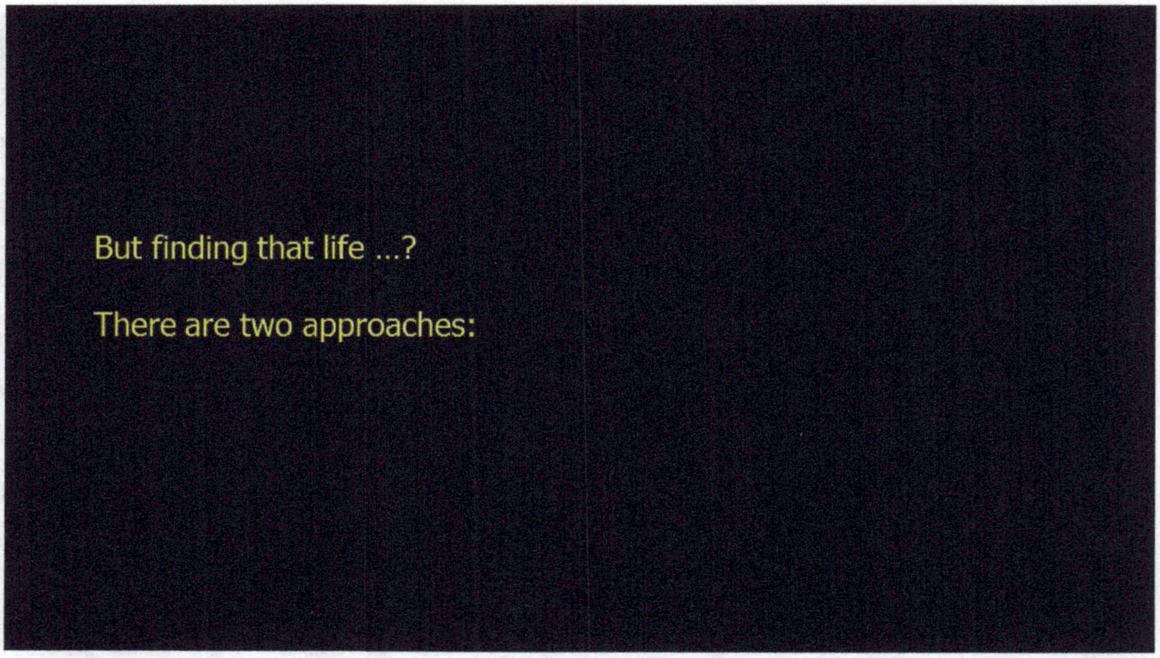
But finding that life ...?
There are two approaches:

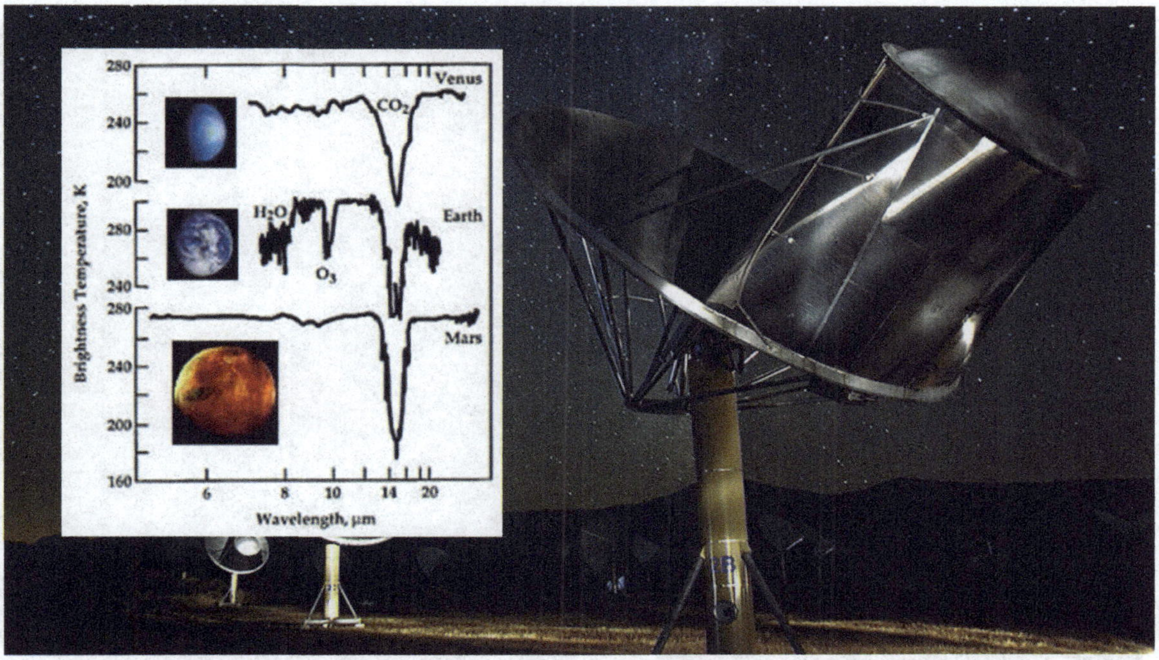
Venus
CO_2
Earth
H_2O
O_3
Mars
Brightness Temperature, K
Wavelength, μm

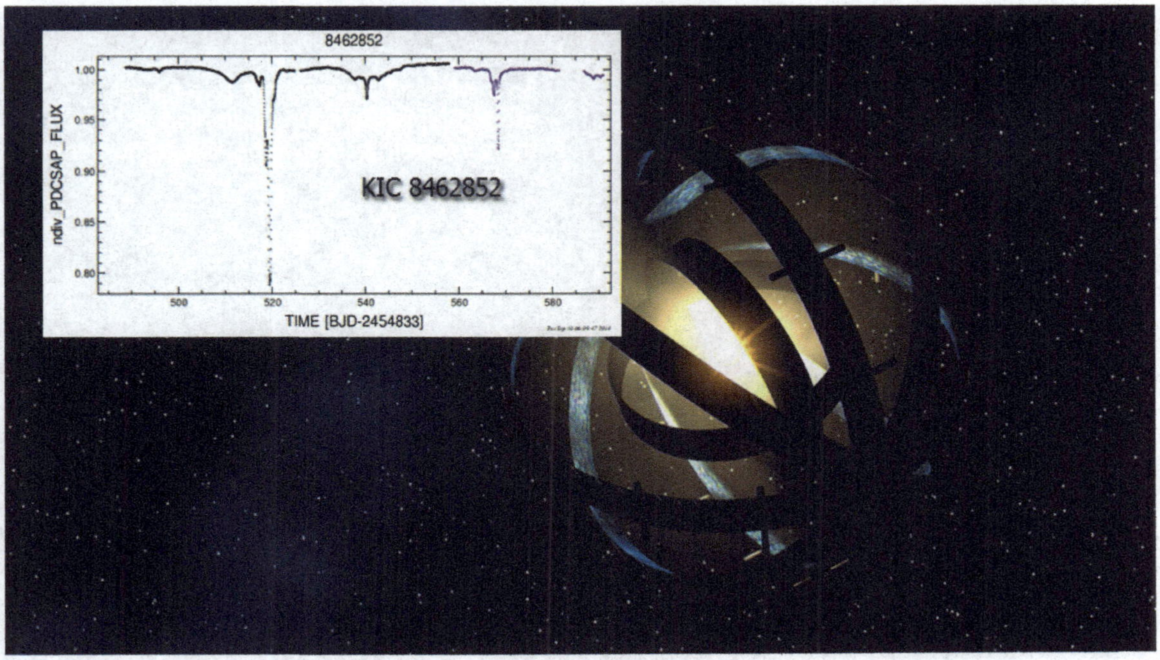
8462852
ndiv_PDCSAP_FLUX
KIC 8462852
TIME [BJD-2454833]

Hard way …
Viking Lander, 1976

So if we send a spacecraft to a planet around Alpha Centauri (assuming there *is* one), would we be able to do better?

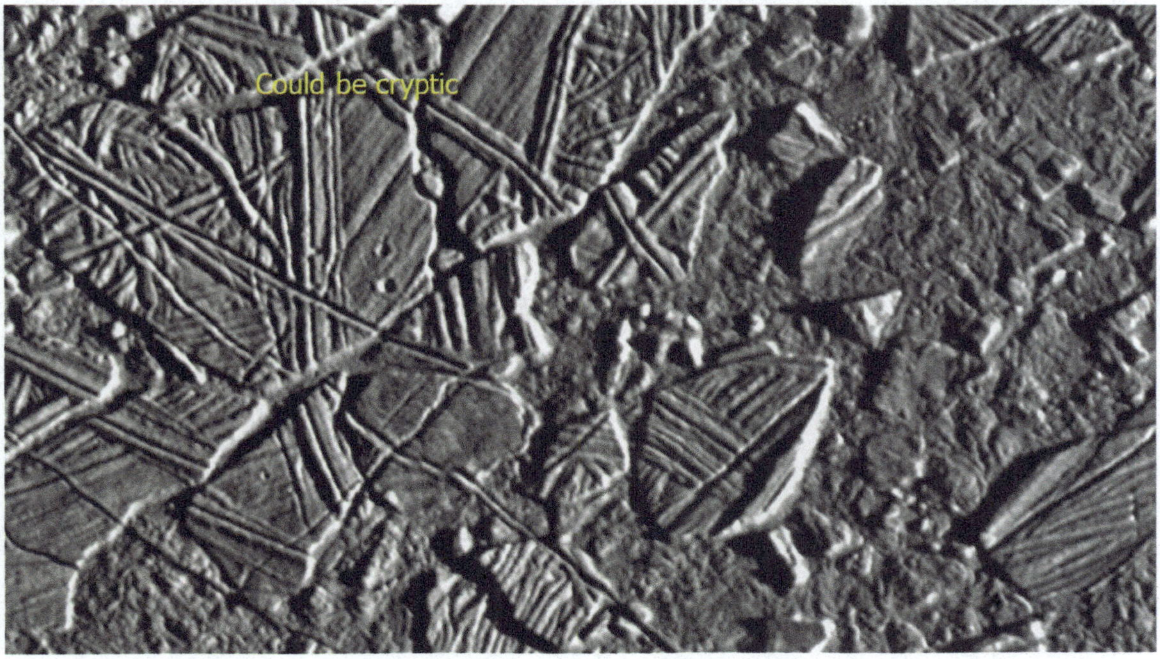

Getting up close may lead to inconclusive results.

So maybe forget the 100YSS, and instead embrace a
100 year telescope project?

Go for the "easy" way to find life, and forget the
unpredictable (and unmodifiable) "hard" way?

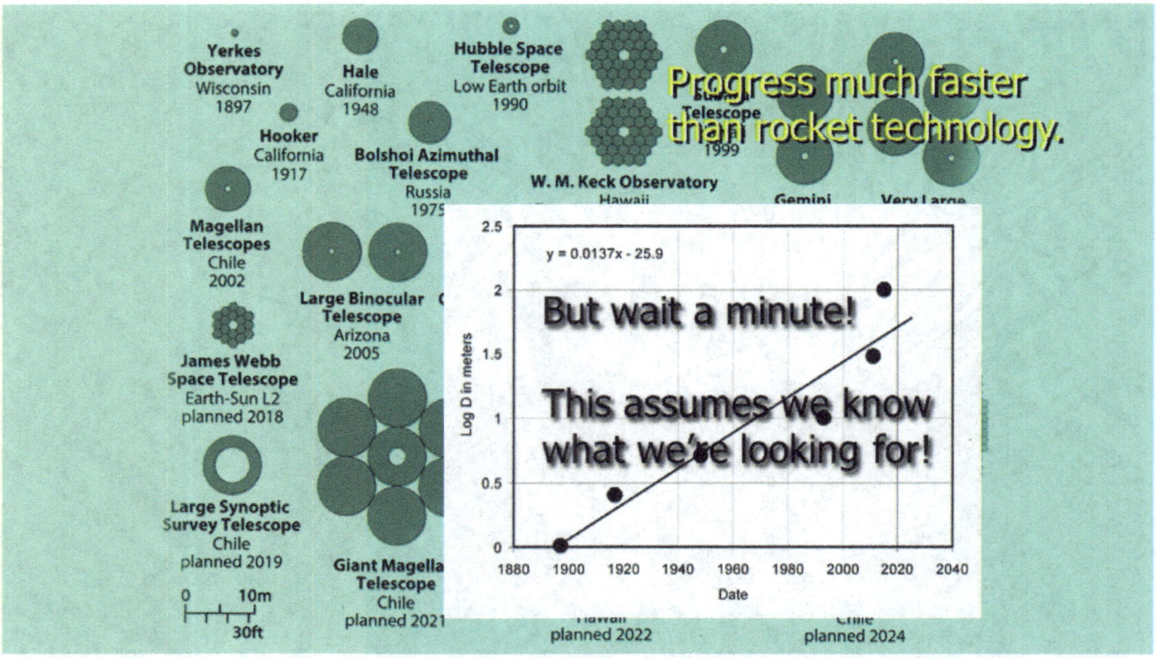
Yerkes Observatory
Wisconsin
1897
Hale
California
1948
Hubble Space Telescope
Low Earth orbit
1990
Hooker
California
1917
Bolshoi Azimuthal Telescope
Russia
W. M. Keck Observatory
Magellan Telescopes
Chile
2002
Large Binocular Telescope
Arizona
2005
James Webb Space Telescope
Earth-Sun L2
planned 2018
Large Synoptic Survey Telescope
Chile
planned 2019
Telescope
Chile
planned 2021
planned 2022
planned 2024
0 10m
30ft
Progress much faster
than rocket technology.
y = 0.0137x - 25.9
But wait a minute!
This assumes we know
what we're looking for!
Log D in meters
Date
2.5
2
1.5
1
0.5
0
1880
1900
1920
1940
1960
1980
2000
2020
2040

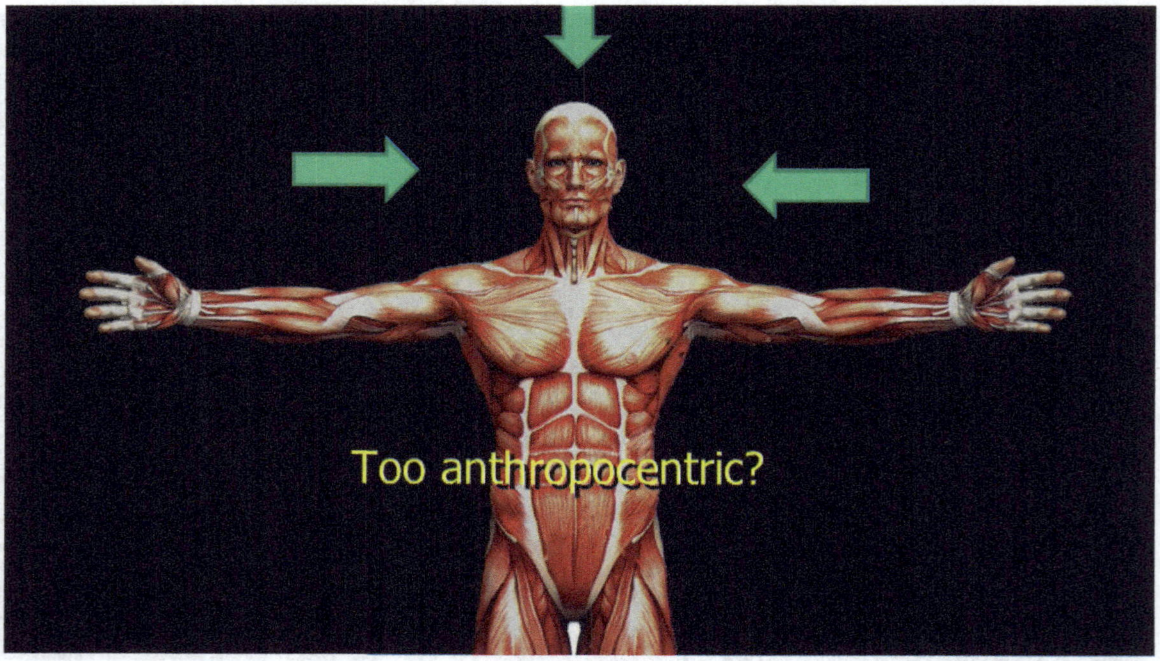
Too anthropocentric?

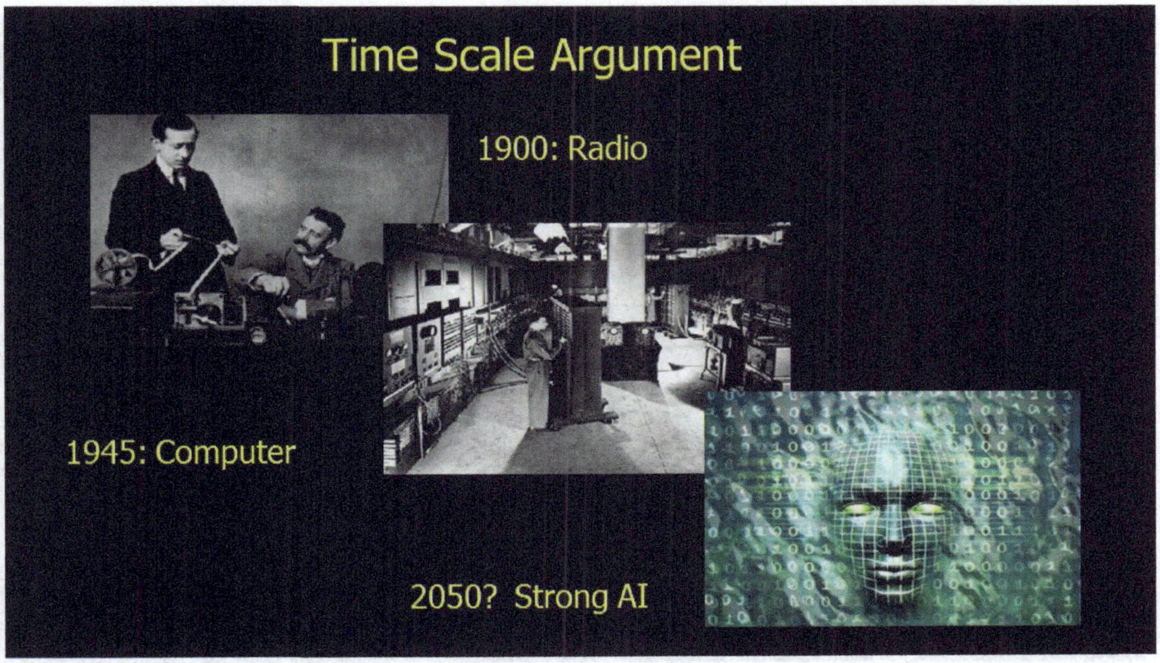
Time Scale Argument
1900: Radio
1945: Computer
2050? Strong AI

It's not going to be these guys!

The real value will not be looking up close for stuff we might better find from far away.
But discovering the truly unknown.

Don't tell him what to find.
Just get him a ship.

Design & Layout

The wide variety of disciplines represented and information contained in the preceding papers provides unique design challenges. As in the previous year, the goal was to create a uniform presentation and easily accessible layout while still maintaining the individual author's needs and desires. The overall design is intended to be conducive to the experience of education and engagement. This year's improvements in design have focused on uniformity between the papers. Formulas, illustrations, and other non-text elements are more consistent (although not absolutely so). The challenge engaged and met is to provide a cross-discipline publication that highlights the variety of thought, research, and effort needed to achieve the audacious dream of interstellar travel.

Primary text is typeset in Adobe Caslon Pro. Headings are set in Helvetica Neue and League Gothic.

Design and layout by Jason D. Batt.

Made in United States
Orlando, FL
05 November 2026

90209471R00267